ИЗДАТЕЛЬСТВО · МИР ·

В. Н. Кобранова

ПЕТРОФИЗИКА

Издательство «Недра» Москва

V. N. Kobranova

Petrophysics

Translated from the Russian
by V. V. Kuznetsov

Springer-Verlag Berlin Heidelberg GmbH

V. N. Kobranova

ISBN 978-3-662-09246-0 ISBN 978-3-662-09244-6 (eBook)
DOI 10.1007/978-3-662-09244-6

Originally published by Springer-Verlag Berlin Heidelberg New York in 1989
Softcover reprint of the hardcover 1st edition 1989

Preface

The chemical, physical and physicochemical processes that are at work in the depths of the Earth, both connected and unconnected with man's activities and coupled to the relevant properties and characteristics of the rocks, began to be intensively studied in the early decades of the present century. Until then little evidence had been available concerning the physical and physicochemical properties of rocks, and the data that existed were one-sided and uncoordinated.

Both in this country and elsewhere an interest in investigating natural processes, the processes taking place in rocks, and the properties and characteristics of rocks arose as a result of the intensive development of oil and gas engineering, the mining of coal and ore, the construction of large projects, railroads, etc. Information on the properties of rocks was needed, in particular, to facilitate progress in engineering, technology, and geological and geophysical methods of prospecting for extracting and processing mineral deposits.

In the late 1920s and early 1930s, methods involving intrinsic and induced polarization were introduced.

Moreover, little information was available concerning the petrophysical and petrochemical quantities characterizing the different contribution of various rocks to electrical processes.

Electrical methods were followed by other methods of applied physics based on the novel electrical, thermal, magnetic, nuclear, elastic and other properties of rocks.

It proved impossible to find data on these properties in the classical physics, physical and colloid chemistry literature, since the natural physical relations of rocks were not reported. Little was known about the significance of the physical quantities obtained for rock types using different procedures and techniques, or about the connections between petrophysical and physicochemical quantities.

It was not until the 1960s that sufficiently clear concepts about the physical and physicochemical processes occurring in rocks were proposed, a sound basis for providing petrophysical data was laid, and essentially new evidence about petrophysical connections was obtained.

Following the creation and development of industries concerned with rocks and applied geophysics both in the USSR and elsewhere, scientific schools appeared that focused on the theoretical and experimental studies of a wide range of chemical, physical and physicochemical properties of rocks. One of such schools at the Moscow Institute of Oil and Gas Engineering was created and led by V. N. Dakhnov.

Apart from Dakhnov, important contributions to petrophysics have been made by A. S. Semenov, A. I. Zaborovskii, N. B. Dortman, V. M. Dobrynin, E. M. Leont'ev, M. L. Ozerskii and others.

These researchers laid the foundations of petrophysics and showed that in order to correctly plan geophysical investigations in the field and interpret the results, one must know how individual petrophysical quantities are distributed within geological bodies. Petrophysics is of particular importance for geophysical studies of boreholes where the spectrum of physical quantities measured in a borehole enables us to identify the rocks found therein, isolate mineral deposits and estimate their reserves.

The 1980s have witnessed the appearance of handbooks, monographs, manuals and, finally, this textbook, which describes the new geological discipline and the problems faced by it, and also a vast bulk of data to be primarily used by geophysicists. The manual will be of great value for the training of mining engineers.

Contents

List of Symbols

A_{da}	—diffusion and adsorption activity
A_f	—filtration of electrochemical activity
$\beta = 1/K$	—coefficient of compressibility
β_g	—compressibility factor of gas
C_{capt}	—flux of captured neutrons
C_{scat}	—flux of scattered neutrons
δ_{dr}	—density of dry rock
$\delta_{dr.cl}$	—density of dry clayey component
δ_s	—density of solid clayey component
$\varkappa$	—specific magnetic mass susceptibility
$d_{gr.eff}$	—effective diameter of rock grains
$d_{gr.av}$	—effective and average diameter of rock grains
$d_{gr.m}$	—grain median diameter
d_p	—potential determining the layer
d_0	—Helmholtz double layer
E_{cd}	—concentration diffuse potential of polarization
$\varepsilon = E_{0av}/E_{av}$	—dielectrical permeability
$\varepsilon_{air,\ sph\ (or\ s)\ w.r}$	—relative dielectrical permeability in solid phase
$\varepsilon_{w.b.av}$	—dielectrical permeability of bound water
ε_{mr}	—dielectrical permeability of moist rock
$\varepsilon_{s1,2}$	—dielectrical permeability of solid rock
H_c	—coercive force (under normal)
$H_{c.t}$	—coercive force (under thermomagnetization)
$H_{c.e}/H_e$	—relative coercive force
H_{ef}	—action of effective magnetic field
H_{ex}	—action of external magnetic field
I_s	—surficial current
$I_{abs.a}$	—absorption active current
$I_{abs.r}$	—absorption reactive current
$\vec{J}_{mr}$	—residual mass magnetization
$\vec{J}_m$	—mass magnetization
$\vec{J}_{mn}$	—residual normal mass magnetization
$\vec{J}_{rs}$	—residual magnetization of saturation
$\vec{J}_V$	—specific volumetric magnetization
$\vec{J}_{Vr}$	—residual isothermal magnetization
$\vec{J}_{Vs}$	—volumetric magnetization of saturation

$\vec{J}_{mn.d}$ —detrital mass magnetization
$\vec{J}_{Vn.d}$ —detrital volumetric magnetization
k_{da} —diffusion-adsorption emf coefficient
$k_{p.cl}$ —closed voids ratio
$k_{p.d}$ —dynamic voids ratio
k_p —voids ratio
k_{perm} —permeability coefficient
$k_{o.p}(k_{por.o})$ —open porosity coefficient
$k_{eff.p}$ —coefficient of effective porosity
$k_{d.p}$ —dynamic porosity coefficient
k_f —coefficient of filtration
$k_{fill.cl.V}$ —filling coefficient
$k_{w.b}$ —water saturation of rock by bound water
$k_{fm\ m}$ —coefficient of mass ferrimagnetics
$k_{w.r}$ —coefficient of residual water
$k_{c.m}$ —coefficient of mass cementation
$k_{cl.m}$ —coefficient of mass clayiness
$k_{p.red}$ —reduced voids ratio
$k_{p.rel.}$ —relative voids ratio
$k_{sw.cl.}$ —swelling clayey component coefficient
$\varkappa_t$ —relative residual magnetization
$K = E/3(1\text{-}20)$ —modulus of overall expansion
$\varkappa$ —specific magnetic volumetric susceptibility
$\varkappa_{av}$ —average magnetic susceptibility
L_s —slowing-down length
L_d —diffusion length
$L_{k.av}$ —average statistical length of ion path in rock sample
λ, μ —Lamé's coefficient $\left(\lambda = \dfrac{\nu}{(1+\nu)(1-2\nu)} E,\right.$ $\left.\mu = \dfrac{1}{2}\dfrac{E}{1+\nu}\ \text{N/m}^2,\ \text{Pa}\right)$
μ_{av} —average magnetic susceptibility
$\mu_B = |e|h/2m(e)$ —Bohr magneton
ν —Poisson ratio
P_p —parameter of porosity
$\vec{P}_{e.av}$ —electron average polarization
$\vec{P}_{a.av}$ —atom average polarization
$\vec{P}_{i.av}$ —ion average polarization
$\vec{P}_{eo.av}$ —electroosmotic average polarization
$\vec{P}_{el.d}$ —elastic displacement polarization
$\vec{P}_{ip}$ —induced polarization
$\vec{P}_d$ —orientation dipole polarization
$\vec{P}_{el}$ —electrolytic polarization
$\vec{P}_{er}$ —electron relaxation polarization
$\vec{P}_{cd}$ —concentration-diffusion polarization
$\vec{P}_{ith}$ —ion relation thermal polarization

P_s	—saturation parameter
$P_{os.p}$	—oil saturation parameter
$P_{ws.p}$	—water saturation parameter
Π	—surface conduction parameter
q_{rm}	—total (specific mass) radioactivity
ϱ_{mr}	—specific resistance of moist rock
ϱ_{op}	—overall geostatic pressure
ϱ_r	—specific resistance of ore
$\varrho_{n\text{-}m.c}$	—specific resistance of non-ore component
$\varrho_{s1,2}$	—specific resistance of solid rock
$\varrho_{ws.r}$	—specific resistance of water-saturated rock at right angles
$\varrho_{wb.av}$	—average effective specific resistance of bound water
$S_{por.}V_{dr}$	—pore space surface in volume
$S_{por.m_{dr}}$	—pore space surface in mass
σ_{air}	—specific electrical conductivity of air
σ_e	—attenuation coefficient calculated for one electron
$\sigma_{f.cl}$	—free electrolyte resistance filling voids space of clayey content
$\sigma_{ws.r}$	—specific conductivity of water-saturated rocks
σ_{scat}	—scattering cross section
σ_{capt}	—capture cross section
t_C	—Curie temperature
$\tau_{d.ws}$	—lifetime of water-saturated rock
$\tau_{d.os}$	—lifetime of oil-saturated rock
η	—dynamic viscosity coefficient
η_{act}	—activation overvoltage
η_{ads}	—adsorption overvoltage
U_{ap}	—potential of applied polarization
U_{ip}	—potential of induced polarization
U_f	—filtration potential difference
V_P	—velocity of propagation of elastic waves
V_{por}	—volume of a rock voids space
$V_{por.d}$	—volume of dynamic porosity
V_r	—volume of rock
V_w	—maximum possible volume of water
$V_{w.b}$	—volume of bound water
$V_{w.c}$	—volume of capillary water
$V_{cl.por}$ ($V_{por.cl}$)	—volume of closed pores
$V_{op.por}$ ($_{por.o}$)	—volume of open pores
$V_{m.r}$	—volume of moist rock
V_{sus}	—volume of suspended water
$V_{w.r}$	—volume of residual water
$V_{w.str.b}$	—volume of strongly bound water
$V_{w.l.b}$	—volume of loosely bound water
$V_{w.fr}$	—volume of free water
w_c	—coefficient of capillary moisture capacity

w_h	—coefficient of hygroscopic moisture
$w_{m.h}$ ($m.\ hyg\ m$)	—coefficient of maximum hydroscopic moisture
w_{sus}	—coefficient of suspended capacity
w_t	—total moisture capacity coefficient
wJ_V	—action of internal molecular field
$\zeta_{\mathrm{liq.0}}$	—potential determining layer in the liquid phase
ζ_s	—potential determining layer in the solid phase

Introduction

The energy of a nation is largely conditioned by the extent of prospective, surveyed and extracted oil and gas reserves and other natural resources. To solve the tasks associated with the need to increase all types of mineral wealth, all those engaged in geological, drilling and prospecting work make extensive use of applied geophysics, in particular of a new geological discipline—petrophysics. Petrophysics which is the study of the physical properties of reservoir rocks, deals with the following:

(1) the physical and physicochemical processes that are at work in rocks during which the latter manifest their physical and physicochemical properties (porosity, density, water-, gas-, oil-, heat- and electroconductivity, magnetization, elasticity, radioactivity etc.);

(2) the petrophysical quantities (voids ratio, coefficients of permeability, electroconductivity, radioactivity etc.) characterizing the nonuniform participation of different types of rocks in individual physical and physicochemical processes; the concepts of these parameters determining their equations and units of measurements;

(3) variation rows of petrophysical quantities and characteristics of these rows (their limiting, mean, median and modal values) for rock types and species governed by the variations in their composition, structure, thermobaric conditions, origination and life-span;

(4) the classification of petrophysical quantities according to their values and also of rocks according to their petrophysical and other characteristics;

(5) the relationship between petrophysical and other (petrochemical, petrophysical-chemical) quantities; their classifications;

(6) the local and regional variations in petrophysical quantities owing to the specific composition and structure of geological bodies or regions, magmatism, sedimentation, subsurface water regime, mineral deposits;

(7) the mechanisms and regularities of variations of petrophysical quantities as observed in borehole cross sections;

(8) the petrophysical regionalization, i.e. the determination of regional boundaries governed by variation row features of petrophysical and petrochemical characteristics of rocks species, petrophysical connections and other features of the areas in question;

(9) the determinations of petrophysical quantities by laboratory methods.

It is out of the scope of the present manual to encompass all the aspects studied by petrophysics. Additional information on various related disciplines can be found in the Bibliography.

It was not until the 1960s that petrophysics emerged as a separate discipline; this occurred after general surveys on the physical properties of rocks had appeared and the core of petrophysics as a self-contained science had been formed.

The development of petrophysics is connected with the names of such scientists as V.N. Dakhnov, A.S. Semenov, A.I. Zaborovskii and others.

Currently petrophysics is also receiving much attention abroad. Important investigations have been conducted by Archie, Gassman, Willey, Fett and many others.

The bulk of the accomplished work that enable petrophysics to emerge as a separate scientific discipline in this country is due to the personnel of a number of leading research institutions and laboratories of various agencies that are engaged in prospecting for and extracting mineral deposits and in construction ventures.

To solve the various problems encountered in geological prospecting one must have a knowledge of the different types of physical properties and petrophysical characteristics. The physical properties of rocks that are studied in the geophysics of oil and gas can be grouped according to the following characteristics:

(1) capacity (porosity, moisture content);

(2) capillary (capillary pressure, wettability);

(3) gaseous- and hydrodynamic (gas-, water- and oil permeability);

(4) density (solid-, liquid-, gaseous phase densities and overall rock density);

(5) electrical (electroconductivity or electrical resistance, dielectrical permeability, natural and induced electrochemical and redox activities);

(6) thermal (heat capacity, thermal conductivity and thermal diffusivity);

(7) magnetic (magnetization, magnetic susceptibility, Curie-point temperature);

(8) nuclear (radioactivity, neutron activity);

(9) elastic (capability for geometry variation, transfer and absorption of elastic vibrations);

(10) strength (compressive strength, rupture- and shear strength);

(11) plastic (plasticity) group.

Each group is characterized by a complex of petrophysical quantities: for example, the capacity group, by coefficients k_p, $k_{o.p}$, $k_{eff.p}$, $k_{d.p}$ of the total, open, effective and dynamic porosity, respectively and coefficients w_t, w_c, w_{sus}, $w_{m.h}$, w_h of the total, capillary, suspended, maximum hygroscopic and hygroscopic moisture capacity, respectively. Thus, a specialist has to deal with more than 60 different characteristics.

The petrophysical investigations that were carried out from the mid-thirties to the present time made the following achievements possible.

1. The techniques and equipment for the determination of many petrophysical parameters have been developed. For example, devices for the determination of magnetic susceptibility, residual magnetization, α-, β- and γ-activity, elasticity characteristics have been produced commercially. Unfortunately, however they are yet far from being perfect. They are cumbersome, bulky, difficult to control, sometiсmes insufficiently reliable and difficult to adjust. What is more, not only the equipment but also modern methods for the determination of petrophysical quantities have not been unified; thus, petrophysical determinations made at different laboratories

are hardly comparable. Currently both the equipment and the determination procedures are being further developed.

2. Physico-mathematical theories of electroconductivity, natural and induced polarization, heat conductivity, radioactivity, elasticity of rocks and many other physical and physicochemical processes have been developed; equations have been derived permitting one to calculate to the first approximation the values of one or another petrophysical quantity from data on the composition and structure of rocks and also from the characteristics of the latter.

3. A multitude of different petrophysical characteristics, their variation limits, average, median and modal values of their rows for rock types and groups have been determined.

4. The utilization of petrophysical parameters has allowed for the development of a more detailed rock classification system based on the properties of the latter.

5. Cross sections and charts have been prepared illustrating the variation in petrophysical quantities or their complexes as a function of depth for quite a number of areas in the USSR, major regions (the map of rock magnetization in the USSR) and within identical-age or identical-type sediments (the map of Mesozoic deposits density in West Siberia).

6. A number of pair correlations between petrophysical characteristics have been established.

Pair correlations are common for coeval, equally transformed and consolidated sediments and rock massifs if the concentration of anomalous (in terms of its properties) element, mineral or component differs from one variety to another. Such correlations have been found for: (a) density δ_{dr} and iron content C_{Fe} for dry ferruginous quartzites in the Kursk Magnetic Anomaly Region (Fig. 1*a*) and a number

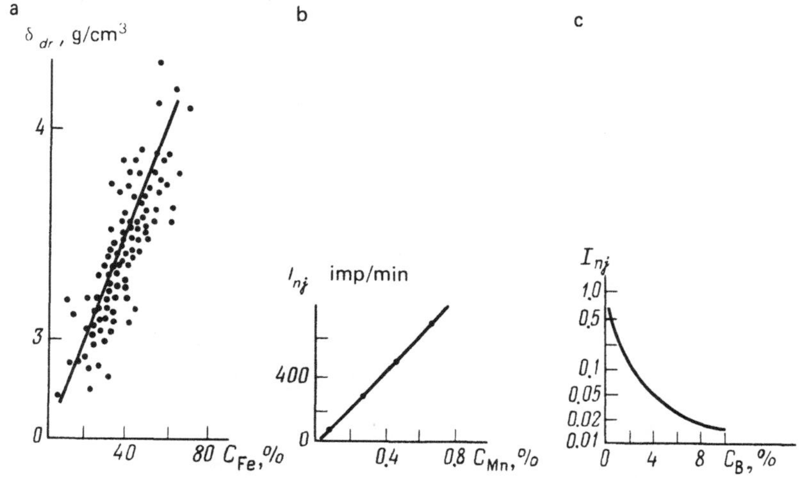

FIG. 1. Dependences of density δ_{dr} on iron content C_{Fe} for ferruginous quartzites (*a*) (after B.A. Andreev), induced activity of Mn, $I_{n\gamma}$, on its content, C_{Mn}, in ore (b), and neutron count rate I_n on boron content, C_B, in magnetite ore (for a weight 0.1 kg) (after E.M. Filippov)

of rock types in iron ore deposits; (b) induced activity $I_{n\gamma}$ and manganese content C_{Mn} in ore (Fig. 1*b*) for different manganese ore deposits; (c) neutron count velocity I_n and boron content C_B in boron bearing ores (Fig. 1*c*); (d) magnetic susceptibility $\varkappa$ and magnetite content $C_{Fe_3O_4}$ for ferrous quartzites; (e) density δ_{dr} and content C of ore components in ore (see Fig. 43). A markedly intimate connection is observed only under conditions of minor and practically invariable porosity of geological bodies.

Pair correlations also exist between petrophysical quantities characterizing the definite origin, composition, and extent of transformation of loose sediments with an unequal content of anomalous-property components (mineralized water). Examples of such connections can be provided by the relationship between specific resistances ϱ_{mr}, thermal conductivity λ_{mr}, velocity v_P of propagation of elastic waves, density $\delta_{dr}(\delta_{mr})$ and the voids ratio k_p (see Figs. 44, 157), and also the less intimately connected specific resistance ϱ_{mr} or porosity parameter P_p and density δ_{mr} (or velocity of propagation of elastic waves v_P), thermal conductivity λ_{mr} (see Fig. 95) and others.

The connections between three petrophysical characteristics are more closely related for both the consolidated and the loose sediments of identical origin and extent of transformation when one of the petrophysical quantities characterizes the ratio between the solid and liquid phases in a rock; the other, its structure; and the third is governed both by one and the other quantity. Moreover, fairly close connections between two of these characteristics can be obtained given a fixed value of the third one (see Figs. 87, 88).

Petrophysical connections provide a basis for the determination of the many petrophysical quantities needed for estimations of mineral deposits by geophysical techniques.

Chapter I

Heterogeneity of Geological Bodies

Rocks occur in the Earth's crust as interrelated bodies of different origin, geometry, size, composition, structure and other properties (single crystals and polycrystals of minerals, beds and facies, shields and plates, continental platforms, continental, geosyncline folded, orogenic systems and belts, continents, oceanic basins, terrestrial crust bodies, mantle, Earth's core). The general and most important property of geological bodies is their heterogeneity, each of geological bodies contains various (in terms of origin, size and other characteristics) heterogeneity types; taken together, it is they that are responsible for inhomogeneity of a body as a whole.

When solving various practical tasks, we identify and study individual heterogeneity types of a geological body, mentally subdividing it into unit volumes to which we assign ever increasing or decreasing sizes, yet much smaller (by two-three or more orders of magnitude) than the size of the object under study. Heterogeneity of a definite species shows itself in variation of average values of its characteristics as we pass from one unit volume to another. Heterogeneity characteristics are provided by petrophysical, petrochemical and other variables determined and averaged in terms of unit volumes.

Given the size of a unit volume and average heterogeneity characteristic values, subregions are identified within a geological body, i.e. heterogeneity elements where heterogeneity characteristics manifest definite constant values.

The degree of detail of study of a geological body can be enhanced by continuously diminishing the size of a unit volume, yet the number of heterogeneity elements will then vary in a stepwise fashion in conformity with discreteness—definite size levels of individual heterogeneities of the geological body. Given that the unit volumes are of major size, the commensurable heterogeneity elements will only be identified when, by using a measuring device, we determine the proper heterogeneity characteristic, a much lesser rock volume is studied (at least by two orders of magnitude) than the heterogeneity element involved.

In this case we will be able to unambiguously isolate heterogeneity commensurable with or larger than the size of the device, whereas all others whose volume is less (by several orders of magnitude) than heterogeneity volume, studied by the measuring device, will not be isolated and their presence will not practically tell, or, if it will, then only inappreciably, on the average value of the variable being determined. Heterogeneity whose volume is close to that to be determined by the measuring device, contribute to the appreciable spread of the variable being determined and may prevent us from being able to isolate the heterogeneity of interest to us. By continuously diminishing unit volumes and, accordingly, volumes to be studied by the use of a measuring device, we distinguish ever decreasing elements and heterogeneity types in a geological body. The larger the geological body, the greater is the number of its heterogeneities and the greater must be the set of unit volumes and measuring devices.

The most elementary and relatively small geological entities (single crystals of minerals), for example, show structural and nonstructural heterogeneity. Structural heterogeneity is lack of identical structural cells throughout the volume filled by them. Structural cells of minerals are different in chemical composition, i.e. the same positions can be occupied in their volume by different atom species, the cells may also contain vacancies (nodes unfilled by atoms), incorporated atoms and dislocations. Other types of mineral structure heterogeneities also occur. To reveal structural heterogeneity, diffraction (X-ray and electron diffractometry) techniques and radio spectroscopy are used.

Nonstructural heterogeneity is nonuniform composition and structure of individual portions of a mineral element. This heterogeneity may be of primary or secondary origin and be due to a number of factors. For minerals it is visually studied, by using a light or an electron microscope, electron microprobe with the purpose of determination of mineral structure and origination conditions.

Bedded bodies, apart from the aforementioned, have lower heterogeneity types or levels. These bodies are filled by multicomponent rocks between polymineral, different-size grains of rocks with structural and nonstructural heterogeneity are incorporated cement of a definite composition, natural waters and sometimes their mixtures with liquid and gaseous hydrocarbons. Along the strike and the dip of a bedded body one can often observe a regular variation in the ratio between their solid and other components and in the mineral and granulometric composition of grains (Fig. 2) and cement (Fig. 3). Heterogeneity associated with lamination and

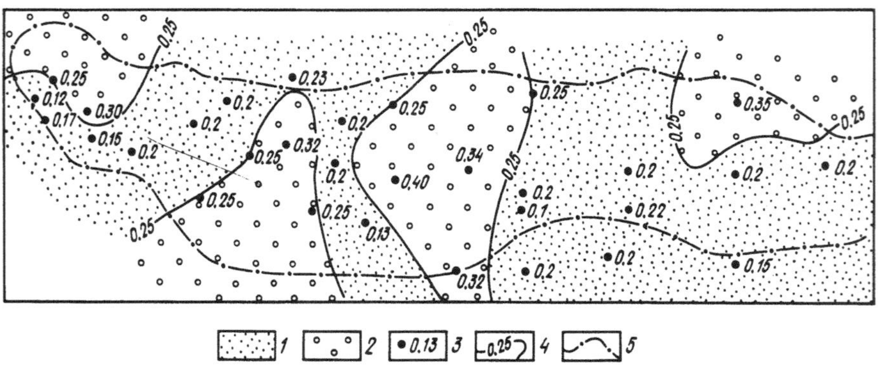

FIG. 2. A diagram of distribution of sand-aleurite reservoir beds having different average diameters of grains in a bed C_1 of the Mukhanovo deposit (after O.A. Chernikov and A.I. Kurenkov).

Regions of distribution of rocks having an average grain diameter: *1*—<0.25 mm; *2*—>0.25 mm; *3* — location of borehole and value of average grain diameter (in mm) in it; *4*—lines of equal values of grain diameter; *5* — boundaries of zones with different grain contents within bed C_1

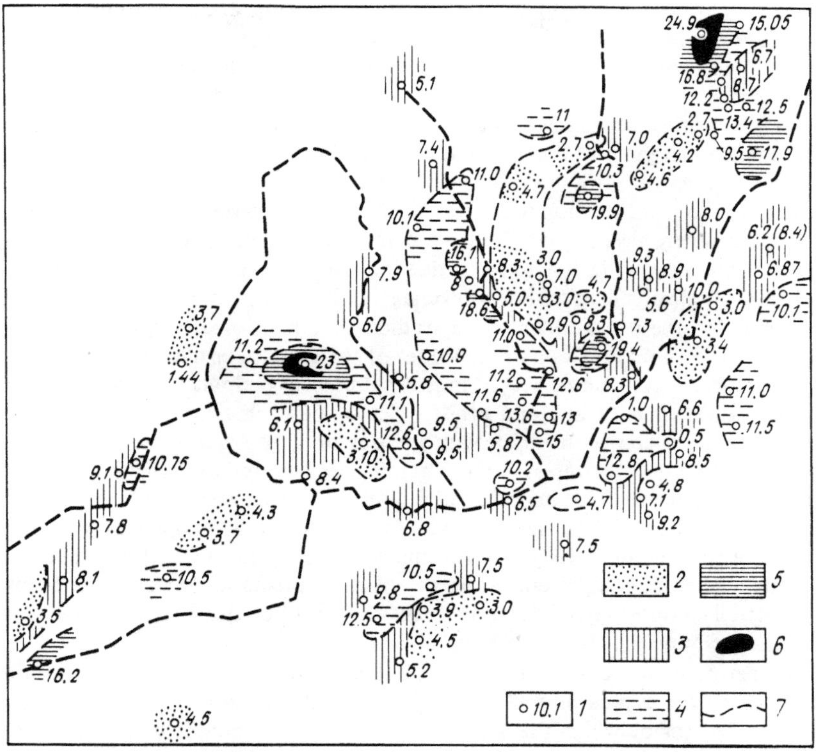

FIG. 3. A diagram showing distribution of reservoir beds having different content of cementing material of bed D_1 of the Tuimaza oil field (after O.A. Chernikov and A.I. Kurenkov). *1* — boreholes and cement content in rock (in %); cement content interval in rock (in %); *2* — 0-5; *3* — 5-10; *4* — 10-15; *5* — 15-20; *6* — >20; *7* — outlines of regions of working

equigranular pattern of these grains is determined macroscopically (visually), by using a light microscope and granulometric analysis; the polymineral character of grains and cement by using a light and an electron microscope; structural and nonstructural grain heterogeneity by the aforementioned techniques.

To identify individual heterogeneity types we take and then investigate specimens from a seam following a definite procedure. The heterogeneity of bedded bodies is studied in terms of their origin and subsequent formation, for search, prospecting and extraction of mineral deposits, for enhancing the effective use of applied geophysics and for solution of a number of other practical tasks.

A much greater variety of different heterogeneity types are contained in such major geological entities as shields and plates, continental platforms, continental geosyncline folded orogenic systems, belts, etc. For continental platforms, apart from those considered for beds, one can single out still lower heterogeneity levels conditioned by the structural multistage pattern of these entities (a structural stage is a series of lithologically different rock strata forming a single complex), the presence of deep faults and other specific features. In this case we refer to methods of applied geophysics, satellite-aided photogrammetry etc. to reveal low levels of heterogeneity.

It may also prove necessary to study heterogeneity of major geological entities to find out their origin and living conditions in the context of search for mineral deposits, construction projects and many other practical applications.

In addition to identification and study of individual heterogeneity types of geological entities petrophysical variables are determined by investigating rock specimens and using geophysical techniques dealing with separate different regions of a bed for lithologically identified or similar-age rock masses etc. Since rock volumes are different for these quantities to be studied they may involve different heterogeneity types, so nonuniform average values of variables determined for specimen collections and a large number of various regions of a geological body are likely. Spread of values of both one and other types of petrophysical characteristics is also due to the difference in physical properties of a geological body within its volume. The difference in values for a whole bedded body is greater the lesser is the volume of specimens (or investigated regions) and the lesser the homogeneity of the geological body. If the size of the measuring device is inappreciably greater than the sizes of heterogeneities of a specimen or rock area under study, then the sought characteristics of different properties of these media are governed not only by the material composition of the constituent elements but also by their volumetric ratio and distribution in the rock volume.

Thus, petrophysical characteristics are not constants; their values are conditioned both by the degree of rock homogeneity and the ratio of specimen sizes or rock and devices by which the values of parameters of interest are determined. They are also conditioned by the temperature, pressure, and frequency of the physical field used for their evaluation, and the origination, living conditions and the age of rocks concerned.

Viewed in the context of prospecting for and exploitation of oil and gas deposits, orebodies, coal-bearing suites it is the general practice to study heterogeneity of individual beds and veins by referring to a wide spectrum of petrophysical variables to be determined for a representative collection of rock specimens. The samples are caused to assume a regular (cylindrical, rectangular) shape and a definite size. Complex geophysical studies of strata into which a borehole has been sunk are also conducted by using devices that differ in type and size. By referring to these data, through the use of statistics, we find limiting, average and median values of petrophysical characteristics for a bed or the whole orebody. The observed results are also represented as petrophysical stratigraphic and other maps showing lateral heterogeneity of individual seams, petrophysical cross sections of individual boreholes, composite area sections and those of regions, territories, coal basins, used for petrophysical regionalization. Petrophysical quantities thus established are also used for identification of correlations between different characteristics. It is not uncommon that such studies of geological bodies permit one to find out definite regularities in spatial variation of averaged characteristics.

Chapter II

Porosity, Clayiness

The solid phase fills part of the volume of a rock; the remaining volume is taken by other (generally liquid and gaseous *) rock constituents. The property of rocks to contain various volumes not filled by the solid phase volumes V_{por} in a definite dry volume ** V_{dr} is known as *porosity*. The volume V_{por} is made up by individual minor volumes, voids, that is why it is often described as a rock void space. Voids may differ in origin, shape, size, interrelation of other voids.

Sec. 1. The Origin, Shape, Size, Interrelation of Voids. Porosity Types

The Origin of Voids

We distinguish between primary and secondary voids. *Primary voids* occur during rock formation and vary in size and shape as rock is packed, cemented or metamorphosed. Primary voids are structural pores between detrital (sandy, aleurite, argillaceous, calcareo-magnesial) rock granules, intercrystalline pores of consolidated igneous, metamorphic, sedimentary rocks, tuffaceous, siliceous sinter voids, cinereous and other voids. *Secondary* (newly formed) *voids* are caverns, cracks, channels in rock bodies upon their dissolution, weathering, crystallization, recrystallization, limestone dolomitization, in the course of tectonic and biochemical process.

Caverns occur upon dissolution (leaching) by acidic waters of fissured carbonate rocks and gypsums. Cracks with a forward and crossing orientations originate in consolidated rocks as a result of abrupt changes in their volumes following marked temperature variations and pressure changes, due to crystallization, recrystallization, dolomitization.

Channel-shaped biochemical voids can be observed in rocks left by decayed root systems, tracks of worms and other animals.

Voids Shape

The shape of voids may be: close to rhombohedral (for loose, well sorted, rounded detrital rocks), close to tetrahedral (for the same, yet compact rock species), slit-like (for clays, micas, laminated texture rocks), irregularly shaped widening and narrow-

---

* Ice and gas hydrate are contained here less commonly.

** A dry rock is a rock dried to form constant mass at a temperature ensuring removal of free and bound water (see Chap. III).

ing channels (for poorly sorted detrital formations); fissure-shaped (for hard igneous, metamorphic, consolidated sedimentary rocks), cavern-shaped (for carbonates and gypsum species), vesicular (for igneous rocks), cellular (for tuffs and siliceous sinters), channel-shaped (for loesses), showing a shape corresponding to crystals of dense igneous, metamorphic, sedimentary rocks.

Voids Size

The size of a pore is characterized by *an effective* (or average) *diameter*, or cross section that may vary for different rock species over a wide range.

In terms of the effective diameter d_{eff} we distinguish: (1) large supercapillary pores with a diameter more than 10^{-1} mm; these are numerous for sorted out weakly cemented detrital and clastic rocks (shingles, gravel, coarse and medium-grained sands, sandstones, clastic carbonate varieties); for leached carbonate rocks where they are appreciable in size and are called megavoids (these are Karst caverns in the carbonate rock and gypsum regolith); (2) capillary pores with the effective diameter 10^{-4}-10^{-1} mm typical of less sorted, coarse, more cemented rocks (fine sands, sandstones, clastic carbonate and other similar formations); supercapillary and capillary pores are often collectively called macropores; (3) thin subcapillary (meso- or transition), small diameter pores whose d_{eff} is in the range from 2×10^{-6} to 10^{-4} mm typical of natural adsorbents (clays, microcrystalline chalky limestones, diatomites, tripoli, ash tuff etc.); (4) minor size (thin) micropores with $d_{eff} < 2 \times 10^{-6}$ mm; these have been found for certain natural zeolite species.

Pore size classification is based on the data on the pattern of bonding between pore liquid and the solid component and its transport in rocks. The fraction of the water bound with the solid component is inappreciable in supercapillary pores, and the bulk of it moves in conformity with the laws of hydrodynamics for pipes in the direction of gravitational force. The content of bound water in capillary pores is greater, and its fall by gravity is prevented by the capillary rise.

In subcapillary pores natural waters are practically stagnant being practically bound with the solid rock component and adsorbed ions. Micropores whose diameter is commensurable with the size of molecules contain solely strongly bound water one molecule thick. The adsorption forces cause the pores to be filled by this water. Water mobility is even less than in mezopores.

The Interrelation of Pores and Porosity Types

Rock pores (or voids) may be both interconnected and isolated. The interconnected pores (these are numerous in well sorted weakly cemented clastic rocks) are known as *open* and isolated pores as *closed*. The latter type is often met with in carbonate igneous rocks manifesting cavernous and vesicular pores.

The total sum of the volumes $V_{op.por}$ and $V_{cl.por}$ of open and closed pores, respectively is the volume V_{por} of all rock pores or its *total porosity*:

$$V_{por} = V_{op.por} + V_{cl.por}$$

Sec. 2. Configuration (or Structure) of Rock Porosity Space

Rock types (sands, sandstones, aleurolites, limestones, granites, gabbro etc.) and rock groups (sandy, aleurolitic, clayey, acidic, intermediate, basic, ultrabasic etc.) have different (in terms of origin, geometry, size, internal bonds) pores that are unequally distributed in the pore space; they are also unequally distributed over the rocks volume. Hence comes a concept of a specific configuration (structure) of pore spaces.

For instance, fissured cavernous limestones and dolomites' solid component crystals demonstrate primary micropores uniformly distributed throughout the volume; their total volume represents a relatively minor fraction of the volume of all the pores. Secondary pores of these rocks—fissures and caverns (both visible and invisible to the unaided eye) show subcapillary, capillary and supercapillary dimensions. It is not uncommon that caverns differ very much in the effective diameters, are open or closed, nonuniformly distributed over the volume of one rock variety or another; their total volume may often take an appreciable fraction of the pore space of rocks in question. They may also differ in the extent of oppenness, length of parallel or crossing cracks; their fraction in the pore space is of minor value.

The structure of pore spaces of clastic rocks showing well sorted rolled and weakly cemented grains (Fig. 4*a*) is much simpler compared with the rocks described above. However, it is far from being similar. According to O.A. Chernikov and A.I. Kurenkov, the size of the pores of the D_1 container bed of the Tuimaza deposit varies from 5 to 50 μm and on the average d_{eff} = 13.5 μm, and for the C_1 bed of the Mukhanovo deposit the same values vary from 5 to 150 μm with the average d_{eff} = 30 μm. Clastic rock structure is rendered more complex under conditions of consolidation and cementation [e.g. by clay (see Fig. 4*b*)], where rocks show minor pores of a complicated geometry.

What is more, the structure of different argillaceous cementing minerals is far from being similar. Crystals of laminated argillaceous minerals with a rigid crystal lattice (caolinite, dickite etc.) represent hexagonal plates with the base spacing from 10 to several hundreds of micrometers and thickness less by an order of magnitude. Ordered pack-

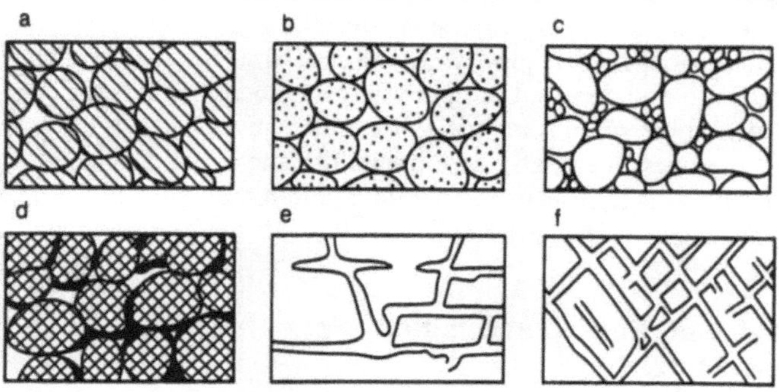

FIG. 4. Different structures of voids space

Clastic rocks: *a* — highly porous with well rounded and sorted grains; *b* — very high-porous with well rounded sorted and porous grains; *c* — low-porous with poorly rounded and sorted grains; *d* — decreased porosity with well rounded sorted but cemented grains; *e* — with solution pores; *f* — with fissure porosity (after O.E. Meinzer)

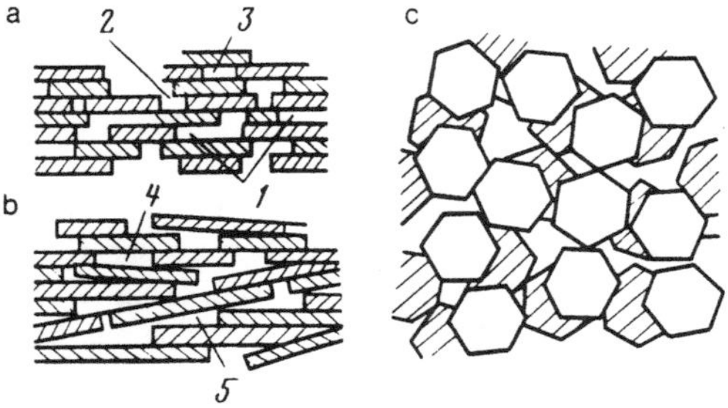

FIG. 5. Types of voids in laminated minerals (after A.P. Karnaukhov).

a, *b* — cross sections of mineral in different directions; *c* — top view; voids: *1* — slit-like; *2* — maze-like; *3* — closed; *4*, *5* — wedge-like

ing of these plates is observed solely along the *C* axis, being randomly oriented along other axes (Fig. 5). The communicating pores of the adjacent layers show a maze-like pattern (Fig. 5*a*, *2*), slit-like pores (Fig. 5*a*, *1*) have the width in each layer equal to that of the particles. Isolated pores may also occur (Fig. 5*a*, *3*). Expanding lattice argillaceous minerals (montmorillonite, vermiculite) show adjacent layers less rigidly interconnected, and polar adsorbate molecules (e.g. of water) are introduced between them to extend them. The pore space of these minerals is formed by primary slit-like variable thickness pores and secondary pores—spaces between laminated crystals.

Non-laminated natural adsorbents, zeolites, have a three-dimensional alumosilicate framework joined together by strong covalent bonds, their pore space being made up by crystalline structure pores whose configuration and size is conditioned by the type of cuboctahedra (crystal's structural elements) and cation type and localization.

The shape of a rock's structural elements is what governs the form of pores in fibrous tubular argillaceous minerals with crystals shaped as fibres, pipes or needles. Halloysite crystals, e.g. represent short hollow tubes. Needle-shaped palygorskite and sepiolite crystals have longitudinal zeolite channels, respectively, 0.37×0.64 and 0.56×1.10 nm in cross section. Secondary pores of all minerals being considered are formed by interstitial space. The size and configuration of secondary pores are governed by the crystal's size and their packing type. The structure of the pore space of individual rock varieties of definite beds differs to a lesser degree. Even if not quite similar in origin, configuration, size, internal bonds of pores, their volumetric ratio and distribution, pore spaces of beds are close in type.

The structure of rock's pore space is identified visually or by the use of electron beam, light, raster microscopes. It can also be evaluated by the rocks' clay content, pore composition, specific surface, effective and modal pore radius, specific volume of all as well as of open and effective pores.

Sec. 3. The Clayiness, Pore Composition, Pore Space Surface and Specific Surface

Clayiness

The clayiness of sedimentary rocks is the property of these latter to contain particles with $d_{eff} < 0.01$ mm (< 10 μm), less often with $d_{eff} < 0.001$ mm (or less than 0.002 or 0.005), as particles of such a small size are mainly responsible for the properties of sedimentary rocks. Their content in rocks, however, is not generally determined.

Clayey particles represent fragments of clayey minerals of kaolinite, montmorillonite, hydromica (illite) groups typically with $d_{eff} \leqslant 0.005$ mm, fragments of quartz, feldspars, micaceous and heavy minerals and contain admixtures of limonite, hematite, carbonates, sulphates (gypsum), sulphides (pyrite, marcasite) and other minerals.

Montmorillonite may be predominant in one rock species, kaolinite in another and hydromicas in still another species. The size of clayey particles, their adsorptional properties and their swelling capacity may differ from one clayey mineral to another. For example, montmorillonite group minerals demonstrate the greatest dispersion and adsorptional, and swelling capacity. The property of rocks to contain a plurality of clayey particles filling the space between larger grains or sharing it is known as *dispersed clayiness* in contradistinction to *bedded clayiness*, i.e. the property of rocks to show a composition with thin clay intercalations.

The property of rocks to contain a different mass $m_{dr.cl}$ (or volume $V_{dr.cl}$) of dry clayey particles for their definite mass m_{dr} (or volume V_{dr}) in a dry state is assessed by a *specific mass* $k_{cl.m} = m_{dr.cl}/m_{dr}$ (or *specific volume* $k_{cl.V} = V_{dr.cl}/V_{dr}$) *clayiness*. Hence

$$k_{cl.V} = V_{dr.cl}/V_{dr} = \frac{m_{dr.cl}/\delta_{dr.cl}}{m_{dr}/\delta_{dr}} = \frac{m_{dr.cl}}{m_{dr}} \frac{\delta_{sol}(1 - k_p)}{\delta_{dr.cl}} \approx k_{cl.m}(1 - k_p)$$

where δ_{sol}, $\delta_{dr.cl}$ and δ_{dr} are densities of, respectively, solid, dry clayey rock components and of dry rock.

The specific mass clayiness of sedimentary rocks may vary in the range from several to 90% and more.

It is better practice to evaluate the specific volume content $k_{sw.cl}$ of the swelling clayey component in a moist rock by using the relationship

$$k_{sw.cl} = V_{sw.cl}/V_{m.r}$$

where $V_{sw.cl}$ is the volume of moist (swelling) clay in a volume $V_{m.r}$ of a moist rock.

The degree of filling the space between sandy aleurolite, carbonate or other grain species by the clayey material is characterized by such petrophysical quantities as *relative clayiness* η_{cl} and *filling coefficient* $k_{fill.cl.V}$.

The relative clayiness η_{cl} is the ratio of the volume $V_{dr.cl}$ of the dry clayey component to the total sum of volumes V_{por} of rock pores and $V_{dr.cl}$ of the dry clayey component (to the volume of pore filler between larger-size rock fragments):

$$\eta_{cl} = \frac{m_{dr.cl}/\delta_{dr.cl}}{m_{dr}/\delta_{dr} - (m_{Ps+Av}/\delta_{sol})} = \frac{V_{dr.cl}}{V_{dr} - V_{Ps+Av}} = \frac{V_{dr.cl}}{V_{por} + V_{dr.cl}}$$

where *Ps* and *Av* are abridged forms standing for a sandy and aleurite fraction of rocks.

Since $V_{dr} = V_{por}/k_p$ and $k_{cl.V} = V_{dr.cl}/V_{dr}$,

$$\eta_{cl} = \frac{k_{cl.V}V_{dr}}{k_pV_{dr} + k_{cl.V}V_{dr}} = \frac{k_{cl.V}}{k_p + k_{cl.V}}$$

The coefficient

$$k_{fill.cl.V} = \frac{V_{sol.cl} + V_{w.sw.cl}}{V_{sol.cl} + V_w}$$

where $V_{w.sw.cl.}$, $V_{sol.cl}$ and V_w are water volumes of, respectively, clayey cement swelling, its solid phase and in swelled clayey rock pores. It characterizes the degree of clay filling of the swelled clayey rock pore space.

A number of petrophysical quantities: the average effective and median diameter of pores and grains, their modal radius, specific surface, specific water-, gas-, electro-, thermoconductivities, diffusion and adsorptional activity etc. are very much governed by their specific volumetric or relative clayiness and by the pattern of clayey particles—dispersed or bedded distribution (see below). The methods of determining clayiness used in petrography are far from invariably satisfying petrophysicists.

The values of clayiness characteristics are exaggerated for excessive rock pounding and transformation of larger particles to clayey particles, and diminished if, during preparation of a rock specimen its sesquioxides have been partially removed, or clayey particles are found not only in the interstitial space between larger grains but also in them proper, e.g. in feldspar sandstone grains in connection with transformation of their minerals to clayey materials.

Pore Composition

Pore composition (pore size distribution) is an essential structural rock characteristic that determines the concentration plurality of various effective-diameter pores in the rock pore volume. The concentration of definite-size pores in their mixture (pore volume) is expressed (by a volume and volumetric method) in fractions of the volume of all rock pores. The pore composition of uncemented clastic rocks is mainly governed by their granulometric composition.

Relatively large and practically similar in size pores are predominant in coarse-, large- and medium-grained well sorted and weakly cemented rocks. Similar, yet small- and fine-grained rocks demonstrate more small and thin pores. In poorly sorted uncemented clastic rocks different in size yet predominantly small and thin pores occur approximately in equal proportion.

The pore composition of cemented clastic rocks is conditioned not only by the granulometric composition but also by the content of such cement species as calcium carbonate, sesquioxides etc. Pores with various effective diameters occur here. At this, the fraction of large pores decreases and that of small and subcapillary pores increases.

The pore composition of many carbonate rocks is diversified, occasionally very complex. Among biogenic limestones there are, for example, coquina and algal limestones composed by practically intact shells and well preserved alga remnants. Fairly large and similar-size pores are predominant in the pore space of such rock varieties. Detrital and biodetrital biogenic limestone varieties are made up by various-size shell fragments. Their pore composition involves more minor size pores. Chemogenic limestones and dolomites are pelitomorphic, fine-, small- and medium-grained and less commonly display an oölitic structure. Similar size pores are predominant in pelitomorphic carbonate rocks, oölitic limestones also have pores that are approximately equal in size yet are much larger. The size of relatively equal pores in other chemogenic carbonate rock varieties is governed by the size of the constituent

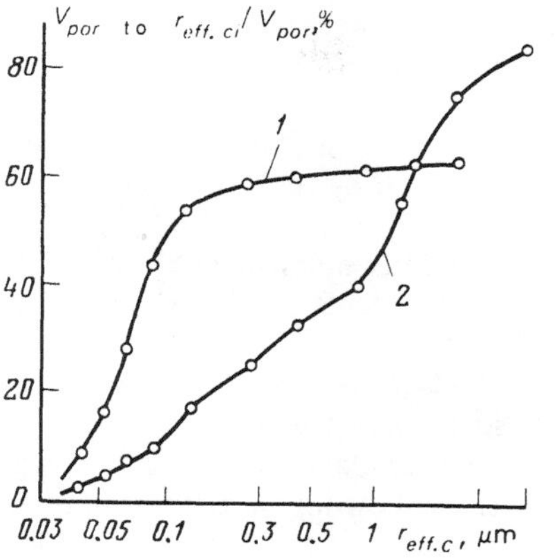

FIG. 6. Cumulative curves of radii, $r_{eff.c}$, of voids of polymictic sandstones of the Uzen deposit (after V.V. Pospelov).

k_p (in %); k_{perm} (in fm^2), respectively: *1* — 27.8; 0.7; *2* — 24.6; 333

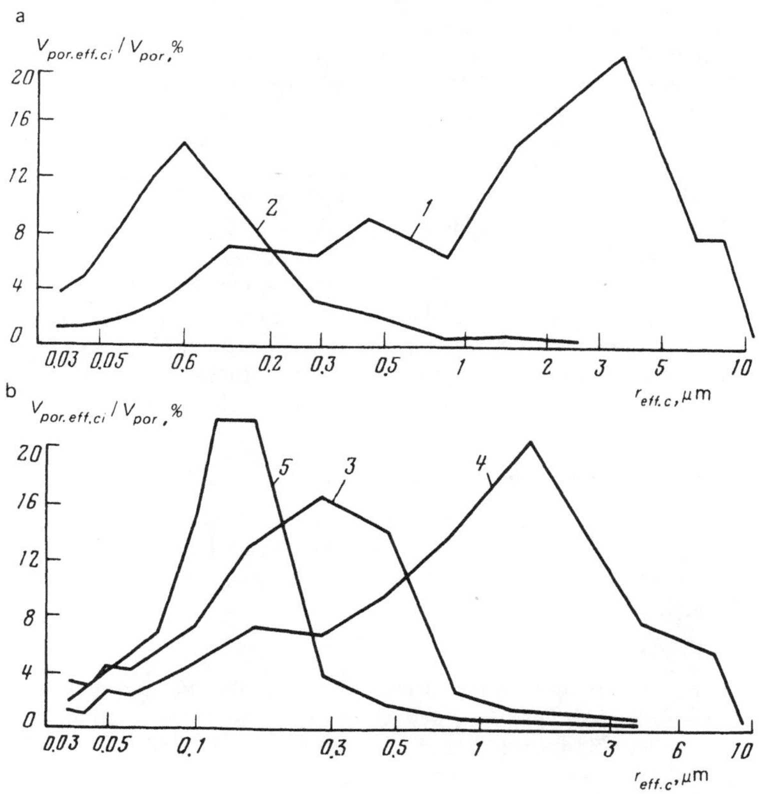

FIG. 7. A polygon of distributions of radii, $r_{eff.c}$, of voids in samples of polymictic sandstones of the Zhetybai deposit (*a*) and Uzen deposit (*b*) (after V.V. Pospelov, 1975).

k_p (in %); k_{perm} (in fm^2), respectively: *1* — 24.6; 333; *2* — 27.8; 0.7; *3* — 21.5; 1.4; *4* — 25.2; 4.2; *5* — 23.8; 1.3

grains. All the aforementioned rocks that have undergone secondary transformations may also contain caverns, sulphatized space, so their pore composition is of a complicated character.

As to crystalline rocks, their pore composition that has been inadequately studied is governed by the form and size of crystals and by their vesiculation and jointing.

The pore composition of rocks is what largely determines many of its other petrophysical quantities: gas and water permeability coefficients, specific electroconductivities, electrochemical and γ-activities etc.

To study pore composition, use is made of methods of capillary mercurial indentation (mercurial porometry), water displacement (aqueous porometry), capillary condensation, adsorption, optical methods involving capillary luminophor impregnation of a sample, electron microscopy and others. Following these studies we obtain porograms, pore size distribution curves characterizing the pore composition of the sample under study. They are also used to calculate the median, average and modal values of the effective diameters (or radii) of pores.

The pore composition is characterized by an integral (cumulata) and a differential (polygon) porogram (Figs. 6, 7). Both these graphs are plotted in a rectangular coordinate system. The pore effective radii r_{eff} are plotted along the abscissa axis of the cumulata, and along the ordinate axis sections whose lengths are proportional to the accumulated frequences of definite fractional contents $V_{por\ \mathrm{to}\ r_{eff.ci}}/V_{por}$ and $V_{por\ \mathrm{to}\ d_{ci}}/V_{por}$ to the effective capillary radius $r_{eff.ci}$ (or diameter $d_{eff.ci}$). Along the polygon's abscissa axis the effective radius $r_{eff.ci}$ (or diameter $d_{eff.c}$) of pores is plotted, and along the ordinate axis their fractional contents $V_{por.\,r_{eff.ci}}/V_{por}$ (or $V_{por.\,d_{eff.ci}}/V_{por}$) corresponding to each of the radii (or diameters) being determined.

It is only by combining adsorption, capillary condensation, mercury porometric, electron microscope, raster and sometimes X-ray and spectroscopy methods that we are able to fairly adequately characterize the structure of the pore space and follow the variation of its structure at any stage of rocks' life.

Specific Surface and Pore Space Surface

The rocks' pore space surface is a surface dividing the solid phase of rocks from all others. The property of rocks to possess different pore space surface is evaluated by *specific surface—volumetric* S_V and *mass surface* S_m. The former bounds the pore space of a unit volume, the latter of a unit mass of a rock:

$$S_V = S_{por.\,V_{dr}}/V_{dr}; \quad S_m = S_{por.\,m_{dr}}/m_{dr}$$

where $S_{por.\,V_{dr}}$ and $S_{por.\,m_{dr}}$ are pore space surfaces in the volume V_{dr} and mass m_{dr} of a rock. The specific surface S_V is measured in m^2/m^3 (or m^{-1}) (or, alternatively, cm^2/cm^3 (or cm^{-1}), and S_m in m^2/kg and more commonly in m^2/g.

The specific surface of sedimentary and clastic rocks is governed by the mineral and granulometric composition, configuration of grains, cement content and type. It decreases with increasing the average diameter of grains and decreasing clayey cement content (Fig. 8). However, the values of S are mainly governed by clayiness of clastic rocks (Fig. 9*a*, *b*). The specific surface also increases with increasing

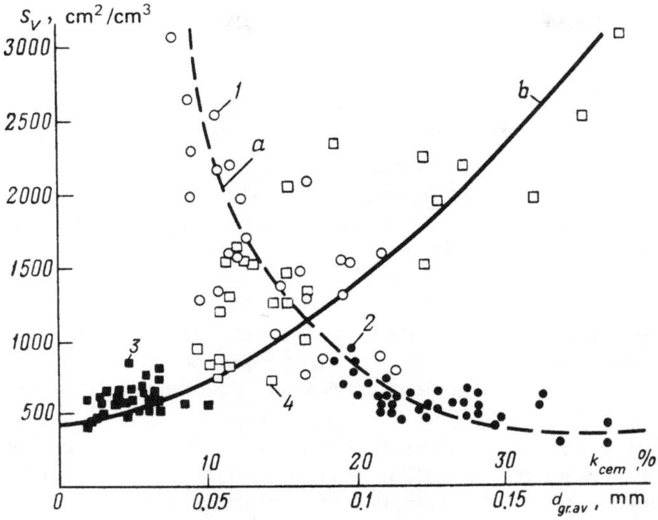

FIG. 8. Dependences of the specific surface S_V on average grain diameter $d_{gr.av}$ (*a*) and clay content k_{cem} (*b*) (after O.A. Chernikov and A.I. Kurenkov).

Average grain diameter (in mm): *1* — 0.1; *2* — >0.1; cement content (in %); *3* — to 10; *4* — >10

residual water saturation $k_{w.r}$ (Fig. 10). The older are rocks and the greater is the depth of their occurrence i.e. the greater is the action on them by the pressure and temperature, the lesser is generally the specific surface.

The greatest specific surface is shown by natural adsorbents: clays, tripoli, some bauxite varieties, ash tuffs and some other similar rock species. Very small particles with $r_{eff.ci}$ from $(18\text{-}19) \times 10^{-4}$ to 0.1-0.2 μm are predominant in their granulometric composition, and S_m is in the range from $(10\text{-}400) \times 10^3$ m²/kg. A much lesser ($(0.5\text{-}2) \times 10^3$ m²/kg) specific surface is shown by well sorted, rolled, weakly

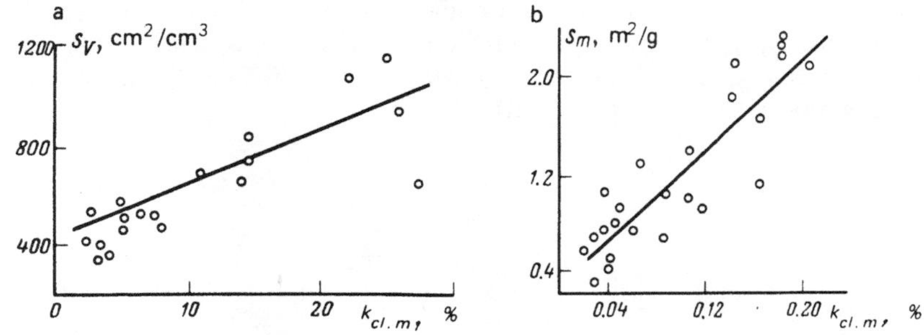

FIG. 9. Dependences of the specific surface S_V (*a*) (after V.L. Komarov) and S_m (*b*) (after B.I. Tulbovich) on the coefficient of mass clayiness $k_{cl.m}$ of terrigenic rocks

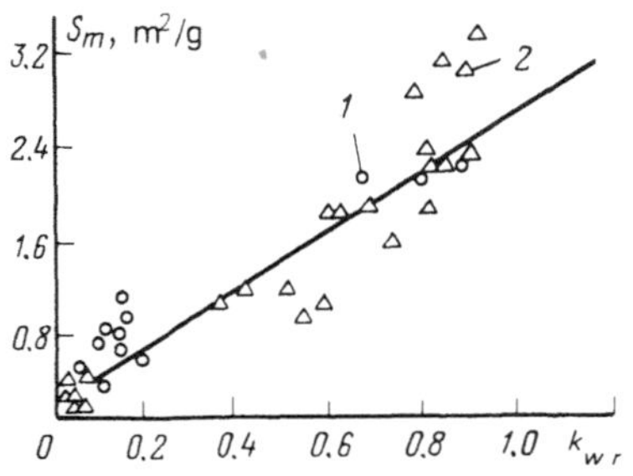

FIG. 10. Dependence of the specific surface S_m on the coefficient of residual water saturation, $k_{w.r}$, of terrigenic rocks (after B.I. Tulbovich).
1, 2 — S_m values for different deposits of the Perm Region

cemented, medium- and large size detrital rocks. Oil and gas traps have particles with both the former and latter radii and often the fragments (pores) of large size are predominant in them. Viewed in this context the limits of their S_m variation are approaching (0.5-2) × 10^3 m^2/kg.

The specific surface, characterizing the structure and adsorptional capacity of rocks' pore spaces, determines the formation and transformation conditions of sedimentary rocks, makes it possible to distinguish their structural types and adsorbent types. It is closely associated with a number of other petrophysical parameters and may not be only determined in a laboratory but also evaluated by the available relationships. Given the specific surface, it is possible, from the relations with other characteristics, to evaluate approximately the latter. Adsorptional, filtrational, based on the capillary effect, optical, electron and microscopic, granulometric and other laboratory methods have been developed to determine specific surface. In addition, the above methods may have sometimes several varieties.

The specific surface and the average effective pore radius can be also assessed from other values that are closely associated with it:

$$r_{eff.av} \approx \sqrt{8k_{sp}/k_p}/3$$

The above relationship has been derived on an assumption that the rocks' pore space provides three systems of parallel and equally spaced cylindrical capillaries each of which is at right angles to one of three mutually perpendicular faces of a rocks's parallelepiped.

The values $r_{eff.av}$ of a rock are also estimated from their connections with the exchange capacity, specific volumetric adsorptional moisture capacity, surface conductivity, diffusional adsorptional activity, permeability coefficient, γ-activity and some other values (see Figs. 14, 18, 51).

Chapter III

Moisture content, moisture capacity, electrical double layer

Water containing soluble salts will invariably occur in rocks. The volume of water is equal to that of pores of a dry rock or is more or less than it. In the latter case the rocks' pore space, apart from water, contains either gas or oil or their mixtures.

The content of rocks' water is conditioned by *the moisture* (*or water*) *content* of rocks. As to the property of rocks, depending on their power and thermobaric conditions to retain one amount of water or another is known as *moisture capacity*. Since the increased moisture content of rocks leads not only to variation in their properties but also to those of the entrapped water in a steplike fashion, we may distinguish hygroscopic, maximum adsorptional, maximum hygroscopic, suspended, capillary and full moisture capacity; not only definite amounts but also water characteristics, in particular, mobility correspond to them.

Sec. 4. Full Moisture Content, Interphase Interaction and an Electrical Double Layer

Full moisture capacity is the property of rocks to retain varying maximum possible volume V_w of water for a definite volume V_{dr} of dry rock. Given this moisture content, 10-20% of the pore volume of unswelling rocks may be taken by the entrapped air. In the case of swelling rock species the volume of water exceeds V_{por} of dry rock. This is accounted for by water penetration (interstitial water) between crystalline lattice packets of some rock minerals, such as montmorillonite, vermiculite, halloysite hydrate varieties etc. An increased water content compared with porosity is also observed upon filling with water of tubular channels between structural units of sepiolite-attapulgite-palygorskite minerals.

Under conditions of full and other types of moisture capacity at the rock's interphase boundaries there occur electrochemical reactions to produce on either side *an electrical double layer* (EDL), a specific electrical charge distribution in the rocks' interphase boundary regions. In the presence of an EDL each phase has a volumetric charge and between them there appears an electrical potential difference. It is solely the EDL at the contact of electrolytes with the solid rock component that has been sufficiently studied. This layer consists of an ionic, adsorptional and orientational EDL types.

The ionic layer type (Fig. 11*a*) is produced due to residual electrical forces at the surface of the rocks' solid component given different chemical potentials of

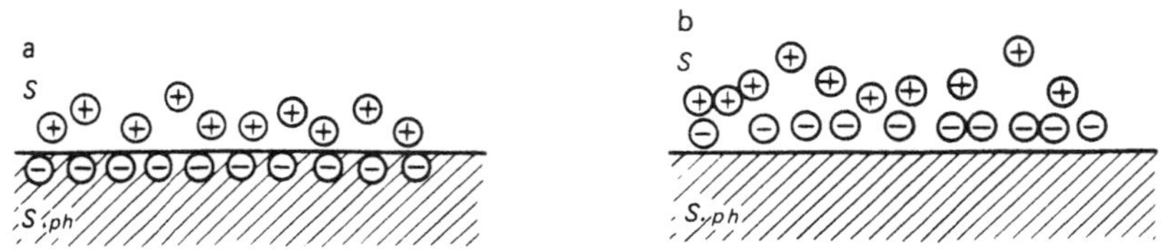

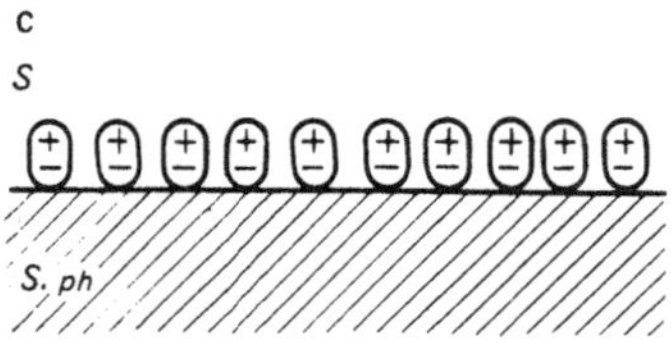

FIG. 11. Structure of the ionic (*a*), adsorptional (*b*) and orientational (*c*) types of an electric double layer.

S — solution; S_{ph} — solid phase of rock

similar ions in different rock phases. It includes a layer of dehydrated, potential determining ions in the solid phase and a layer of hydrated counterions in the liquid phase. Potential determining and counterions have the opposite signs, so individual phases are charged, and on the whole the rock is neutral. The potential determining ions are diffusely located at the very interphase boundary with density decreasing with greater depth of the solid phase. They impart to the latter a definite (positive or negative) potential (charge sign), hence the term "potential determining". The sign and volume of the solid phase charge are governed by the chemical composition of the contacting phases, the extent and state of the interphase surface.

The layer of hydrated counterions adjoins the rocks' solid phase, being retained here by potential determining ions. Yet thermal molecular motion in the liquid phase acts to wash out this layer, so this has an immovable and diffuse part. The immovable or dense part of the counterion layer is located in the pore space and adjoins the solid phase followed by the diffusional part where ion density decreases toward deeper reaches of the liquid phase.

The electrical double *adsorption layer* (see Fig. 11*b*) results from selective adsorption of indifferent electrolyte ions on the solid phase's uncharged surface. The indifferent electrolyte has no ions shared by the solid phase. What is adsorbed is that ion that has a greater valency and lesser hydrate radius; it closer approaches the adsorbent and is stronger held here by the van der Waals forces. The latter appears at the solid phase's surface under the action of the adsorbing ion. As it approaches the adsorbent, it causes its surface atoms to be polarized. At this, it polarizes its electron shells, the atoms being transformed to momentous dipoles capable of retaining adsorbing ions. Indifferent electrolyte ions polarize solid phase atoms to a greater degree the larger is their electrical field. In particular, the cations deform adsorbent surface atoms the more the greater is their charge and the less is their hydrate radius. The degree of deformation increases as the atoms and ions increase in size. Anions are of a greater size than cations, so their polarization degree is more appreciable.

The structure of the adsorptional EDL type does not differ from that of the ionic EDL, but adsorbed ions and counterions of this layer are to be found in the electrolyte.

The orientation type of EDL is made up by dipole water molecules. Part of them are oriented and held in the pore space close to positive and negative noncompensated or van der Waals charges on the rocks' solid phase surface forming a layer of strongly bound water. This layer's field and residual electrostatic forces cause additional water amounts forming its loosely bound layer to be oriented. Loosely bound water includes also water oriented near ions of the diffuse part of EDL. The two layers form an orientation type electrical double layer—*bound water* (see Fig. 11*c*).

Two other types of the electrical double layer also appear at the solid phase's surface, for pore waters generally contain not only electrolyte with an ion common for the two phases but also indifferent electrolyte in appreciable concentration. Depending on the properties of the solid phase and electrolyte one or two types of the electrical double layer predominantly develop.

An electrical double layer including all three EDL types seems to originate in the pore space of calcite and pure (without clayey particles) limestones saturated by chloride, mainly sodium, waters of the composition (Cl $81SO_4 17$)/[(Na + K) 83 Ca 10 Mg6] (after Kurlov's formula) and mineralization 10 g/l.

The ionic layer type is then formed due to $CaSO_4$ and $Ca(HCO_3)_2$. The former enters the composition of the aforementioned waters and the latter forms upon CO_2 containing water dissolving $CaCO_3$ in conformity to the reaction:

$$CaCO_3 + H_2O + CO_2 = Ca(HCO_3)_2$$

In water $CaSO_4$ and $Ca(HCO_3)_2$ decompose to form hydrated ions Ca^{2+}, SO_4^{2-}, HCO_3^-. Calcium is dehydrated, is extracted from the rock's liquid component by the solid phase to complete its crystal lattice and be located there diffusely (with a density dropping with deeper penetration into the solid phase) near the very boundary of the liquid phase, ions SO_4^{2-}, HCO_3^- forming a counterion layer with a dense and diffuse parts; in the latter counterions are arranged also diffusely (with density decreasing toward free solution). *The free solution* is a liquid in the central pore regions. It is practically unaffected by the solid phase surface electrical forces. Thus the solid phase is positively charged, and the electrolyte negatively.

NaCl, KCl, $MgSO_4$ and H_2O concentration in the pore electrolyte solution is much greater compared with $CaSO_4$ and $Ca(HCO_3)_2$ concentration (particularly, that of the latter, due to inappreciable solubility of $CaCO_3$ in the presence of CO_2), so, according to the mass action law known alternatively as Guldberg and Waage law and owing to the effect of the induced van der Waals forces two other EDL types appear at the solid component's surface as well. In so doing, ions Cl^- and SO_4^{2-} are found in the adsorptional type layer. In a solution containing ion HCO_3^- the limestone adsorbs it to get a negative charge.

It appears that the same mechanism is involved in the origination of the three types of electrical double layer in pure (not containing clayey, carbonate or ferruginous cement) limestones saturated by chloride, mainly sodium waters with the chemical composition Cl 98/Na 99 (after Kurlov's formula) and mineralization

330×10^3 kg/m^3 (330 g/l). The interaction of quartz and water results in the appearance at its surface of silanol (—SiOH) groups according to the reaction

$$2(\text{—SiO}) + H_2O \rightarrow 2\,(\text{—SiOH}) + O$$

This surface compound (polysilicic acid) is capable of partial dissolution in water of the acidic type: $-SiOH \rightarrow SiO^- + H^+$. The H^+ (in fact, H_3O^+) ions go to the pore electrolyte to form here a counterion layer which is held close to the solid phase owing to electrostatic interaction with ($-SiO^-$) ions at the quartz's surface. The electrolytic H^+ and OH^- are potential determining ions that condition ($-SiOH$) group dissociation degree. The OH^- enhance and the H^+ increases dissociation. Consequently, with increasing pH the surface negative potential and zeta (or electrokinetic) potential (see what follows) increase accordingly.

The adsorptional layer is formed of NaCl and the orientation one, of water dipoles. The Cl ion and strongly bound water molecules are located close to the solid phase, and the Na ion in a counterion layer. Water dipoles surrounding Na as well as ones caused to be oriented by the strongly bound water layer and residual van der Waals forces are what forms a loosely bound water layer.

The mechanism of origination of an electrical double layer in clays is much the same. Negative ions produced by dissociation of the alumosilicate component of rocks are mainly located at the surface of clayey particles, these ions attract positive ions from the solution. As a result, dense and diffuse portions of the electrical double layer are formed (Fig. 12). The adsorbed ions cause the water dipoles that are in the immediate vicinity to be oriented.

Recently use has been extensively made of a hypothesis that in many minerals the electrical double layer appears due to isomorphic substitution for one type of atoms (or ions) at the nodes of their crystal lattice by other species. For example, the Si atom of some alumosilicate lattice minerals can be replaced during the formation of these latter by an Al atom. Thereupon the solid component acquires an electron and becomes negatively charged.

The electrical double layer varies in thickness, has unequal potential drop at this thickness and at the movable portion of the diffuse layer. The thickness *d* of the electrical double layer is the distance between the centres of gravity of charges

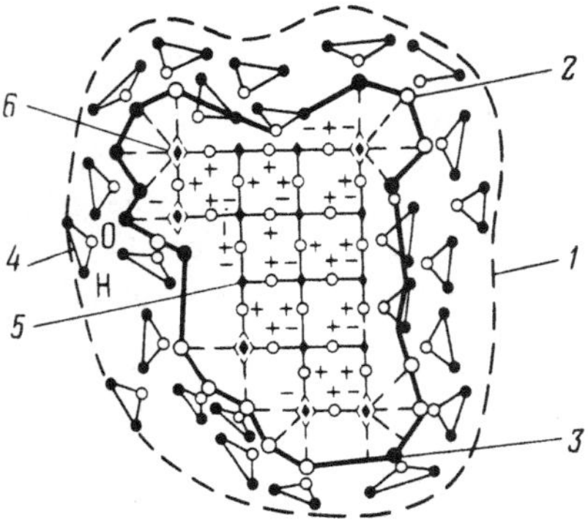

FIG. 12. Structure of an electric double layer near a clay particle (after Pauli and Valko).

1 — boundary of particle with EDL; *2*, *3* — exchange, respectively, monovalent (Na^+, K^+ etc.) and divalent (Ca^{2+}, Mg^{2+} etc.) cations; *4* — water molecules; *5* and *6* — anions of crystal lattice having saturated and nonsaturated bonds of the potential determining and diffuse layers. It is made up of the thickness of the potential determining layer d_p (the distance from the centre of gravity of the charges' potential determining layer to the interphase boundary), the Helmholtz double layer d_0 (the distance from the solid phase to the plane passing through centres of gravity of the closest counterions—plane of the greatest closeness (Fig. 13*a*) several decimal fractions of the nanometre in thickness depending on the degree of ion dehydration) and thickness λ of the diffuse portion of the electrical double layer equal to the distance from the plane of the maximum approximation to the centre of gravity of counterions of the diffuse portion of EDL.

The thickness $d \approx d_0 + \lambda$ (see Fig. 13*b*) can vary from several decimal fractions of the nanometre to several hundreds of micrometres. The greater the rock electrolyte concentration, the less is the thickness of the electrical double layer. The electrical double layer contracts as a result of decrease in the thickness of the loosely bound water film and the diffuse portion of the counterion layer. Under conditions of pronounced pore electrolyte concentration these layers may be nonexistent which is due to the great osmotic pressure of highly mineralized electrolyte molecules and ions at which thermal molecular motion is unable to tear ions and water dipoles from the interphase surface and diffusely distribute them in the electrolyte.

A potential drop at the thickness of the electrical double layer (galvanopotential) is known as *the Nernst potential* $\mathscr{E}$. There is no way of measuring it since the electrical charges whose distribution at the interphase surface determines $\mathscr{E}$ are discrete and connected to the rock material. As the charged particle passes from the solid to the liquid stage or vice versa, not only electrical work is performed but also work determined by the chemical potential difference in the two phases. The Nernst potential is composed of the ζ_{sol} potential discontinuities at the potential determining layer in the solid phase $\zeta_{liq.0}$ and $\zeta_{liq.d}$ respectively on the dense and diffuse portions of the counterion layer in the liquid phase. Consequently, $\mathscr{E} = \zeta_{sol} + \zeta_{liq.0} + \zeta_{liq.d}$.

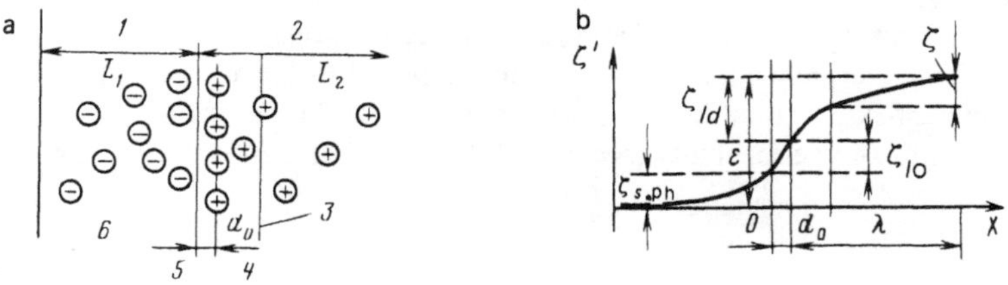

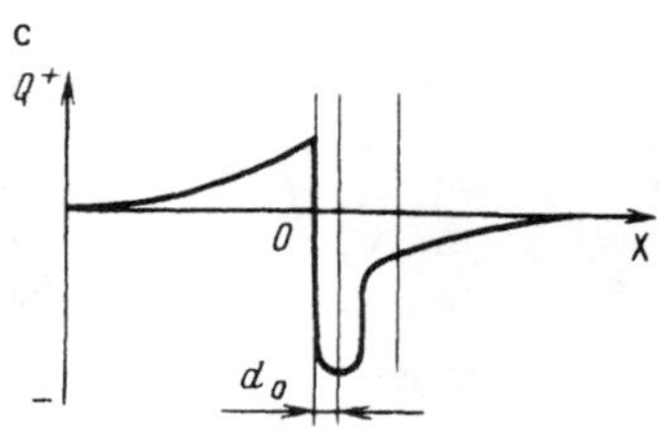

FIG. 13. Structure of an electrical double layer (*a*), distribution of potential ζ' (*b*) and charge Q (*c*) at the interphase boundary of the solid phase and electrolyte (after V.A. Myamlin and Yu.V. Pleskov).

1 — solid body: *2* — electrolyte; *3* — plane of sliding of liquid; *4* — Gouy's layer; *5* — Helmholtz layer; *6* — region of volumetric charge

The $\zeta_{liq.d}$ potential discontinuity consists of its discontinuities on the hydrodynamically movable (zeta or electrokinetic potential ζ) and immovable ($\zeta_{liq.d} - \zeta$) portions of the diffuse layer (see Fig. 13*b*).

Let us consider the zeta potential representing a potential difference between points on the outer boundary of the immovable portion of the counterion layer and in the electrolyte immediately outside the electrical double layer on its portion moving together with the free solution with respect to the solid phase under the action of the electrical field (or applied pressure).

The values of this potential can be determined since both of its boundaries are found in the liquid phase; it is determined from the velocity of electrophoresis or electroosmosis. The zeta potential is conditioned by the number of the charges near the interphase boundary and their distribution in the electrical double layer. The zeta potential values of rocks vary from -35 to 35 mV.

The properties of water and ions in the electrical double layer differ from those in a free solution. Strongly bound water is denser compared with free water (its density δ_w varies from 1.2 to 2 g/cm^3), displays greater viscosity, shear strength, elasticity and lesser electroconductivity compared with distilled water. Bound water freezes in various rock species at different temperatures: in kaolinite at $t = -20$ °C, in montmorillonite at -193 °C. In addition, the freezing temperature is conditioned by the specific surface and pore size. In ultramicroscopic pores it freezes with more difficulty. The amount of frozen water is also governed by the composition of the exchange ions. At $t = 6$ °C the amount of unfrozen water in Na, K and Mg clays proved to be roughly the same, and in Ca clays less. As strongly bound water is formed appreciable amount of heat is liberated. This water will not dissolve either salts or sugar which provides a basis for determining its volume V_{aq} in a rock. Loosely bound water has properties approaching those of free water, yet the viscosity of the former exceeds the latter's and it can slowly move from one rocks' solid phase region to another. The freezing temperature of this water is -1.5 °C, and its ability to dissolve generally soluble materials is inappreciable.

In the laboratory we determine the volume of all water remaining at the specified pressure gradient (residual water) without dividing it into strongly and loosely bound water. This water may not contain the entire amount of loosely bound water. The latter's content is governed by the pores' radius (Fig. 14), exchange capacity (Fig. 15), voids ratio and permeability (see Figs. 20, 55), and also by clayiness coefficient (Fig. 16).

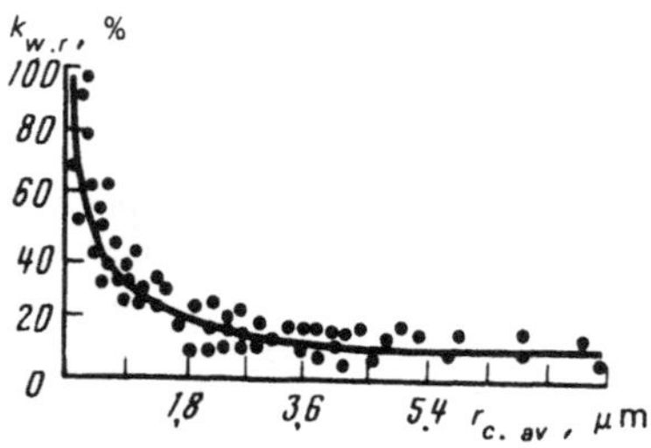

FIG. 14. Dependence of coefficient of residual water saturation, $k_{w.r}$ on the average radius of capillaries, $r_{c.av}$ of foraminiferous algal limestones for a group of various deposits of the Perm Region (after B.I. Tulbovich)

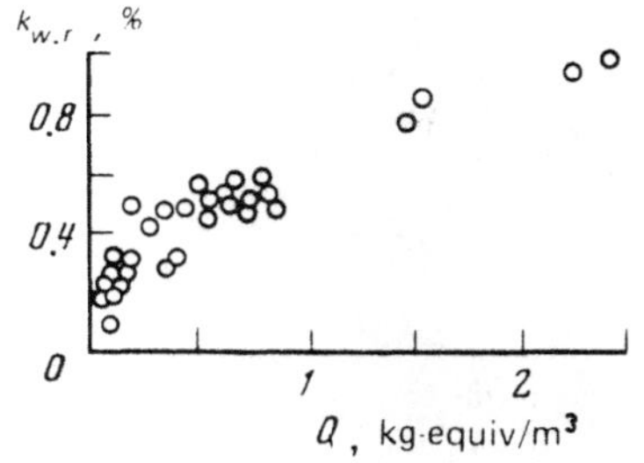

FIG. 15. Dependence of coefficient of residual water saturation, $k_{w.r}$, on the capacity, Q, of exchange of sandy-clayey Carboniferous sediments of the Eisk-Berezanskii region of the Krasnodar province (after L.K. Tankayeva)

Counterions are replaced by greater valency ions with a smaller hydrate radius according to the series $Fe^{3+} > Al^{3+} > Ba^{2+} > Ca^{2+} > Mn^{2+} > Mg^{2+} > K^{+} >$ $> Na^{+} > Li^{+}$. The exchange energy of hydrogen is higher than that of univalent and bivalent ions. Exchange is facilitated by the growth of concentration of electrolyte coming to the rock. The exchange capacity of rocks is characterized by exchange capacities: mass capacity $Q_{0.1m}$, volume capacity Q_V and reduced capacity q_V. *The mass exchange capacity* $Q_{0.1m}$ is the number of milligram equivalents of exchange atoms for 0.1 kg of a dry rock, it varies in the range of $<10 - >1\,500$ mg-eq/kg. *The volume capacity* is the number of milligram equivalents of exchange ions for 1 m³ of rock:

$$Q_V = \frac{Q_{0.1m}}{0.1}(1 - k_p)\delta_{sol} = \frac{Q_{0.1m}}{0.1}\delta_{dr}$$

where δ_{dr} and δ_{sol} are, respectively, density of a dry rock and of its solid component, kg/m³; k_p is the voids ratio.

The reduced exchange capacity is

$$q_V = Q_V/k_p$$

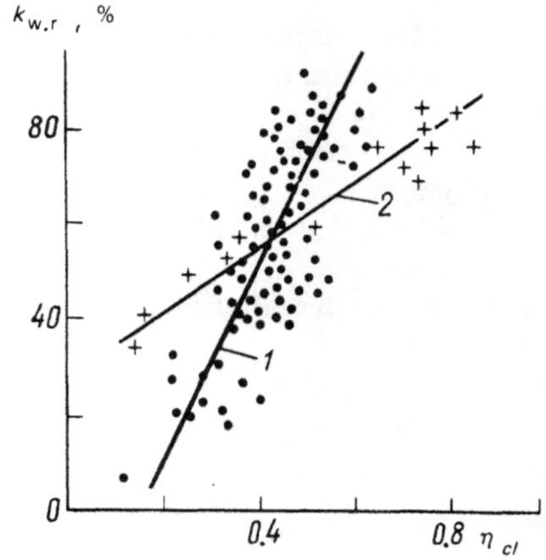

FIG. 16. Dependence of coefficient of residual water saturation, $k_{w.r}$, on the relative clayiness, η_{cl}, of productive sandy-clayey beds of West Siberian deposits (*1*) (after E.I. Leont'ev) ($k_{perm} > 1$ fm²) and Triassic carbonate sediments of the Beli Isvor and Glavaci (*2*) (after I.B. Nikolaeva)

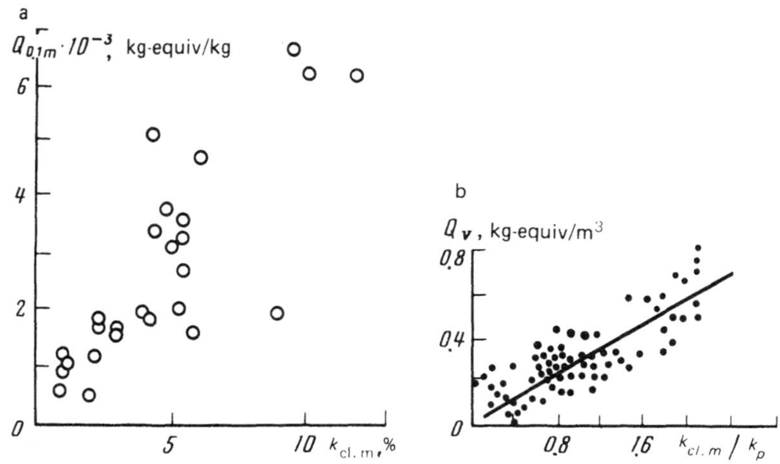

FIG. 17. Dependences of exhange capacities Q on mass, $k_{cl.m}$, and reduced, $k_{cl.m}/k_p$, clayiness.

a — for carbonate rocks (Eocene) from borehole 4 of the area of the Chisten'kaya village of the Crimean region (after A.A. Vedenin); *b* — for sediments of Kartamysh and Araucarite formations of the Dnieper-Donets depression (after M.M. Ellanskii)

It varies for sandy-aleurite and clayey rocks in the range of (2-140) × 10^6 mg-eq/m^3.

The greatest exchange capacity is common for natural adsorbents, such as clays, tripolis, diatomites, opokas, ash tuffs etc. The difference in the exchange capacity is governed by the mineral composition, structure of clayey and other rock minerals, degree of its perfection, isomorphic substitutions and dispersion extent. The most important exchange cations in clays are Ca^{2+}, Mg^{2+}, H^+, Na^+, K^+*, arranged according to the order of their mutual exchange. The most important among anions are PO_4^{3-}, SO_4^{2-}, Cl^-, NO_3^- (anion exchange is poorly understood).

In the exchange complex of clays either Na or Ca prevails. If Na is predominant, clays will swell, if Ca, they will not. The swelling of Na in montmorillonite clays may occur attaining the ratio of the original volume to the increased volume 1:8.

The exchange capacity of sandy-aleurite and clayey rocks increases with their mass $k_{cl.m}$ or relative ($k_{cl.m}/k_p$) clayiness (Fig. 17*a*, *b*) and is governed by the mineral composition and diameter of grains (Fig. 18). Given that d_{gr} = const, it is the smallest for kaolinite cement rocks and the greatest for montmorillonite cement rocks.

Counterions of the electrical double layer diffuse portion at distances from the solid phase more than two or three water molecules in thickness are movable. They

* Potassium is substituted by Na^+, Na^+ by H^+ etc.

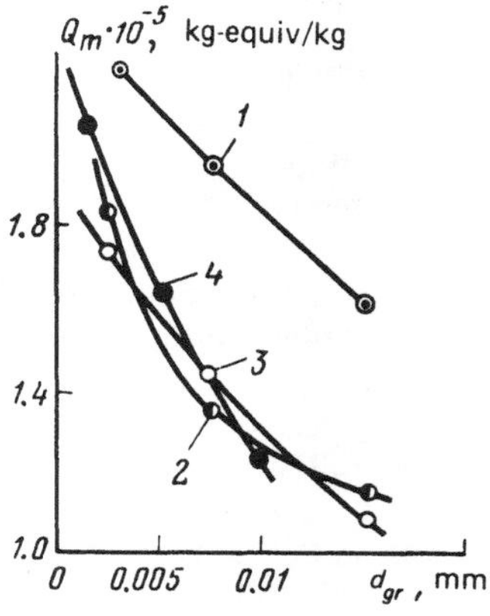

FIG. 18. Dependence of exchange capacity Q_m on grain diameter d_{gr} of the clayey fraction (after V.L. Komarov).

Cement: *1* — hydromica-kaolinite; *2* — kaolinite; *3* — mixed laminated hydromica-montmorillonite, kaolinite; *4* — argillite of the Tula horizon (Tuimaza oil field); quantity Q_m is the number of kilogram-equivalents per kg of dry rock

are transported in a rock jointly with its water under the action of the tangential field E (in V/cm) or pressure gradient. The linear speed (in cm/s) of their motion is

$$u_0 = \varepsilon \zeta E / 4\pi\eta$$

where η is the viscosity of the solution; ε is dielectrical permeability; ζ is zeta potential.

However, the speed of movement (or mobility) of ions in the electrical double layer is much less than outside it, in the free solution. As the ion moves in a double layer of rocks it is being acted on by the force of the external electrical field conditioned by friction in bound water, the forces governed by the electrostatic interaction between the mobile ions (relaxation effect and electrophoresis forces), and the force conditioned by the ionic electrostatic field of the solid phase. All of them, excepting the force of the E field act to break the ion motion. A number of theories and relevant relationships are available determining cation mobility upon its removal from the solid phase to a maximum value outside the electrical double layer.

The electrical double layer, in particular, a bound water layer, solely for clays, tripolis, diatomites, ash tuffs, zeolites etc. and rocks showing an appreciable adsorbent content (clay-sandy, calcareous and magnesial rocks) occupies an appreciable fraction of the pore space. Mesopores are predominant in such rocks, and the specific surface is appreciable (10-100) $\times$ 10^3 m^2/kg. When other rock species are involved, the fraction of the pore space taken by the electrical double layer is inappreciable. The voids space represented mainly by supercapillary and capillary pores is almost entirely taken by free water. It varies in composition and mineralization for different-age and different type rocks occurring at different depths.

Free water together with bound or practically entirely bound water is to be found in subsurface water reservoirs representing a self-contained geological body where water is entrapped (say, a bed).

An artesian basin represents a higher-order hydrogeological reservoir. The overburden of an appreciable thickness is composed here by alternating oil or gas traps and aquicludes.

The foundation contains jointing or vein jointing structure reservoirs. The complex foundation reservoirs include hydrogeological massifs, i.e. outcrops to the earth's surface or tectonic uplifts of typically crystalline rocks that display aquiferous "zones" of jointing and cracks. There also exist other reservoir types of underground waters.

Within continental platforms major artesian basins are predominant and hydrogeological massifs are less in number. As to folded regions, an opposite pattern is to be observed.

In stratal water basins the ion salt composition and mineralization of subsurface waters vary from top downward. We distinguish here the following hydrochemical zones: *A* is zone of fresh waters with mineralization up to 1 g/kg (1 g/l) with subzones of extremely fresh, normally fresh and hard fresh waters; *B* is zone of salt water with mineralization from 1 to 35 g/kg (36 g/l) containing subzones of brackish, weakly saline and strongly saline waters; *C* is zone of brines with mineralization from >35 (36 g/l) to >420 g/kg (>500 g/l) containing subzones of very weak, weak, strong, very strong and extremely saturated brines.

In the vertical cross section of stratal water basins the hydrochemical zones and subzones are located in a different sequence. Their particular combination reflecting the hydrochemical cross section of the entire thickness of the sedimentary cover of a basin is known as a *hydrochemical belt*.

The belts are classified as a single-zone (*A*), a two-zone (*AB*) and a three-zone (*ABC*) belt. A single-zone belt generally surrounds marginal parts of artesian basins, a two-zone belt will be found in the internal field, and the central part will manifest all the three zone types. Different water zones vary in thickness over a wide range. Subsurface waters mineralization often agrees with their ion salt composition. Fresh waters are generally hydrocarbonate, salt waters are sulphate and brines are chloride.

We refer to a normal hydrochemical cross section such a section that has mineralization increasing from top downward and shows this trend in ion salt composition variation: HCO_3^- (Ca^{2+}, Na^+) → SO_4^{2-} (Ca^{2+}, Na^+) → Cl^- (Na^+, Ca^{2+}) or zone variation $A \to B \to C$. It is frequent that we observe a different hydrochemical zone distribution sequence said to be a hydrochemical inversion. A drop in waters mineralization with depth is observed in recent-origin artesian basins of Trans- and Ciscaucasia, in the Middle Asia and in some West Siberian deposits.

Full moisture content is manifested by rocks occurring below the groundwater level. Above this level we generally observe a zone of incomplete water saturation of rocks that includes a capillary, a suspended (or perched) water subzone, a maximum hygroscopic and a hygroscopic moisture content subzone. The zone of incomplete water saturation of rocks may be absent in humid regions where the groundwater level in clayey rock mass more often than not agrees with the ground surface. The depth of this zone in other regions varies from one place to another since the groundwater level fluctuates depending on the climatic conditions and properties of rocks.

Sec. 5. Capillary, Suspended, Maximum Hygroscopic and Hygroscopic Moisture Capacity

Capillary Moisture Capacity

Capillary moisture capacity is the property of rocks to retain different volume $V_{w.c}$ of bound and capillary head water per definite volume V_{dr} of a dry rock. Capillary head water is one that is contained in capillary pores communicating with the groundwater level. The particular type of moisture capacity is observed immediately above the groundwater level. In the immediate vicinity of this latter all rocks' voids, except supercapillary pores, are filled by bound and free capillary head water. With increasing the distance from the groundwater level the number of capillary pores containing free water will decrease; completely filled by water will be ever thinner pores where the capillary rise of water is greater. This will continue until the suspended moisture capacity subzone is reached. Consequently, capillary moisture capacity is a variable quantity.

The height of the capillary subzone is 0.3 to 0.6 m for large and medium grain sandy rocks; for not so well sorted varieties where the aleurite fraction predominates it is 1 m, and for high clay content rocks it is up to 2-3 m.

Suspended Water Capacity

Suspended moisture (water) capacity is the property of rocks to retain a different volume $V_{w.b}$ of bound or $V_{w.sus}$ of capillary suspended water for a definite volume V_{dr} of a dry rock.

Under conditions of suspended moisture content the pore space of coarse-grained rocks contains a volume $V_{w.b}$ representing a sum of volumes $V_{w.str.b}$ and $V_{w.l.b}$, of, respectively, strongly bound and loosely bound water, the remainder being occupied by the air.

Medium-grained rocks with the particles' diameter 0.01-0.05 to 1 mm (e.g., sands and sandstones), given that the diameter of capillaries decrease, apart from capillary bound water, contain suspended water maintained by the difference in meniscus pressure. This type of water is not communicating to the groundwater level or that of capillary head water; it completely fills the pores some distance from the upper level of the capillary zone. However, under definite conditions capillary suspended water percolates downward, then such rocks contain solely bound water.

Fine-grained rocks have the pores' gap so small that the pores prove to be filled solely by sorptional bound water (which is in this case also suspended).

Suspended moisture capacity is revealed in strata composed by clastic rocks at a depth smaller compared with those of capillary moisture capacity where the water evaporation rate is more intensive than that of capillary rise, and the pores in such types of strata contain less water and more air than does the capillary moisture capacity subzone.

Maximum Hygroscopic Moisture Capacity

The maximum hygroscopic moisture capacity subzone of the same sediments following the suspended moisture content subzone contains even less water and more humid air, the water occurring quite close to the earth surface, and water evaporation rate is more pronounced here.

Maximum hygroscopic moisture capacity is observed at the relative air humidity in the rocks' pores at least 94%. In addition, a maximum volume $V_{w.str.b}$ of strongly bound water and a volume $V_{w.an.por}$ of pore angles (interstitial water) are held in the vicinity of the solid phase, i.e. $V_{w.m.h} = V_{w.str.b} + V_{w.an.por}$.

The mechanism of origination of pore angle water appears to be as follows. If the pore space of a dry rock is to be filled by air whose relative humidity exceeds 80% and does not exceed 94%, then a layer of loosely bound water appears above strongly bound water which is the first to settle. However, on flat pore surfaces it fails to be held and is caused to be drawn to the pores' angles by the meniscus forces. Vapour elasticity above concave menisci in the pores' angles is less than above relatively flat surfaces of grains (or pores). This results in a more intensive water vapour condensation rate and additional accumulation of water at the particles' joints. The latter process is known as capillary condensation which is particularly appreciable for rocks with pore size less than 0.0002 mm where menisci show an appreciable curvature rate.

Pores' angle water refers to loosely bound water type. It is mobile solely within collars, does not transfer hydrostatic pressure and its molecules may be oriented about solvent ions. Compared with free and loosely bound water it is strongly bound and is held in pores' angles more reliably, so, as rocks dry up, it is the last to be removed.

Thus, maximum hygroscopic moisture capacity is the property of rocks to absorb from the air at its relative humidity >94% and retain different volumes $V_{w.m.h}$ of strongly bound and interstitial water for a definite volume V_{dr} of a dry rock.

Hygroscopic Moisture Capacity

Hygroscopic moisture capacity is observed in daytime in sun-desiccated rocks of the upper layers of the aeration zone of a desert and semidesert; at this, the water content in pores may be even less than it is necessary to provide maximum amount of strongly bound water formed at relative air humidity about 55% and corresponding maximum adsorptional moisture capacity of rocks.

Thence it follows that maximum adsorptional moisture capacity is the property of rocks to absorb and retain different volumes $V_{w.str.b}$ (or mass $m_{w.str.b}$) of strongly bound water for a definite volume V_{dr} (or mass m_{dr}) of a dry rock.

Chapter IV

Effective and Dynamic Porosity. Voids Ratio and Moisture Capacity Coefficient

Sec. 6. **Effective and Dynamic Porosity**

Effective porosity is the maximum volume $V_{por.eff.}$ of voids in water-, oil- or gas-bearing strata, oil and gas traps that may contain free water, oil- or natural gas:

$$V_{por.eff} = V_{por.o} - V_{w.str.b} - 1/nV_{w.l.b} = V_{por.o} - V_{w.r}$$

where $n = 1$ for water saturated strata and more than 1 for oil and gas-bearing strata; $V_{w.r}$ is the volume of residual water.

This porosity is responsible for geological reserves of hydrocarbons. It is assumed in this case that oil or gas accumulated in an oil and gas trap may squeeze out $1/n$ fraction of loosely bound water.

The volume $V_{w.r}$ for water-saturated traps may be, for example, ascertained from capillary pressure curves (Fig. 19) obtained by the technique of displacement from a rock sample of water wetting it:

$$V_{w.r} = k_{w.r}V_{por.o}$$

where $k_{w.r} = V_{w.r}/V_{por.o}$ is the saturation coefficient of a rock by residual water $V_{w.r}$ which is not practically decreased under conditions of the experiment (see Fig. 19); $V_{por.o} = V_{dr}k_{por.o}$ and is evaluated from the values of the volume V_{dr} of a dry rock and coefficient $k_{por.o}$ of open porosity (both of these quantities are determined by a separate experiment).

The volume of residual water may also be attained by centrifugation. The fraction of loosely bound water displaced from traps during the origination of extremely saturated oil or gas fields is governed by the properties of gas, oil, natural waters and structure of the oil and gas trap.

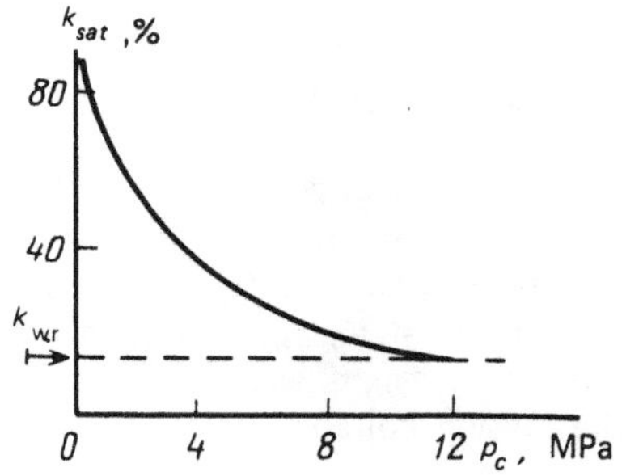

FIG. 19. Dependence of the coefficient of water saturation, $k_{sat} = V_{sat}/V_{por}$, on capillary pressure p_c for a sandy-clayey rock

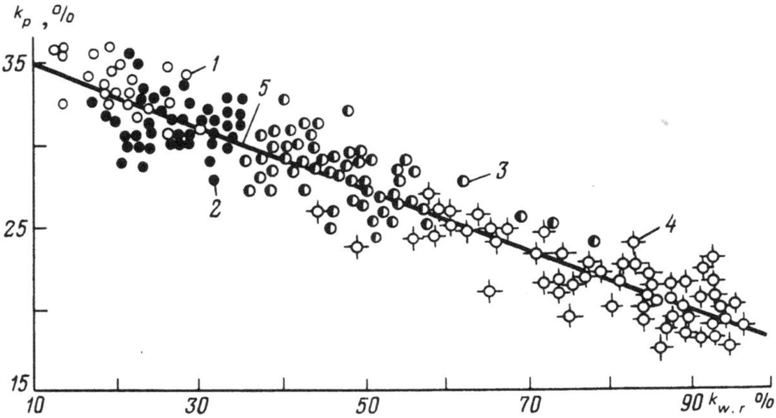

FIG. 20. Dependence of the voids ratio, k_p, on coefficient of residual water saturation, $k_{w.r}$, of polymictic sandy-clayey rocks of the PK bed of the Russkoe deposit (West Siberia) (after E.I. Leont'ev).

k_{perm} (in fm²): *1* — >500; *2* — 100-500; *3* — 10-100; *4* — <10; *5* — regression line

The coefficient $k_{w.r}$ and, consequently, $V_{w.r}$ increase, for example, with decrease of the pore diameter (see Fig. 14), with increasing relative clayiness (see Fig. 16), specific surface (see Fig. 10), exchange capacity of sandy-clayey and carbonate rock samples (see Fig. 15). It also increases with decreasing the voids ratio of carbonate rocks (Fig. 20). However, the pattern of increase differs for traps showing dissimilar structure.

Dynamic porosity is the volume $V_{por.d}$ of voids occupied in the voids space of an oil or gas trap or a rock sample, the homogeneous or nonhomogeneous fluid seeping down at a definite gradient (grad) of pressure p_i. For water-saturated rocks this porosity is generally determined from the difference of the volume $V_{por.o}$ of open pores and the volume $V_{w.r.\,\mathrm{grad}\,p_i}$ of residual water available in the rock's pores during seepage through it of the liquid at a pressure gradient p_i:

$$V_{por.d\,\mathrm{grad}\,p_i} = V_{por.o} - V_{w.r\,\mathrm{grad}\,p_i}$$

$$= V_{por.o} - V_{w.str.b} - 1/nV_{w.l.b\,\mathrm{grad}\,p_i} - 1/mV_{w.fr\,\mathrm{grad}\,p_i}$$

where $1/n$ and $1/m$ are fractions of loosely bound and free water not seeping down the voids space at a pressure gradient grad p_i.

Given an appreciable gradient of pressure $n > 1$, $m \gg 1$, and given a small pressure gradient, $n = 1$, $m > 1$. $V_{por.d\,\mathrm{grad}\,p_i}$ may be equal to or less than $V_{por.eff}$. $V_{por.d\,\mathrm{grad}\,p_i} = V_{por.eff}$ if the experiments to determine these quantities have used the identical maximum pressure gradients. $V_{por.d\,\mathrm{grad}\,p_i} < V_{por.eff}$ when pressure gradients applied to a sample are less than those in experiments to determine $V_{por.eff}$. In this case not only a fraction of the loosely bound water is displaced but also that of the free water, the latter fraction being the greater the lesser the pressure gradient. In the case of an oil-saturated rock

$$V_{por.d\,\mathrm{grad}\,p_i} = V_{por.o} - V_{w.r\,\mathrm{grad}\,p_i} - V_{o.r\,\mathrm{grad}\,p_i}$$
$$= V_{por.o} - V_{w.str.b} - 1/nV_{w.l.b\,\mathrm{grad}\,p_i} - 1/mV_{w.fr\,\mathrm{grad}\,p_i}$$
$$- 1/pV_{o.b\,\mathrm{grad}\,p_i} - 1/qV_{o.fr\,\mathrm{grad}\,p_i}$$

where $V_{o.r\,\mathrm{grad}\,p_i}$ is the volume of all residual oil at pressure gradient p_i; $V_{o.fr\,\mathrm{grad}\,p_i}$ and $V_{o.b\,\mathrm{grad}\,p_i}$ are the volumes, respectively, of free and bound oil at pressure gradient p_i.

Given appreciable pressure gradients, in this relationship $m \gg 1$, $q \gg 1$, $n > 1$ and $p > 1$, and it takes on this form:

$$V_{por.d\,\mathrm{grad}\,p_i} = V_{por.o} - V_{w.str.b} - 1/nV_{w.l.b\,\mathrm{grad}\,p_i} - 1/pV_{o.fr\,\mathrm{grad}\,p_i}$$

Given that grad p_i are small, in the above relationship $n = 1$, $p = 1$, and m and $q > 1$. Thus, for oil-saturated rocks $V_{por.o} > V_{por.eff} > V_{por.d\,\mathrm{grad}\,p_i}$. The volume $V_{por.d\,\mathrm{grad}\,p}$ is what determines the amount of oil (or gas) to be extracted. When it is desired to determine $V_{por.d\,\mathrm{grad}\,p_i}$, first of all we estimate the pressure gradient at which homogeneous or nonhomogeneous fluid is seeping into the borehole, the established value being applied for experiments to determine dynamic porosity.

Sec. 7. **The Voids Ratio and Moisture Capacity Coefficient**

The property of rocks to demonstrate different volumes of various pore or moisture types for a definite volume of a dry rock is characterized by a number of quantities known as *the voids ratio* and *the moisture capacity coefficient.* These provide proportionality coefficients in the following relations:

$$V_{por} = k_p V_{dr}; \qquad V_w = w_t V_{dr}$$
$$V_{por.o} = k_{p.o} V_{dr}; \qquad V_{w.c} = w_c V_{dr}$$
$$V_{por.cl} = k_{p.cl} V_{dr}; \qquad V_{w.sus} = w_{sus} V_{dr}$$
$$V_{por.eff} = k_{p.eff} V_{dr}; \qquad V_{w.m.h} = w_{m.h} V_{dr}$$
$$V_{por.d} = k_{p.d} V_{dr}; \qquad V_{w.h} = w_h V_{dr}$$

and are called, respectively, *a total* k_p, *an open* $k_{p.o}$, *a closed* $k_{p.cl}$, *an effective* $k_{p.eff}$, *a dynamic* $k_{p.d}$ *voids ratio* and *a total* w_t, *a capillary* w_c, *a suspended* w_{sus}, *a maximum hygroscopic* $w_{m.h}$ and *a hygroscopic moisture capacity coefficient* w_h. They suggest that, even though volumes of pores (voids) and water (moisture) for different porosities and moisture capacity magnitudes may increase with increasing the volume V_{dr} of a dry rock, different volumes of the quantities in question (different voids and water type volumes) may be obtained for various rock species for the same volume. The above coefficients are specific volumetric quantities as distinct from specific mass moisture capacity coefficients often used as moisture capacity characteristics of rocks.

Specific mass moisture capacity coefficients $w_{t.m}$, $w_{c.m}$, $w_{sus.m}$, $w_{m.hm}$, w_{hm} are proportionality coefficients in the relations connecting the water masses m_w, $m_{w.c}$,

$m_{w.sus}$, $m_{w.m.h}$, $m_{w.h}$, given different moisture capacity types, with the mass m_{dr} of a dry rock. As porosity characteristics use is made of *reduced voids ratios*

$$k_{p.red} = V_{por}/V_s$$

as well as *relative voids ratios*

$$k_{p.rel} = V_{por}/m_{dr}$$

All the aforementioned quantities are dimensionless except $k_{p.rel}$ which is measured in m^3/kg (or cm^3/g). For oil saturated rocks $k_p > k_{p.eff} > k_{p.d}$.

Sec. 8. A Total Voids Ratio of Minerals, Primary Sediments, Sedimentary Rocks, Their Altered and Metamorphosed Varieties, Igneous Rocks and Ores

The total voids ratio of minerals has been insufficiently studied and does not generally exceed 1% (Table 1), sometimes attaining a value 2-3%. The voids ratio k_p

TABLE 1. The Total Voids Ratio of Several Minerals (after B.P. Belikov, K.S. Aleksandrov and T.B. Ryzhova)

Mineral	k_p, %	Mineral	k_p, %
Garnet (grossular)	0.095	Biotite	1.56
Garnet (pyrope)	0.232	Phlogopite	1.65-3.17
Diopside (baikalite)	0.152	Microcline-perthite	0.622
Augite	2.026	Albite	0.421
Aegirine	0.575	Oligoclase	0.31
Diallage	1.2	Labradorite	0.402
Hornblende	0.89	Nephelinite	0.151
Muscovite	2.14		

of primary sediments, newly formed sedimentary rocks, their varieties altered in the course of diagenesis, katagenesis and metamorphism varies from values close to 1 up to practically zero (Fig. 21, Table 2). The greatest values of the total voids ratio in the range of ones in excess of 50-80% to ones greater than 30% are manifested by arenaceous, aleurite, argillaceous, calcareous, diatomite, spongy, radiolarian, sapropel, ferriferous, bauxite and other mud types. The total voids ratio of these primary sediments deposited on the river, lake, sea and oceanic beds is the greater the less is the roundness, the median diameter (Fig. 22) and the higher is sortedness of their grains governed by the mineral composition of the ablation sources and the conditions of particles transfer. It is a problem to accurately establish the effect of the geometry of fragments on the values of the voids ratio of sediments since they are made up by fragments that differ in configuration. Given equal size of

TABLE 2. Limiting Values of Quantities Characterizing Total Porosity and Density of Primary Sediments and Dry Rocks

Primary sediment or rock	k_p, %	δ_{dr}, g/cm^3
Sediments, sedimentary and metamorphic rocks and ores derived from them		
Arenaceous mud	30-70	1.27-1.94
Sand	4-50	1.3-2.3
Sandstone	0.5-40	1.3-3.6
Sandy shale	1-15	1.8-2.9
Quartzite and ferruginous quartzite	0.2-20	2.19-4.4
Aleurolite mud	60-70	—
Loess	30-60	1.14-1.93
Loessial rock	—	1.13-2.25
Aleurolite	1-50	1.51-3.2
Algillaceous mud	50-60	0.8-1.8
Clay	4-75	1.2-3.18
Argillite	1-30	1.6-3.35
Phyllite	0.5-0.6	2.4-2.76
Slate:		
clay shale	1-25	1.3-3.2
roof slate	0.4	2.79-2.8
slate	0.9-1.1	2.77-2.78
mica schist	—	2.5-2.8
Paragneiss	0.4-2.6	—
Diatomite	30-80	0.4-1.57
Tripoli	28-66	0.5-1.94
Opoka	25-66	0.27-2.13
Spongolite	1-17	0.18-2.66
Jasper	—	2.32-2.97
Flint	1-6	2.32-2.80
Calcareous mud	65-87	—
Chalk	10-55	1.85-2.6
Limestone	0.5-48	1.3-3.5
Calcareous tuff	20-30	2.3-3.0
Marmorized limestone	0.4-6	1.9-3.5
Marble	0.1-2.2	2.3-3.0
Dolomite	0.1-37	1.3-3.5
Dolomitic meal	33-55	1.48-2.08
Marmorized dolomite	0.4-2.4	—
Marl	0.7-48	1.2-3.0
Marl	—	2.0-3.1
Magnesial paraamphibolite	$<$0.1-27	2.0-2.98
Gypsum rock	1-37	—
Anhydrite rock	0.2-17	2.3-3.0
Rock salt	0-5	2.1-2.3
Sylvinite	—	1.9-2.95
Carnallite rock	—	1.6-2.15

TABLE 2 (*continued*)

Primary sediment or rock	k_p, %	δ_{dr}, g/cm^3
Ores:		
bauxite	0-44	1.3-3.4
ferruginous laterite	39-56	—
martite and bog iron ore	0.2-28	3.8-4.3
siderite	1-28	2.45-3.58
iron ore of leptochlorite formation	—	1.6-2.46
manganese, oölitic, sooty ore	—	1.8-4.5
apatite	—	3.18-3.41
phosphorite	—	2.8-3.0
peat	Up to 95	0.45-0.80
coal	1-40	1.1-1.8
brown coal	12-40	1.2-1.3
hard coal	3-12	1.2-1.35
anthracite	1-2	1.4-1.8
graphite	—	2.2-2.8
graphite gneiss	—	2.1-2.57
Magmatic and metamorphic rocks and ores formed from them		
Dunite	0.3-6.7(0.3-2)	2.43-3.6(2.43-3.0)
Peridotite	0.3-10(0.3-2)	2.43-3.3(2.43-3.0)
Pyroxenite	0.2-4.5(0.3-2)	2.78-3.6(2.78-3.0)
Serpentinite	0.8-12(—)	1.96-3.6(1.96-2.5)
Gabbro	0.2-6.2(0.2-3)	2.5-3.3(2.5-2.7)
Eclogite	0.2-1.5(0.2-1.5)	3.2-3.5(3.2-3.5)
Anorthosite	0.3-3.0(0.3-3.0)	2.66-2.9(2.66-2.9)
Labradorite	0.4-12(0.4-3.0)	2.24-2.98(2.6-2.98)
Diorite	0.4-5.0(0.4-3.0)	2.47-3.03(2.7-3.03)
Syenite	0.5-9.0(0.4-4.0)	2.55-3.0(2.55-3.0)
Granite	0.3-12(0.3-4)	2.2-2.98(2.55-2.68)
Alaskite	0.2-7.4(0.2-4)	2.48-2.8(2.55-2.63)
Nepheline syenite	0.6-5.0(0.6-4)	2.55-2.7(2.55-2.61)
Basalt	0.4-40(—)	2.1-3.06(2.1-3.06)
Diabase	0.2-18(0.2-3)	2.3-3.3(2.7-3.06)
Andesite	0.4-22(—)	2.17-2.72(2.17-2.72)
Porphyrite	0.4-18(0.4-6)	2.34-2.94(2.6-2.94)
Trachite	1.2-22(1.2-22)	2.2-2.3(2.2-2.3)
Liparite	4.5-36(4.5-36)	1.9-2.84(1.9-2.84)
Quartz porphyry	0.5-10(0.5-10)	2.32-2.85(2.5-2.84)
Obsidian	1-12(1-12)	—(—)
Welded tuff	6-60(6-60)	0.4-2.62(0.4-2.62)
Tuff:		
volcanic	3-57(3-57)	1.51-2.86(1.51-2.86)
basaltic	3.9-53(3.9-53)	0.60-2.6(0.60-2.6)
porphyric	—(—)	2.19 av (2.19 av)

TABLE 2 (*concluded*)

Primary sediment or rock	k_p, %	δ_{dr}, g/cm³
Tuffites	—(—)	1.51-2.82(1.51-2.82)
Slates:		
quartz-chlorite	1.5-6.6(1.5-6.6)	—(—)
chlorite	0.1-0.8(0.1-0.8)	2.5-2.92(2.5-2.92)
talk schist	—(—)	2.6-2.86(2.6-2.86)
quartz-sericitic schist	—(—)	2.6-2.77(2.6-2.77)
sericitic	—(—)	2.61-2.72(2.61-2.72)
crystalline (ortho)	—(—)	—(—)
Orthoamphibolite	0.1-27(0.1-6)	2.67-3.27(2.67-2.27)
Secondary quartzite	7.5-11(7.5-11.0)	2.3-2.68(2.3-2.68)
Orthogneiss	0.2-14(0.2-6.0)	2.3-3.1(2.3-3.1)
Skarn	0.3-21(0.3-7.0)	2.9-4.0(2.9-4.0)
Hornblende	0.2-13(0.2-5.0)	2.55-2.8(2.55-2.8)
Ore formation rocks and ores:		
magnetite skarn	0.3-12(0.3-5)	2.8-4.8(3.1-4.8)
titanum magnetite	—(—)	3.1-4.3(3.1-4.3)
magnesioferrite	—(—)	3.3-4.0(3.3-4.0)
skarn		
chromite	—(—)	2.76-4.5(3.1-4.8)
sulphide copper-nickel	—(—)	2.8-4.9(2.8-4.9)
tin	—(—)	1.88-4.65(1.88-4.65)
stanniferous pegmatites	—(—)	2.65-2.7(2.65-2.7)
cassiterite-quartz	—(—)	2.65-3.82(2.65-3.82)
cassiterite-silicate and		
cassiterite-sulphide	—(—)	2.32-4.27(2.32-4.27)
stanniferous skarns	—(—)	2.34-4.27(2.34-4.27)
tungsten	—(—)	2.68-2.99(2.68-2.99)
polymetallic	0.2-9(0.2-3)	2.7-4.5(2.7-4.5)
copper	0.2-7(0.2-2)	2.67-5.07(2.67-5.07)
gold ore	—(—)	2.4-3.4(2.4-3.4)
zinc	—(—)	2.5-3.29(2.5-3.29)
lead	—(—)	2.73-4.19(2.79-4.19)
plutonic graphite		
crystalline	—(—)	2.35-2.8(2.35-2.8)
actimonite	—(—)	2.8-2.96(2.8-2.96)
corundum	—(—)	3-4(3-4)
barite	—(—)	4-4.5(4-4.5)
fluorite	—(—)	2.9-3.3(2.9-3.3)

Note. For plutonic rocks the limits of variation of k_p and δ_{dr} are presented, in parentheses for unweathered rocks.

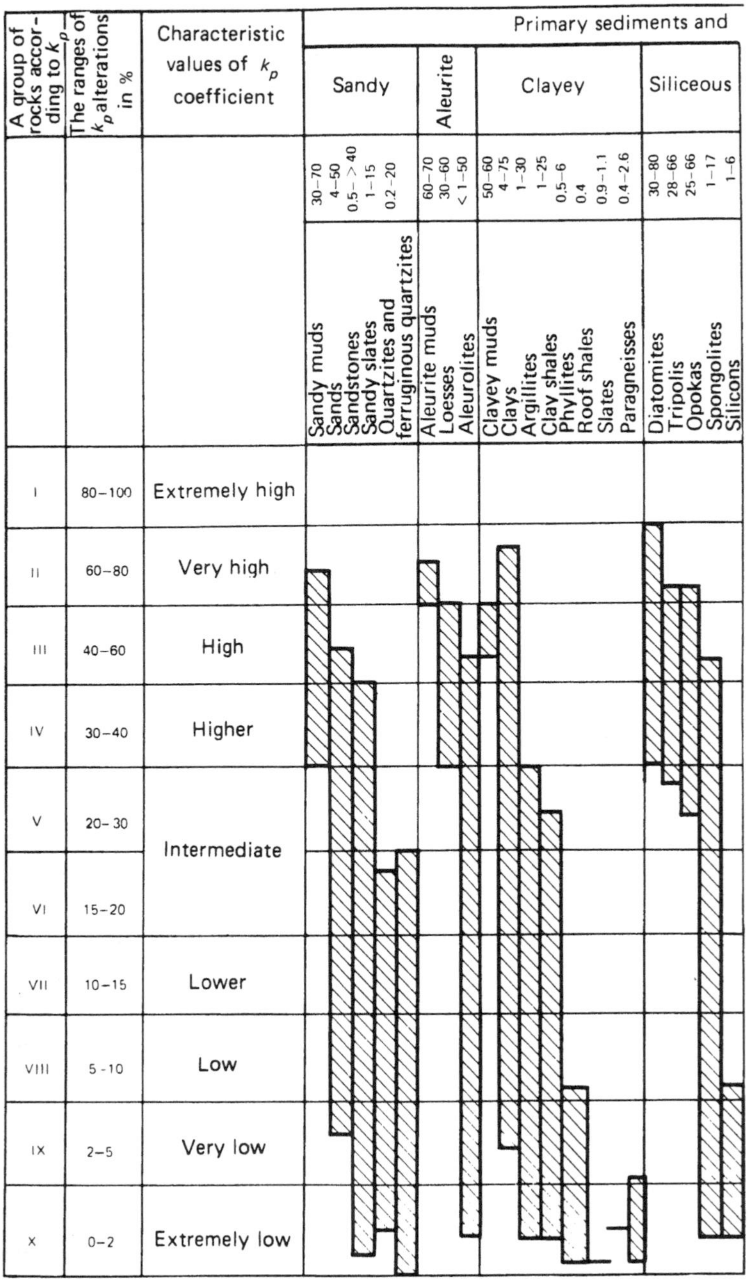

FIG. 21. Values of the voids ratio, k_p, of primary sediments, sedimentary and derived metamorphic rocks and ores

sedimentary and metamorphic rocks originated from the previous ones

Group	Subgroup	Rock	
Calcareo-magnesial	Calcareous	Lime mud	65–87
		Limestones	< 0.5–48
		Limestones: biomorphic	0.5–35
		biogenic and detrital	< 0.5–30
		detrital	< 0.5–30
		slime	0.8–22
		pelitomorphic	0.1–40
		microgranular	4–33
		fine-granular	10–20
		medium-granular	10–20
		oölitic	11–45
		clastic	0.8–30
		Chalk	10–55
		Tufa	(20–30) m
		Marmorized limestones	0.4–6.0
		Marble	0.1–2.2
	Dolomite	Dolomites	< 0.1–37
		Dolomites: pelitomorphic	1–40
			< 0.5–30
		fine-grained	< 0.5–30
		clastic	< 0.8–30
		Dolomitic meal	33–55
		marbled	0.4–2.4
Marls		Marls: dolomitic	0.7–48
		Papaamphibolite	11–23
		magnesite	< 0.1–27
Salt		Gypsum rocks	< 1–37
		Anhydrite rocks	< 0.2–17
		Rock salt	0–5
Aluminious		Bauxites: residual	33–40
		kaolinite	15–34
		loose	9–44
		stone	0–25
Ferruginous		Ferruginous laterites	39–56
		Martites and limonites	0.2–28
		Siderite rocks	1–28
Peat and fossil coal		Peat	up to 95
		Coals: brown	18–45
		candle	11–18
		gas	7–11
		rich	5.4–7

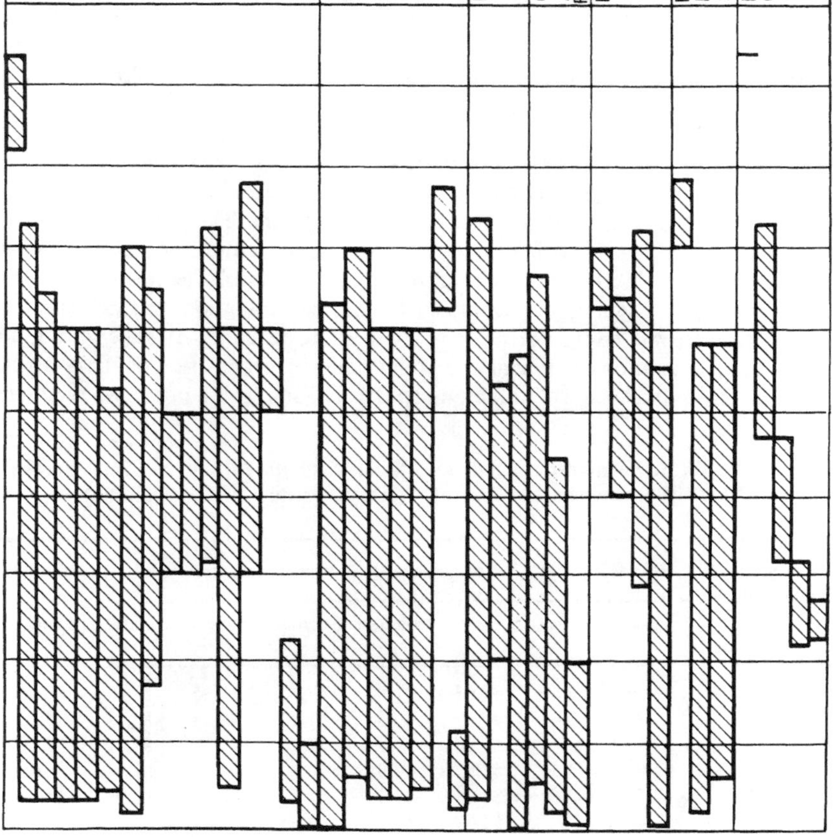

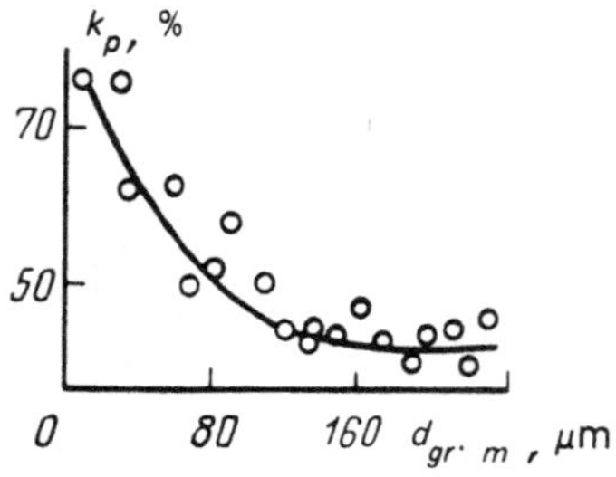

FIG. 22. Dependence of the voids ratio k_p on the median diameter of grains $d_{gr.m}$ for modern sediments of the Mediterranian (after Füchtbauer and Reinecke)

quartz, orthoclase, mica fragments, their occurrence jointly with angular and lamellar fragments is more porous than with rounded grains (Table 3). The angular grain pattern facilitates arch formation and increases porosity.

The appreciable porosity of fine-grained argillaceous sediments is, besides, ascribed to their special cellular or honey-comb structure, and that of arenaceous, aleurite, calcareous, diatomite and other mud types not only to the angular shape of particles, their sufficient roundedness, but also to the presence of films of bound water near the surface of their solid phase.

During diagenesis muds are consolidated, dehydrated transforming to rocks showing a porosity that is less than the original one. As rocks are sinking, with increasing pressures and temperatures there occur processes of early (to a depth of 2-3 km) and then later katagenesis and metamorphism. They differently transform sedimentary rocks derived from muds giving rise to a multitude of their types forming groups (families) whose parental materials are muds. The aforementioned and other geological factors contribute to further consolidation and dehydration of rocks, their cementation, partial recrystallization, leaching of salts and solid phase material, alteration of mineral and granulometric composition of rocks which often causes them to transform to metamorphic varieties. These processes facilitate

TABLE 3. The Values Relationship Between k_p Coefficient (in %) and the Total Porosity of Minerals from the Roundness and Size of Grains (according to V.V. Okhotin)

Diameter of fraction, mm	Rounded quartz		Angular quartz		Angular orthoclase		Mica (muscovite)	
	loose	compact	loose	compact	loose	compact	loose	compact
2-1	36.06	33.40	47.63	37.90	47.50	45.46	87.0	80.46
1-0.5	36.30	33.63	47.10	40.61	51.98	46.88	85.18	75.20
0.5-0.25	39.66	33.42	46.98	41.09	54.76	49.18	83.71	72.17
0.25-0.1	44.80	34.35	52.47	44.82	58.46	51.62	82.74	66.30
0.10-0.06	44.53	39.60	54.66	45.31	61.22	52.72	82.98	68.98
0.06-0.01	—	—	55.99	45.68	62.53	—	—	65.33

both decreasing and increasing the voids ratio of rocks. So, for example, clayey muds ($k_p > 80\%$) are the initial material of the clayey rocks group. At depth of 3 m they transform to plastic clays with $k_p = 75\%$, at 5 m their k_p drops to 73%. Toward the end of diagenesis these rocks become clayey species showing $k_p = 45\text{-}35\%$, whereas at the last stage of katagenesis and during metamorphism their k_p drops to 1%, and they turn into argillites and slates.

The concept of variations in the value of the voids ratio of clastic rocks as a result of shifting their grains during consolidation is accounted for by the porosity theory fairly valid only for well rounded and weakly cemented rocks. A rock is considered to be a system of equal-diameter spheres whose structural element is provided by a rhombohedron cutting out of the eight adjacent sphere portions that, taken together, form one sphere. The voids ratio of the rhombohedron of the entire system of spheres can be found from a relationship:

$$k_p = \frac{V_r - V_{sp}}{V_r} = 1 - \frac{\pi}{6(1 - \cos\theta)\sqrt{1 + 2\cos\theta}}$$

where V_r and V_{sp} are, respectively, volumes of the rhombohedron and the entire sphere made up of its portions cut out by the rhombohedron from the system of spheres; θ is the angle of the rhombohedron varying in the range 60-90°.

Since at $\theta = 90°$ the rhombohedron converts to a cube, the relation to calculate k_p takes on this simpler form:

$$k_p = \frac{d^3 - \pi d^3/6}{d^3} = \frac{6 - \pi}{6} = 0.48 \text{ or } 48\%$$

For $\theta = 60°$ $k_p = 26\%$. Thus during displacement of particles the voids ratio may vary in the range from 48 to 26%.

The total voids ratio of sandy and aleurite rocks varies from values less than several decimal fractions to values in excess of 50%; the total voids ratio of clayey carbonate and siliceous rock groups from a few decimal fractions to 50-70%. The k_p values of coals and peat varies over the same wide range. The range of variation of the voids ratio of saliferous rocks is narrower, and for individual types of sedimentary rocks k_p varies still less, yet sometimes over quite a wide range (see Fig. 21, Table 2).

The appreciable variations of the total voids ratio of rocks' groups and types is accounted for by the original differences in the mineral composition and structure of their individual varieties, by the effect of pressure, temperature, time and other geological factors on the degree of filling the volume by the solid phase.

Low and very low values of k_p of clastic rocks (sands, sandstones, aleurites, aleurolites, calcarenites etc.) are often due to the poor sorting out of fragments (occurring originally or with lapse of time under the pressure of the overburden) and cementation of their pore space by siliceous, carbonate, ferruginous, argillaceous and other cement varieties. For sandy-silt-rocks, e.g. a drop in values of k_p has been established with decreasing the coefficient G_f (*B*) of sorting out of grains and increasing the clayiness coefficient $k_{cl.v}$ (Fig. 23), the sum of coeffi-

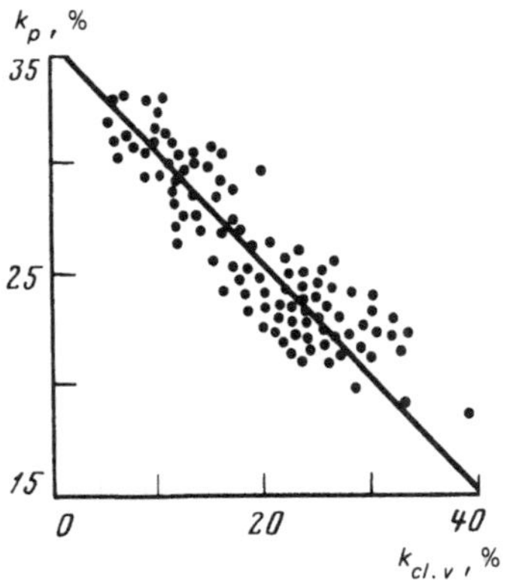

FIG. 23. Dependence of the voids ratio k_p on the coefficient of volumetric clayiness $k_{cl\ v}$ of Mesozoic sandy-clayey sediments of the Nadym area of bed PK of the Russkoe deposit (West Siberia) (after E.I. Leont'ev)

cients $k_{cl.m}$ of clayiness and $k_{c.m}$ of cementation*, coefficient $k_{w.r}$ of residual water saturation (see Fig. 20) and median size of grains.

The voids ratio of uncemented and unconsolidated sandy-argillaceous rocks decreases until the entire space between larger (sandy and aleurite) grains is filled by the cementing (clayey or clayey soluble) portion [(Fig. 24), the left branches of the design** and the experimental curves $k_p = f(k_{cl.m})$]. Then, with increasing the clayey fraction's content, k_p increases again attaining greater values compared with

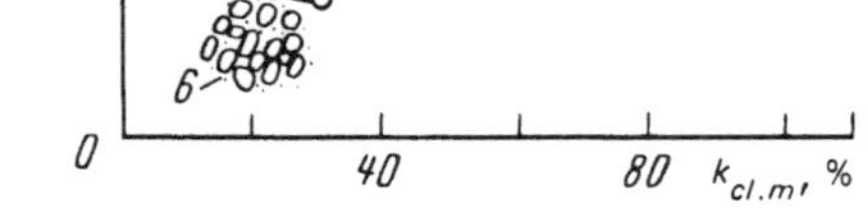

FIG. 24. Dependences of the voids ratio k_p on the coefficient of the mass clayiness of sandy-silt-rocks.

1—design curve; *2-4*—experimental curves (after V.V. Okhotin) obtained upon addition to fraction 16-8 fractions 1-0.05 (in mm); *3* — 2-1, and *4* — 4-2; *5* — *7* are structures of rocks, respectively, in the absence of a clayey fraction, upon the filling of all voids between sandy-aleurite particles by a clayey material and in the presence of predominance of clayey material over sandy-aleurite material

---

* The cementation coefficient shows the specific mass content in a rock of not only clayey fraction but also of other cement types (carbonates, sesquioxides) etc.

** The following equation corresponds to the left branch of the $k_p = f(k_{cl.m})$ design curve

$$k_p = \frac{(m_{Ps+Av}/\delta_{drPs+Av})k_{pPs+Av} + (m_{Pl}/\delta_{drPl})k_{pPl} - (m_{Pl}/\delta_{drPl})}{m_{Ps+Av}/\delta_{drPs+Av}}$$

where m_{Ps+Av}, m_{Pl} are masses of sandy-aleurite and clayey particles in 1 000 mm^3 of a rock; $\delta_{drPs+Av}$, $\delta_{dr.Pl}$, k_{pPs+Av} and k_{pPl} are densities and voids ratios of pure sandy-aleurite and clayey rocks.

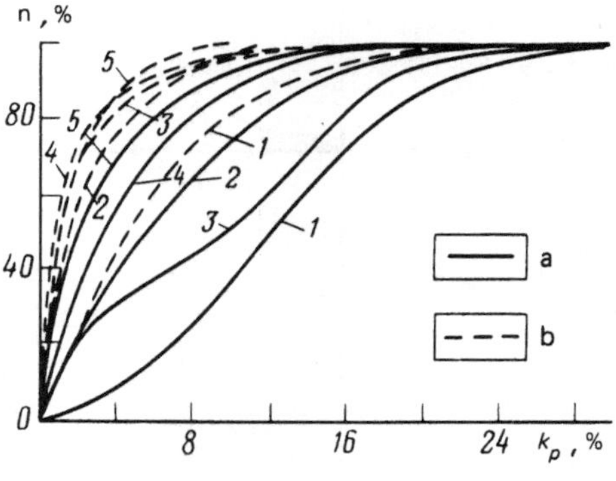

FIG. 25. Cumulative curves of distribution of values of the voids ratio k_p of carbonate rocks of the Middle Carboniferous of the Perm Kama Region (after Sh.V. Abashev and Yu.N. Reshetnyakova).

1 — dolomites; *2* — calcareous shell sandstones; limestones; *3* — biomorphic and detrite-biomorphic; *4* — micrograined; *5* — detrite; areas of: *a* — platform; *b* — Cis-Ural trough; *n* — accumulated frequencies of k_p values

weakly cemented rock species where the sandy-aleuritic fraction predominates [see Fig. 24, the right-hand branches of the design and experimental curves $k_p = f(k_{cl.m})$*]. As can be seen from Figs. 23 and 24, the clayey fraction, to a definite content, causes the voids ratio of rocks to be drastically decreased. Rock cementation by the clayey soluble fraction results in a greater drop in the k_p value. The limits of variation and distribution of k_p values of individual rock types of different sandy-silt-rock sediments are governed by their origin and transformation and are therefore far from being equal.

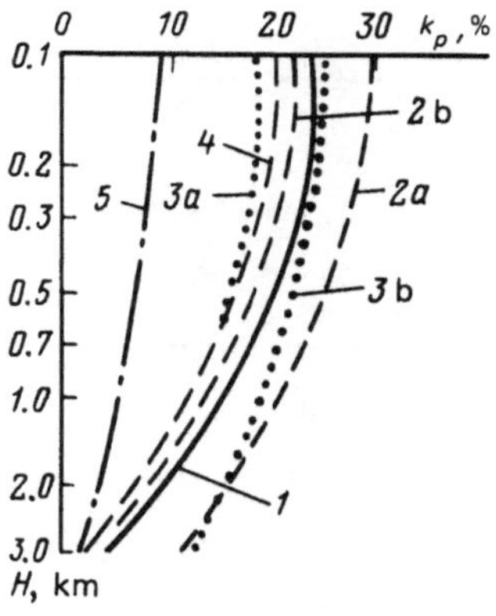

FIG. 26. Dependences of the voids ratio k_p on depth H for the Middle and Lower Carboniferous sediments (after M.L. Ozerskaya and S.G. Semenova).

1 — clay; *2a* and *2b* — aleurolite; *3a* and *3b* — sandstone; *4* - marl; *5* — limestone

* The right-hand branch of the design curve is obtained on an assumption that the volume of the clay fraction increases the volume of the pore space between sandy and aleurite particles (if they are brought into contact) and that sand grains swim, as it were, in a mass of clay particles. Then

$$k_p = \frac{(m_{Pl}/\delta_{drPl})k_{pPl}}{m_{Pl}/\delta_{drPl} + m_{Ps+Av}/\delta_{sPs+Av}}$$

where δ_{sPs+Av} is the density of the solid phase of the sandy-aleurite fraction of rock.

A group of rocks according to k_p	The ranges of alternation of coefficeint, k_p, %	The characteristics of values of k_p
I	80–100	Extremely high
II	60 – 80	Very high
III	40–60	High
IV	30–40	Higher
V	20–30	Intermediate
VI	15–20	
VII	10–15	Lower
VIII	5–10	Low
IX	2–5	Very low
X	0–2	Extremely low

FIG. 27. Values of the voids ratio k_p of magmatic and metamorphic rocks and magmatogenic ores.

Rocks: *1* — nonweathered or nonserpentinized; *2* — weathered or serpentinized

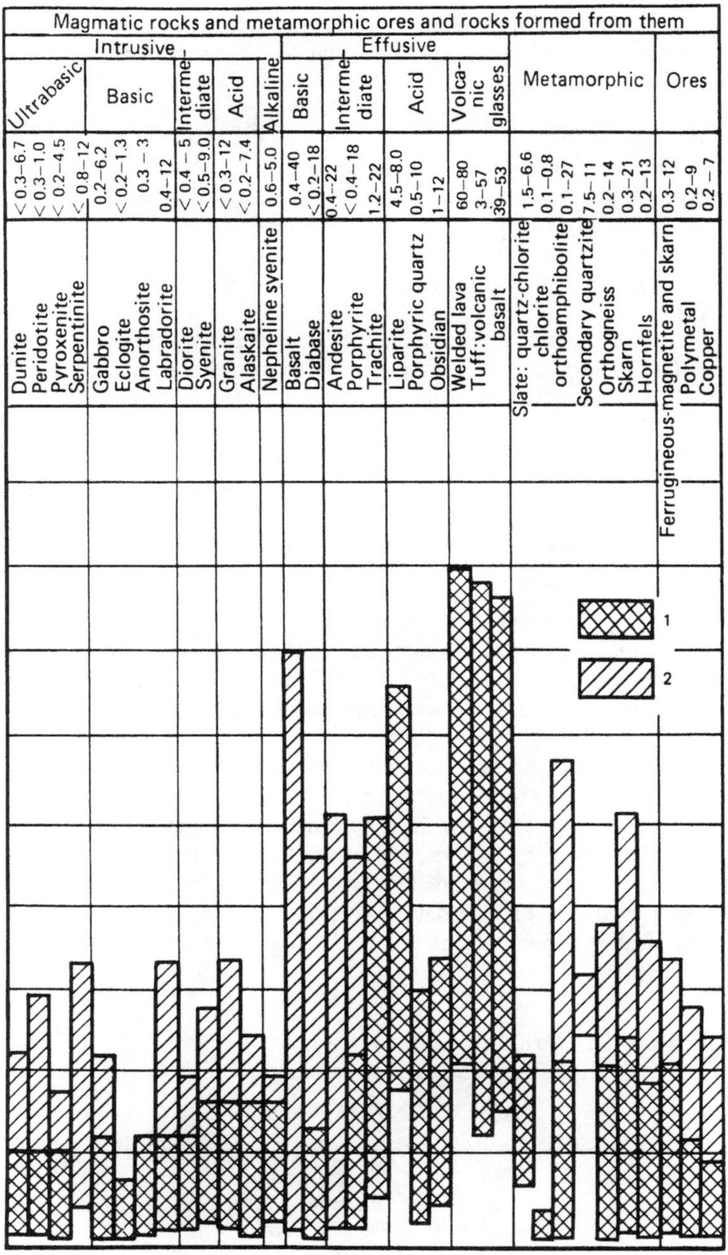
Magmatic rocks and metamorphic ores and rocks formed from them
Intrusive
Effusive
Ultrabasic
Basic
Intermediate
Acid
Alkaline
Basic
Intermediate
Acid
Volcanic glasses
Metamorphic
Ores
< 0.3–6.7 Dunite
< 0.3–1.0 Peridotite
< 0.2–4.5 Pyroxenite
< 0.8–12 Serpentinite
0.2–6.2 Gabbro
< 0.2–1.3 Eclogite
0.3 – 3 Anorthosite
0.4–12 Labradorite
< 0.4 – 5 Diorite
< 0.5–9.0 Syenite
< 0.3–12 Granite
< 0.2–7.4 Alaskaite
0.6–5.0 Nepheline syenite
0.4–40 Basalt
< 0.2–18 Diabase
0.4–22 Andesite
< 0.4–18 Porphyrite
1.2–22 Trachite
4.5–8.0 Liparite
0.5–10 Porphyric quartz
1–12 Obsidian
60–80 Welded lava
3–57 Tuff: volcanic
39–53 basalt
1.5–6.6 Slate: quartz-chlorite
0.1–0.8 chlorite
0.1–27 orthoamphibolite
7.5–11 Secondary quartzite
0.2–14 Orthogneiss
0.3–21 Skarn
0.2–13 Hornfels
0.3–12 Ferrugineous-magnetite and skarn
0.2–9 Polymetal
0.2 – 7 Copper
1
2

For different carbonate rock types (dolomites, biogenic, and detrital and biogenic limestones, calcareous shell sandstones, microgranular and detrital limestones) they are also governed not only by their original structure but also by transformation processes which, in turn, are associated with the living conditions of the rocks. For example, cumulates of values of k_p of the carbonate rocks of the Middle Carboniferous of the Permian Kama River region are distinctly differentiated. However, this is not typical of k_p value distribution curves of the Cis-Urals depression sediments (Fig. 25). The limits of variation of k_p in the former case are greater and the greatest fraction of k_p values of different carbonate rocks is taken by the high values of this quantity. This is not characteristic of the areas of the Cis-Urals depression (see Fig. 25).

The coefficients k_p of similar sedimentary rocks are generally the smaller, the greater is the depth of their occurrence and their lifetime due to the aforementioned dia- and epigenetic processes. The fashion the voids ratio decreases with depth H is shown by $k_p = f(H)$ curves (Fig. 26) obtained by different sedimentary rock types and their deposition and life conditions. Particularly dramatic is the decrease of the voids ratio of fine-grained sedimentary rocks (clays, aleurolites). Voids ratio values of limestones decrease with depth not so drastically. The variation of k_p of single-type and similar-age rocks with depth can be approximated by the equation

$$k_{p.avH} = k_{p.max} e^{-0.045H/H_0} \qquad (1)$$

where $k_{p.max}$ is the value of k_p for H_0; H_0 is the depth of occurrence of unweathered rocks (here H_0 is taken equal to 0.1 km); H is the depth for which it is desired to determine $k_{p.avH}$. In abnormal pressure zones deviations from Eq. (1) are observed.

Less commonly k_p values increase with depth owing to tectonic processes and leaching of the solid phase. Thus with increased extent of transformation of sedimentary rocks and with passage of them to other rock types the total voids ratio generally diminishes (see Fig. 21), yet, when found in leaching and weathering zones, rocks can deconsolidate. The voids ratio values of weakly altered intrusive, orthometamorphic rocks, ores of magmatogene origin do not, as a rule, exceed 2-5%. Moderately weathered varieties of the same rocks may show k_p values as

TABLE 4. Rock Classification According to Values of k_p

A group of rocks	k_p, %	Description of rocks
I	>20	Highly porous
II	15-20	Extremely porous
III	10-15	Middle porous
IV	5-10	With reduced porosity
V	<5	With low porosity

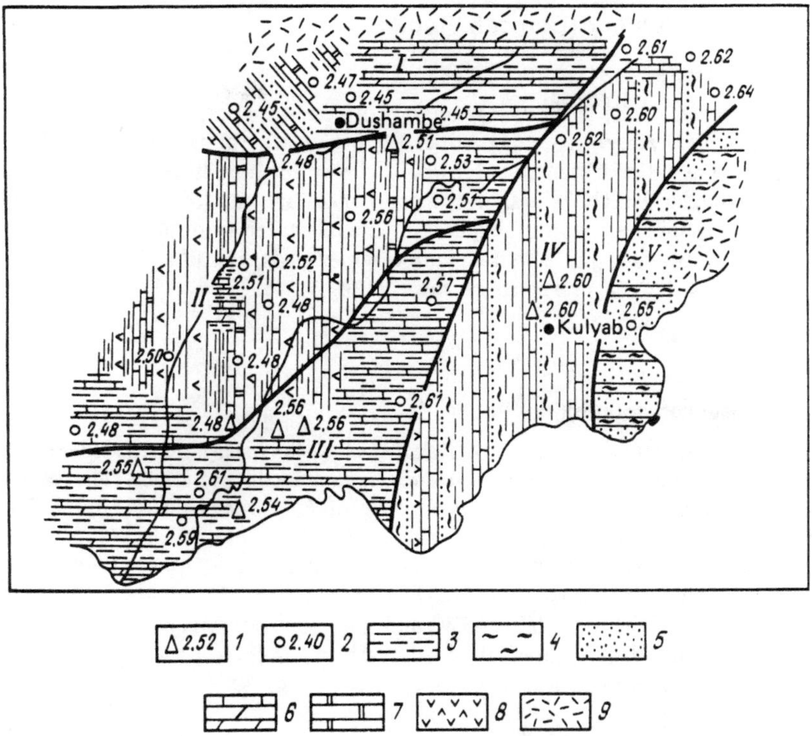

FIG. 28. Variations of the voids ratio k_p and density δ_r of rocks of Alai strata of the Paleogene of the Tajik depression (After S.N. Dembitskii).

1 — boreholes and values of δ_r (in g/cm^3); *2* — the same, in outcrops; *3* — clays; *4* — aleurolites; *5* — sandstones; *6* — limestones; *7* — dolomites; *8* — gypsums and anhydrites; *9* — Paleozoic framework; zones with corresponding δ_r (in g/cm^3) and k_p (in %): *I* — 2.45-2.47, and 6—7; *II* — 2.50-2.55 and 5.5-7.4; *III* — 2.55-2.60 and 3.8-5.5; *IV* — 2.60-2.65 and 1.9-3.8; *V* — >2.6 and 2

high as 10% and more (Fig. 27) whereas k_p values of weakly altered effusive rocks may attain more than 20% (e.g. for basalts, diabases, porphyrites). Particularly great voids ratio values are manifested by ash tuffs (30-60%) and by Arctic welded lava. In the hypergene zone plutonic and metamorphic rocks may give rise to gruss and clayey sediments exhibiting appreciable k_p values.

Presented below is a rock classification according to their voids ratio (Table 4) proposed by oil prospecters. There exist other classifications of rocks according to their values.

Definite spatial regularities in the variation of k_p values of similar-age identical or different type rocks (Fig. 28) governed by the conditions of their origin, occurrence and transformation have been established.

Sec. 9. Moisture Capacity Coefficients of Minerals, Sedimentary (Detrital) Rocks

Minerals

Data on the values of the moisture capacity of minerals are practically unavailable. One may assume *à priori* that these values are in the general case less than those of the total voids ratio of minerals. However, clayey minerals have the coefficient $w_t \geqslant k_p$. Particularly high values of the total moisture capacity coefficient are shown by clayey minerals with an expanding crystal lattice (montmorillonite, vermiculite), their volume V_w in a wetted state often exceeds that in a dry state, and, as a result, $w_t = V_w/V_{dr}$ proves to be more than

$k_p = V_{por}/V_{dr}$*

Sedimentary (Detrital) Rocks

The values of moisture capacity coefficients of sedimentary detrital rocks are governed by the mineral composition, structure and texture of rocks. The values of w_t vary over about the same range as those of k_p if rocks do not swell. For swelling rocks w_t exceeds k_p.

The values of the capillary moisture capacity coefficient w_c are evaluated from the $w_c = f(H)$ curve (Fig. 29) obtained in the capillary rise subzone, they are found between the levels of the water table (GW) and the upper boundary of the capillary zone (CZ). Moisture distribution in bedded overburdens corresponding to the two probable cases is illustrated by curves presented in Fig. 30.

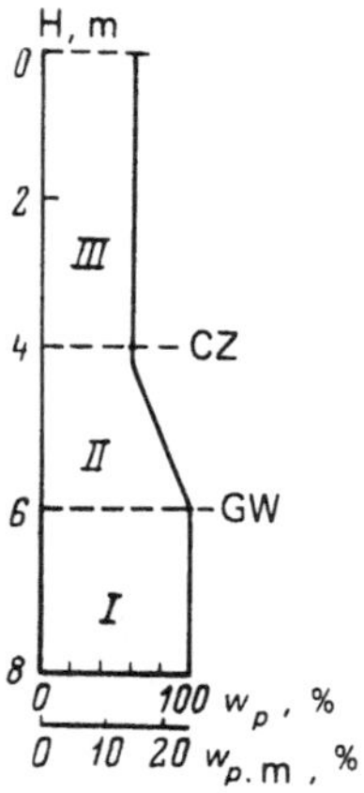

FIG. 29. Curve of distribution of moisture in the rock mass following their through wetting, given a deep occurrence of groundwater (after A.A. Rode).

GW — groundwater table level; CZ — upper boundary of capillary zone; zones: *I* — saturated; *II* — capillary; *III* — suspended moisture capacity

* When evaluating the moisture capacity coefficients in fractions of the volume of dry rock, if the latter's moisture capacity is expressed in fractions of the volume of wetted rock, this will not be valid, and w will not be >1.

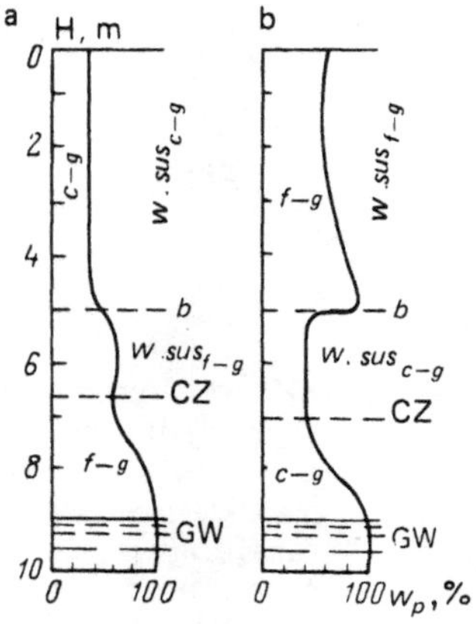

FIG. 30. Curve of the distribution of moisture in bedded rock masses (after A.A. Rode).

a — given coarse-grained bed *c-g* underlain by fine-grained *f-g*; *b* — given fine-grained bed *f-g* by coarse-grained bed *c-g*; *b* — boundary of beds; CZ — upper boundary of the capillary zone; GW — groundwater table level; suspended moisture capacity of bed: $w.sus_{c\text{-}g}$; $w.sus_{f\text{-}g}$

The coefficient of suspended moisture capacity w_{sus} in sands is equal to 3-5%, for unsorted, as a rule, sandy-aleurite rocks it attains 10-12%, and for unsorted, generally, aleurite and clayey rocks and clays it increases, respectively, to 12-18-22%.

No reliable laboratory methods of determination of w_{sus} have yet been suggested; as for the field methods, the interested reader is referred to works of A.A. Rode. Approximate values of suspended moisture capacity of sedimentary rocks of oil deposits with the coefficient $k_{w.sus} = V_{w.sus}/V_{por}$ are listed in Table 5.

Analysis of data in Table 5 suggests that the greatest amount of suspended water is demonstrated by fine-grained rocks, viz. sand, sandstones and clays. Much suspended water is commonly contained in chalky limestones. The mass maximum hygroscopic moisture capacity coefficient of sedimentary rocks varies from 0.24 to 50% and more*.

TABLE 5. The Moisture Capacity of Different Rocks of Oil Deposits (after P. Jones)

Rocks	w_{sus}, %, from V_{por}, given k_{perm} in fm^2			
	10	100	1 000	10 000
Coarse-grained limestones and sands	12	8	6	4
Medium-grained sands and limestones	35	25	18	8
Fine-grained sands and limestones	55	40	30	15

Note. Rocks of oil deposits have suspended moisture capacity indicated in the table whereas the remaining voids space is filled by oil or gas: more commonly, in addition to bound, they contain capillary water.

* The determination of $w_{m.hm}$ uses a weighing technique: a dry rock sample is held over a 10% solution of H_2SO_4 (relative humidity of the air in the vessel where the sample is placed is 94%). By knowing the mass of the sample prior to and following wetting, we calculate $w_{m.hm}$.

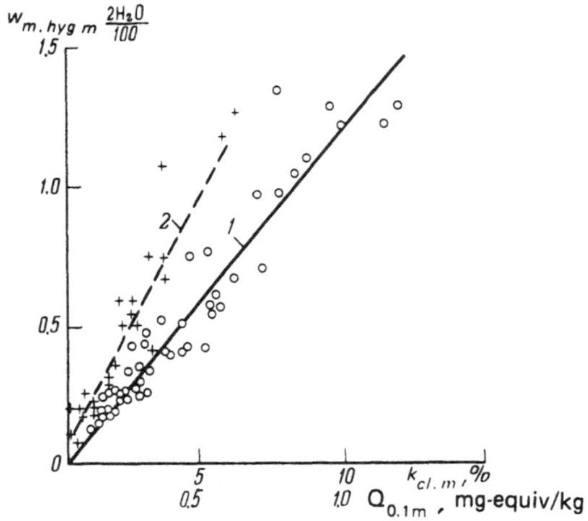

FIG. 31. Dependence of the maximum hygroscopic moisture capacity $w_{mh.m}$ on the coefficient of mass clayiness $k_{cl.m}$ (*1*) and exchange capacity $Q_{0.1m}$ (*2*) for carbonate rocks of the Eocene from borehole 4 of the area of the Chistenkaya village of the Crimean Region (after A.A. Vedenin)

These limits of variation of $w_{m.hm}$ are in good agreement with the specific surface of the aforementioned rocks. The maximum hygroscopic moisture capacity of fine-grained rocks with angular poorly rounded grains, clays and strongly clayey rocks is more than that of other varieties of sandy-aleurite and clayey and other detrital rocks since they have an appreciable specific surface and an increased SiO_2/R_2O_3 ratio which is often linked with the presence of expanding crystal lattices in clay minerals.

As a result, the values of $w_{m.hm}$ of limestones increase with growth of clayiness and exchange capacity (Fig. 31). Maximum hygroscopic moisture capacity is also governed by the composition of the absorbed ions which is accounted for by the orientation of unequal amounts of water dipoles around ions that differ in charge and radius. Some of the researchers either reject altogether or underestimate this mechanism of hygroscopic water accumulation.

The values of the mass maximum adsorptional moisture capacity $w_{m.adm}$* are in the range 0.1-50% and do not generally exceed 13% (for clays). They also increase with higher clayiness and exchange capacity and, consequently, with greater specific surface of rocks.

---

* It is determined from wetting heat volume of the fluid in which salts and sugar are not dissolved.

Chapter V

Oil and Gas Saturation. Chemically Bound Water

Sec. 10. **Oil and Gas Saturation of Rocks**

Oil, water and gas may be present in the pore space of oil- and gas-saturated rocks in different ratios and show various distribution; the latter is governed by the capacity of the solid phase of the rock to be wetted (hydrophility) or fail to be wetted (hydrophobic nature) by water.

The position of water and oil (or water and gas) in a single void and, consequently, in a rock, given different water content, can be diagramatically represented for rocks with *a hydrophylic and hydrophobic surface* of the solid phase as is illustrated in Fig. 32. Under conditions of total water saturation in a single void whose surface is wetted with water, the latter takes all its volume although in some cases minor amounts of gas, oil or both can remain in its central part (Fig. 32*A*, *a*). In the case of a hydrophobic oil (gas) trap a layer of oil or bitumen is found between the solid phase and water contained in the central part of voids (Fig. 32*B*, *a*).

With decrease of water saturation, the area occupied by unwetting phases (oil, gas or oil and gas) grows (Fig. 32*A*, *b*) and, given definite inappreciable water saturation extent, attains a critical value (Fig. 32*A*, *c*). Then the water assumes the form of a film that completely covers the surface of the solid phase.

In the case of a hydrophobic oil trap the water accumulates in the central parts of a void. With decrease of water saturation, the water content first drops in thin interconnected channels whereas in larger channels it is still preserved (Fig. 32*B*, *b*, *c*). The wetting water is then located on separate surfaces of voids (Fig. 32*A*, *d*) and water, which is not wetting the solid phase, is present as veins passing through centres of major voids (Fig. 32*B*, *d*).

Hydrophylic oil traps where water wets the solid phase mainly occur in nature. Hydrophobic traps have also been found in part. These, for example, include oil traps of the Carboniferous period found in the Radayevskaya area of the Kuibyshev Region of the Volga basin, productive deposits that seem to be hydrophobic due to the presence in oil of resins and sulphur of the Carboniferous period to be found in the Arlan area in Bashkiria with coal inclusions, bituminous Athabbaka sandstones in Canada whose grains are covered by a continuous bitumen layer*, Wilcox sandstone in Ocklahoma City and Springer sandstone in the Sholem Aleichem deposit.

Hydrophobity of oil traps is governed by the content of unwettable inclusions (e.g. coal ones); moreover, the pattern of distribution of water, oil, gas in the pore space of rocks is believed to be conditioned by their compostion and properties.

---

* These rocks approach fully hydrophobic oil and gas traps.

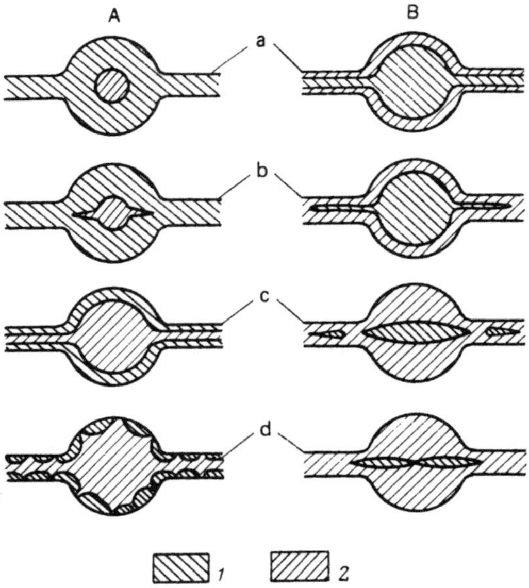

FIG. 32. Diagram of location of water and oil (or water and gas) in a separately taken void of hydrophilic (*A*) and hydrophobic (*B*) reservoir beds, given different degrees of their saturation.

Saturation of reservoir bed by water: *a* — total; *b* — higher than critical; *c* — critical; *d* — lower than critical; *1* — water; *2* — oil

An important role is played by organic acids contained in oils that impart to oil surface active properties, the ability to wet the solid phase.

However, the nature of hydrophobity of oil and gas traps has not yet been sufficiently understood. A number of petrophysical quantities (e.g. specific electric resistance, permeability coefficient etc.) are governed not only by the water, oil and gas content but also by their distribution pattern in the oil and gas traps.

In the general case of oil-, gas- and water-saturated rocks the sum total of volumes of oil V_o, gas V_g and V_w is equal to the volume of the pore space of rocks V_{por}:

$$V_o + V_g + V_w = V_{por}$$

and, consequently,

$$(V_o/V_{por}) + (V_g/V_{por}) + (V_w/V_{por}) = 1$$

The V_o/V_{por}, V_g^*/V_{por} and V_w/V_{por} ratios, expressed in fractions of a unity or in percent are defined, respectively, as k_o, k_g, k_w, and are said to be *coefficients of oil-, gas- and water saturation* which are used to evaluate the degree of saturation

* At atmospheric pressure.

of the rocks' pore space.

For oil-saturated rocks $V_o + V_w = V_{por}$, consequently,
$(V_o/V_{por}) + (V_w/V_{por}) = k_o + k_w = 1$

for gas-saturated rocks $V_g + V_w = V_{por}$ and

$(V_g/V_{por}) + (V_w/V_{por}) = k_g + k_w = 1$

The following relationships are true for oil-, water- and gas-saturated rocks:

$k_o = 1 - (k_g + k_w), \quad k_g = 1 - (k_o + k_w)$

The coefficient k_o measured on samples is the closest to its true value if during taking a core sample an oil-base drilling fluid has been made and solely bound water was available in the trap.

If the drilling uses a clayey solution, then an appreciable amount of oil is displaced by the solution filtrate, therefore, oil saturation of the cores is decreased, usually not higher than 20-40%.

The same *saturation* known to be *residual saturation* and defined as k_{sr} is also observed in clayey solution filtrate-washed zones of permeable oil-saturated strata immediately adjacent to the borehole's walls.

Under natural conditions oil saturation of rocks is 95% and more (hydrophobic oil traps). There are, however, oil and gas traps with oil saturation 50-60% yielding pure oil. These are represented by small-grained and clayey varieties of quartz and feldspar sandy-aleurite and clayey rocks (coarse- and fine-grained aleurolites) bounding much water. The k_o values are governed by the mineral composition and hydrophility (hydrophobic nature) of oil traps. The k_o coefficient grows with increasing the average radius $r_{eff.ch.av}$ of pore channels of the terrigenous oil and gas traps (Fig. 33*a*).

Extremely high k_o or k_g values are common for water floating oil and gas deposits of substantial thickness (e.g., in West Siberia).

Only oil or gas and bound water ($k_{w.r} = k_{w.b}$) is found in the limiting (maximum) saturation zone. The amount of bound water is governed by the specific surface of the rock, the greater the latter, the higher is $k_{w.b}$ and the lesser is $k_o(k_g)$.

The limiting saturation zone is followed in the direction of the borehole by a transition zone that stretches to the surface below which the stratum is completely saturated. In addition to bound and capillary water, the transition zone contains moving water. Here $k_o(k_g)$ varies from maximum values to zero. The thickness of the transition zone in the region of Tataria varies from 1.5 to 2 m, in the Kuibyshev region from 6 to 7 m, for the West Tebuk deposit in the Komi Autonomous Republic up to 11 m, in West Siberia it is as great as 30-40 m.

An insufficient saturation subzone can be singled out in the transition zone that is in contact with the limiting saturation zone. The insufficient saturation zone is characterized by the inflows of water-free oil (or gas). Here $k_o(k_g)$ varies from the maximum (>83%) to the critical (~55%) value. This portion of the transition zone contains less residual water than its lowermost portion where, apart from bound water, capillary water is contained, and the effective relative permeability for water varies from 0 to 1 and in the course of exploitation there occur inflows of oil with water, or gas and water.

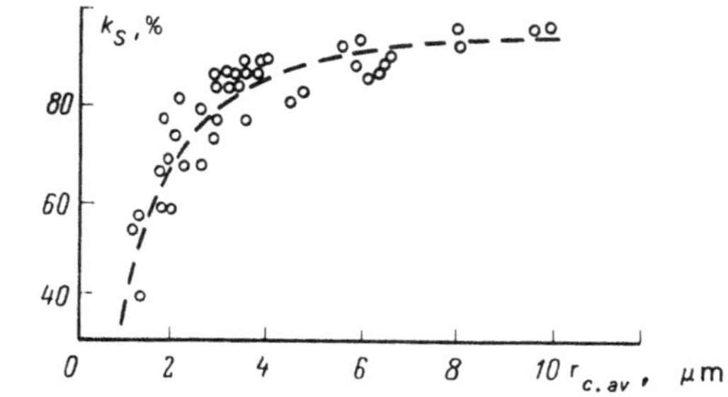

FIG. 33a. Dependence of the coefficient of oil saturation $k_{o.s}$ on the average radius, $r_{ch.av}$ of voids channels for terrigenic reservoir beds of the Baklanovskoe and Moskude deposits (after B.N. Tulbovich)

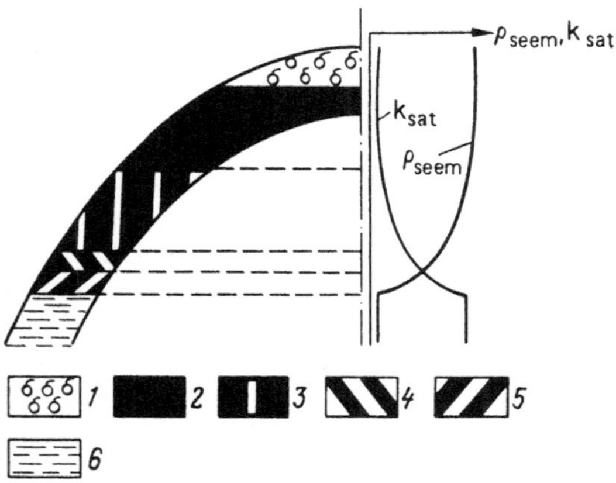

FIG. 33b. Diagram of the distribution of hydrocarbons according to the height of the oil deposit with a gas cap

Zones: *1* — gaseous, *2* — limiting oil saturation; *3* — undersaturation; *4*, *5* — transition (*4* — subzone of two-phase motion of fluids; *5* — subzone of residual oil saturation); *6* - aquiferous; dotted lines show zone boundaries; ϱ_{seem} — seeming specific electric resistance; k_{sat} — coefficient of water saturation (after Yu.V. Konoplev)

A water oil contact (WOC) or a water gas contact (WGC) is currently represented by an assumed surface of a stratum above which it yields water-free oil (or gas) and below which to the bottom of the transition zone, oil and water or gas and water (see Fig. 33*b*).

The determination of the gas- and water saturation coefficients of samples uses a device designed by S.L. Zakhs. Hydrophobity is established by observing the pattern of oil and water impregnation by a rock sample.

Sec. 11. **Chemically Bound Water**

Apart from free and adsorbed water, rocks contain crystallization water, and the heating of some rock species gives rise to constitutional water. Crystallization and constitutional waters are *chemically bound water* types.

Crystallization water makes part of the crystal lattice of many minerals; *constitutional water* forms during the heating of a number of mineral species commonly containing hydroxyl ions. That is why it is also known as hydroxyl water. Unlike free, adsorbed and interlayer waters (swelling waters) that are released from minerals without rupturing their structure, most commonly at temperatures less than 110 °C, removal and synthesis of chemically bound water are, as a rule, likely at higher temperatures. In this process their crystal lattice is ruptured.

Crystallization Water

Crystallization water is present in the crystal lattice as H_2O molecules and is commonly removed upon heating a rock to 300 °C and less commonly to higher temperatures. As a result, a water-free compound forms or a compound with lesser water content.

Minerals containing crystallization water represent compounds of weak bases and strong acids and of strong bases and weak acids.

Strong bases and weak acid compounds include, for instance, sodium carbonate $Na_2CO_3\ 10\ H_2O$ from which crystallization water is removed at laboratory temperature ($t \approx 20$ °C)*. Gypsum $CaSO_4\ 2H_2O$ and epsomite $MgSO_4\ 7H_2O$ are weak base and strong acid compounds. Gypsum loses crystallization water at t up to 400 °C.

Crystallization-water minerals include also hydroboracite $CaMgB_6O_{11}\ 6H_2O$, boronatrocalcite $NaCaB_5O_9 \times 8H_2O$, dioptase $Cu_6(Si_6O_{18})\ 6H_2O$, mirabilite $Na_2SO_4\ 10H_2O$ and others. Water is removed from these minerals, respectively, at temperatures 200-325, 115-500 and 200-600 °C; mirabilite loses crystallization water at laboratory temperature in dry air.

Constitutional Water

Constitutional water is formed by the OH^-, H^+ ions and oxonium (or hydronium) H_3O^+ ion entering the crystal lattice of minerals.

These minerals are classified as basic salts of weak bases and as bases if their composition contains the OH^- ion: in this case water is removed from minerals at t = 300-1 300 °C. Minerals involving constitutional water may be strong-base acidic salts and acids, then they contain the H^+ ion, from which water forms at temperatures up to 300 °C.

Among minerals basic salts of weak bases are predominant, and, upon heating them, water is generally produced from the OH^- ions.

This type of water is shown to occur upon the heating of rocks by chemical analysis.

Minerals the heating of which gives rise to synthesis of constitutional water include, for example, hydrargillite $Al(OH)_3$, talk $Mg_3(OH)_2(Si_4O_{10})$, serpentine $Mg(OH)_4(Si_2O_5)$, brucite $Mg(OH)_2$, diaspore $Al(OH)O$, allophane $Al_2O_3nSiO_2$, kaolinite $Al_2(OH)_4Si_2O_5$, halloysite $Al_2(OH)_4(Si_2O_5)H_2O$, montmorillonite $(AlMg)_2(OH)_2(Si_4O_{10})H_2O$ and others.

The temperature for constitutional water to be removed for talc, serpentine, brucite and diaspore is, respectively, equal to 930, 670, 410 and 510 °C.

On being heated, allophane loses interlayer water and hydroxyl groups from the crystal lattice gradually over the range t = 110-1 000 °C (Fig. 34, curve *1*). As a result 10-15% water is formed. In the case of kaolinite the loss of OH groups from the crystal lattice forming hydroxyl water occurs over the range of tempera-

---

* Temperature at which crystallization water is lost is very low for sodium carbonate and mirabilite, it is unusual for other minerals.

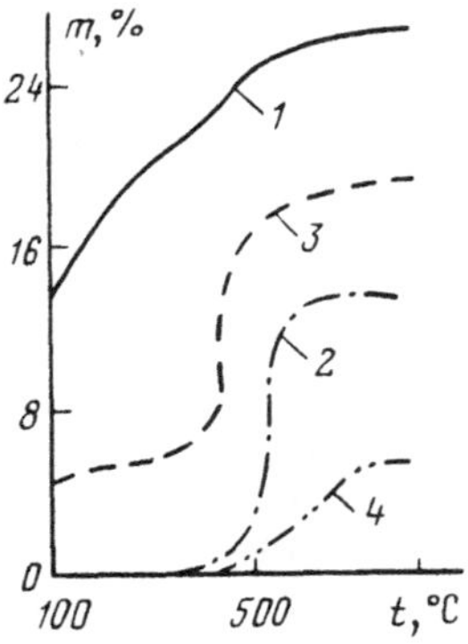

FIG. 34. Curves of losses of the mass *m* of clays at different temperature *t* (after K.S. Ross, P.F. Kerr and S.B. Hendriks).

1 — allophane (Moorfield, Kentucky); *2* - kaolinite (Ion, California); *3* - halloysite (Liège, Belgium); *4* — montmorillonite

tures 400-525 °C (Fig. 34, curve *2*). Halloysite (Fig. 34, curve *3*) loses hydroxyl groups mainly over the range of temperatures from 400-430 and up to 500 °C. Hydroxyl water is the principal water type for nonaqueous halloysites, where 20% water forms, whereas as little as 2-3% interlayer water is contained in them. By contrast, aqueous halloysites contain more low temperature interlayer water.

The moment the hydroxyl groups entering the crystal lattice of montmorillonite start to be lost and hydroxyl water starts to form is difficult to establish since dehydration curves show no abrupt breaks (Fig. 34, curve *4*). It is assumed that it starts at 300 °C. For common montmorillonites a rapid loss of hydroxyl water starts at about 500 °C. Dehydration terminates at 800 °C (Fig. 34, curve *4*).

As far as plastic rocks are concerned, constitutional water mainly forms from the aforementioned minerals of clays. Particularly much hydroxyl water is formed of allophane, kaolinite and montmorillonite.

Apart from the aforementioned types still another water variety exists, viz. constitutional crystallization water. It is formed from salts of H_3O^+ oxonium ion typically upon the heating of minerals to a temperature 300 °C. Oxonium is to be found in potassium-containing laminated structure silicates. Upon weathering potassium is substituted by the oxonium ion. An example of an oxonium-containing mineral is provided by hydromuscovite $(K, H_3O)Al_2(OH)_2[Al_2Si_3O_{10}]$ derived from mica.

Chapter VI

Density

Sec. 12. **The Definition and Physical Basis**

Density is the property of a rock to have a definite mass of a unit volume different from the specific mass of its other varieties. The density* is

$$\delta_r = m_r / V_r \tag{1}$$

where m_r is the mass of rock; V_r is the volume of rock.

In the general case the mass of a rock m_r consists of the masses m_s, m_l and m_g, of its solid, liquid and gaseous phase, respectively.

Hence,

$$\delta_r = \frac{m_r}{V_r} = \frac{m_s + m_l + m_g}{V_r} = \frac{\delta_s V_s + \delta_l V_l + \delta_g V_g}{V_r} \tag{2}$$

where δ_s, δ_l, and δ_g are densities of, respectively, the solid, liquid and gaseous phase of the rock; they determine the masses of the volumes V_s of the solid, V_l of the liquid and V_g of the gaseous phase in a volume V_r of the rock. The unit of measurement of density is kg/m^3. Alternatively a fractional unit, g/cm^3, may be used.

Given that $V_r \approx V_{dr}$ (as for weakly clayey and other nonswelling water-saturated rocks) and $V_{dr} = 1$

$$k_r = V_{por}/V_{dr} = V_{por}/V_r = V_{por}; \qquad V_g = 1 - V_{por} = 1 - k_r$$

$$w = \frac{V_l}{V_{dr}} = \frac{V_l}{V_r} \simeq V_l; \qquad V_g = V_{por} - V_l = k_r - w$$

Here V_{dr} and V_{por} are the volumes of, respectively, a dry sample and its pore space; k_r and w are the total voids ratio and the total moisture capacity, respectively.

A substitution of the latter relationships into Eq. (2) yields

$$\delta_r = (1 - k_r)\delta_s + w\delta_l + (k_r - w)\delta_g \approx \delta_{dr} + w\delta_l \tag{3}$$

where $\delta_{dr} = (1 - k_r)\delta_s$ is the density of a dry rock.

The latter is true since $\delta_g = \delta_{air} \simeq 1.2 \times 10^{-3}$ g/cm^3 and for $w \to 0$ $\delta_{air} k_p$ is also very small (δ_{air} is the density of the air).

Consequently, rock density is governed by the density of the rock's phases and specific volumes of the voids and water. Given $k_r \to 0$ and $w \to 0$, $\delta_r \to \delta_s$; and given $k_r \to 1$ and $w \to 1$, $\delta_r \to \delta_l$.

---

* Literature on field geophysics generally defines density as σ.

The theory is well corroborated by the experimental evidence for weakly clayey and other, nonswelling rock species. The rock density values are determined from samples taken from rock outcrops and boreholes. For analysis of experimental data rock density is evaluated from the density of the rock's phases, values of the voids ratio and moisture capacity coefficient. The phase densities are also used for solving other geological problems: determining the substance composition of waters, oil and gases, evaluating the purity extent of individual mineral varieties, calculating the volume of a unit cell of mineral crystals as a characteristic of the chemical and mineral composition of dense igneous, metamorphic and sedimentary rocks etc.

For clayey rocks and clays Eq. (3) is not valid due to their shrinkage, i.e. decrease of their pore volume upon drying. The density $\delta_{dr} = (1 - k_r - \Delta k_r)\delta_s$ determined on samples of dry clayey rocks is much greater than the actual contribution of the solid phase to the density of these rocks.

To find the density of maximum wet clayey rocks, to the uncorrected values of δ'_{dr} of clayey rocks a correction is given

$$\Delta\delta = \delta_r - \delta'_{dr} = \frac{m_s + m_l}{V_{por} + V_s} - \frac{m_s}{V_{por} + V_s - \Delta V_{por}}$$

$$= \left(1 - \frac{\delta'_{dr}\Delta V_{por}}{k_m V_{por}}\right)(k_m k_r) \approx q k_m k_r \approx qw$$

where $q = (1 - \delta_{dr}\Delta V_{por}/k_m V_{por})$ and ΔV_{por} is shrinkage, decrease in the pore volume following drying a sample.

Consequently, the density of clayey rocks is $\delta_r = \delta'_{dr} + q k_m k_r \approx \delta'_{dr} + qw$. The value of q is estimated from the content of clayey fractions in clays and the experimental relation $q = f(k_{cl.m})$ for the particular region determined by R.V. Kalinina (Fig. 35).

In order to give a correction for shrinkage to δ'_{dr} of clayey rocks and clays, I.Kh. Yudborovskii also proposes the following equation obtained experimentally:

$$\delta_{dr} = 1.4\delta'_{dr} - 0.96 \qquad (4)$$

in case the above relation cannot be established for the regions where samples are to be taken. The error of density determination without taking into account shrinkage is about 0.2 g/cm^3.

The difference between the density values determined from humid samples and calculated from Eqs. (3) and (4) is on the average ±0.03 g/cm^3.

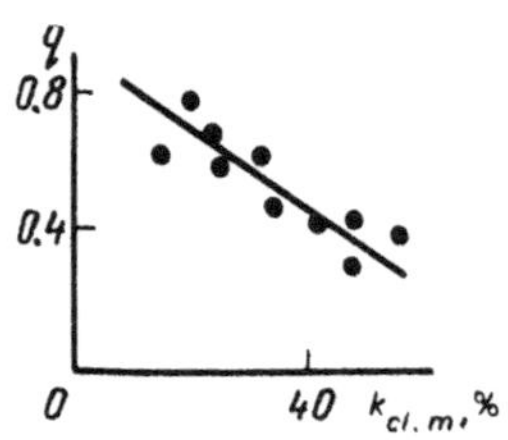

FIG. 35. Dependence of the coefficient q on the coefficient of mass clayiness $k_{cl.m}$ of Gzhel clays

By proceeding in like manner, for oil-, water-saturated rocks we can write:

$$\delta_r = \delta_s V_s + \delta_w V_w + \delta_o V_o$$

but since $V_{por} = k_p$, then $V_w = k_w k_p$; $V_o = k_o k_p$ and

$$\delta_r = (1 - k_p)\delta_s + \delta_w k_w k_p + \delta_o k_o k_p = (1 - k_p)\delta_s + \delta_w(1 - k_o)k_p + \delta_o k_o k_p$$

The densities of gas and water saturated and oil-, water- and gas saturated rocks are found, respectively, from these equations:

$$\delta_r = (1 - k_p)\delta_s + \delta_w k_w k_p + \delta_g k_g k_p$$

and

$$\delta_r = (1 - k_p)\delta_s + \delta_o[1 - (k_w + k_g)]k_p + \delta_w[1 - (k_o + k_g)]k_p + \delta_g[1 - (k_w + k_o)]k_p$$

These relations are obtained much as the previous one.

As can be seen from the theory of density, there is no way of analytically determining this quantity without knowing the densities of the phases of rocks, their voids ratio and moisture capacity coefficient. Let us consider the densities of the phases to estimate the contribution of each of them to the densities of rocks.

Sec. 13. **The Gaseous and the Liquid Phase**

The Gaseous Phase

The density of the air at 20 °C and atmospheric pressure 0.1 MPa is 0.0012 g/cm^3, that of hydrocarbon gases varies from 0.000715 (methane) to 0.00317 (pentane) g/cm^3 and is equal on the average to the density of the air.

The Liquid Phase

The natural water filling the voids whose density at 20 °C, depending on the mineralization, varies from 1.010 g/cm^3 (fresh water) to 1.240 g/cm^3 (extremely strong brine) as well as mixtures of this water with oil represent a liquid phase. If this latter includes oil and water mixtures, its density is governed by the densities δ_o and δ_w of its constituents, oil and water, and their volume ratio. The density of oils may vary, depending on the chemical composition, from 0.72 to 1 g/cm^3 and only seldom does it exceed that limit. The oil and water ratio in the liquid phase may differ very much (see Sec. 10).

For measurements of the density of natural waters and oils one generally use areometers and, less frequently, other apparatus.

Sec. 14. **Minerals**

The density of the solid phase of a rock is what governs its mineralogical composition. Let us therefore consider the density δ_s of the same component of individual minerals. Since the voids ratio of these media is commonly small (generally not more than 1%), their density δ_s is approximately equal to the density δ_m of the entire mineral.

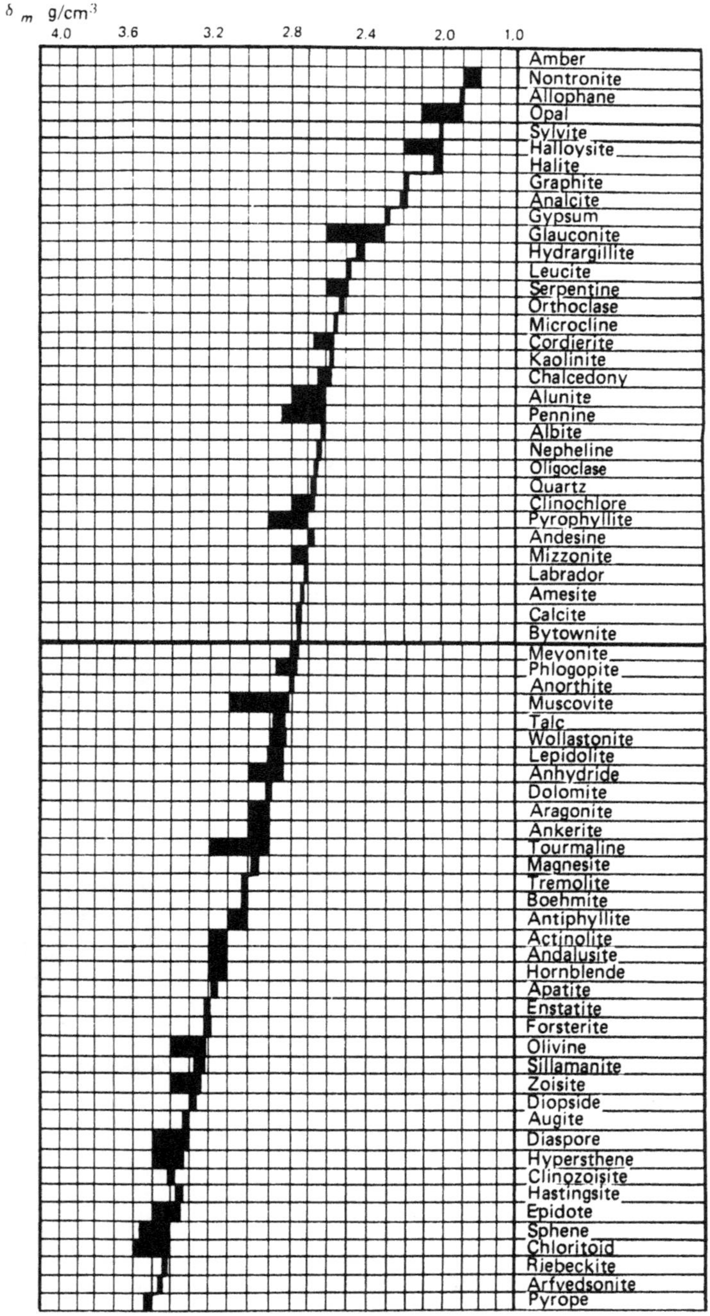
δ m g/cm3
4.0
3.6
3.2
2.8
2.4
2.0
1.0
Amber
Nontronite
Allophane
Opal
Sylvite
Halloysite
Halite
Graphite
Analcite
Gypsum
Glauconite
Hydrargillite
Leucite
Serpentine
Orthoclase
Microcline
Cordierite
Kaolinite
Chalcedony
Alunite
Pennine
Albite
Nepheline
Oligoclase
Quartz
Clinochlore
Pyrophyllite
Andesine
Mizzonite
Labrador
Amesite
Calcite
Bytownite
Meyonite
Phlogopite
Anorthite
Muscovite
Talc
Wollastonite
Lepidolite
Anhydride
Dolomite
Aragonite
Ankerite
Tourmaline
Magnesite
Tremolite
Boehmite
Antiphyllite
Actinolite
Andalusite
Hornblende
Apatite
Enstatite
Forsterite
Olivine
Sillamanite
Zoisite
Diopside
Augite
Diaspore
Hypersthene
Clinozoisite
Hastingsite
Epidote
Sphene
Chloritoid
Riebeckite
Arfvedsonite
Pyrope

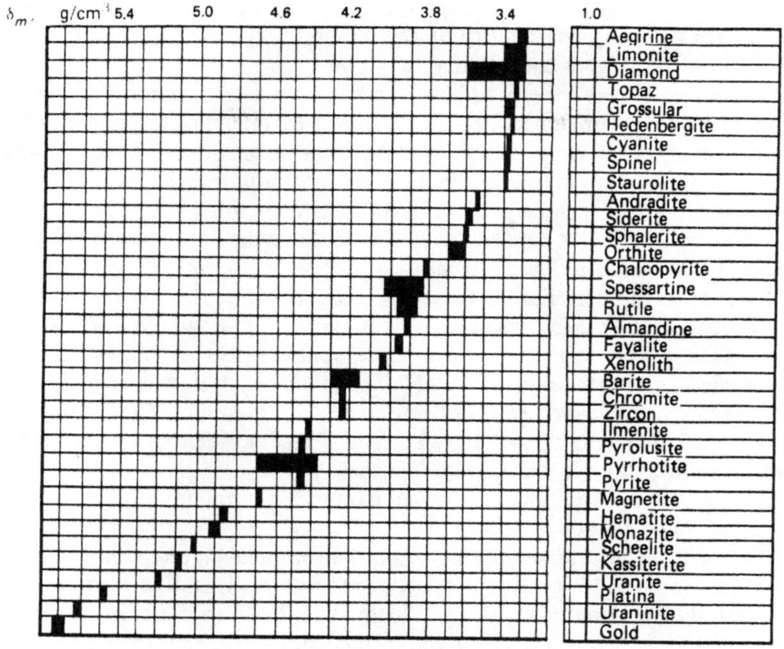

FIG. 36. Values of densities of minerals

The more is the number of atoms of an appreciable relative atomic mass and the more the atoms are packed in a unit volume, the higher is the density of a mineral. Packing is governed by atomic (ionic) radii, valencies and the bonding type of the particles. The factor that mainly affects the density of ore minerals is the weighted average relative atomic mass, and what affects the density of rock-forming minerals is packing.

The most important rock-forming minerals (calcite, quartz, oligoclase, nepheline etc.) comparatively little differ in terms of the weighted average relative atomic mass, and it is comparatively small, so differentiation of densities of these media is mainly governed by their structure—particles' packing density in single crystals. The effect of the radius of the particles can be easily shown by comparing the density of ionic crystals of sylvine and halite. Even though the composition of sylvine includes a potassium atom with a greater relative mass than a sodium atom found in a halite molecule, the density of sylvine (1.99 g/cm^3) is less than that of halite (2.1-2.2 g/cm^3) owing to the difference in ionic radii of K and Na equal to 0.0133 and 0.098 nm, respectively.

The dependence of the density on valency of ions that compose minerals can be accounted for as follows. Large ions (for example, the O^{2-}, S^{2-}, Cl^- ions) a number of minerals give rise to the densest packings. Cations of a smaller radius are located in-between. At this, the number of cations among anions may be the greater, and, consequently, the density of the mineral may be higher the lesser is their valency and the higher the valency of anions.

The relationship between the density of the mineral's structure and the density of packing of its atoms or molecules can be shown by referring to minerals that are identical in chemical composition but different in structure. Even though graphite and diamond have the same chemical composition, they differ drastically in structure. A laminated structure is characteristic of graphite, with the distance between the layers being 0.34 nm and interatomic distances in a layer 0.142 nm. By contrast, diamond has a denser structure with the interatomic distances 0.154 nm. This shows up the densities of these minerals, equal to 2.2 and 3.5 g/cm^3, respectively. The difference in structure is also typical of aragonite and calcite. Aragonite has a denser structure which is responsible for its higher density (2.85-2.94 g/cm^3) compared with calcite (2.73 g/cm^3). The density as determined for most minerals varies from 0.98 g/cm^3 for ice to 22.5 g/cm^3 for iridic osmium. Minerals are classified into dense ($\delta_s > 4$ g/cm^3), medium-density ($\delta_s = 2.5$-4.0 g/cm^3) and small-density ($\delta_s < 2.5$ g/cm^3) species.

Most minerals have a medium density. An inappreciable fraction of all minerals (10-15%) show small density.

Dense minerals include: selenium, tellurium, native metals (gold, silver, copper etc.), sulphides, with rare exception (pyrite, pyrrhotite etc.), half of oxides and hydroxides (pyrolusite, hematite etc.), rare silicate varieties (thorite, zircon etc.), rare phosphate varieties and their analogues (monazite, carnotite, xenotime etc.), some carbonates (cerussite, viterite etc.), wolframates (scheelite, wolframite etc.), some sulphates (barite, anglesite etc.), chlorides and bromides: cerargyrite, bromyrite, calomel etc. The composition of dense minerals shows an important concentration of particles with a great relative atomic mass (lead, mercury, silver, copper etc.), these materials also have a relatively small atomic (or ionic) radius; such minerals often display the densest atom packing—cubical (face-centred) and hexagonal.

The highest content of dense atoms with relatively small atomic radii is displayed by native metals and their natural alloys that have maximum densities (Fig. 36).

Small-density minerals include the following species. Of native nonmetals, these are graphite and sulphur; some of oxides and hydroxides (ice, opal); many silicate class minerals (halloysite, montmorillonite); some borate species (borax); hydrous carbonates (sodium carbonate known also as soda); hydrous sodium sulphates (mirabilite), some chlorides, such as halite, sylvite (see Fig. 36).

Small-density minerals comprise elements with a small relative atomic mass (hydrogen, boron, carbon, oxygen, sodium, silicium, sulphur, chlorine, potassium etc.); atoms of these elements often have large atomic or ionic radii; many of them (ice, graphite, some silicates, calcite etc.) display a loose texture.

The medium-density mineral group includes most of the remaining minerals (see Fig. 36). The variable density of minerals is generally due to impurities. The good differentiation of mineral density permits this quantity to be used for their identification.

Density may, in particular, serve to isolate individual members of isomorphic series, for example, plagioclase series minerals that represent isomorphic mixtures* of albite ($NaAlSi_3O_8$) and anorthite ($CaAl_2Si_2O_8$) with densities from 2.61 to 2.77 g/cm^3.

The density of rock-forming minerals of sedimentary rocks varies from 1.85 (for allophane) to 5.18 g/cm^3 (for pyrite), most of them being medium-density minerals. The density of ore minerals attains great magnitudes varying from 2.3 (hydrargillite) to 7.5 g/cm^3 (galenite). The density of rock-forming minerals of igneous rocks attains 4 g/cm^3 and more.

Sec. 15. **The Solid Phase**

The density of the solid phase of rocks is the weighted mean density of its constituent minerals, i.e.

$$\delta_s = \sum_{i=1}^{i=n} V_{si}\delta_{si} = V_{s1}\delta_{s1} + V_{s2}\delta_{s2} + \dots + V_{sn}\delta_{sn}$$

where n is the number of minerals constituting the solid phase; δ_{s1}, δ_{s2}, ..., δ_{sn}, V_{s1}, V_{s2}, ..., V_{sn} are, respectively, densities and volumes taken by individual minerals in unit volume of the solid phase.

Apart from hydroargillites and bauxites, it is the density of the solid phase of massive ores that varies most and generally from values >3.5 to <5.5 g/cm^3, since it displays appreciable concentrations of dense minerals (magnetite, hematite and others in the solid phase of iron ores; pyrrhotite, pentlandite, chalcopyrite and others in the solid phase of sulphide copper nickel ores; galenite, sphalerite, pyrite, chalcopyrite in the solid phase of polymetallic ores etc.). The density δ_s of sparsely impregnated ores may drop to values close to that of the solid phase of the enclosing rocks that mainly is composed of silicates (quartz, serpentine, olivine, tourmaline etc.) with a density from 2.5 to 3.0 g/cm^3 and somewhat higher. The density δ_s of the solid phase of individual varieties of iron ores of a similar mineral composition shows the following ratio with the C_{Fe} (in %) content of iron in an ore, with an average ξ (in %) content of iron in an ore mineral and average densities $\delta_{s.o}$ and $\delta_{s.en}$ of, respectively, ore and enclosing minerals:

$$\frac{1}{\delta_s} = \frac{C_{Fe}}{\xi\delta_{s.o}} + \frac{\xi - C_{Fe}}{\xi\delta_{s.en}} \tag{5}$$

Given constant values of densitites $\delta_{s.o}$ and $\delta_{s.en}$ and a constant average mass iron content in ore minerals Eq. (5) (Fig. 37) for single-mineral iron ores of different composition are well corroborated by the experimental data. They make it possible to determine the iron content in individual ore varieties of a deposit from the density of their solid phase as well as the density of the solid phase of the enclosing rocks.

* An isomorphic mixture is a mixture of two or more minerals that are similar in composition and in crystalline form.

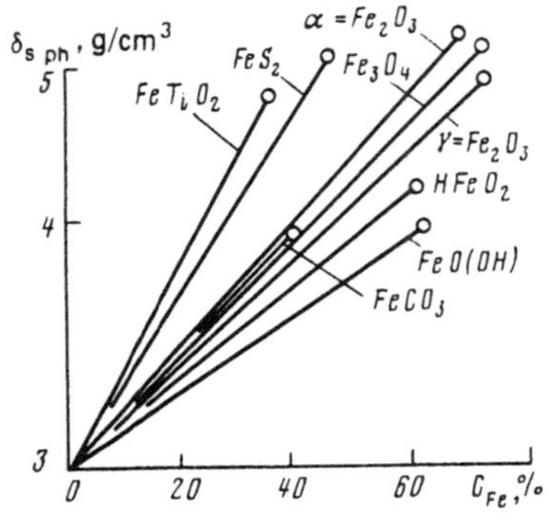

FIG. 37. Dependences of the density of the solid phase, δ_s, of iron ores bearing a single ore mineral, on their iron content, C_{Fe} (after A.K. Weinberg)

The density of the solid phase of igneous, metamorphic, calcareous and magnesial rocks is generally less than δ_s of ores and varies in the range from 2.5 to 3.5 g/cm^3. For igneous rocks, δ_s increases from acidic (granites etc.) to basic (gabbros etc.) and ultrabasic (peridotite, pyroxenite) as in this direction the content of the above rocks of ferruginous and magnesial minerals (olivine, pyroxenes, micas) or their components (FeO, Fe_2O_3, CaO, MgO) increases in the solid phase.

The highest solid phase density values refer to limestones and dolomites containing ferruginous mineral admixtures (pyrite, glauconite, siderite, chamosite etc.); the least values by rocks with admixtures of opal, kaolinite and other less dense minerals compared with calcite and dolomite. The most probable are values of δ_s of limestones from 2.71 to 2.74 g/cm^3, and those of dolomites from 2.8 to 2.9 g/cm^3.

The density of the solid phase of sandy-aleurite and clayey rocks drops to 2.2 g/cm^3. Peak values (in excess of 3.5 g/cm^3) of the considered density of these rocks do not practically differ from δ_s of calcareous and magnesial rocks. This is accounted for by that the composition of these rocks, in addition to rock-forming minerals (quartz, feldspars, clayey minerals etc.) includes impurity and accessory minerals of a relatively appreciable density, such as garnet, zircon, rutile, disthene, limonite, ilmenite for quartz-garnet sands, cassiterite for quartz-cassiterite sands, ilmenite for quartz-ilmenite sands, zircon for quartz-zircon sands, ilmenite and magnetite for ilmenite- and magnetite-quartz sands, hornblendes for hornblende sands, micas for

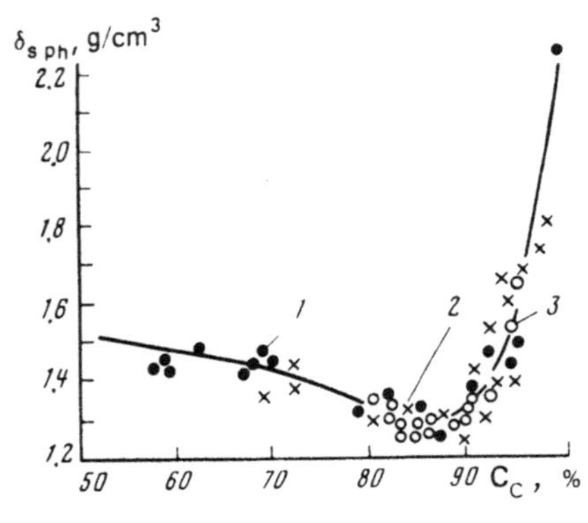

FIG. 38. Dependence of the density, δ_s, of dry ashless coals on their carbon content, C_C, (degree of metamorphism).

1 — after Dulhunty and Penrose; *2* — after Franklin; *3* — after Zwittering

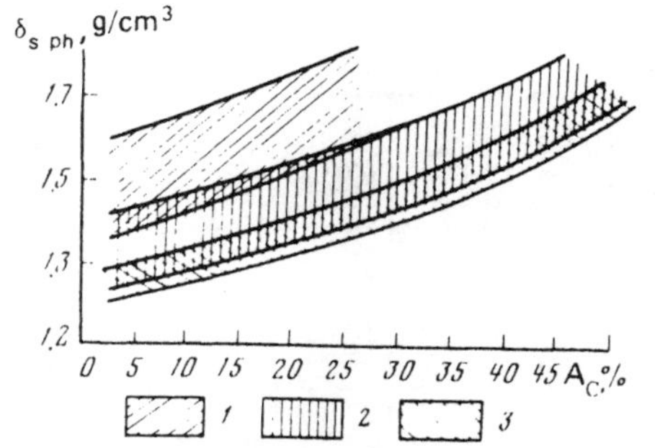

FIG. 39. Dependence of the density, δ_s, of various metamorphosed coals on the ash content A_C.

Coals: *1* — anthracites; *2* — hard coals; *3* — brown coals

micaceous sands; pyrite, glauconite, siderite, chamosite and micas for sandstones, aleurolites and clays. Relatively low densities of the solid phase of the aforementioned rocks are connected with the presence of opal, kaolinite, allophane and other clayey minerals showing small density. Due to the fact that it is not uncommon that the content of secondary and accessory minerals may vary even for rocks of the same type and age, the density of their solid phase varies over a wide range.

The density of the solid phase of saline rocks may be even smaller. For mirabilite, e.g. it is about 1.5 g/cm³, for rock salt 2.1-2.2 g/cm³, for gypsums 2.3 g/cm³. However, the maximum values of the solid phase's density of this rock group may be in excess of 3 g/cm³ (for example, for anhydrites with admixtures of pyrite, marcasite and other high-density minerals).

The density of the solid phase of siliceous rocks varies from 2 (for diatomite, whose rock-forming mineral is opal) up to >2.65 g/cm³ (for jasper and flints with rock-forming chalcedony or its mixture with opal or quartz).

The lowest δ_s values have been established for graphite and coals. The δ_s values of these rocks are in the range from 1.25 to 2.27 g/cm³. Low values of δ_s are shown by pure coals containing 85% C (Fig. 38). Pure coals have few mineral admixtures (quartz, clayey minerals, pyrite). At this, the density $\delta_{s.org}$ of their ashless organic solid phase may be evaluated from the relation

$$\delta_{s.org} = \frac{(100 - \xi)\delta_s\delta_{s.m}}{100\delta_{s.m} - \xi\delta_s}$$

where ξ is the mineral admixture content, %; δ_s and $\delta_{s.m}$, are, respectively, densities of the ash coal solid phase and admixtures.

The highest values of solid phase density are shown by high-ash content and metamorphosed coals (Fig. 39).

Sec. 16. **Dry Sediments and Rocks**

In terms of their δ_{dr} density values as determined on a very great number of samples, dry sediments and rocks can be divided into the following five groups.

Group I includes sediments and rocks with a very low density δ_{dr} (from 0.5 to 1.5 g/cm³). This group comprises sediments and rocks of various groups with

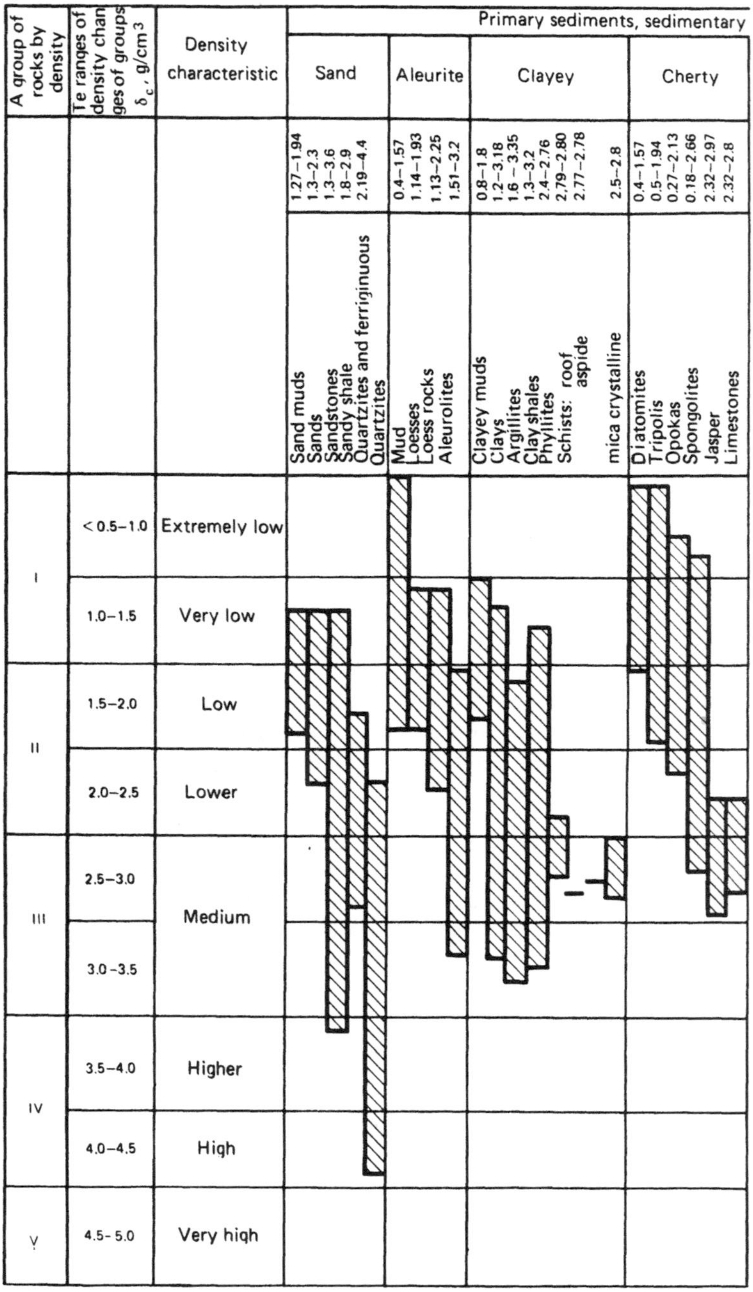

FIG. 40. Values of the density δ_{dr} of primary sediments, derived from sedimentary and metamorphic rocks and ores

rocks and metamorphic rocks and ores formed from them		
Calcareo-magnesial: calcareous	Limestones:	1.3–3.5
	Limestones: biogenic	1.9–2.95
	biogenic-detrital	1.27–2.95
	detrital	1.91–2.65
	slime	2.14–2.43
	pelitomorphic	1.64–2.43
	microgranular	2.52–2.62
	fine	2.56–2.75
	oölitic	2.2–2.4
	clastic	1.97–2.72
	Chalk	1.2–2.6
	Tufa	1.85–2.6
	Marmorized limestones	2.3–3.0
	Marble	2.3–3.0
	Dolomites	1.9–3.5
Calcareo-magnesial: Dolomitic	Dolomitic meal	1.48–2.08
Marls	Marls	1.2–3.0
	Dolomitic marl	1.4–2.5
	Paraamphibolitic magnesite	2.0–3.1
Salts	Gypsum rocks	2.0–2.98
	Anhydrite rocks	2.3–3.0
	Rock salt	2.1–2.3
	Sylvinite	1.9–2.95
	Carnallite rock	1,6–2,41
Aluminiferous	Bauxites: loose	1.3–2.4
	stony	1.68–3.47
	clayey	1.34–2.04
	red, greasy to touch	2.5–2.8
Ferruginous	Martites and limonites	3.8–4.3
	Iron ores of leptochlorite formation	1.6–2.19
	Siderite rocks	2.45–3.58
Manganese	Oölitic, soft, sooty	1.8–4.5
Phosphate	Apatite ores	3.18–3.41
	Phosphorites	2.8–3.0
Peat, coals and graphites	Peat	0.45–0.8
	Coals: brown	1.2–1.30
	hard	1.2–1.35
	anthracites	1.4–1.8
	Graphite	2.23–2.27

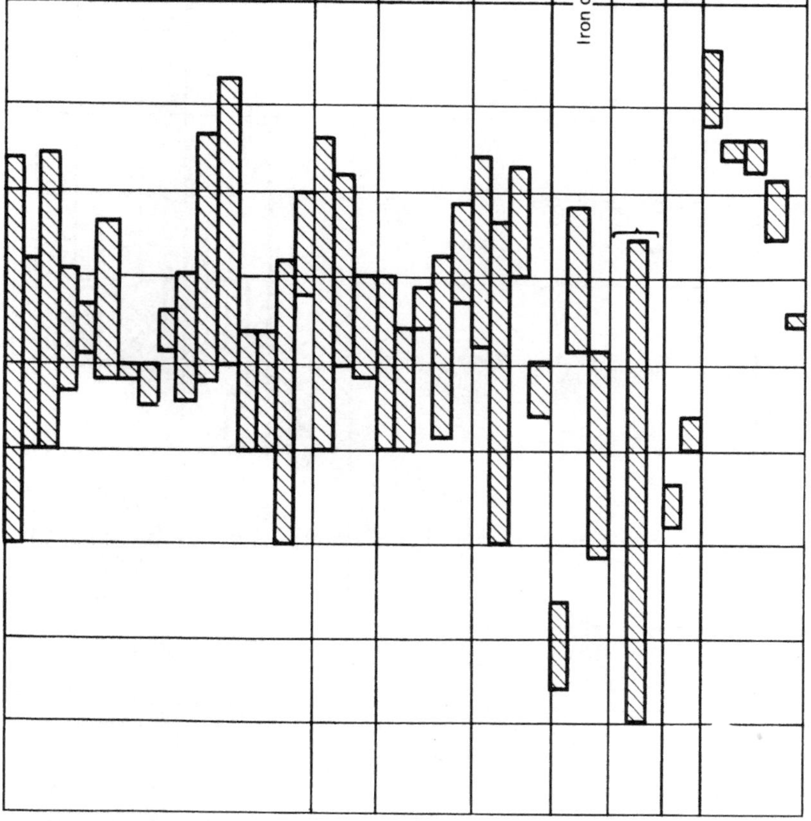

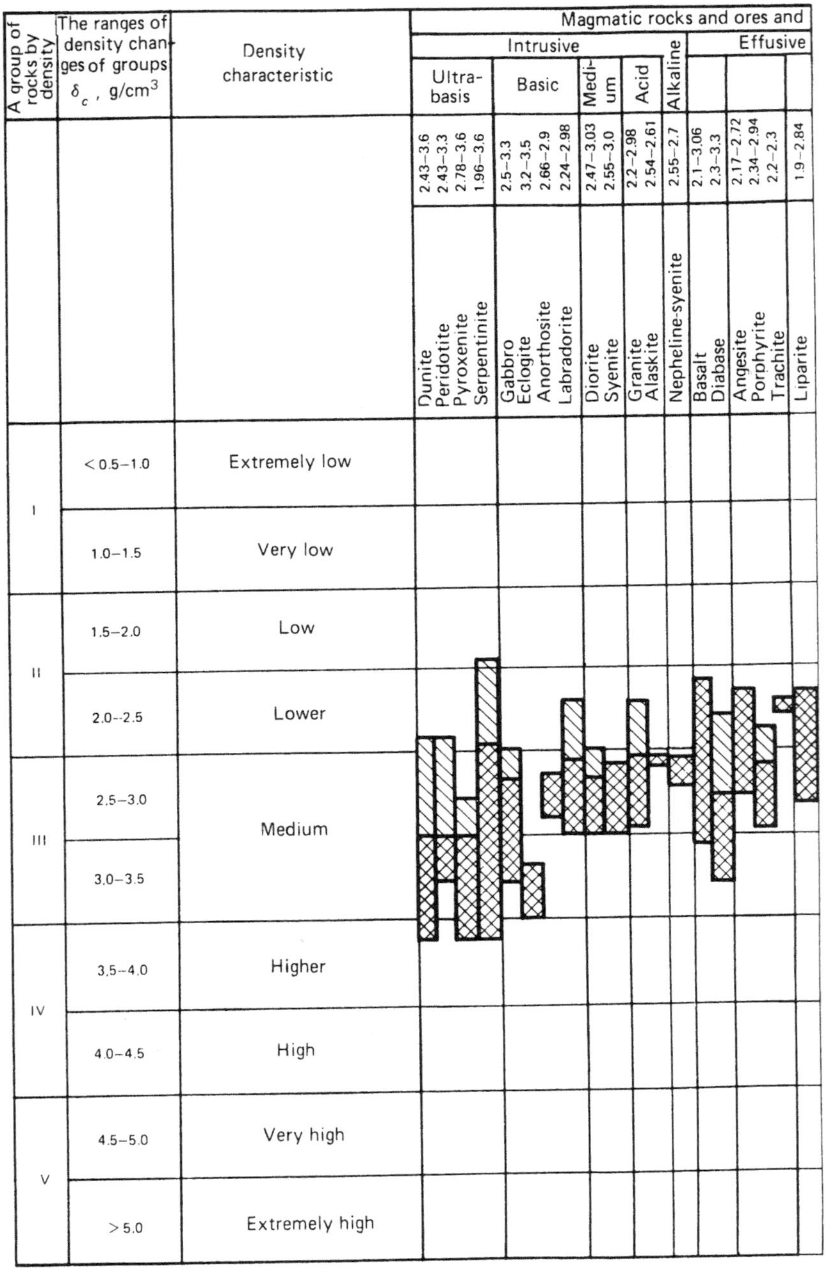

FIG. 41. Values of the density δ_{dr} of magmatic, metamorphic rocks and ores of magmatogenic origin.

Rocks: *1* — nonweathered or nonserpentinized; *2* — weathered or serpentinized

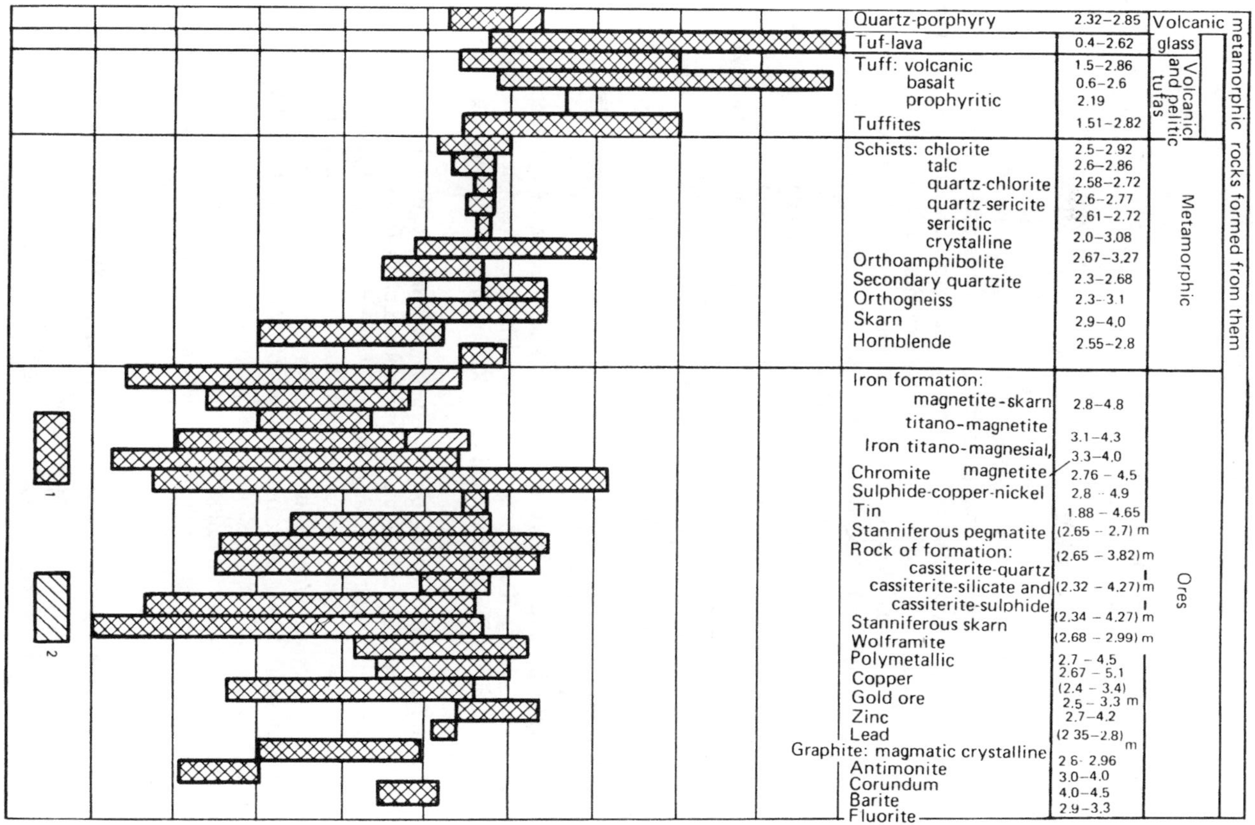
metamorphic rocks formed from them
Volcanic
glass
Volcanic and pelitic tufas
Metamorphic
Ores
Quartz-porphyry 2.32–2.85
Tuf-lava 0.4–2.62
Tuff: volcanic 1.5–2.86
basalt 0.6–2.6
prophyritic 2.19
Tuffites 1.51–2.82
Schists: chlorite 2.5–2.92
talc 2.6–2.86
quartz-chlorite 2.58–2.72
quartz-sericite 2.6–2.77
sericitic 2.61–2.72
crystalline 2.0–3.08
Orthoamphibolite 2.67–3.27
Secondary quartzite 2.3–2.68
Orthogneiss 2.3–3.1
Skarn 2.9–4.0
Hornblende 2.55–2.8
Iron formation:
magnetite-skarn 2.8–4.8
titano-magnetite 3.1–4.3
Iron titano-magnesial, magnetite 3.3–4.0
Chromite 2.76 – 4.5
Sulphide-copper-nickel 2.8 – 4.9
Tin 1.88 – 4.65
Stanniferous pegmatite (2.65 – 2.7) m
Rock of formation:
cassiterite-quartz (2.65 – 3.82) m
cassiterite-silicate and (2.32 – 4.27) m
cassiterite-sulphide (2.34 – 4.27) m
Stanniferous skarn (2.68 – 2.99) m
Wolframite 2.7 – 4.5
Polymetallic 2.67 – 5.1
Copper (2.4 – 3.4)
Gold ore 2.5 – 3.3 m
Zinc 2.7–4.2
Lead (2.35–2.8) m
Graphite: magmatic crystalline 2.6–2.96
Antimonite 3.0–4.0
Corundum 4.0–4.5
Barite 2.9–3.3
Fluorite
1
2

6*

very high voids ratio values (extremely porous clayey, sandy, calcareous, diatomite and other muds, extremely porous varieties of clayey, sandy, aleurite rocks, some varieties of chalk, diatomites, tripolis, opokas, ash tuffs etc.) as well as rocks both with high and low voids ratio values yet showing an abnormally small density δ_s (peat and differently metamorphosed coals excepting high density varieties of anthracites and high-ash content coals (see Figs. 40, 41, Table 2).

Group II includes low- and decreased-density δ_{dr} rocks (from 1.5 to 2.5 g/cm^3). These comprise highly metamorphosed and ash coals, sedimentary rocks (clayey, aleurite, sandy, calcareous and magnesial, salt, gypsums, bauxites of loose and clayey varieties, some hard bauxite varieties) excepting their very high density varieties with ore and other hard inclusions and cement, porous and weathered varieties of tin ore, polymetallic and other ores, some graphite rock varieties.

Group III comprises medium-density δ_{dr} rocks (from 2.5 to 3.5 g/cm^3). The group includes dense sedimentary, igneous and metamorphic rocks, such as dense sandstones, aleurolites, argillites, limestones, dolomites with ore impregnations and dense cement, anhydrites, dunite, gabbro, granite, slates and schists, hornfels and others, bauxites' hard and red and smearing varieties, some graphite rock species, impregnated sulphide copper-nickel, copper, iron and polymetallic ores.

Group IV includes increased- and high-density δ_{dr} (from 3.5 to 4.5 g/cm^3) rocks. It is represented by massive unweathered tin, sulphide copper-nickel, iron, chromite and polymetallic ores excepting some varieties of sulphide copper-nickel, tin, polymetallic and copper ores.

Group V includes rocks with very high and extremely high density δ_{dr} values (from >4.5 g/cm^3). The group is represented by some very dense varieties of tin, sulphide copper nickel, polymetallic and copper ores with a very high ore mineral content.

It should be noted that δ_{dr} of some rocks (for example, galenite ores) may attain values in excess of 6 g/cm^3. Some density group rock varieties may exhibit δ_{dr} values falling within the limits indicated for one, two or more groups due to variations in the composition and structure of their individual varieties. The most probable δ_{dr} values of various types of the most important sedimentary rock groups increase in the following order: sandstones (2.1-2.4 g/cm^3) < aleurolites (2.1-2.5 g/cm^3) < clays (2.2-2.4 g/cm^3) < limestones (2.4-2.6 g/cm^3) < dolomites (2.5-2.6 g/cm^3) < anhydrites (2.8 g/cm^3 and more).

Thus, loose varieties are the most likely to occur for sandy aleurite and clayey rocks.

Sec. 17. The Relationship Between Density δ_{dr} and Other Petrophysical and Petrochemical Quantities

Since it is not uncommon that massifs, orebodies or rock beds whose origin, age and transformation extent have been established should contain igneous, metamorphic or sedimentary rock varieties of the same type that may change in composition and structure it is possible to find fairly close correlations between the density δ_{dr} and rock composition and structure characteristics. The density δ_{dr} of rocks of individual crystal-

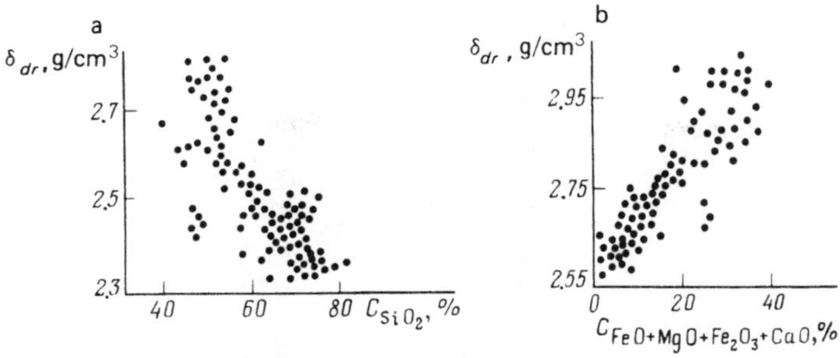

FIG. 42. Dependence of the density δ_{dr} of crystalline rocks of the Pre-Cambrian foundation of the eastern Russian platform mainly represented by gneisses on the contents of the principal components C_{SiO_2} (*a*) and $C_{FeO+MgO+Fe_2O_3+CaO}$ (*b*) (after N.V. Podoba)

line massifs increases with increasing *C* concentration in their separate component varieties ($FeO + Fe_2O_3 + CaO + MgO$) of dense ferruginous and magnesial minerals and decreasing the SiO_2 content (Fig. 42).

The δ_{dr} value of rocks of definite orebodies is greater the higher their content in individual varieties of ore metal(s), their oxides or ore minerals with abnormally high

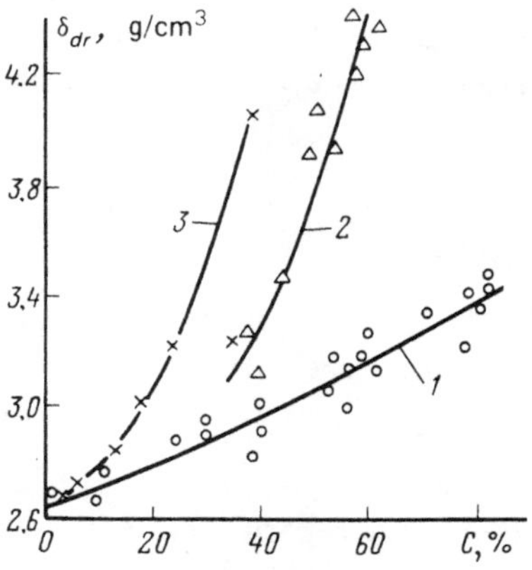

FIG. 43. Dependences of the density δ_{dr} on the mass content, *C*, of sulphides in sulphide-cassiterite ores (*1*) (Ternistoe deposit), Cr_2O_3 in chromite ores (*2*) (Almaz-Zhemchuzhina deposit) and CU + Fe + Pb + Zn in copper ores (*3*) (Djeskazgan deposit, after B.M. Urazaev)

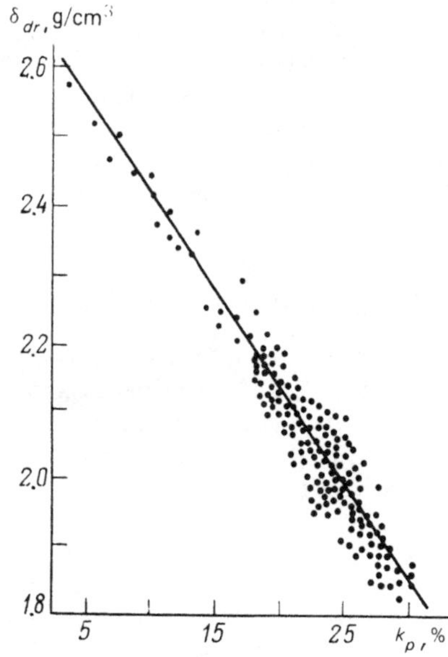

FIG. 44. Dependence of the density δ_{dr} on the voids ratio k_p for sandy and clayey rocks of bed AB_1 of the Samotlor oil field in West Siberia (after E.I. Leont'ev)

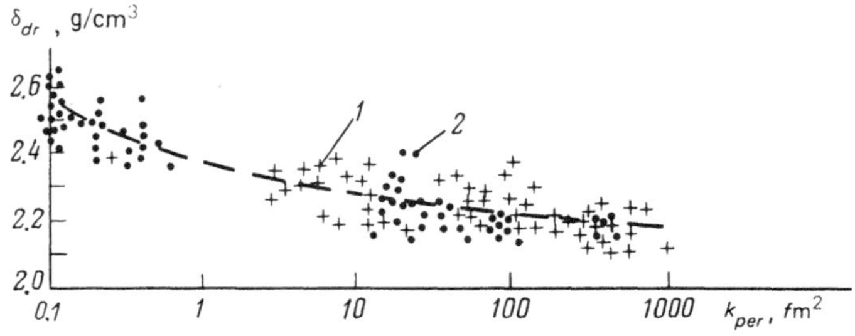

FIG. 45. Dependence of the density δ_{dr} on the coefficient of gas permeability, for sandy-clayey rocks of the Jurassic and Lower Carboniferous sediments of various areas in West Cis-Caucasia (after V.K. Popov and D.N. Silkina).

1 — sandstones; *2* — aleurolites

density, for example, iron (see Fig. 1), the Cu + Fe + Pb + Zn sum, Cr_2O_3 and sulphides (Fig. 43).

The density of individual graphite ore varieties increases with decreasing their content of graphite, and the density of coals with increasing their ash content, of ultrabasites with decreasing their serpentinization. The density δ_{dr} of rocks of many sandy, aleurite, carbonate and other sedimentary deposits, and of rocks of crystalline magmatogene massifs drops drastically with increasing their voids ratio k_p (Fig. 44) since the volumetric content of air in rocks increases during this process. The density δ_{dr} of rocks of sandy and clayey sediments decreases with increasing their permeability coefficient k_{perm} (Fig. 45). The density values of dry δ_{dr} and moist δ_m rocks also decreases with decreasing their clayiness coefficient $k_{cl.m}$ (Fig. 46). These connections are conditioned by the fact that both δ_{dr} and k_{perm} as well as $k_{cl.m}$ are to an appreciable degree governed by the coefficient k_p.

Sec. 18. **Maximum Moisture-content Rocks**

The density δ_m of maximum moisture-content highly porous rocks is higher than their density δ_{dr}. The densities δ_m and δ_{dr} of low-porosity rocks differ inappreciably. The $\delta_m = f(w_t)$ graphs are flatter than $\delta_{dr} = f(k_p)$ graphs. This is in good agreement with density differences of water and air.

The density δ_m of maximum moisture-content rocks can be readily found from Eq. (3) or special transparencies, given the total moisture capacity coefficient w_t, density δ_{dr} and density δ_w of water. The density of oil-saturated rocks is lower than that of maximum moisture-content rocks, since the density of oil is less than that of water; it can be found from Eq. (4), given the densities δ_{dr}, δ_w and δ_o, respectively, of dry rock, water and oil, the voids ratio k_p and coefficient of water saturation $k_{w.s} = (1 - k_o)$.

Due to the fact that the voids ratio of detrital rocks is governed by their mineral composition, fragment configuration, degree of sorting and compaction of fragments, so are these characteristics responsible for the density of rocks.

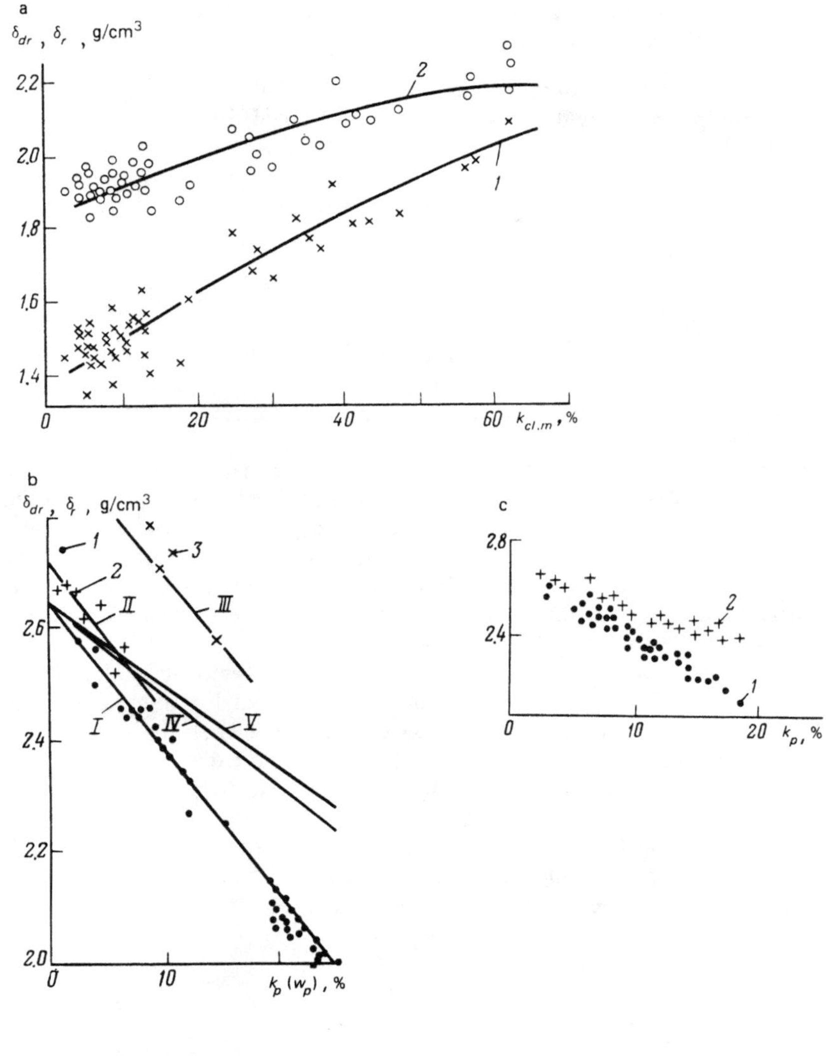

FIG. 46. Dependences of densities δ_{dr} and $\delta_{m(r)}$ on coefficients of mass clayiness $k_{cl,m}$ (*a*) on the total voids ratio k_p and total moisture capacity coefficient w_t (*b*, *c*).

a — dependences of δ_{dr} (*1*) and $\delta_{m(r)}$ (*2*) for terrigenic rocks of the Moscow coal basin on $k_{cl.m}$ (after I.T. Kozel'skii); *b* — dependences of $\delta_{dr} = f(k_p, \delta_s)$ for dry rocks having δ_s (in g/cm^3): *I* — 2.65; *II* — 2.73; *III* — 2.9; dependences of $\delta_{m(r)} = f(w_t, \delta_s, \delta_l)$ for moist rocks, given δ_s = 2.65 g/cm^3, and δ_l (in g/cm^3): *IV* — 1; *V* — 1.16; rocks: *1* — sandy and clayey; *2* — limestones; *3* — siderite-clastic; *c* — dependences of δ_{dr} (*1*) and $\delta_{m(r)}$ (*2*) on k_p for sandstones and aleurolites of Pre-Zhivetian sediments of areas of Bashkiria and Tataria (according to data obtained in the petrophysics laboratory of the Moscow Institute of Oil and Gas

Sec. 19. The Density of the Sedimentary Cover of the Earth's Crust

The densities, δ_{dr} and $\delta_{m(r)}$, of individual types of sandy, aleurite, clayey, calcareous and other groups of sedimentary rocks increase with the depth of their occurrence and age (Fig. 47*a*, *b*). This is mainly connected with irreversible postsedimentation porosity decrease. The density values of individual sedimentary rock types can to a definite depth be estimated from the following relations:

$$\delta_{dr.av} = \delta_s(1 - k_{p.max} \times 10^{-2} e^{-0.045H/H_0})$$

and

$$\delta_{r.av} = \delta_s + k_{p.max} \times 10^{-2} e^{-0.045H/H_0}(1 - \delta_s)$$

where H is the depth of occurrence of the rock concerned; H_0 is the depth of the determination equal to 0.1 km for the above equations; $k_{p.max}$ is the initial voids ratio—maximum value of k_p of the given rock type (in %) close to the earth's surface at a depth H_0.

The initial value of $k_{p.max}$ and the value of δ_s at a definite depth H is conditioned by the original substance composition and structure of the rock, tectonic conditions of the region at the initial geological period and depth of occurrence of rocks. Characteristic values of δ_s and k_{pmax} have been established for 64 oil and gas bearing regions of the USSR, and density sections have been plotted for them (see Fig. 47). These illustrate the distribution of density values for different types of sedimentary rocks in the depth interval from 100 to 3 000 m.

Definite regularities have also been established in the regional areal variation of the average density of series of the same age conditioned by the variation of the composition and structure of the rocks composing them.

So, for example, according to E.E. Fotiadi, the average density of carbonate Paleozoic rocks of the Russian platform decreases from east westwards, amounting to 2.7 g/cm^3 in the eastern regions and to 2.3 g/cm^3 in the centre of the region. This regularity is accounted for by the decrease from east to westwards of the dolomitization degree of rocks. The appreciable dolomitization of limestones in the eastern Russian platform is connected with the presence of deep mineralized waters and

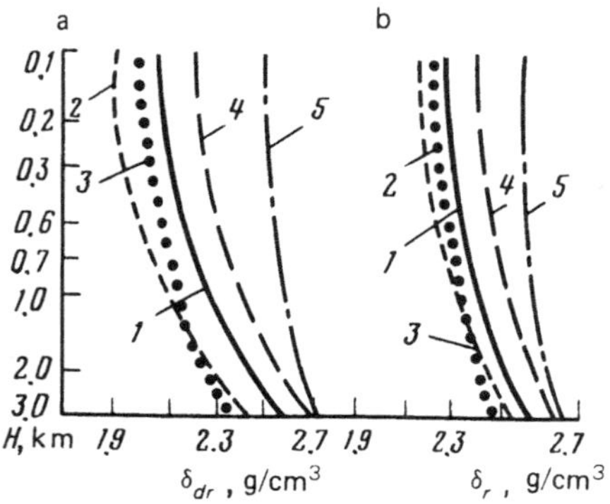

FIG. 47. Dependences of δ_{dr} (*a*) and $\delta_{m(r)}$ (*b*) of gas- and water-saturated rocks on the depth of occurrence H in the Russian platform (after M.L. Ozerskaya and S.G. Semenova).

1 — clay; *2* — aleurolite; *3* — sandstone; *4* — marl; *5* — limestone

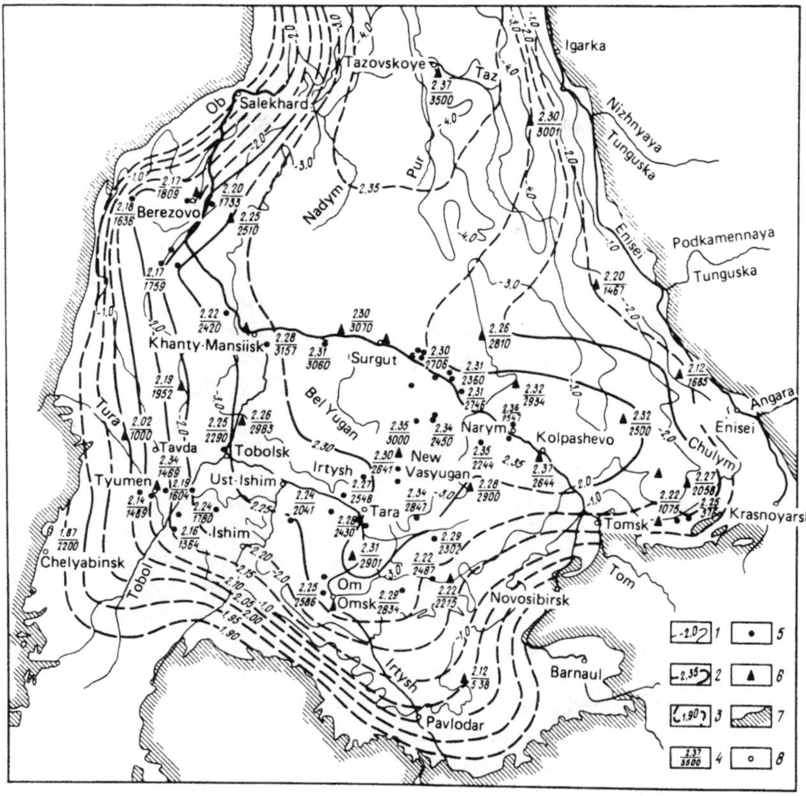

FIG. 48. Schematic map showing densities of rocks, δ_r, of the Mesozoic-Cenozoic cover of the West Siberian plate (compiled by N.A. Tuezova and N.I. Bryuzgina and simplified by V.N. Kobranova).

1 — isohypses over the roof of the foundation; *2* — density isolines; *3* — insufficiently certain density isolines; *4* — average density values (numerator) and average thickness values (denominator); boreholes: *5* — exploratory; *6* — reference; *7* — Pre-Jurassic foundation; *8* — populated points

proximity to the Urals folded region. The variation pattern of the average density of the Meso-Cenozoic series on the area of the West Siberian plate has been also established. Schematic maps of the equal density of different rock types and age complexes (Fig. 48) testify that the average density of the Meso-Cenozoic series increases from the fringe to the centre (porosity decreasing). The density of the sedimentary cover varies from 1.87 g/cm^3 near the western fringe of the lowland to 2.37 g/cm^3 in its centre; the increase in density toward the centre of the lowland is accounted for by the growth in the same direction of the thickness and depth of occurrence of rocks of the Meso-Cenozoic complex. According to A.T. Donabedov, the average density of sandy and clayey rocks of the coal-bearing mass of the middle Carboniferous within the area of the Greater Donbas coal basin decreases from 2.7 g/cm^3 in the centre to 2.3 g/cm^3 at its northern portion. This regularity results from the increase, as the centre of the depression is reached, of the thickness of the rock mass, degree of catagenesis of rocks and drop in their porosity. The extent of metamorphism is responsible for the type and composition of coals. The same regularity has been also found in the density variation for other coal-bearing regions.

The average density of terrigene rocks often decreases with approaching the local positive structure dome owing to the trend for detrital rocks to become coarser in this direction. As the dome of local structures is approached, the clayiness of detrital rocks generally decreases and their sorting and porosity increase which leads to a drop in rocks' density. An opposite regularity in porosity and density distribution within the structures being considered is observed less commonly.

A drop in the density of terrigene rocks with approaching the structures' dome has also been detected in terrigenous rock mass in West Siberia. This is attributed to an increased thickness of siliceous rocks in the structures' dome. A decrease in the density of the structures' domes has also been discovered in the Upper Tertiary zone of the Apsheron Peninsula. This zoning is accounted for by the facial features of this mass, but also by jointing that may sometimes appear in the dome.

The density of rocks of calcareous and magnesial positive structures is intimately connected with their position on the structure. Predominantly limestones are developed on a structures' dome, with dolomite content increasing toward limbs which contributes to the increased density.

A definite regularity in density distribution is observed in carbonate-clayey masses. As these are formed carbonate material is located in elevated, and clayey material in lower portions of the paleorelief of the bed. This results in increased density at areas where carbonate rocks develop.

Zoning has been discovered in density variation for many local structures of the Volga-Urals province. A drop in the density of carbonate rocks due to increased jointing is commonly observed when approaching the structures' dome. One may well refer to other examples of a regular variation in the density of rocks within the sedimentary cover of the earth's crust.

Chapter VII

Permeability

Sec. 20. **Absolute Permeability**

The capacity of rocks for transmitting liquids, gases or their mixtures in the presence of a pressure gradient $\Delta p/l$ is said to be *permeability* (or perviousness). We distinguish *an absolute* (physical) and *an effective* permeability. The absolute permeability is the permeability of a rock with respect to a dry gas or a single-component liquid. The greater the cross section F, and acting pressure gradient $\Delta p/l$ (l is the length of a rock volume) of the rocks, and the less viscosity μ of the infiltrating substance, the greater is the quantity of the dry gas, Q, or homogeneous single-component liquid, that is transmitted by the rocks for a unit period of time.

However, the gas (or a single-component fluid) discharge for different rocks, given a specified cross section, pressure gradient and viscosity, varies from one case to another which governs the proportionality coefficient k_{perm} in the equation

$$Q = k_{perm} \frac{\Delta p F}{\mu l} \tag{6}$$

known as the permeability coefficient which has area, L^2, and is measured in SI system, in m^2. However, use is still made of a non-system unit, darcy (D). The permeability of 1 D is shown by a rock sample 1 cm^2 in cross section, 1 cm long, given a pressure gradient 0.1 MPa and viscosity of a homogeneous infiltrating substance equal to 1 sP; 1 D $\approx 10^{-12}$ m^2 = 1 μm^2, and 1 mD $\simeq$ 1 fm^2. The coefficient k_{perm} of absolute permeability for water is less than that for gas since a certain fraction of the voids space of moist rocks is taken by strongly and loosely bound water, dead-end or very thin capillary water with a mobility less than that of free water or with no mobility.

The difference in the value of the coefficients under consideration is particularly great as far as the infiltration of fresh waters through clayey rocks is concerned. The coefficient of gas permeability of rocks varies over a wide range from <0.001 fm^2 to $>5\,000$ μm^2.

Sec. 21. The Analytical Relationships Between the Coefficient of Absolute Permeability and Structural Characteristics of Rocks

Let us imagine rock samples as being cubical and hexagonal packings of cemented capillaries r in radius and n in number for a unit area of the model cross section (Fig. 49). The discharge of a dry gas or homogeneous fluid flowing through such simplest media will be equal to the sum of the discharges of the infiltrating substance in individual capillaries. Suppose that infiltration through capillaries obeys Poiseuille's law. Then the true velocity v_{tr} of the flowing substance (liquid or gas) in a capillary is

$$v_{tr} = \frac{d_c^2 \Delta p}{32\mu l} \tag{7}$$

and the discharge (volume rate of flow) of it through a capillary d_c in diameter is

$$Q' = \frac{\pi d_c^4 \Delta p}{4 \times 32\mu l}$$

The discharge of a fluid through n capillaries is

$$Q = n \frac{\pi d_c^4 \Delta p}{4 \times 32\mu l}$$

Since for the model in Fig. 49*a*

$$k_p = \frac{nF\pi d_c^2 l}{4Fl} = \frac{n\pi d_c^2}{4}$$

and

$$k_p = \frac{\pi d_c^2}{4d_c^2} = \frac{\pi}{4}$$

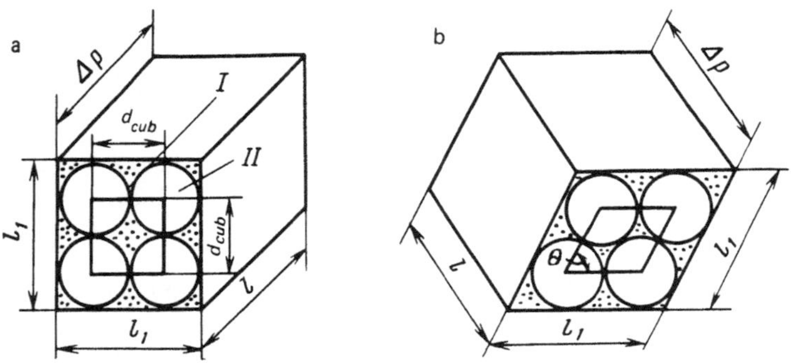

FIG. 49. Diagrams of elementary rock samples.

Cylindrical capillary packing: *a* — cubic; *b* — hexagonal; *I* — solid phase of models; *II* — capillaries filled by an infiltrating agent: l_1 — faces of model the size of which is unity; θ is the angle of a rhombus or cube formed by connecting the centres of capillaries then

$$n = \frac{4k_p}{\pi d_c^2} = \frac{4\pi}{4\pi d_c^2} = \frac{1}{d_c^2}$$

$$Q = \frac{1}{d_c^2} \frac{\pi d_c^4 \Delta p}{4 \times 32\mu l} = \frac{\pi d_c^2 \Delta p}{128\mu l} \quad (8)$$

By equalizing the right-hand sides of Eqs. (6) and (8), given $F = 1$, we get

$$k_{perm} = \frac{\pi d_c^2}{32 \times 4}$$

For a hexagonal packing of capillaries (for $\theta = 60°$)

$$k_p = \pi/(4 \sin \theta)$$

$$n = \frac{\pi/(4 \sin 60)}{\pi d_c^2/4} = \frac{1}{d_c^2 \sin 60}$$

$$Q = \frac{1}{\sin 60 d_c^2} \frac{\pi d_c^4 \Delta p}{4 \times 32\mu l}$$

and

$$k_{perm} = \frac{\pi d_c^2}{27.67 \times 4}$$

or

$$k_{perm} = \frac{\pi d_c^2}{32 \times 4} = \frac{k_p d_c^2}{32} \quad \text{and} \quad k_{perm} = \frac{k_p d_c^2}{27.7}$$

i.e. the values of k_{perm} are governed by the pattern of packing of capillaries and are in direct proportion to their squared diameter, k_p representing a proportionality coefficient.

Let us complicate the model. Let us assume that a cylindrical rock sample S in cross section and l in length (Fig. 50a) can be divided into n structural elements l in length and ω_i in cross section such that each of them has inside an infiltration channel with a variable cross section $\omega_{ch.i}$ and length $l_{ch.i}$.

Let us assume the element with an averaged statistical cross section $\omega_{av} = \frac{1}{n}\sum_{i=1}^{i=n} \omega_i$, channel cross section $\omega_{ch.av} = \frac{1}{n}\sum_{i=1}^{i=n} \omega_{ch.i}$ and channel length $l_{ch.av} = \frac{1}{n}\sum_{i=1}^{i=n} l_{ch.i}$ to be identical to a "tube" section (Fig. 50b). Let us show that, after transformation, Poiseuille's equation (7) is valid for a circular cross-sectional pipe (Fig. 50c).

We reduce Eq. (7) to a form applicable to pipes with any cross-sectional configuration as follows: substitute the diameter by a hydraulic radius $r_{hydr} = V_t/S_t$, where V_t is the pipe volume, and S_t is its inner lateral surface.

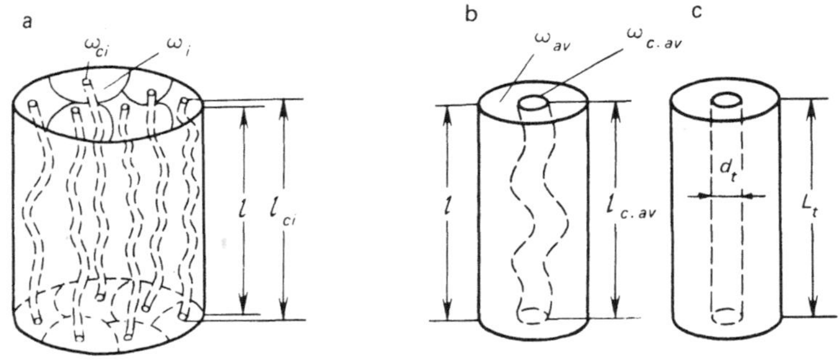

FIG. 50. A diagram illustrating the derivation of the Kozeni-Carman equation.
a — rock sample with meandering channels; *b* — unit structure element of rock sample; *c* — section of a round pipe

For a circular cross-sectional pipe

$$r_{hydr} = \frac{\pi r^2 L_t}{2\pi r L_t} = \frac{r}{2} = \frac{d}{4}$$

and, consequently, $d = 4r_{hydr}$.

After substituting d by r_{hydr} Eq. (7) takes on this form

$$v_{tr} = \frac{r_t^2}{2} \frac{1}{\mu} \frac{\Delta p}{L_t}$$

Calculations and experience show that equations for pipes of whatever cross section are close to the latter relationship. They differ by a constant factor in the denominator whose value varies form 2 to 3 and is generally close to 2.5. Let us denote this factor in the denominator, which is known as the Kozeny constant, as f.

Hence

$$v_{tr} = \frac{r_{hydr}^2 \Delta p}{f \mu L_t} = \frac{v_t^2 \Delta p}{f S_t^2 \mu L_t}$$

If an element of the model (see Fig. 50*b*) has a volume $V_{dr} = 1$, then its $k_{p.d} = V_{por.d.av}/V_{dr} = V_{por.d.av.}$, where $V_{por.d.av}$ is the average volume of the infiltration channel.

By defining its lateral surface as S_{fV}, we have

$$v_{tr} = \frac{k_{p.d}^2 \Delta p}{S_{fV}^2 l_{c.av} \mu f} \tag{9}$$

By taking into account that v_{tr} as much greater than the infiltration rate $v_f = k_{perm} \Delta p / \mu l$ as the cross section ω_{av} is greater than the cross section $\omega_{c.av}$, one can write:

$$v_{tr} = \frac{\omega_{av}}{\omega_{c.av}} v_f = \frac{\omega_{av}}{\omega_{c.av}} k_{perm} \frac{\Delta p}{\mu l}$$

but since

$$k_{p.d} = \frac{\omega_{c.av} l_{c.av}}{\omega_{av} l}$$

and

$$\frac{\omega_{av}}{\omega_{c.av}} = \frac{l_{c.av}}{k_{p.d} l}$$

then

$$v_{tr} = k_{perm} \frac{\Delta p l_{c.av}}{\mu l^2 k_{p.d}} \tag{10}$$

By equalizing the right-hand sides of Eqs. (9) and (10), we get

$$\frac{k_{p.d}^2 \Delta p}{f S_{fV}^2 \mu l_{c.av}} = \frac{k_{perm} \Delta p l_{c.av}}{\mu l^2 k_{p.d}}$$

but $l_{c.av.}/l = T_{hydr}$ is hydraulic meandering. Hence

$$k_{perm} = \frac{k_{p.d}^3}{S_{fV}^2 T_{hydr}^2 f} \tag{11}$$

Consequently, according to the Kozeny-Carmen equation, based on a model sufficiently approximating actual rocks, the permeability coefficient k_{perm} is in direct proportion to the cubed dynamical voids ratio k_p and is inversely related to the squared specific surface of infiltration S_{fV} and hydraulic meandering T_{hydr} as well as to the coefficient f taking into account the configuration of the pore channels. Let us further show that the coefficient

$$k_{perm} = F(k_{w.r},\ \tau_{av},\ k_p,\ T_{hydr} \text{ and } f)$$

Here $k_{w.r}$ is the coefficient of rock saturation by residual water and τ_{av} is the thickness of its film.

Let us assume that each of the model meandering voids channels contains not only free water but also a film of bound water whose thickness is τ_{av} and content $k_{w.r}$. Then the following is true:

$$\tau_{av} S_{fV} = k_{w.r} k_p$$

Hence

$$S_{fV}^2 = k_{w.r}^2 k_p^2 / \tau_{av}^2$$

By substituting into Eq. (11) for $k_{p.d}$ the product $k_p(1 - k_{w.r})$ and for S_{fV}^2 the ratio $k_{w.r}^2 k_p^2/\tau_{av}^2$ we get

$$k_{perm} = \frac{(1 - k_{w.r})^3 \tau_{av}^2 k_p}{k_{w.r}^2 T_{hydr}^2 f} = \frac{(1 - k_{w.r})^3 k_p}{k_{w.r}^2 (T_{hydr}/\tau_{av})^2 f} \tag{12}$$

This relationship determines the character of dependence of k_{perm} on the quantities $k_{w.r}$, k_p, T_{hydr}, τ_{av} and f, which are jointly responsible for the volume of the movable fluid in a rock, the force causing it to be held in the voids and average paths of individual streams of the infiltrating component.

Since for sediments of a definite type $f \approx$ const, T_{hydr} = const, $\tau_{av} \approx$ const, then k_{perm} should mainly change depending on the coefficients k_p and $k_{w.r}$. The higher the saturation coefficient of rock by residual water the greater is the decrease in the k_{perm} of the bound water.

The permeability of rock composed of fragments of approximately identical shape diminishes with a decrease in their cross sections and a decrease in the voids ratio. It is easy to show the dependence of k_{perm} on d_{gr} and k_p.

If spherical grains of the same size with diameter d_{gr} show cubical packing, then the rock's voids ratio is

$$k_p = (d_{gr}^3 - \pi d_{gr}^3/6)/d_{gr}^3 = 1 - \pi/6 \tag{13}$$

and the specific surface is

$$S_V = \pi d_{gr}^3/d_{gr}^3 = \pi/d_{gr} \tag{14}$$

As follows from Eqs. (13) and (14),

$$S_V = 6(1 - k_p)/d_{gr}$$

Since

$$S_V^2 = k_p^3/k_{perm} f T_{hydr}^2$$

then

$$k_{perm} = k_p^3 d_{gr}^2/36(1 - k_p)^2 T_{hydr}^2 f \tag{15}$$

Consequently, the coefficient of permeability of dry gas-saturated rocks is power-related to the average diameter of their grains and increases in direct proportion to the voids ratio.

Sec. 22. The Dependence of the Coefficient of Absolute Permeability on Different Petrophysical Quantities from Experimental Data

The experimental data point to: (1) growth of the coefficient of permeability with increasing the quadratic average radius (diameter) of capillaries (Fig. 51); (2) connections for different types of sedimentary deposits—direct between the coefficients k_{perm} and k_p (Fig. 52); (3) inverse connections between k_{perm} and S_V (see Fig. 51, curve *2*); k_{perm} and T_{hydr} (Fig. 53). These are not close connections. The former fails to take into account the difference between individual rock samples in terms of the specific pore volume, the latter, the resistance of the voids channels offered to the flow of the infiltrating fluid, the third type, the specific volume or also the surface of the voids channels; (4) fairly close relationships can be found between k_{perm}, k_p and S_V (Fig. 54) that are connected through an equation

$$k_{perm} = A k_p^a / S_V^b$$

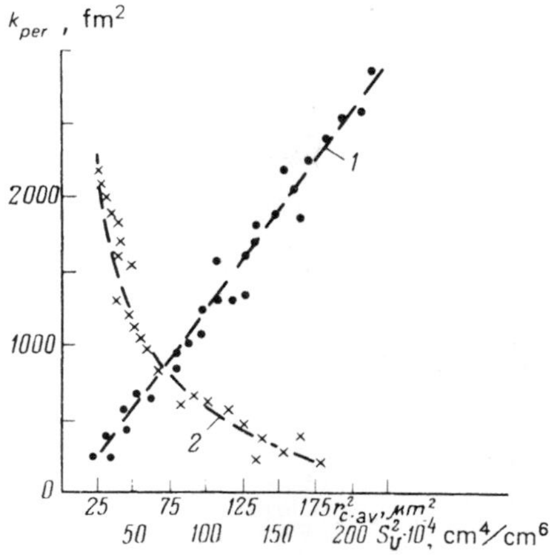

FIG. 51. Dependences of the coefficient of gas permeability k_{perm} on the quadratic average radius of pore channels $r^2_{ch.av}$ (*1*) and quadratic specific volumetric S_V^2 (*2*) for slightly cemented sandstones and aleurolites of bed D_1 of the Tuimaza oil field (after A.A. Bolotov, B.A. Belinskii, L.L. Sinii).

Cross section of borehole 1397, depth interval 1650.8-1923.9 m; k_{perm} = 200-2000 fm²; k_p = 20-26%; $r_{ch.av}$ = 5-15 μm; S_V = 500-1400 cm^{-1}

where $a \approx 3$; $b \approx 3$; A is a constant for a definite type of sediments equal approximately to 10^2 (when measuring k_p in %); (5) inverse connections between k_{perm} and $k_{w.r}$ ($k_{w.r}$ is a coefficient of residual water saturation) (Fig. 55) and a closer connection between k_p, log k_{perm} and $k_{w.r}$.

The factors determining k_{perm} values of sedimentary rocks include also the composition of their phases, shape, granulometric composition of grains and its statistical characteristics (the median and effective diameters, coefficients of homogeneity and sorting), the specific content of clayey and clayey dissoluble cementing component. The above features and properties of rocks are what is responsible for the interphase in voids of rocks, their specific volumetric surface S_V, total k_p and open specific voids volume, the composition of voids of these media and its statistical characteristics, specific volumetric concentration of clayey $k_{cl.m}$, dissoluble k_{cem} and clayey dissoluble $k_{cl.cem}$ component, specific volumetric filling of voids by bound and residual water.

The water permeability coefficient of rocks is also very much affected by the composition of the water and exchange complex of the solid phase determining the structure of the electric double layer and the thickness of the film of bound water. The coefficient of permeability is minimal if the Na ion is predominant in the counterions layer. Then the voids of the rock contain a great amount of bound

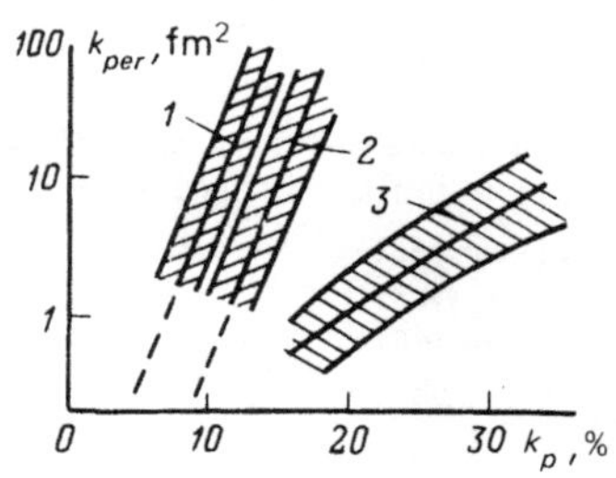

FIG. 52. Relationship between the coefficient of permeability k_{perm} and voids ratio k_p for calcareous sediments of different type.

1 — oölite limestone smakover, Magnolia; *2* — crystalline, secondary-porosity limestone of Saint Andrews, Vessen; *3* — Devonian chalky limestone crossite, Texas; shaded areas are the areas of possible spread of points

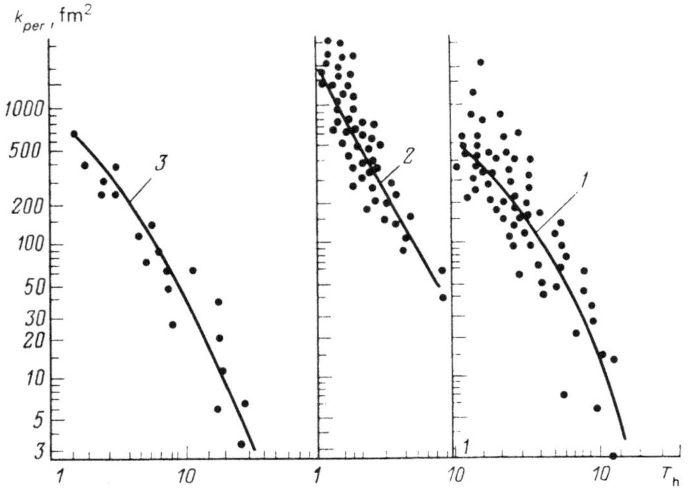

FIG. 53. Dependence of the coefficient of permeability k_{perm} on meandering rate T_h of sandy-aleurolite rocks (after O.A. Chernikov and A.I. Kurenkov).

Deposits: *1* — Uzen (horizon XVII); *2* — Mukhanovo (bed C_1); *3* — Tuimaza (bed D_1); $T_h = k_{o.p}/k_{p.eff}$; effective voids ratio determined from microsections: $k_{p.eff} = (F - F_1)/F$, where F, F_1 are areas, respectively, of the entire microsection and that taken by grains and cement

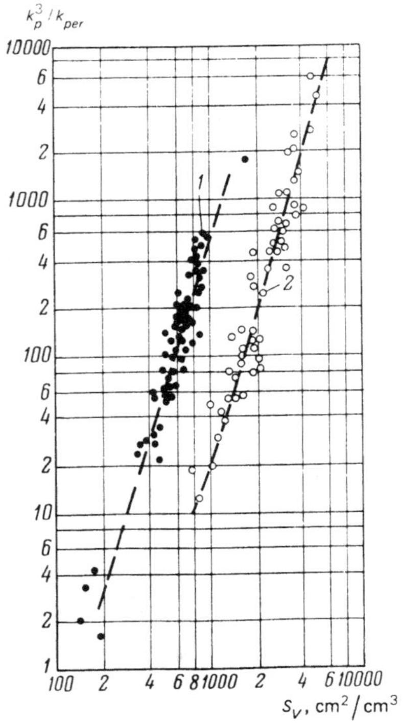

FIG. 54. Dependences of the k_p^3/k_{perm} ratio on the specific volumetric surface S_V.

1 — Grozny area sandstones and aleurolites (after A.M. Nechai); *2* — sands and aleurolites of the lower division of the productive rock mass of the Neftyanye Kamni oil field (after F.I. Samedov and L.A. Buryakovskii)

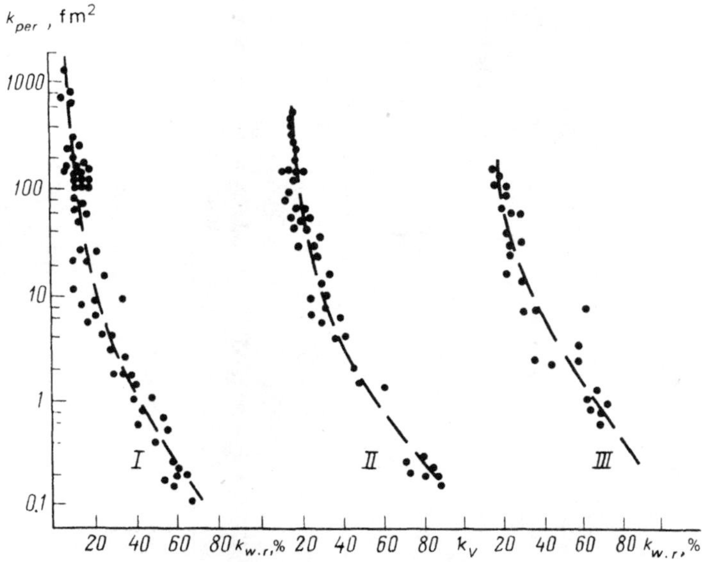

FIG. 55. Dependences of the coefficient of gas permeability k_{perm} on the coefficient of residual water saturation $k_{w.r}$ for carbonate rocks of different origin and structure (after B.I. Tulbovich).

Limestones: *I* — biomorphic; *II* — blob-like and cloddy; *III* — detrital, slime-detrital

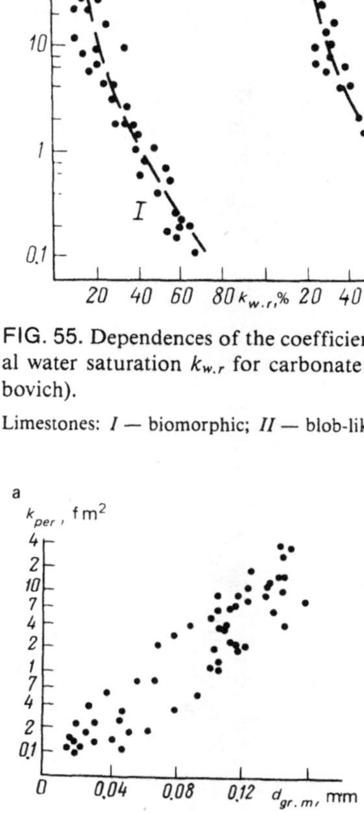

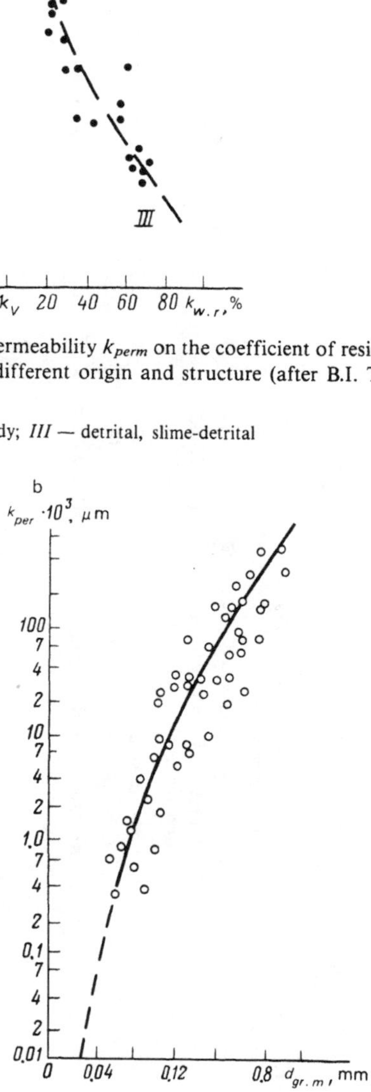

FIG. 56. Dependences of the coefficient of permeability, k_{perm}, of sandy-clayey rocks on the grains median diameter $d_{gr.m}$ of sandstone (*a*) (after M.K. Kalinko) and polymictic sandstones and aleurolites (*b*)

TABLE 6. Filtration Properties of Principal Rock Types (After N.A. Plotnikov's Data Augmented by E.E. Kerkis)

Rock group in terms of permeability	Principal rock types	Permeability coefficient, μm^2	Filtration coefficient*, m/s	Voids ratio, %	Principal regularities in the filtration properties
I (very high)	1. Pebbles, gravel without filling material (washed)	>5 000 (500-5 000)	>350 (350-3 500)	25-30	Permeability is generally uniformly distributed over the area
	2. Massive and crushed rocks in screes and stone falls without a filling material	>500 (500-10 000)	>350 (350-6 500)	20-35	Permeability unchanged
	3. Strongly karsted	>100 (100-5 000)	>70 (70-3 500)	2-15	Permeability is very much variable and decreasing with depth
	4. Neovolcanic, mainly basalts, andesito-basalts and andesites	(100-2 000)	70-1 300	2-25	Permeability is conditioned by primary cracks of jointing and sometimes cavernosity; rather constant in plan and cross section
	5. Strongly fissured intrusive gneisses and crystalline schists	>100 (sometimes to 1 500)	>70	1-5	Permeability is very variable, attains its maximum in discharge zone and in zones of recent-origin faults, decreases rapidly with depth, gneisses and crystalline schists show a pronounced filtration anisotropy
II (high)	1. Pebbles and gravel with various grained sand filling material; sand and gravel sediments, coarse sands	20-100	15-70	25-35	Permeability is variable, lamination and filtration anisotropy often occur
	2. Pure various and middle sands	10-80	7.0-55	25-35	Ditto
	3. Karsted	10-100	7.0-70	1-8	Permeability is very much variable, decreases with depth
	4. Fissured igneous (intrusive and ancient-origin effusive), gneisses and crystalline schists	10-50	7.0-35	0.5-2.0	Regularities similar to ones in rocks of Group I and item 5; permeability low at depth

	5. Cavernous dolomites and limestones	10-100	7.0-70	2-20	Permeability is variable, decreases with depth
III (intermediate)	1. Pebbles and gravel with filling small and middle sand	1-10	0.7-7.0	25-30	Permeability is variable; bedded sediments have a filtration anisotropy
	2. Small and middle sands	1-10	0.7-7.0	25-40	Ditto
	3. Porous cemented (sandstones, gravellites, conglomerates etc.)	1-10	0.7-7.0	6-20	Permeability is often rather constant; filtration anisotropy is likely
	4. Weakly karsted and paleokarsted	1-10	0.7-7.0	0.5-1.0	Permeability is generally very variable, particularly that of paleokarst rocks
	5. Fissured intrusive, ancient-origin effusive and metamorphic with a medium rate of jointing	1-10	0.7-7.0	0.2-0.5	Similar to regularities shown by rocks of Group II, item 4
	6. Fissured detrital, cemented	1-10	0.7-7.0	0.2-2.0	Permeability is variable, decreases with depth; bedded rock masses show a filtration anisotropy
IV (low)	1. Fine-grained, powdered and clayey sands, light clay sands	0.1-1.0	0.07-0.7	25-40	Similar to regularities of rocks of Group III, item 2
	2. Porous cemented (aleurolites, sandstones etc.)	0.1-1.0	0.07-0.7	5-20	Similar to regularities of rocks of Group III, item 3
	3. Weakly fissured igneous, metamorphic and sedimentary	0.1-1.0	0.07-0.7	0.1-1.0	Permeability is variable, at a small depth decreases almost to zero
	4. Anthracite and dry coals	0.1-1.0	0.07-0.7	0.1-1.0	A small filtration anisotropy is possible
	5. Slightly decayed peat	0.1-1.0	0.07-0.7	50-70	An appreciable moisture capacity; a sufficiently high initial gradient is observed
	6. Loess, loess-like clayey sands	0.1-1.0	0.007-0.07	25-50	Macroporosity, filtration anisotropy; permeability is often rather constant

TABLE 6 (*concluded*)

Rock group in terms of permeability	Principal rock types	Permeability coefficient, μm^2	Filtration coefficient*, m/s	Voids ratio, %	Principal regularities in the filtration properties
V (very low)	1. Sandy clays, heavy clayey sands	0.01-0.1	0.007-0.07	25-50	Permeability may be rather constant over the area of distribution of rocks
	2. Sandy-argillaceous chlorite schists etc., phyllites	0.01-0.1	0.007-0.07	2-4	Filtration anisotropy is often appreciable; permeability rapidly diminishes with depth almost to zero
	3. Densely cemented detrital having minor porosity (aleurolites, sandstones, sandy slates etc.)	0.01-0.1	0.007-0.07	8-10	Similar to regularities in rocks of Group IV, item 2
	4. Various very weakly fissured	0.01-0.1	0.007-0.07	0.01-0.1	Permeability decreases almost to zero at a very small depth
	5. Well decayed peat	0.01-0.1	0.007-0.07	100-150 and more	Moisture capacity is very great; high initial gradient
VI (close to zero)	1. Clays, marly clays, argillites, heavy sandy clays	<0.01	<0.005	15-45	Permeability is often variable in a direction at right angles to bedding
	2. Halides, gypsum, anhydrite below the upper fissured and karsted zone	~ 0	~ 0	~ 0	Permeability becomes zero due to the closing of fissures and other cracks under the effect of geostatic and tectonic pressure
	3. Various hard and semihard rocks below upper fissured zone	<0.01	<0.05	<0.01	Similar to regularities of rocks of Group V, item 4

* Given for fresh water at temperature of 10 °C. The values of filtration coefficient is rounded.

water and the cross section of filtering pores is very much diminished. Given calcium diffuse layers, k_{perm} of the oil and gas trap is greater since the swelling of their clayey fraction is less.

The pattern of decreasing k_{perm} with diminishing the median diameter $d_{gr.m}$ of grains of sandy-clayey rocks can be inferred not only from Fig. 56 but also from the data presented in Table 6. The rate of increasing permeability with the diameter of particles is variable. This is due to the specific features of the voids ratios, configuration and meandering rate of the voids, mineral composition of rocks of dissimilar sediments. The k_{perm} values decrease with deteriorating the degree of sorting and increasing rock cementation (Fig. 57).

Given a constant voids ratio, the permeability of rocks is mainly governed by the absolute or relative content of the clayey or the clayey-dissoluble component in their voids space. The greater the content of the latter in the rock, the less is the permeability magnitude.

A drop in the permeability, which is at first very appreciable with the increasing of the clayey and dissoluble component in the rock (see Fig. 57), is accounted for by the concurrent decrease in porosity and increase in the specific surface. This drop is then less marked, given that the volume of cementing material exceeds that between larger rock grains, and porosity begins to grow.

The variation of k_{perm} with increasing the cement content is affected by its mineral composition. The coefficient k_{perm} of quartz sands drops more drastically and faster attains the values approaching zero if they are cemented by montmorillonite rather than kaolinite showing less swelling (see Fig. 57*b*).

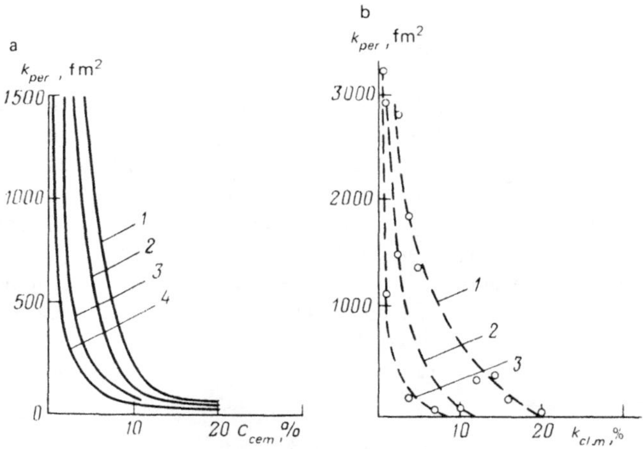

FIG. 57. Dependences of the coefficient of permeability k_{perm} on the content in rocks of heterogeneous cement C_{cem} and clayey cement $k_{cl.m}$.

a — for different sandy-clayey sediments (after A.A. Khanin and N.V. Smirnova); sandstones: *1* — medium- and fine-grained Pre-Zhivetian age of Pachelma; *2* — small-grained of the Uger suite of Bilce-Wolica; *3* — small-grained suites of the Goryachi Klyuch, Stavropol Region; *4* — small-grained of the coal-bearing formation of Archeda; *b* — for quartz sands (after M.A. Fedorova); clay: *1* — kaolin; *2* — polymictic; *3* — montmorillonite

Since the density δ_{dr} and coefficient of permeability k_{perm} are, as a rule, governed by their voids ratio k_p, sometimes rather a close connection between these quantities can be observed. Thus, k_{perm} is governed by the origin and conditions of the ensuing life of rocks, since they are responsible for the initial pattern of the pore space and its change with time.

Sec. 23. **The Permeability of Jointy Rocks**

These rocks demonstrate an intergranular and crack permeability.

The crack permeability of slates, limestones, dolomites, aleurolites and sandstones varying from 15 to 40 fm^2, is much more than intergranular permeability which does not generally exceed 0.1 fm^2.

The calculation of the coefficient of crack permeability k_{perm} may use the relation

$$k_{perm} = 8.45 \times 10^6 b^2 k_p$$

where b is the opening of cracks, μm; $k_p = Tb$ is the coefficient of crack porosity (here T is the volumetric density of cracks, m^2/m^3), %.

The crack permeability in a microsection is also determined under microscope. The use of this method makes it possible to obtain: (1) the openness b (in μm) of microcracks from measurements in a microsection; (2) the total length l (in cm) of microcracks in a microsection; (3) the area of the microsection $S_{m\text{-}s}$ (in cm^2).

The coefficient of crack permeability k_{perm} can alternatively found from the relation

$$k_{perm} = Ab^3 l/S_{m\text{-}s} = (A/B) b^2 k_p$$

where $k_p = Bbl/S_{m\text{-}s}$; A and B are numerical coefficients that take on different values depending on the geometry of crack systems.

An indirect calculation of k_{perm} uses a quantity $L = 1/D$, where D is the density of cracks, i.e. the number of cracks for a unit length of the perpendicular erected to the plane of cracks in the microsection. The parameter D is calculated from the total length (extent) l of cracks cutting the section and the area $S_{m\text{-}s}$ determined in the section by using the equation $D = Cl/S_{m\text{-}s}$, where C is a numerical coefficient assuming different values depending on the geometry of crack systems.

Since the opening of cracks in lithologically different jointy rocks is approximately constant (14-15 μm), the distance between cracks L may serve for an approximate estimation of the permeability of rocks. This distance varies from 8 to 140 μm. Accordingly, the permeability of jointy rocks also varies; for instance, it equals 0.5 fm^2 for dolomites with $L = 14$ cm and 24 fm^2 for marls with $L = 0.8$ cm.

Sec. 24. **The Classification of Rocks in Terms of the Permeability Coefficient**

Referred to the permeability coefficient, rocks are divided into permeable, semipermeable and practically impermeable species.

Permeable rocks include coarse detrital rocks (pebbles, gravel), weakly cemented and well sorted sandy-silt-rocks, cavernous and, particularly, karsted and fissured, calcareous and magnesial rocks, fissured igneous rocks.

The pore space of permeable rocks takes, as a rule, an appreciable fraction of the rock volume (20, 30 40% and more) and is generally composed by a relatively small number of supercapillary, large capillary or capillary pores often uniformly distributed through the bulk of the rock. Such rocks contain little bound water. The permeability coefficient of these rocks may vary from 10^{-2} to several thousand μm^2.

Semipermeable rocks include less sorted clayey sands, some aleurite, sandstone and aleurolite varieties as well as a number of carbonate rocks, in particular, small-crack chalky limestones and dolomites. The voids space of these rocks is largely represented by subcapillary pores, the content of bound water is increased. The permeability coefficient varies from 10^{-4} to 10^{-2} μm^2.

Practically impermeable (or impervious) rocks include rocks with $k_{perm} < 10^{-4}$ μm^2: clays, argillites, clay slates, subcapillary pore marles, strongly cemented sands, sandstones and aleurolites, dense chalk and chalky limestones, unweathered crystalline carbonates and igneous rocks, close porosity rocks etc.

The voids space of clays, chalky limestones may often take as much as 50% of their volume. For argillites, slates, marls, crystalline unweathered carbonates and igneous rocks it does not generally exceed 6-8%. For some rocks (crystalline unweathered carbonates and igneous rocks) the lack of permeability is due to isolated voids. Practically all water in most voids is bound and cannot move at naturally occurring pressure gradients. The permeability of rocks is higher in the direction of stratification compared with the direction at right angles to it (the phenomenon of anisotropy).

Table 6 presents a more detailed rock classification. It covers practically all rock types and, referring them to six groups with definite limits of variation of k_{perm}, k_f, k_p (k_f is an infiltration coefficient, $k_{perm} = k_f\mu/\delta_w$), takes into account the conditions of their occurrence. It should be noted that this classification associates the highly productive sandy-clayey oil and gas traps with rocks having average and even low, rather than very high, permeability, unlike well-known oil specialists.

Sec. 25. The Classification of Oil and Gas Traps in Terms of Their Most Important Characteristics

There exist several classifications of oil and gas traps in terms of k_{perm}, k_p and other rock characteristics.

A classification of sandy-silt-rocks containing oil and gas proposed by A.A. Khanin is particularly popular. It takes into account the granulometric and pore composition, the bound water content, the effective porosity and permeability.

The classification is based on the dependence of the effective gas permeability coefficient on the effective voids ratio $k_{p.eff}$ for different types of sandy-silt-rock oil and gas traps (Fig. 58). According to A.A. Khanin, these rocks fall into six classes, each having a different permeability and capacity (Table 7). Individual classes are represented by four rock types with their respective effective gas permeability (see what follows) and porosity.

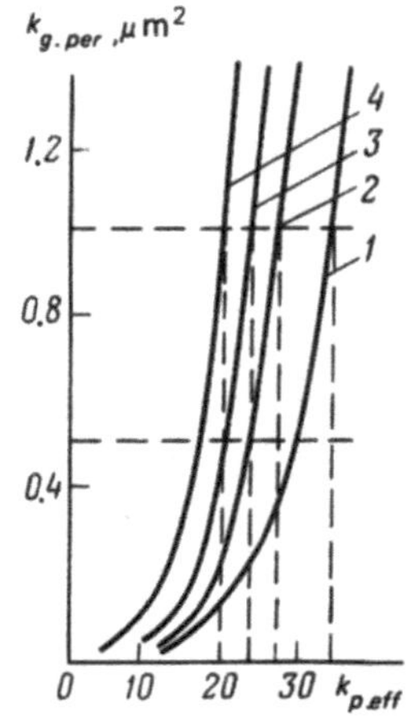

FIG. 58. Dependence of the coefficient of effective gas permeability $k_{perm.g}$ on the effective voids ratio $k_{p.eff}$ for sandy and aleurite reservoir bed groups (after A.A. Khanin).

For reservoir beds showing predominance of the following fraction: *1* — aleurite 0.05-0.01 mm (aleurite small-grained reservoir bed group); *2* — aleurite 0.1-0.05 mm; *3* — small-grained sandy; *4* — medium-grained sandy

For example, Class I of zone rocks with a very high permeability and capacity are represented by middle sandstones with $k_{perm.g} > 1$ μm² and corresponding $k_{p.eff} > 17\%$ (see Fig. 58, curve *4*). The traps of this class may also include fine sandstones. However, due to a more complex structure of the voids space of these rocks their $k_{p.eff}$ must exceed 20% (see Fig. 58, curve *3*).

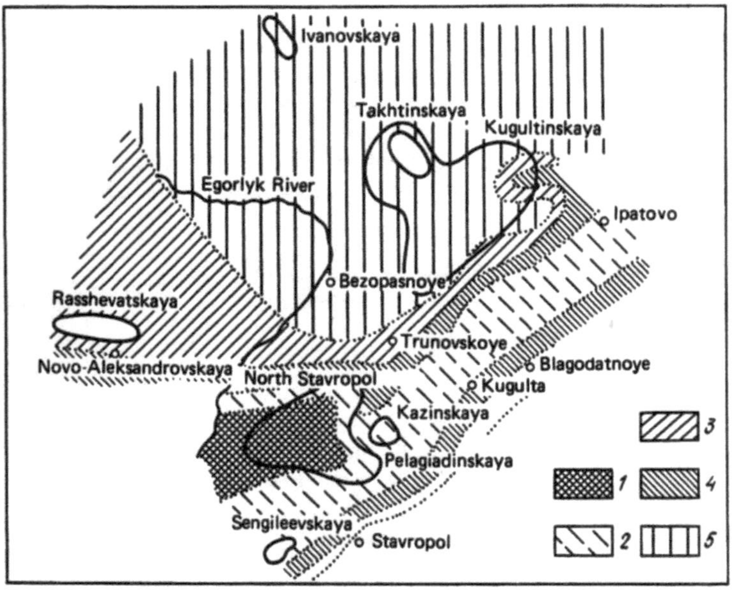

FIG. 59. A diagram showing the distribution of gas traps (relative to $k_{perm.g}$ and $k_{p.eff}$ values) developed in the Khadum horizon at the area of Central Cis-Caucasia (after A.A. Khanin).

$k_{perm.g}$ (in fm³) and $k_{p.eff}$ (in %): *1*—100->1000 and 25; *2*—100-1000 and 23; *3*—10-500 and 19; *4*—1.0-100 and 15; *5*—<5 and 13; $k_{p.eff} = k_{p.o} - k_{w.r}k_{p.o}$

TABLE 7. Classification of Sandy and Aleurite Rocks Composing Oil and Gas Traps (after A.A. Khanin and M.I. Koloskova)

Trap class	Oil and gas trap group (in terms of predominance of granulometric fraction)	Effective voids ratio, %	Size of main filtering pores, μm	Content of main filtering pores of total voids space, %	Total content of voids <2 μm of total voids space, %	Residual water saturation of total voids space, %	Permeability coefficient, μm^2	Permeability and capacity of oil and gas trap
I	Middle sandstones	⩾17	50-150	40-80	0-17	-		Very high
	Fine sandstones	⩾20	20-100	40-80	0-20	5-25	>1	
	Coarse aleurolites	⩾23.5	-	-	-	-		
	Fine aleurolites	⩾30	-	-	-	-		
II	Middle sandstones	15-17	40-100	25-50	5-25	-		High
	Fine sandstones	18-20	18-60	36-60	5-30	10-35		
	Coarse aleurolites	21.5-23.5	12-30	40-80	10-35	-	0.5-1	
	Fine aleurolites	26.5-30						
III	Middle sandstones	11-15	20-50	25-40	15-40	-		Middle
	Fine sandstones	14-18	16-40	15-50	15-40	10-45	0.1-0.5	
	Coarse aleurolites	16.8-21.5	10-30	25-65	20-45	-		
	Fine aleurolites	20.5-26.5						
IV	Middle sandstones	5.8-11	15-40	15-30	20-45	-		Decreased
	Fine sandstones	8-14	10-35	15-45	20-50	30-60	0.01-0.1	
	Coarse aleurolites	10-16.8	3-20	20-50	20-56	-		
	Fine aleurolites	12-20.5						
V	Middle sandstones	0.5-5.8	-	-	-	-		
	Fine sandstones	2-8	5-16	20-40	50-85	50-95	10^{-3}-10^{-2}	Low
	Coarse aleurolites	3.3-10	3-8	20-50	50-90	-		
	Fine aleurolites	3.6-12				-		
VI	Middle sandstones	<0.5	-	-	-	-		Rather low (such oil and gas traps do not have generally economic importance)
	Fine sandstones	<2	-	-	-	-	$<10^{-3}$	
	Coarse aleurolites	<3.3	-	-	-	-		
	Fine aleurolites	<3.6	-	-	-	-		

A higher value of the effective voids ratio $k_{p.eff}$ (>23.5%) should be demonstrated by coarse aleurite oil and gas traps (see Fig. 58, curve *2*) to ensure the same permeability (1 μm^2), given a more involved fine and meandering voids space of the rock where both paths of individual gas (or oil) flows are longer and the forces holding gas (or oil) in the trap are greater. The greatest effective porosity ($k_{p.eff} > 29\%$) for the same permeability (1 μm^2) should be manifested by fine aleurite rocks composing the trap (see Fig. 58, curve *1*).

Class I sandy gas and oil traps contain from 40 to 80% of relatively large (from 20 to 150 μm in size) principal infiltering pores, few (from 0 to 20%) voids with the size <2 μm and a little (from 5 to 25%) bound water.

Class II gas and oil traps having a high permeability in the range from 0.5 to 1 μm^2 also include four types of sandy and aleurite rocks; the limits of values of $k_{p.eff}$ much as for Class I gas and oil traps, are determined from the respective relations $k_{perm.g} = f(k_{p.eff})$ (see Fig. 58 and Table 7) etc.

For Class V low-permeability and water capacity oil and gas traps, not only low $k_{p.eff}$ values are characteristic (see Table 7) but also a relatively small (20 to 50%) content of principal filtering pores of small size (3 to 16 μm) and an appreciable content of pores less than 2 μm in size and of bound water (>50-95%).

The physical permeability and porosity characterize the pattern of oil and gas traps. Viewed in this context, it is recommended to determine k_{perm} for as great volume of rocks as possible. It is also important to study the distribution of this quantity in space within the limits of definite strata and beds, particularly, productive ones (Fig. 59).

Sec. 26. **The Effective and Relative Permeability**

The effective permeability is the capacity of rocks for transmitting through them definite components (gas-water, oil-water, gas-oil-water). Effective permeabilities are characterized by the respective *coefficients for gas* ($k_{perm.g}$), *water* ($k_{perm.w}$) and *oil* ($k_{perm.o}$) which are proportionality coefficients in the equations:

$$Q_g = k_{perm.g} \frac{\Delta p/l}{\mu_g} F; \quad Q_w = k_{perm.w} \frac{\Delta p/l}{\mu_w} F; \quad Q_o = k_{perm.o} \frac{\Delta p/l}{\mu_o} F \tag{16}$$

where Q_g, Q_w and W_o are discharge rates of individual components during infiltration of a heterogeneous liquid; $\Delta p/l$ is the pressure gradient; μ_g, μ_w, and μ_o are viscosities of gas, water and oil; F is cross section of the cylindrical sample

$$k_{perm.g} = K_{perm.g} k_{perm}; \quad k_{perm.w} = K_{perm.w} k_{perm};$$
$$k_{perm.o} = K_{perm.o} k_{perm}$$

Proportionality coefficients in the relations that follow are called *relative permeability coefficients* for gas, water and oil, respectively; their values are governed by the volumetric ratio of the components (gas, oil, water) in the infiltrating mixtures which are conditioned by the quatities k_g, k_w and k_o.

The dependences of $K_{perm.o(g)}$ and $K_{perm.w}$ on k_w have been sufficiently studied solely for water-oil and water-gas two-component mixtures (Fig. 60). Studies that have been made show the following.

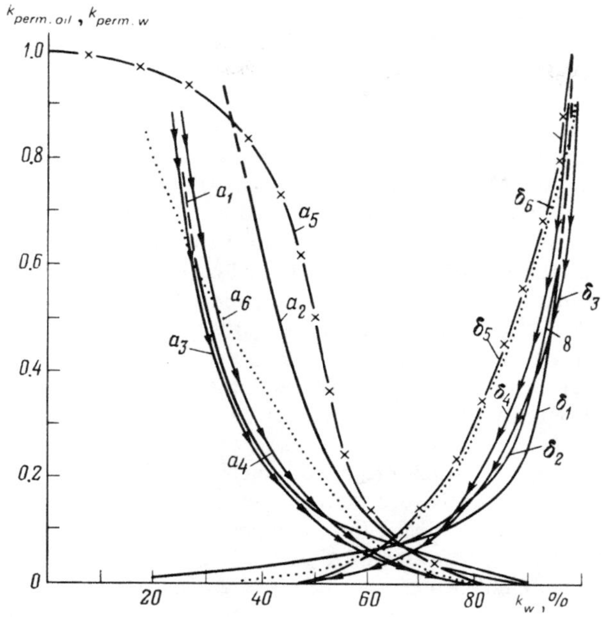

FIG. 60. Dependences of the coefficients $k_{perm.oil}(a)$ and $k_{perm.w}(b)$ of relative permeability of rock concerning oil or water depending on the coefficient of water saturation k_w (after the data submitted by various authors).

a_1, b_1—Samotlor oil field, bed AB_{2-3} (k_{perm} = 800 fm^2, $k_{o.p}$ = 25.8%); a_2, b_2—the same, bed BC_8 (k_{perm} = 492 fm^2, k_p = 25.2%) (after A.G. Kovalev); a_3, b_3—Samotlor field, bed BC_8; a_4, b_4—Ust-Balyk oil field, BS_1 bed (after data of the Siberian Research Institute of Oil Engineering); a_5, b_5—Trekhozernoe oil field, P horizon (k_{perm} = 70 fm^2, k_p = 24%) (after G.N. Pokrovskaya); a_6, b_6—Shagirt deposit, Yasnopolyanskie sediments (after B.I. Tulbovich)

1. Given $k_w < 20$-50%, $K_{perm.w} \approx 0$ ($k_{perm.w} \approx 0$). This is accounted for by the fact that until 20%-water saturation for relatively well sorted and large-grained rocks and until $k_w = 50\%$ for their poorly sorted and small-grained varieties water is held in small and blind pores at locations of grain contacts and does not participate in the infiltration process. The pellicular and microdrop water of the rocks, adsorbed by their solid phase, is also immovable in this case.

With increasing water saturation, the water begins to infiltrate and $K_{perm.w}$ attains as much as 70-90%, given k_w = 97-98%. Thus the value of the water saturation coefficient, given which appreciable water infiltration starts depends on the structure of the voids space of rocks. However, it is also governed by the physicochemical properties of the infiltrating heterogeneous fluid. Alkaline water facilitates the separation of oil films from a rock, and thus relative permeabilities over the whole water saturation interval are great both for oil and confined water.

2. The values of $K_{perm.o}$ and $K_{perm.g}$ are close to zero for $k_w > 75$-90%. They increase with decreasing k_w to attain 70% (for oil) and 87-97% (for gas), given

k_w = 20-30%. The value of relative permeability for oil and gas is also conditioned by the structure of the voids space and the nature of the aqueous and non-aqueous components of a heterogeneous fluid.

3. Depending on the degree of water saturation a two- and single-phase flow is possible.

4. The discharge rate of a heterogeneous fluid composed of two non-mixing components determined by the sum of the effective permeability coefficients is generally less than that of a homogeneous fluid evaluated by the absolute permeability coefficient, and relative permeability varies from 0 to ~1.

5. During the infiltration through a rock of gas-saturated water the gas discharge rate attains values close to that of a homogeneous fluid, given values k_w < 17-37%; for k_w greater than specified values, the gas discharge rate is much less compared with the discharge rate of a homogeneous fluid.

6. The discharge rate of oil through a rock attains values close to those of a homogeneous liquid, solely given k_w < 10%, and is much less for higher water saturation degrees.

During the infiltration through a rock of gas, oil and water mixtures a one-, two- and three-component flow is likely to occur. The region of combinations of k_w, k_g and k_o coefficient values for which a three-component flow is possible is very much restricted. Note that: (1) the value of $K_{perm.w}$ is mainly governed by the coefficient k_w and is practically independent of the oil to gas ratio in a rock; (2) values of $K_{perm.o}$ and $K_{perm.g}$ vary depending on the saturation of the voids space by each of the components of a heterogeneous liquid (Fig. 61); (3) the effective permeability for gas during the infiltration of a three-component heterogeneous liquid is much less than for the infiltration of a gas-saturated liquid, given the same gas saturation of a rock; (4) given the same degree of oil saturation of a rock, the effective permeability for oil in a three-component flow can be greater or less than that of oil for a two-component flow; (5) the bound water in amounts to 20-50% decreases the effective permeability for oil to a lesser degree than does the same amount of free water.

The graphs illustrating the dependence of $K_{perm.o(g)}$ on k_w obtained on a core of a definite sediment type are used for an evaluation of effective permeability

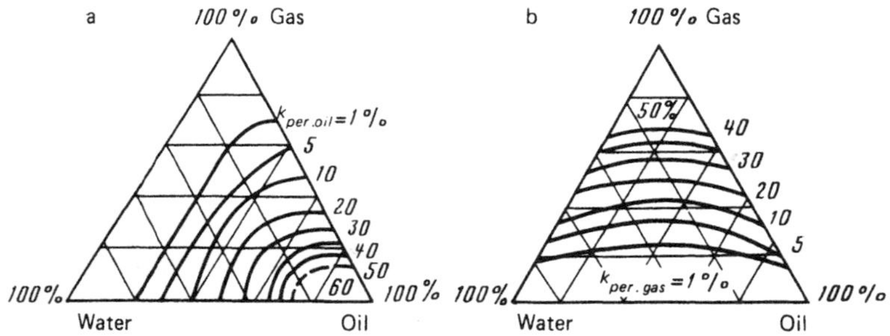

FIG. 61. Isolines of equal relative permeabilities $K_{perm.oil}$ for oil (*a*) and $K_{perm.gas}$ for gas (*b*) during infiltration through sands of water, oil and gas mixtures

coefficients for oil and gas. For sandstones such curves are presented in Fig. 60. By using both them and the values of absolute permeability coefficients and water saturation coefficients, it is possible to evaluate the effective permeability of a given rock type.

In like manner we can determine the effective permeability coefficients of rocks for oil or gas from the values of k_w, k_g, k_o, k_{perm} and diagrams of the type represented in Fig. 61 which make possible the evaluation, first, of $K_{perm.o}$ and $K_{perm.g}$, and then $k_{perm.o}$ or $k_{perm.g}$.

During the infiltration in oil traps of a two- or three-component mixture the discharge rates Q_g, Q_w, Q_o are governed not only by the effective permeability coefficients of the components but also by their viscosity.

Chapter VIII

Polarization, Electrical Conductivity, Losses and Their Characteristics

Sec. 27. Polarization Types and Their Characteristics

Being acted on by electromagnetic fields, rocks are polarized, conduct electrical current and a fraction of the field's energy is lost converting to heat (Joule heat losses in a direct field and dielectrical losses in an alternating field). In addition, the rocks are polarized and conduct current as a liquid is being pressed through them and if they are in contact with other moist rocks or electrolytes.

The mechanism of polarization, electrical conductivity and losses for rocks heterogeneous in terms of materials, phase and structure is very much involved and diversified as far as ion and electron conducting rocks are concerned, and is dependent on the field frequency, temperature and pressure.

Polarization in an electrical field reduces to the separation in rocks of unlike charge carriers and origination of an electrical dipole moment in any rock volume. In rocks, the field gives rise to certain processes facilitating their definite polarization which for ion-conducting rocks (sands, sandstones, limestones etc.) in a stationary field is due to the following processes: elastic displacement (of electrons, atoms, ions), orientation (of dipole, strongly bound molecules), relaxation (orientational dipole and ionic thermal), migration (volumetric polarization), concentration redistribution, electroosmosis.

Electron and ion-conducting rocks (coals, chalcopyrite, iron ores and other ores) are polarized much as are ion-conducting rocks. However, apart from the aforementioned ones, in the given case polarization is also affected by electrolytic processes. Elastic displacement polarization, relaxation and migration polarization appear and disappear rather rapidly. Other polarization processes develop and terminate at a much slower rate (slow polarization types). Due to the increased frequency of the polarizing current ever fewer polarization types are observed.

Displacement (elastic) polarization. It appears in rocks containing charged and interconnected particles that can shift with respect to one another under the effect of a field. Displacement polarization falls into electron, atomic, ionic and elastic dipole types.

Purely *electron polarization* can occur for nonpolar atoms and molecules of rocks (argon, helium, xenon atoms, nitrogen, oxygen, hydrogen and other gas molecules), minerals with valency crystals made of similar atoms (sulphur, diamond, graphite, selenium). Apart from ion, electron polarization takes place for ion crystals as well. At this, the electron shells of nonpolar atoms and ions of the medium (Fig. 62 *Ia*) are displaced with respect to the nuclei in the electric field

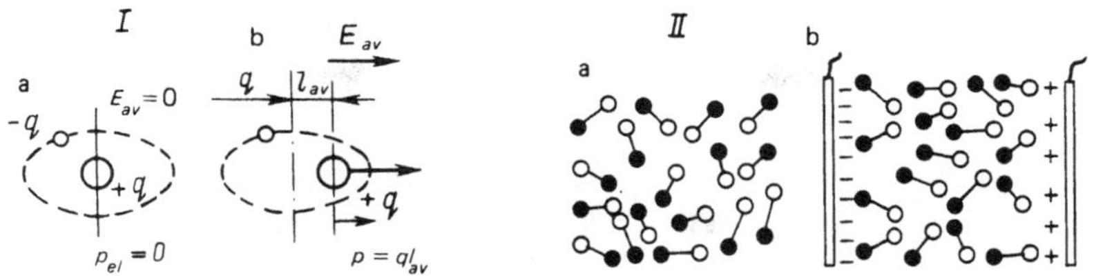

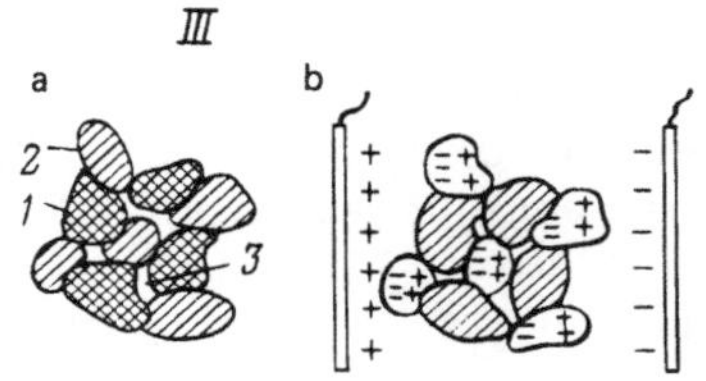

FIG. 62. Diagrams illustrating different polarization types in the absence (a) and presence (b) of an external field.

Polarization: *I*—electron; *II*—orientational; *III*—migrational

E, and in a volume ΔV there appears a dipole moment-polarization vector:

$$\vec{P}_{e.av} = \lim_{\Delta V \to 0} \frac{\Sigma \vec{p}_e}{\Delta V}$$

where $\vec{P}_e = ql_{av}$ is the dipole moment of the atom—vector directed from a negative to a positive charge (Fig. 62 *Ib*) (q is the charge of the electron; l_{av} is the average distance between the dipole's poles).

Electron polarization terminates within 10^{-14} to 10^{-15} s when polarizing field frequences ranging from zero to optical are involved.

Atom polarization occurs for rocks with valency crystals of dissimilar atoms. The latter are bound to form molecules by forces of valency electron exchange interaction (covalent bond), and outer shell electrons are redistributed between atoms asymmetrically, being shifted toward atoms with stronger bonds; that is why atoms acquire charges of unlike polarity. Field-induced polarization appears as a result of a relative displacement in molecules of dissimilar charged atoms. Atom polarization $\vec{P}_{a.av} \ll \vec{P}_{e.av}$ and sets in within a time interval 10^{-11} to 10^{-13} s, which is somewhat more compared with the time interval taken for electron polarization to set in.

Ion elastic displacement polarization is assumed to occur for quartz, corundum, calcite. It is also possible for other ion crystals (halite, sylvite etc.), whose crystal lattice contains densely packed dissimilar ions. Polarization $P_{i.av}$ in an electric field reduces to displacement of unlike ions from their equilibrium position in the crystal lattice. It sets in within time interval 10^{-13} to 10^{-14} s.

Ion polarization weakly depends upon frequency (until the frequency of the infrared region of the spectrum). With increasing temperature polarizability enhances due to lessening of the elastic bond between ions. Polarization grows for minerals whose ions have a great value of the α_i/r_i^3 ratio (α_i is polarizability of the ion crystal molecule; r_i is the ion radius).

Elastic displacement polarization $\vec{P}_{el.d}$ of polar molecules is characteristic of dipole dielectrics with strongly bound molecules which are able to rotate under the effect of the field solely through minor angles.

Relaxation (*thermal*) *polarization*. It occurs for rocks containing weakly bound particles which can change equilibrium state under conditions of thermal movement. We distinguish between orientation dipole, ion thermal and electron-relaxation polarization.

Orientation dipole polarization $\vec{P}_{d.av}$ is characteristic of rocks whose composition contains substances (water, gases) with dipole polar relatively weakly and elastically bound molecules. Polar molecules with sufficient thermal motion energy rupture elastic bonds with other similar molecules and are brought from one equilibrium state to another. Polarization is expressed in the preferable orientation of axes of dipole molecules in the direction of the field (see Fig. 62 *IIb*). Thermal molecular motion prevents this moment by disorienting molecules. As a result, the dipole moment axes before the action of the field have a random spatial orientation and the resultant dipole moment of the rock equals zero.

The time τ for setting in of relaxation polarization (relaxation time) of dipole molecules of polar liquids (oil, water) is

$$\tau = A \exp (U/kT)$$

where U is the height of the potential barrier separating two equilibrium states; A is a constant weakly depending on the absolute temperature T; k is the Boltzmann constant.

Rocks with orientation polarization manifest electron and sometimes atom or ion polarization connected with their solid phase.

As temperature and frequency of the field grow orientation polarization decreases. It takes 10^{-10} to 10^{-7} s to develop.

Polarization $\vec{P}_{d.av}$ is observed not only for the liquid and polar gaseous rock phase but also for minerals with a ring and carcass type lattice and loosely packed particles (crystallohydrates, clay minerals, zeolites). They all contain dissimilar type of water. Representatives of these mineral groups are: cordierite, beryll, spodumene, muscovite, biotite, gypsum, talc, etc.

Ion relaxation thermal polarization $\vec{P}_{ith.av}$ is possible for ion crystals with weakly bound ions the appearance of which is caused by defects and specific structure of the crystal lattice. Minerals and rocks containing these crystals have also electron and ion polarization. Under conditions of thermal motion, weakly bound ions are detached from the locations where they are fixed in place to be displaced in crystals at distances commensurate with interatomic distances, negotiating the potential barrier. The external field creates an excess ion transfer in its direction which leads to rock polarization. This process is compensated for by inverse ion diffusion.

Relaxation time and trajectories of polar molecules and weakly bound ions, under conditions of thermal polarization, are governed by the structure and temperature of rocks.

Electron relaxation polarization $\vec{P}_{er.av}$ appears due to excess "defect" electrons or "holes" alternatively known as electron holes which are activated by thermal energy.

Interlayer (macrostructure, volumetric) *polarization*. This polarization is assumed for rocks whose conducting components (see Fig. 62 *IIIa*, *2*) are alternated by nonconducting ones (see Fig. 62, *IIIa*, *1*) or air (see Fig. 62, *IIIa*, *3*). In the particular case within a short time interval positive ions of conductive inclusions are displaced along the field and negative ones in an opposite direction and are retained within inclusions near the interphase surface since the second component of a rock does not practically conduct electricity (see Fig. 62, *IIIb*). Given this polarization type, electron migration toward the anode is possible as is positive ion accumulation at an opposite side of a grain with electron conductivity included in the nonconductive phase.

As a resutl of the above processes conducting rock particles are polarized (layer polarization*) and acquire a dipole moment P_m much as a large molecule. Growth of polarization charge stops if its field is fully compensated for by the applied field. Under conditions that have set in the field and current are absent from conductive particles. Interlayer polarization terminates within 10^{-6} to 10^{-3} s close to the time taken by orientation (dipole) polarization to set in.

The time required for these polarization types to set in is comparable with the frequency of electromagnetic fields involved. That is why they are called slow compared with electron and ion types. Electrochemical polarization types considered in what follows are even slower.

Concentration and diffusion polarization (concentration redistribution polarization). This polarization type $P_{cd.av}$ of ion conductive rocks in an electric field occurs under conditions of dramatic heterogeneity of cross sections of their voids channels filled by electrolyte and with the electric double layer near the interphase boundary. Clearly, in so doing, counterions and coions in the expanded and contracted capillary portions make up unequal fractions of the total number of ions of electrolyte, therefore the effective (average) mobilities and ion transfer numbers also change here.

If the voids space of a rock represents a bunch of parallel and negatively charged capillaries of the above configuration (Fig. 63) then rock counterions are represented by cations, and coions by anions. Then, given the direction of the electric field shown in Fig. 63, the cations, and, in particular, counterions, while travelling toward the cathode (C) will be supplied from a narrow channel *II* (Fig. 63) to a volume

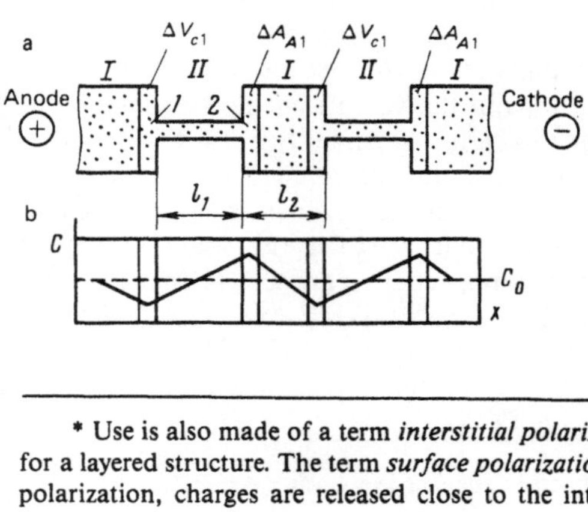

FIG. 63. A model of a capillary of an ion-conducting rock (*a*) and adopted in it distribution of electrolyte concentrations C_0 and C, respectively before and following passage of current (*b*).

I and *II*—wide and narrow portions of the capillary; l_1 and l_2 are their lengths: *1*—minor volume ΔV_{CI} in a wide capillary facing the cathode; *2*—the same, with ΔV_{AI} facing the anode

* Use is also made of a term *interstitial polarization* since the theory was first elaborated for a layered structure. The term *surface polarization* is sometimes used since, given migration polarization, charges are released close to the interphase boundary of different phases.

ΔV_{a1} of a wide channel I from its anode (A) side proceeding further out of it in the same direction.

Masses m_{cII} and m_{cI} (in moles) of, respectively, cations entering a volume ΔV_{aI} and leaving it are found from the relations:

$$m_{cII} = \frac{q}{Fz_c} t_{cII} = \frac{i\tau}{Fz_c} t_{cII} = \frac{Fn_{eq.c}}{Fz_c} t_{cII} = \frac{n_c z_c}{z_c} t_{cII};$$

$$m_{cI} = \frac{q}{Fz_c} t_{cI}$$

where q is the amount of electricity passed through a capillary; i is current; τ is time taken by current to flow; F is Faraday constant; z_c is the charge (effective valency) of a cation; $t_{cII} = u''/(u'' + v'')$; $t_{cI} = u'/(u' + v')$ are transference numbers (fraction of current transferred by cations), respectively, in a narrow and a wide channel; u', v', u'' and v'' are effective (average) mobilites of the cation and anion, respectively, in a wide and narrow channel; $n_{eq.c}$ is the number of gramm-equivalents of transferred cations; n_c is the numbr of moles of transferred cations.

Since in a narrow channel counterions are on the average more mobile than anions of the mainly potential determining electric double layer and constitute an appreciable fraction of all electrolyte ions of this portion of the voids channel, it is their transference number $t_{cII} \rightarrow 1$ that proves greater than that t_{aII} of coions whose value vanishes.

In a wide channel the effective mobilities of different ions and their transference numbers t_{cI} and t_{aI} differ much less both from one another and from their values in a free solution*. The electric double layer constitutes here an inappreciable fraction of the total amount of electrolyte. Given a great difference in cross sections of portions of capillaries I and II, $t_{cII} \gg t_{cI}$ and $\Delta m_{c\Delta V_{aI}} = m_{cII} - m_{cI} = (q/Fz_c)(t_{cII} - t_{cI}) > 0$, i.e. in a volume ΔV_{aI}, cation concentration increases (see Fig. 63). So does anion concentration increase here to the same degree. Their mass m_{aI} (in moles) flowing from a volume of the channel I to a volume ΔV_{aI} on the path to the anode is found from the equation: $m_{aI} = qt_{aI}/Fz_a$, and the mass of ions leaving this volume for the volume of the capillary II, by the equation: $m_{aII} = -qt_{aII}/Fz_a$, but $t_{aI} > t_{aII}$, since in the volume of capillary II the effective mobility v'' of anions is much less than that v' of the same ions in the volume of the capillary I and constitutes a much lesser fraction of the total effective mobility of ions at a portion of the channel II compared with the portion I. In this connection the difference between the entering and leaving the volume ΔV_{aI} masses of the anion $\Delta m_{a\Delta V_{aI}} = (q/Fz_a)(t_{aI} - t_{aII})$ is also greater than zero, i.e. the anion mass in a volume ΔV_{aI} increases. In the last relationship t_{aI} and t_{aII} are transference numbers of ions in capillary portions I and II. Note that $\Delta m_{c\Delta V_{aI}} = \Delta m_{a\Delta V_{aI}}$ according to the condition of electroneutrality of solution in a volume ΔV_{aI}. Thus, electrolyte concentration in a volume ΔV_{aI} of the wide portion of the channel increases.

* A solution found in a wide vessel rather than in the voids space of a rock.

It is also possible to prove that in a minor volume ΔV_{cI} of a wide portion of a channel electrolyte concentration (in moles) drops from its cathode side inversely to its decrease in a volume ΔV_{aI} (see Fig. 63).

Appreciable concentration changes throughout the length of a channel (see Fig. 63*b*) are ruled out after disconnecting the field due to relatively long (about 10 s and more) diffusion processes. In so doing, the resultant of local emf (potentials E_{cd}) is what is responsible for induced concentration-diffusion polarization $\vec{P}_{cd}$ of ion conducting rocks.

Polarization P_{cd} and E_{cd} potentials also exist as a current is passed through the rock, and the sign of E_{cd} corresponds to that of the field applied.

This polarization type, given a greater field frequency, is nonexistent. It is most likely to occur for rocks where macro-, micro- and millimicrovoids (rocks with inter-crystal and crack voids, enhanced clayiness or organic substance content) intermit.

Electrolytic polarization. Electrolytic polarization $\vec{P}_{el}$ appears jointly with concentration and diffusion polarization as an electric current is passed through electron and ion-conducting rocks (graphite, coals, sulphide, iron ores and other ores). Electrolytic polarization is much greater than concentration-diffusion type, therefore, the latter can be ignored when electron and ion-conducting rocks are involved. Before a current is passed the electron conducting component and the electrolyte of rocks are continuously exchanging ions (or electrons)—an exchange current appears, and, as a result of selective adsorption, an electric double layer forms. Owing to this fact, electron conducting rock components have nonequilibrium stationary electrode potentials.

The stationary potential of electron conducting minerals is governed by their composition and structure, the state of the surface, and type and concentration of ions and gases in the electrolyte that it is in contact with.

If gradually increasing voltage is applied to an electron and ion-conducting rock the recharging of the electric double layer and electrolysis cause the stationary potentials from the anode and cathode sides of electron conducting inclusions to unequally vary to convert them to dipoles. The mechanism of electrolysis is as follows. Acted on by the field, electrolyte cations, while rushing to the cathode, approach the electron conducting grain from its side facing the anode (the cathode of grain) and their definite sort, povided it has sufficient energy, reacts electrochemically with the inclusion borrowing electrons and discharges itself (reduces). In addition, the inclusion potential from its side facing the anode (the cathode of grain) become more positive compared with the stationary one. Electrons come to the grain's cathode from its anode. (The anode of grain is its side facing the cathode.) At the anode the potential becomes more negative than the stationary one, since electrons resulting from the oxidation reaction are supplied to the grain. At the locations of current input and output in the electrolyte the products of electrolysis are accumulating: atoms, molecules, lesser valency ions, and other substances. For example, the following reactions are likely to occur in a diluted sulphuric acid at the surface of pyrite grain as a result of appreciable deviation of anode and cathode grain potentials from their stationary value: from the anode side of pyrite (at its cathode) $15FeS_2 + 60H^+ + 60e \rightarrow 15Fe + 30H_2S$; $2H^+ + 2e \rightarrow H_2$, from the

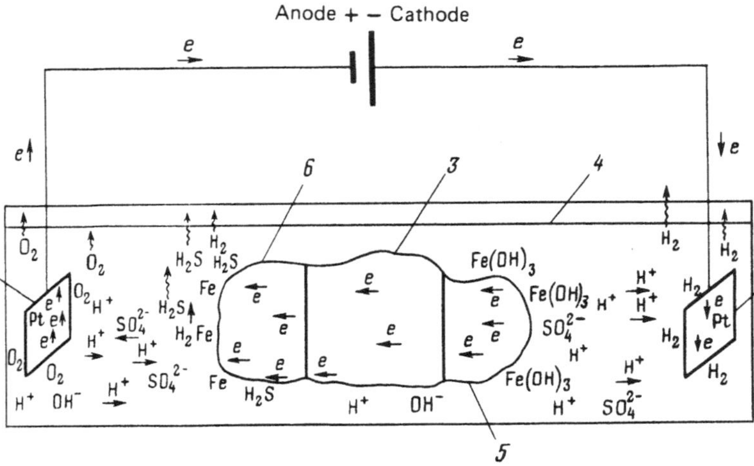

FIG. 64. A diagram illustrating the electrolytic polarization of a pyrite grain in a sulphuric acid solution.

1—Pt anode; *2*—Pt cathode; *3*—a pyrite grain, *4*—level of diluted sulphuric acid solution (H_2SO_4); *5*—anode of pyrite grain, one of its sides facing the Pt cathode *2* (oxidation reaction occurs here: $4FeS_2 + 44H_2O \rightarrow 4Fe(OH)_3 + 8H_2SO_4 + 60H^+ + 60e$; $Fe(OH)_3$ adsorbs on the pyrite anode; H_2SO_4 dissociating into H^+ and SO_4^{2-} ions, and H_2SO_4 concentration in the vicinity of the grain increasing; hydrogen ions move toward the Pt cathode, and here, given the required overvoltage at the cathode, there occurs an electrolytic reduction reaction: $60H^+ + 60e \rightarrow 30H_2$. At this, the electrons needed for to occur are taken from the current source; H adsorbs at the cathode to be liberated into the solution, then to the atmosphere; the electrons emerging during the reaction at the pyrite anode are supplied to the pyrite grain); *6*—the pyrite grain cathode from one of the sides facing the Pt anode (here, given a sufficient overvoltage at the pyrite grain, the following reactions occur: $15FeS_2 + 60H^+ + 60e \rightarrow 15Fe + 30H_2S$ and $2H^+ + 2e \rightarrow H_2$; there are electrons in pyrite for the reaction; Fe precipitates on the cathode, H_2S and H_2 precipitates in the solution and then into the atmosphere, and may adsorb on the grain; concurrently the following oxidation reaction takes place on Pt anode: $30H_2O + 30SO_4^{2-} \rightarrow 30SO_3 + 30H_2O + 15O_2 + 60e \rightarrow 30H_2SO_4 + 15O_2 + 60e = 60H^+ + 30S_4^{2-} + 15O_2 + 60e$, whose products cause the H_2SO_4 concentration to be increased at the anode; then adsorb on the latter or go to the solution and atmosphere (O_2 molecules) and come to the anode (electrons *e*) and move toward the current source

cathode side at its anode $4FeS_2 + 44H_2O \rightarrow 4Fe(OH)_3 + 16H^+ + 8SO_4^{2-} + 60H^+ + 60e$.

The products of their anode side at the cathode of pyrite are hydrogen sulphide, metallic iron and atomic hydrogen, and at the cathode side, at the anode of pyrite—iron hydroxides, SO_4^{2-} ions, hydrogen ions and electrons (Fig. 64).

Electrolysis of the solution at the input and output of current from the ore grain is possible if the difference between external U and stationary U_0 voltage $U - U_0 = \eta = IR + \eta_{cC} + \eta_{cA} + \eta_{actC} + \eta_{actA} + \eta_{adsC} + \eta_{adsA}$, where R is the internal resistance of electron conducting grains; I is current passing through them; η is voltage (overvoltage) necessary to obtain at electron conducting inclusions of rock for electrolysis to begin. It exceeds the stationary potential of grains by the sum of cathode, anode overvoltages and ohmic voltage drop at the inclusions.

Anode overvoltage is the difference between the anode and stationary voltages of an electron conducting inclusion. The cathode potential is more positive than the stationary value by the amount known as cathode overvoltage. Overvoltage occurs because, as current flows through the specimen at the input and output of current from an electron conducting inclusion, concentration of the ion involved in the reaction immediately changes (decreasing at the cathode and increasing at the anode). Since the number of ions possessing an energy sufficient for their discharge is in direct proportion to their concentration, to prevent the current from being changed, the potential must be increased. The quantity η_c is called concentration overvoltage or concentration polarization since the ions are brought to or removed from the grain due to the concentration difference and phenomenon of diffusion. Another resultant of the cathode and anode overvoltages is activation overvoltage η_{act} which is determined by energy expenditures for the ion discharge. Activation overvoltage or discharge overvoltage is the potential to be applied to a rock to enable definite ions of its electrolyte to negotiate the interphase potential barrier and discharge on ore grains.

The third of the principal components of cathode and anode electrolytical polarization (overvoltage) is adsorption or chemical polarization η_{ads}. This overvoltage type occurs due to the presence at the surface of electron conducting grains of a strongly adsorbed layer of intermediate and final products of electrochemical reactions. In this process, at sites of current input and output the composition and, consequently, overvoltage of grains varies. Note that electrode reactions generally have several successive stages each of which is characterized by the formation of a definite product, rate and overvoltage.

Intermediate products of electrolysis are accumulated until the rates of the intermediate and final reaction stage are equal and continuous current appears.

Thus, cathode overvoltage η_c at the cathode of electron-conducting inclusions (given a specified current density) is made up by concentration, activation and adsorption overvoltages: $\eta_C = \eta_{cC} + \eta_{actC} + \eta_{adsC}$. The value η_{cC} is governed by transfer of material in the electrolyte filling the voids, η_{actC} by transfer of electrons across the electrolyte—inclusion interphase. They can be easily identified since η_{actC} grows rapidly upon connection and drops upon disconnection of the polarizing current; η_{cC} changes rather rapidly being dependent on diffusion. The same is true of anode overvoltage at the anode.

Overvoltage depends on the composition of electron conducting grains and electrolyte. Overvoltages at electron conducting inclusions sum up to form the emf $\mathscr{E}_{el}$ and give rise to polarization $\vec{P}_{el}$ of these rocks.

Electrolytic polarization develops rather slowly to drop after the field polarizing the rock is disconnected. Its intensity is conditioned by a number of features of electron and ion-conducting rocks as well as the field's frequency, temperature and pressure.

Activation overvoltage (polarization) sets in 10-50 ms after a field has been applied and is conditioned by the capacity of the electric double layer equal to 1-4 $\mu f/cm^2$. Adsorption overvoltage relaxation time is 100-200 ms, that of concentration polarization is from a few seconds to several minutes.

Electroosmotic polarization. Denoted as $\vec{P}_{eo}$, it results from electroosmosis, which is the movement of the electrolyte through a rock due to the electrical field gradient E. Under conditions of electroosmosis the potential gradient $U_{eo} = E/l$

(l is the rock specimen's length) displaces relative to its solid phase a portion of counterions of the diffuse part of the electric double layer, a fraction of loosely bound water and free water jointly with ions found in it toward the respective electrode (cathode if counterions are provided by cations) to produce in the electrolyte a pressure gradient $\Delta p/l$ similarly directed. Upon disconnection of current, the latter gradient gives rise to infiltration of the pore liquid in an opposite direction and facilitates the origination in the rock of an electroosmotic filtrational potential difference U_{eo}. The mechanism of the origination of polarization $\vec{P}_{eo}$ and potential difference U_{eo} (see Eq. 19) and the quantity it depends on are considered in Sec. 30.

Sec. 28. **Total Polarization and Dielectrical Permeability**

Total polarization is made up by different polarization types possible for the particular mineral or rock at a definite electric field frequency, temperature and pressure. Its greatest values are shown by electron and ion-conducting and structure-heterogeneous rocks in a constant, sufficiently intensive and long standing electrical field, at low temperatures and pressures. The average resulting polarization (algebraic sum of individual polarization types) of electron and ion-conducting rocks is

$$\vec{P}_{av} = \vec{P}_{e.av} + \vec{P}_{a.av} + \vec{P}_{i.av} + \vec{P}_{ed.av} + \vec{P}_{d.av} + \vec{P}_{ith.av} + \vec{P}_{er.av} + \vec{P}_{m} + \vec{P}_{cd} + \vec{P}_{el} + \vec{P}_{eo.av} = \sum_{i=1}^{i=n} \vec{P}_{i}.$$

In particular, given a very great field frequency exceeding 10^{11} Hz, rock polarization is solely associated with rapid polarization types (electron, atom and ion) and that is why it is minimum. Given field frequency 10^{2}-10^{11} Hz, such polarization types are possible to occur as relaxation and migration (volumetric) types, as a result, polarization of the rock in question increases. Finally, in a constant field of long standing other slow polarization types may occur due to which total polarization attains maximum values.

Total polarization is proportional to the average electrical field intensity $\vec{E}_{av}$, i.e. $P_{av} = \alpha_0 n E_{av}$, where n is the number of dipoles in a rock unit volume; α_0 is the average conventional coefficient (polarizability) of rock whose values diminish with increasing rock homogeneity and growing field frequency, temperature and pressure.

All polarization types that have succeeded in developing within the time τ of processing the rocks by a current facilitate the appearance therein of an internal field oriented in a direction opposite to that of the applied field. As a result, the intensity E_{av} of the electrical field applied to a rock is attenuated compared with the intensity E_{0av} of an average field in a vacuum. The attenuation is characterized by a dimensionless quantity $\varepsilon = E_{0av}/E_{av}$ called *dielectrical permeability*. It shows the times the intensity of the electrical field in a dielectric (rock) is attenuated compared with its intensity in a vacuum.

Given the specified value E_{av}, the dielectrical permeability governs polarization of solely rather weakly conducting media. In well conducting materials the charges are liable to migrate until their potentials are equalized, i.e. after the potential difference U between the charges is zero. This lessens charge interaction and leads to the fact that in naturally occurring conductors and well conducting semiconductors $E_{av} = U/d \approx 0$ (d is the path covered by the charge), and ε tends to infinity.

Sec. 29. Induced Potentials and Induced Electrochemical Activity

Induced potentials. After stationary electrical field is disconnected, an inverse process viz. rock discharge begins. The first to disappear practically are rapid polarization types (electron, atomic, ion, elastic, dipole types) followed by slower polarization types: orientation thermal, ion thermal, volumetric (migration) types. The last to disappear are slower polarization types: concentration-diffusion and electrolytic types. In addition, electroosmotic and infiltration polarization types associated with infiltration processes due to the potential or pressure gradient also slowly disappear.

Induced polarization growing during the charging process and decreasing during the discharging process is revealed by recording the growth and drop of voltage U within a time period τ. The connection of the current at a time moment τ_1 (Fig. 65) at $U = f(\tau)$ graphs is marked by a dramatic jump of potential difference (see Fig. 65*b*, time moment τ_1). It corresponds to the ohmic potential drop at a rock specimen and at electrolyte layers adjacent to it and receiving electrodes. Then we observe first a rapid (see Fig. 65*b*, curve portion *1-2*), then decreasing and attaining the limit (see Fig. 65*b*, curve portion *2-3*) growth of potentials of induced polarization. The disconnection of the polarizing current in the $U = f(\tau)$ plot shows itself by a drastic jump of ohmic potential difference (see Fig. 65*b*, curve portion *3-4*), and in the $I = f(\tau)$ graph by a total drop of the polarizing current (time moment τ_2). Then we can observe in the $U = f(\tau)$ plot first rapid, then ever slowing potential drop U_{ip} of induced polarization.

Different stages of charging and discharging processes of minerals and rocks can also be followed in the $U_{ip} = f(j)$ curves (j is the density of the polarizing current). Given small polarizing densities, potentials of induced polarization of metals and electron conducting minerals in solutions of different electrolytes generally increase in proportion to current density. Given current density exceeding several tens or, less often, hundreds of microamperes per cm^2, metals and electron conducting minerals are characterized by complex $U_{ip} = f(j)$ curves.

The process of the charging and discharging of minerals and rocks having electron and ion conductivity is also governed by the shape of the $U_{ip} = f(j)$ curves obtained under conditions of a gradual increase and consequent decrease of current

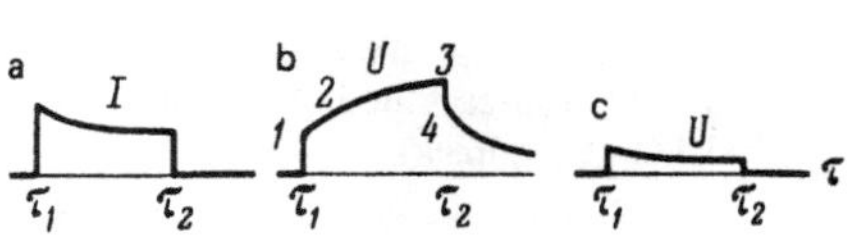

FIG. 65. Current $I(a)$ and potential difference $U(b)$ at the reception circuit with the sample located between the measuring electrodes and $U(c)$ outside them.

τ_1 and τ_2 are time moments the current is switched on and off

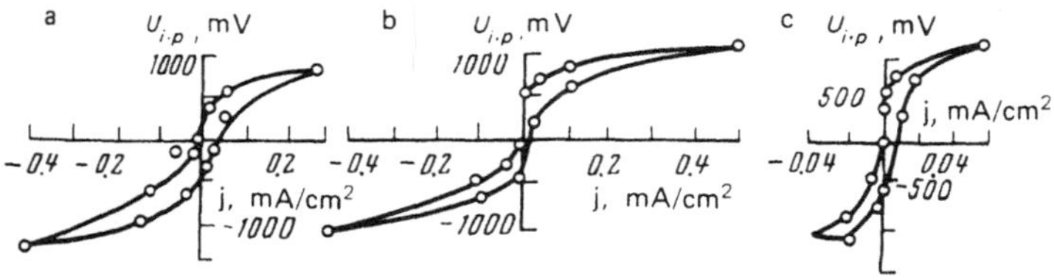

FIG. 66. Dependences of the induced potentials U_{ip} on the density j of the excitation current for electron-conducting minerals and rock (after M.G. Latysheva).

a—chalcopyrite; *b*—sphalerite; *c*—anthracite

densities in a forward and opposite direction (Fig. 66*a-c*). The hysteresis shape of these curves is a result of the effect of noninstantaneous discharge of these rocks.

Induced polarization potentials of rocks manifesting ion conductivity often increase linearly (Fig. 67*c*) to attain sufficiently great polarizing current density values.

Potentials of induced polarization of electron and ion-conducting rocks in electrolyte solutions increase with increasing the content and degree of dispersion of electron conducting inclusions and increased rock moisture content.

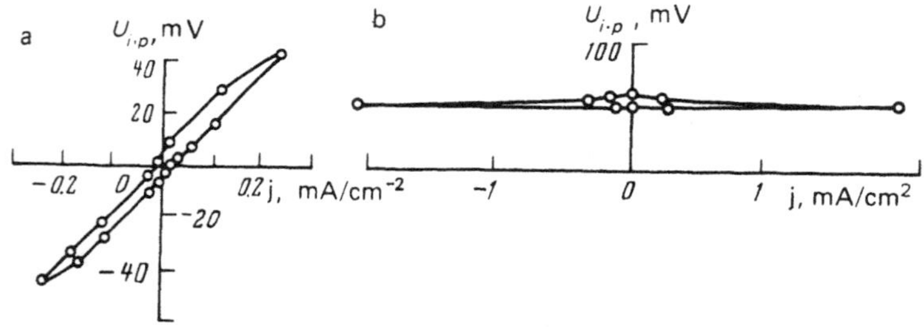

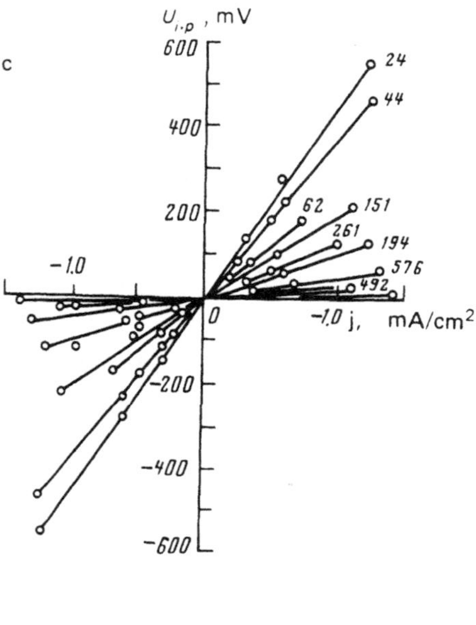

FIG. 67. Dependences of the induced potentials U_{ip} on the current density j for ion-conducting rocks (after M.G. Latysheva).

a—limestone; *b*—clay, *c*—Devonian sediment sandstones of the Tuimaza area

In sandy and clayey rocks potentials of induced polarization increase with increasing the clayey fraction to a definite limit* depending on the clay species and cation type in the diffuse part of an electrical double layer. These potentials increase with decreasing rock moisture content and mineralization of pore moisture.

The longer is the duration of the action of the polarizing field and its intensity and the higher is the specific resistance of the electrolyte of a rock, the slower is the rate of the decreasing of induced polarization. The discharge rate is maximum under conditions of appreciable concentrations of solutions filling the voids.

For quartz sands (with the radius of sand grains $>10\ \mu m$) polarization decreases slower the smaller is permeability of rocks and the smaller the radius of their voids, i.e. the smaller are their particles.

According to M.G. Latysheva, the discharging proceeds following these laws:

(1) for rocks showing electron conductivity

$$U_{ip} = U_{ip0} e^{-\tau/\lambda}$$

where U_{ip0} is the induced potential difference at a time moment $\tau \rightarrow 0$; λ is a constant;

(2) for rocks showing ion conductivity

$$U_{ip} = U_{ip0}/(\tau + c)^m$$

where m is the power index determining the gradient of the drop whose values for sandy clayey rocks vary from ≈ 0.8 to 1.4; c is a constant equal to 1 s.

V.N. Dakhnov believes that discharging of rocks proceeds exponentially. However, in rocks of a complex mineral composition the discharging is characterized by a sum of exponents each of which has its constant λ. The same conclusion is arrived at by I.I. Rokityanskii as well who has established for wet quartz sands the presence of two exponents—one rapidly and the other slowly decreasing.

Note that within the first seconds the effect of induced polarization appreciably diminishes and the remaining fraction of U_{ip} attains zero values after a large time interval (sometimes of the order of tens of minutes).

Induced electrochemical activity. Induced electrochemical activity is the property of rocks to polarize and possess various residual electrochemical polarization (different residual potentials of induced polarization) for a definite time moment upon disconnection of the field. Since all polarization types except concentration diffusion or electrolytical types disappear very rapidly, residual induced polarization potentials recorded by conventional methods, are conditioned, in fact, by them only. Structural heterogeneity of rocks and sometimes the presence of ore inclusions in them are what determines the great difference in their induced electrochemical activity. Rocks are compared in terms of induced electrochemical activity by using the quantity

$$A_v = E_{ip}/E$$

---

* This is a limit at which mineralization of pore solution will not prove appreciable due to the increased content of clayey fraction in the rock, suppressing growth of induced polarization associated with an increase of the specific surface.

where E and E_{ip} are field intensities of, respectively, the polarizing field and one induced by slow polarization types.

The intensity E_{ip} is in direct proportion to rock polarization, but $E = U/l$ and $E_{ip} = U_{ip}/l$, so

$$A_\upsilon = U_{ip}/U \quad (17)$$

Here U and U_{ip} is voltage drop at a specimen l in lenght, related, respectively, with the applied field and one of induced polarization. The induced polarization potential U_{ip} is usually recorded in 0.5 s upon disconnection of the polarizing current and is denoted as $U_{ip0.5}$, therefore Eq. (17) takes on this form:

$$A_\upsilon = U_{ip0.5}/U$$

For electron and ion-conducting rocks this relationship appears as follows:

$$A_\upsilon = U_{ip0.5}/(U + U_{ip})$$

since the potential difference induced in an electron and ion-conducting rock sample is appreciable and, in turn, causes the rock to be polarized.

The quantity A_υ is a dimensionless variable.

Sec. 30. Infiltrational Polarization, Potentials and Electrochemical Activity

As the electrolyte contained in the rock voids infiltrates through a rock sample l in length, ion fluxes appear directed toward the pressure gradient $\Delta p/l$. Since counterions move at a faster rate than do coions, at the ends of rock capillaries there appear a free charge difference, potential difference and a convective surficial current I_s. Yet, as time passes, owing to the fact that counterion concentration at a definite end of the sample grows appreciably, their flux markedly slows down, whereas that of coions speeds up. This gives rise to potential difference U_f throughout the length of the sample and an opposite current I_υ throughout its volume which increases to attain a stationary value $I_V = I_S$, and *the infiltration potential* difference (that of current) U_f becomes constant.

Given a cylindrical capillary with a radius r and length l,

$$I_S = 2\pi \int_0^r u(r)\delta_{layer}(r) r dr = \varepsilon\zeta_0 F_{ef} \Delta p / 4\pi\eta l$$

where $u(r)$ is the linear speed of liquid infiltration in a capillary (to be obtained from the Navier-Stokes hydrodynamics equations); δ_{layer} is the excess charge density in an electric double layer found from Poisson's equation; ε is dielectrical permeability; ζ_0 is the zeta potential of a flat capillary wall; F_{eff} is effective capillary cross section; η is the dynamical viscosity coefficient; Δp is a pressure difference at the ends of a capillary.

Countercurrent in the volume of capillary is

$$I_V = F_{eff} \sigma_w U_f / l$$

where σ_w is the specific electrical conductivity of the fluid in the capillary.

This equation has been derived from the relations $I_V = U_f/R$ and $R = (1/\sigma_w)(l/F_{ef})$.

Given a current has been set in, the following Helmholtz-Smoluchowski equation for a current potential in a capillary is valid:

$$U_f = \zeta_0 \varepsilon \Delta p / 4\pi\eta\sigma_w \tag{18}$$

If we think of the voids space of a cylindrical specimen l in length and F in cross section as being composed of "k" groups of capillaries r in radius parallel to one another and perpendicular to the sample base and allow for the effect of their size and form we can obtain, as did M.A. Belyakov, the following relationship for the infiltration (seepage) potential:

$$U_f = \frac{\varepsilon\zeta_0\varrho_w\Delta p}{4\pi\eta} \frac{\sum_{i=1}^{i=k} \frac{n_i r_i^2}{l_{layer\,i}} (r_i - l_{layer\,i}) \ln\left(1 + \frac{l_{layer\,i}}{r_i - l_{layer\,i}}\right)}{\sum_{i=1}^{i=n} n_i r_i^2} \tag{19}$$

where n_i is the number of capillaries of i-th radius for unit area; ϱ_w is the specific electric conductivity of water saturating a rock's model sample; $l_{layer\,i}$ is the thickness of the electrical double layer in a capillary with an i-th radius.

Since

$$A_f = \frac{\varepsilon\zeta_0}{4\pi\eta} \frac{\sum_{i=1}^{i=k} \frac{n_i r_i^2 (r_i - l_{layer\,i})}{l_{layer\,i}} \ln\left(1 + \frac{l_{layer\,i}}{r_i - l_{layer\,i}}\right)}{\sum_{i=1}^{i=k} n_i r_i^2} = \frac{U_f}{\varrho_w \Delta p}$$

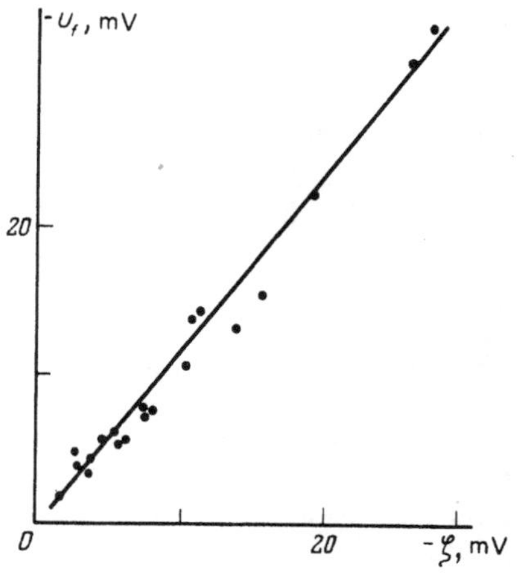

FIG. 68. Dependence of filtration potential U_f on the zeta potential ζ of rock (after M.A. Belyakov).

Specific resistance of water ϱ_w saturating 7 Ohm·m samples; pressure is p = 0.25 MPa

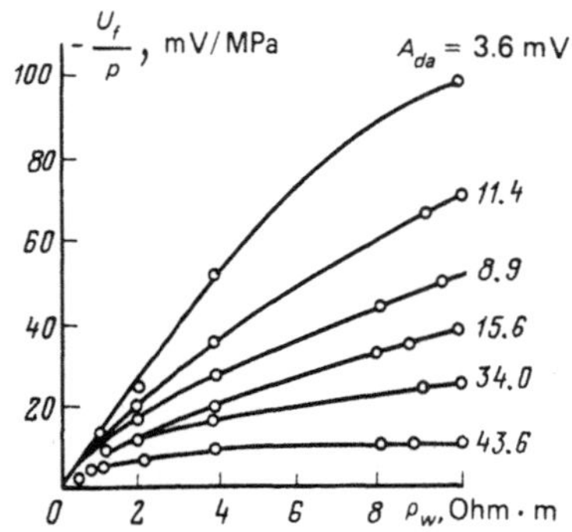

FIG. 69. $U_f/p = f(\varrho_w)$ relationships (after M.A. Belyakov)

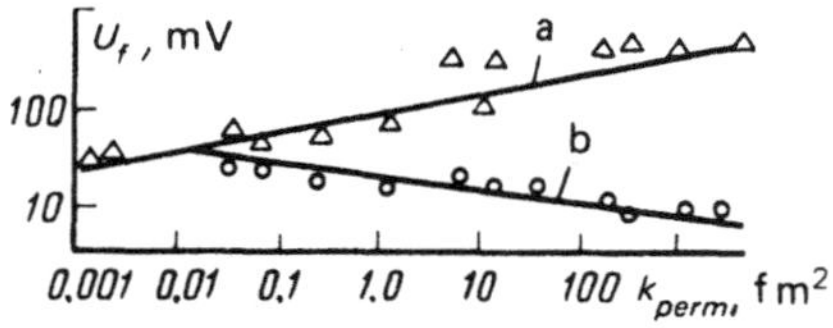

FIG. 70. $U_f = f(k_{perm})$ relationships (after M.A. Belyakov) upon squeezing through a sample of NaCl solution (*a*) and not treated clayey solution (*b*).

$\varrho_w = \varrho_f = 7$ Ohm·m; $p = 5$ MPa

then

$$U_f = A_f \varrho_w \Delta p$$

Here A_f is a quantity known as *infiltration electrochemical activity.* It shows that for a unit of applied pressure difference Δp of specific electric conductivity ϱ_w of electrolyte in the voids of different rocks unequal infiltration potential differences U_f can be obtained, and, the greater is the rock's zeta potential (Fig. 68), its dielectrical permeability, the less is the viscosity coefficient of the fluid saturating the voids. It can also be seen from the latter relationship that A_f is governed by the thickness of the structure of the electric double layer of the voids space of the rock which is characterized by radii of various type capillaries as well as by their ratio in the sample's volume. Experimental studies have established direct relationships $U_f/p = f(\varrho_w)$ (Fig. 69) and $U_f = f(k_{perm})$ (Fig. 70) and an inverse but not close connection $U_f = F(k_{w.r})$.

Sec. 31. Electrochemical Diffusion-adsorption Polarization, Potentials and Activity

At the contact of a moist rock with electrolyte or of different moist rocks there occur electrical potentials called diffusion and adsorption ones. By clarifying the mechanism of the given polarization type let us consider the fashion in which diffusion and adsorption potential originate and differ from diffusion potentials.

Diffusion potentials and diffusion emf coefficient. These potentials are observed for an immediate contact of electrolytes of different concentration or chemical composition. They are conditioned by the different mobility of ions diffusing from a higher-concentration to a lower-concentration solution. In the case of a contact between binary electrolytes differing solely in concentration (e.g. centi- and decimolar solutions of sodium chloride, potassium chloride, sodium sulfate, calcium chloride etc.) then the emerging potential jump (in millivolts) is determined by the Nernst equation

$$\mathscr{E}_{dC_0/C} = a' \cdot 10^3 \frac{RT}{F} \frac{\nu_c u - \nu_a v}{\nu_c z_c u + \nu_a z_a v} \log \frac{a_0}{a}$$

$$= a' \cdot 10^3 \frac{RT}{F} \frac{\nu_c u - \nu_a v}{\nu_c z_c u + \nu_a z_a v} \log \frac{C_0}{C}$$

$$= 58 \frac{\nu_c u - \nu_a v}{\nu_c z_c u + \nu_a z_a v} \log \frac{C_0}{C} = K_d \log \frac{C_0}{C}$$

where $a' = 2.30$ is the module of conversion of Naperian logs to decimal ones; $R = 8314.31$ J/(kmole·degree) is the universal gas constant; T—291.15 K is absolute temperature; $F = 9.64914 \cdot 10^7$ coul/mole is the Faraday constant; ν_c and ν_a are numbers of cations and anions onto which the electrolyte molecule dissociates; z_c and z_a are valencies (charge numbers) of anions and cations; u and v are electrolyte mobilities (or molar ion electric conductivities) of, respectively, anions and cations, m^2/(Ohm·mole); C_0 and C are molar concentrations of contacting electrolytes (C_0 is a constant and greater quantity than C that can take on any specified values less than C_0); a_0 and a* are activities of, respectively, a higher and a lower concentration electrolyte solution;

$$K_d = 58 \frac{\nu_c u - \nu_a v}{\nu_c z_c u + \nu_a z_a v}$$

is the diffusion emf coefficient.

For a binary monovalent electrolyte, e.g., NaCl, $\nu_c = \nu_a = 1$, $z_c = z_a = 1$, and

$$\mathscr{E}_{dC_0/C} = a' \cdot 10^3 \frac{RT}{F} \frac{u - v}{u + v} \log \frac{C_0}{C} = K_d \log \frac{C_0}{C} \tag{20}$$

where the diffusion emf coefficient is

$$K_d = a' \cdot 10^3 \frac{RT}{F} \frac{u - v}{u + v}$$

For contacting sodium chloride solutions at 18 °C

$$\mathscr{E}_{dC_0/C} = 58 \left(\frac{u}{u + v} - \frac{v}{u + v} \right) \log \frac{C_0}{C} = 58(t_+ - t_-) \log \frac{C_0}{C} = -11.6 \log \frac{C_0}{C}$$

where $t_+ = u/(u + v) = 0.39$ and $t_- = v/(u + v) = 0.61$ are transference numbers of, respectively, anions and cations; $(t_+ - t_-) = 0.22$.

Given contacting solutions of different salts, the calculation of diffusion potentials uses the Henderson equation or this is done in laboratory conditions. The design and experimental values of $\mathscr{E}_{dC_0/C}$ show a sufficiently good fit (Fig. 71, curves *1* and *2*).

Coefficients K_d for NaCl, $CaCl_2$, $MgCl_2$, $NaHCO_3$ for 20 °C and concentrations of these electrolytes from infinite dilution to 0.1-1.0 n have, respectively, values

---

* In place of activities use can be made of concentrations only when dealing with diluted solutions since C do not take into account ion interactions. However, practical calculations of $\mathscr{E}_d$ and $\mathscr{E}_{da}$ generally use concentrations or specific resistances ($C = 1/\varrho$), since activities are hard to define.

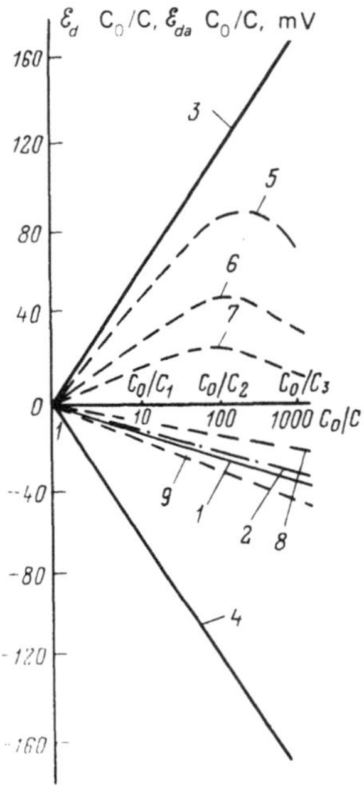

FIG. 71. Dependences of the diffusion and the diffusion-adsorption emf on the solution concentration ratio.

$\mathcal{E}_{dC_0/C} = f(\log C_0/C)$: *1, 2*—design and experimental for NaCl solutions; $\mathcal{E}_{daC_0/C} = f(\log C_0/C)$: *3, 4*—design for NaCl solutions and *5-9* experimental values for sandy-clayey rocks (from curve *5* to curve *9* rock clayiness decreases)

from (−11.6) to 13 mV, from (−19.6) to (−22) mV, from (−22.5) to (−23.5) mV, from 2.2 to 4.2 mV.

Diffusion-adsorption potentials and diffusion-adsorption emf coefficient. The values of electrochemical diffusion-adsorption potentials $\mathcal{E}_{daC_0/C}$ at the contact of different moist rocks or a rock and an electrolyte are governed both by adsorption and diffusion processes. Adsorption facilitates formation in the voids of moist rocks near the interface of an electric double layer. Since mobility of counterions and, in particular, coions of the electric double layer is averagely less than that of ions of the free electrolyte filling the voids, the average or effective mobilities of anions and cations of the electrolyte in the voids is less than their mobility in a free solution. The difference between these mobilities is more appreciable the greater is the fraction of the pore electrolyte taken by the volume occupied by the electric double layer.

The difference between electrolytic mobilities of anions and cations is also greater in the pore electrolyte compared with a free solution. If free electrolyte is practically nonexistent in a rock (as, for example, in clays), the potential-determining ions totally lose mobility, and it is solely counterions that take part in diffusion at a clay-electrolyte or clay-sand etc. contact as far as clay is involved. In this limiting case diffusion and adsorption potentials are the most different in magnitude and sign from diffusion potentials; they can be estimated (in millivolts) for a binary monovalent electrolyte (e.g. NaCl) in the voids of the rock and at its contact by using this relationship:

$$\mathscr{E}_{da\,C_0/C} = 2.3\cdot10^3 \frac{RT}{F}\frac{u' - v'}{u' + v'}\log\frac{C_0}{C}$$
$$= 2.3\cdot10^3 \frac{RT}{F}\left(\frac{u'}{u' + v'} - \frac{v'}{u' + v'}\right)\log\frac{C_0}{C} = K_{da}\log\frac{C_0}{C} \qquad (21)$$

where u' and v' are effective (average) electrolytic mobilities of electrolyte ions; $K_{da} = 2.3\cdot10^3(RT/F)(u' - v')/(u' + v')$ is the diffusion and adsorption emf coefficient.

If the voids of a moist rock are almost completely occupied by an electric double layer (as for natural adsorbents), and the anion is in the potential determining layer ($v' \approx 0$), then Eq. (21) takes on a simpler form:

$$\mathscr{E}_{da\,C_0/C} = 2.3\cdot10^3 \frac{RT}{F}\log\frac{C_0}{C} = 58\log\frac{C_0}{C} \qquad (22)$$

but the potential determining layer can contain also a positive ion (as is the case of some pelitomorphic limestones), then $u' = 0$, and

$$\mathscr{E}_{da\,C_0/C} = -2.3\cdot10^3 \frac{RT}{F}\log\frac{C_0}{C} = -58\log\frac{C_0}{C} \qquad (23)$$

A comparison of diffusion and diffusion and adsorption potentials calculated from Eqs. (20), (22), (23), for NaCl solutions and four C_0/C ratios of their concentrations (see Fig. 71, curves *3*, *4*) shows an appreciable difference.

Given average ion mobilities, generally characteristic of rocks, Eq. (21) does not determine $\mathscr{E}_{da}$ since we do not know these mobilities. An approximate calculation

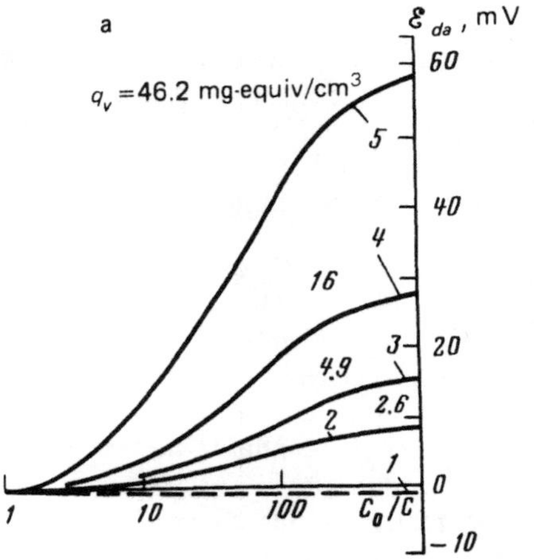

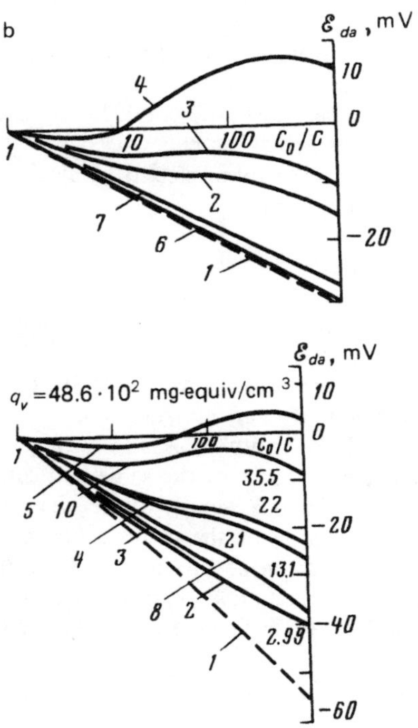

FIG. 72. $\mathscr{E}_{da} = f[\log(C_0/C)]$ relationships for different rocks (C_0 = 0.01 H) saturated by K^+ (*a*), Na^+ (*b*) and Ca^{2+} (*c*) cations (after B.Yu. Wendelstein).

1—$\mathscr{E}_d = f[\log(C_0/C)]$ relationship; *2-10*—$\mathscr{E}_{d.a} = f[\log(C_0/C)]$ relationships, respectively, for Urals refractive clay, kaolin, Gzhel clay, Crimean kill, marshallite, bauxite, Ukhta green clay, Djebol bentonite and askangel of $\mathscr{E}_{da}$ can be made by using more complex equations proposed by D. Shapiro, H. Hill and J. Milburn, B. Wendelstein and others; they enable one to obtain graphs of the dependence of $\mathscr{E}_{da}$ on C_0/C for any adsorptional capacities of rocks that are characterized by the quantities $\alpha = A(S_V/w_s)$ (where S_v/w_s is the relative specific surface; A is a coefficient indicative of the effect on α of the phase composition of rocks) or reduced exchange capacity q_V.

Analytical equations roughly determine $\mathscr{E}_{da\,C_0/C}$ and reveal only general regularities of the variation of this quantity depending on the variation of C_0/C, α, Q_V, q_V, etc. Owing to this fact, $\mathscr{E}_{da\,C_0/C}$ is generally measured in laboratory on rock samples.

Figure 71 represents graphs of experimentally found dependences of $\mathscr{E}_{da\,C_0/C} = f[\log(C_0/C)]$ for rocks (sandy-clayey or calcareo-magnesial) whose clayiness diminishes from curve *5* to curve *9*. Given the anion adsorption type (counterion layer—cations), experimental values of diffusion and adsorption potentials for specified C_0/C ratios are the more different from diffusion ones for the same concentrations, the higher is the clayiness of rocks since it is the clayiness that mainly governs S_V/w_t and q_V, and, consequently, average ion mobility.

However, given equal relative specific surfaces, one can get dramatically different values of $\mathscr{E}_{da\,C_0/C}$—maximum in the presence in the layer of counterions and free solution of potassium ions, and minimum in the presence in the same parts of the pore electrolyte of the Ca ion (Fig. 72). In this connection, in the general case the experimental dependences $\mathscr{E}_{da\,C_0/C} = f[\log(C_0/C)]$ also validate the fact that $\mathscr{E}_{da\,C_0/C} = f(A, S_V/w_r) = f(\alpha)$.

Electrochemical diffusion and adsorption activity. The capacity of rocks to be polarized at a contact with electrolyte or another moist rock and to produce under such conditions different diffusion and adsorption potentials is known as diffusion and adsorption activity.

The measure of electrochemical activity of a given rock sample is provided by the quantity

$$A_{da\,C_0/C_i} = K_{da\,C_0/C_i} - K_{d\,C_0/C_i} = \frac{\mathscr{E}_{da\,C_0/C} - \mathscr{E}_{d\,C_0/C_i}}{\log(C_0/C_i)}$$

where C_0 and C_i are electrolyte concentrations, respectively, in the rock's voids and at its contact.

A_{da} values are governed by the composition of the rock's phases, their structure, temperature and pressure.

The quantity $A_{da\,C_0/0}$, however, is not used for comparing electrochemical activities of rocks of the same and different types, since it is not constant even for a definite rock sample (decreasing with the growth of the C_0/C ratio) due to the imperfection of the method for measuring $\mathscr{E}_{da\,C_0/C}$ (see Fig. 71) and because C_0 concentration of different rock samples is, in practice, indeterminate.

A comparative characteristic of the diffusion and adsorption electrochemical activity of different rocks uses the quantity

$$A_{da} = K_{da} - K_d = \frac{\Delta\mathscr{E}_{da\,C_2\,\mathrm{to}\,C_1} - \Delta\mathscr{E}_{d\,C_2\,\mathrm{to}\,C_1}}{\log C_1/C_2} \tag{24}$$

It is found in this sequence:

(1) finding the increment

$$\Delta\mathscr{E}_{da\,C_2\,to\,C_1} = \mathscr{E}_{da\,C_0/C_2} - \mathscr{E}_{da\,C_0/C_1}$$
$$= K_{da}\log C_0/C_2 - K_{da}\log C_0/C_1 = K_{da}\log (C_1/C_2)$$

of diffusion and adsorption potentials at the same region of the experimental curve $\mathscr{E}_{da\,C_0/C} = f[\log (C_0/C)]$ where K_{da} is constant (region between concentration ratio C_0/C_2 and C_0/C_1, see Fig. 71);

(2) calculating the increment

$$\Delta\mathscr{E}_{d\,C_2\,to\,C_1} = \mathscr{E}_{d\,C_0/C_2} - \mathscr{E}_{d\,C_0/C_1}$$
$$= K_d\log (C_0/C_2) - K_d\log (C_0/C_1) = K_d\log (C_1/C_2)$$

of diffusion potentials at the corresponding region of the curve $\mathscr{E}_{d\,C_0/C} = f[\log (C_0/C)]$ (see Fig. 71);

(3) determining coefficients

$$K_{da} = \frac{\Delta\mathscr{E}_{da\,C_2\,to\,C_1}}{\log (C_1/C_2)} = \frac{\mathscr{E}_{da\,C_0/C_2} - \mathscr{E}_{da\,C_0/C_1}}{\log (C_1/C_2)}$$

$$K_d = \frac{\Delta\mathscr{E}_{d\,C_2\,to\,C_1}}{\log C_1/C_2} = \frac{\mathscr{E}_{d\,C_0/C_2} - \mathscr{E}_{d\,C_0/C_1}}{\log (C_1/C_2)}$$

(4) subtracting from the value of the coefficient K_{da} the quantity K_d.

Thus, diffusion and adsorption activity A_{da} is physicochemical quantity determining the capacity of rocks to produce unequal difference between the increments of the diffusion and adsorption and diffusion potentials per unit log of concentration ratios (or, better, activities) or specific electric resistivity of two contacting with the rock electrolytes, given different composition, temperature and pressure. For the anion adsorption type (C_0 = const, $C_1/C_2 = 10$) and a rock saturated and being in contact with NaCl solutions, the limiting value of the diffusion-adsorption activity for a rock with very high S_V/w_t (e.g. plastic clay)

$$A_{da} = \frac{\Delta\mathscr{E}_{da\,C_2\,to\,C_1} - \Delta\mathscr{E}_{d\,C_2\,to\,C_1}}{\log (C_1/C_2)} = K_{da} - K_d = \frac{58 - (-11.6)}{\log (C_1/C_2)} = 70 \text{ mW}$$

Given the same conditions and cation adsorption (a cation is found in the potential determining layer) $A_{da} = -46$ mW.

Sec. 32. Electrical Conductivity (Electric Resistance), Losses and Their Characteristics

Electrical conductivity (or conductivity) is the capacity of rocks for passing a current (electrical charges), *electrical resistance* is their capacity for preventing a current to flow through them. The process of conductivity reduces to the movement (drift, migration) of charge carriers (ions, electrons, holes) toward the corresponding electrodes through the agency of which an electric current is supplied to the medium concerned. We distinguish these current types: (a) conductivity (let-through), (b) absorption (associated with relaxation polarization types), (c) capacitive current.

Given through current, I_{thr}, free and weakly bound in an electric double layer ions, electrons and holes (if semiconductors or conductors enter the rock's composition) pass through the volume of the rock between the electrodes imparting them their charge directly (electrons and holes) or through electrochemical reactions (ions). Absorption nonthrough, decreasing with time, fairly appreciable current I_{abs} is due to the appearance of all slow (relaxation and migration) polarization types. At this, electrons or ions, having passed through a rock a definite distance, discontinue their directed motion. Absorption currents facilitate the heating of the sample and energy expenditure of the applied field; they have an active $I_{abs.a}$ and a reactive $I_{abs.r}$ components. As a result of displacement polarization and charging of a geometrical interelectrode capacity a purely capacitance current I_{dp} also flows through the rock. The total current is the resultant of the aforementioned current types.

In a constant field a standing current does not practically have as its components an absorption or displacement current. Losses in this case are associated with through conductivity. A constant current I is proportional to the intensity E of the electric field and cross section of the rock's sample:

$$I = \sigma F E$$

where the proportionality coefficient σ (in Cm/m) is *the specific electrical conductivity* characterizing the unequal capacity of different rocks for conducting a current. Since $E = U/l$ (where U is the electrical potential difference at a length l of a rock's sample), $I = \sigma(U/l)F$ and the resistance R (in Ohms) of the sample

$$R = U/I = l/\sigma F = \varrho l/F$$

Thus the electric resistance R of rock samples is in direct proportion to their length and in inverse proportion to the cross section. The proportionality coefficient ϱ (in Ohm·m) shows that given identical size, samples may have different conductivities. *The specific resistance* is a quantity determining the different capacity of dissimilar rocks for preventing a current from passing through a rock.

The number of charges carried across the unit cross section of a current conductor in time unit is found from the relation

$$\vec{j}_{av} = nq\vec{v}_{av}$$

where $\vec{j}_{av}$ is the average current density; n is the number of charge carriers in a unit volume; q is their charge; $\vec{v}_{av} = uE_{av}$ is a vector referring to the average velocity of a directed movement of charged particles ($u = q\lambda/2mv$ is the mobility of the charged particles; λ is the average length of the free path of electrons; m is the electron mass; v is the average speed of the thermal motion of the electron inside the conductor).

Since the current density $\vec{j}_{av} = \sigma E_{av}$, the specific electrical conductivity $\sigma = nqu = qn\lambda/2mv$ depends on the type and state of the conducting rock. However, current density $\vec{j}_a$ in relatively poor conducting rocks that is set up in an alternating electromagnetic field is a vector sum of its active $\vec{j}_a = \vec{i}_{thr} + \vec{j}_{abs.a}$ and reactive $\vec{j}_r = \vec{j}_{dp} + \vec{j}_{abs.r}$ components, i.e. $\vec{j}_n = \vec{j}_a + \vec{j}_r$ (Fig. 73). At this, the active current density component $\vec{j}_a$ (density $\vec{j}_{thr}$ and $\vec{j}_{abs.a}$ of, respectively, active

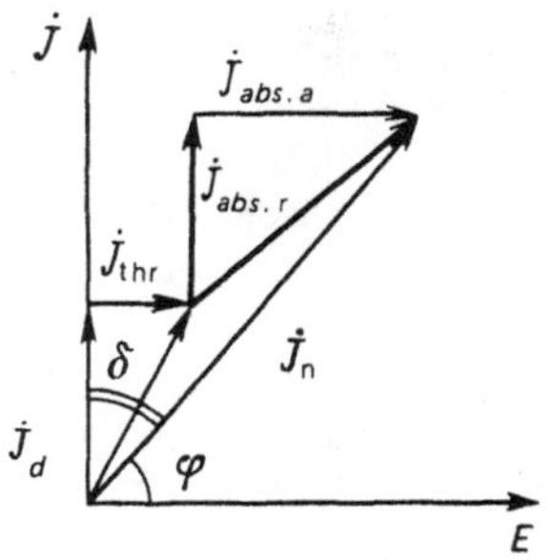

FIG. 73. Vector diagram of current and stress distribution in rock

through and relaxation currents) coincides in phase with the intensity E of the field applied to the rock's sample, and the reactive current density components $\vec{j_r}$ (densities $\vec{j_{dp}}$ and $\vec{j_{abs.r}}$ of, respectively, capacitance and reactive currents) outstrip it by $\pi/2$ (Fig. 73).

In d.c. fields as in a.c. fields Ohm's law is valid, which takes on, however, a differentiated form

$$\vec{j_n} = \sigma^* E$$

where

$$j_n = I_d/F \quad \text{and} \quad \sigma^* = \sigma_a + i\omega\varepsilon_0\varepsilon = \sigma_a + i\omega\sigma_a\tau_0 = \sigma_a + i\sigma_r \qquad (25)$$

is the average value of the complex specific electric conductivity. The quantity σ^* takes into account that an alternating field includes a through current and a displacement current. In the latter equation $\tau_0 = \varepsilon_0\varepsilon/\sigma_a$ is the time of rock relaxation determining the speed of attaining electrical equilibrium in a medium following field superposition; ε and ε_0^* are the dielectrical permeability of the medium and the electrical constant equal to 8.854 pF/m; ω is the current angular frequency; $i = \sqrt{-1}$ is an imaginary unit.

Thus the complex specific electrical conductivity σ^* of rocks is governed not only by a through current but also by currents conditioned by all polarization types that have developed.

Rocks demonstrate all current components since all conducting (e.g. grains of native metals) alternate with insulators (e.g. quartz) and electrolyte (native water).

Upon superposition upon a rock of an alternating electrical field a fraction of its energy is lost (converted to heat). These losses often much exceeds similar losses in a stationary field and are called dielectrical losses. These latter fall into losses due to conductivity, relaxation, governed by thermal polarization and associated with other polarization types. *Dielectrical losses* are characterized by a ratio of active current to reactive current or the value of the angle δ of dielectrical losses in a current and voltage triangle (see Fig. 73). These values are more often used as

---

* $\varepsilon_0 = C_0/G_0$ where C_0 and G_0 characterize the geometrical dimensions of rocks and are found from relations $C = \varepsilon C_0$ and $G = \sigma G_0$. Here C is capacitance and G is conductivity.

electric characteristics for rocks in alternating fields than the complex electrical conductivity σ^*. The ratio of active to reactive current density components defines

$\tan \delta = \vec{j}_a / \vec{j}_r = (\vec{j}_{thr} + \vec{j}_{abs.a})/(\vec{j}_{dp} + \vec{j}_{abs.r})$

The angle δ is complementary to 90° relative to the phase shift angle between full alternating current flowing through a capacitor with a dielectrical sample and voltage across its plates.

Thus, rocks let an electric current flow through them, are polarized due to its effect and at the contact with electrolyte or upon seepage of the latter through these media. Passage of a current (notably, alternating) through rocks is followed by energy losses. The different participation of rocks in these processes are characterized by: specific electrical conductivity σ (or specific resistance ϱ), dielectrical permeability ε, tan angle of dielectrical losses tan δ, induced A_i, diffusion-adsorption A_{da} and infiltration (seepage) A_f activity. The value of all of these quantities vary over a wide range conditioned by the original composition, structure, thermobaric conditions of the formation of rocks, variations of these features and conditions of occurrence of rocks during their lifetime.

Electrical quantities are closely connected with one another and with a number of other crucial petrophysical quantities. That is why it is not uncommon that we can evaluate sufficiently accurately one electrical quantity from another one.

Sec. 33. **The Gaseous Phase**

Let us now consider: (1) electrical properties of the gaseous, liquid and solid phases of rocks; (2) effect of the substance and phase composition, distribution in the volume of the components, their interaction, thermobaric conditions on the values of the electrical quantities of rocks; (3) connections between the electrical and other petrophysical characteristics.

If the gaseous phase is represented by air, an idea of its specific electrical conductivity σ_{air} can be inferred by considering data on the conductivity of the atmosphere which is mainly governed by the presence in the latter of ions:

$$\sigma_{air} = e\left(\sum_i n_i u_i + \sum_j n_j u_j\right)$$

where e is the electron charge; n_i, n_j, u_i and u_j are concentrations and mobilities, respectively, of the positive and negative ions. The value of σ_{air} is also affected

TABLE 8. Dielectrical Permeability of the Air

Temperature, °C	Pressure, MPa	Frequency of polarizing current	ε
0	1	$<3\cdot10^6$	1.00590
19	1	$3\cdot10^6$	1.000576
19	20	$>3\cdot10^6$	1.0108

by the humidity and turbulence of the atmosphere. The average values of the specific electrical resistance of the atmosphere near the earth's surface is 10^{14} Ohm·m. Gaseous hydrocarbons also show similar values of the same quantity.

The dielectrical permeability of the air and gaseous hydrocarbon mixtures is close to unity (Table 8). This is due to the small density of gases and the fact that their molecules are nonpolar, therefore, ε of gases is mainly governed by the inappreciable electron polarization independent of the frequency.

The induced electrochemical activity is not determined in gases since their electron polarization disappears before it can be recorded in laboratory and in field conditions.

Sec. 34. **The Liquid Phase**

The specific resistance of oil is very great (attaining as much as 10^{14} Ohm·m). As to waters filling the voids space, the values of this quantity differ over a very great range depending on the mineralization and temperature, and to a lesser degree by the composition of the waters. The mineralization of these latter occurs following dissolution in water with the specific conductivity 2×10^5 Ohm·m of halite, sylvine, bischofite etc. The dependence of the specific resistance ϱ_w of natural solutions on their mineralization and temperature is easy to follow for aqueous halite solutions (Fig. 74), halite being the most common of soluble minerals with $\varrho_m = 10^{14}$Ohm·m. As can be seen from these data, the specific conductivity of a NaCl solution with a concentration $C_V = 200$ kg/m^3 ($\varrho_{w18\,°C} = 0.055$ Ohm·m) is 1 000 times less than the resistance of a solution with a concentration 0.1 kg/m^3 ($\varrho_{w18\,°C} \approx 55$ Ohm·m).

The effect of temperature on ϱ_w is less. Over the range 20-200 °C ϱ_w can, for example, for a NaCl solution of a specified concentration decrease sixfold (Fig. 75). The decrease of ϱ_w with temperature is attributed to the decreased viscosity of the pore water (see Fig. 75) and hydration of ions of its electrolyte enchancing its mobility and, consequently, conductivity of the electrolyte.

The composition of the waters affects ϱ_w only in a minor way. This is especially true if the waters do not contain OH^- or H^+ ions with anomalously high mobility. Given identical concentration, normal NaCl, KCl, $CaCl_2$ and $MgCl_2$ solutions have an almost identical specific resistance. However, given the same concentration, the specific resistance of H_2SO_4 solutions is much less (see Fig. 74). Even though the specific resistance of solutions decreases with increasing their concentration, this decrease is the less the higher is electrolyte solution concentration, since the ion mobility diminishes in this process as does dissociation degree for weak acid salts. The pattern of the dependence of the specific resistance of salt solutions on their chemical composition is governed by the relation

$$\varrho_w = 10/\Sigma(C_a v f_a + C_c u f_c)$$

where C_a and C_c is the number of gram-equivalents of anions and cations in the solution; v and u are their electrolytic mobilities; f_a and f_c are electrical conductivity coefficients governed by dissolved salt concentration and chemical composition.

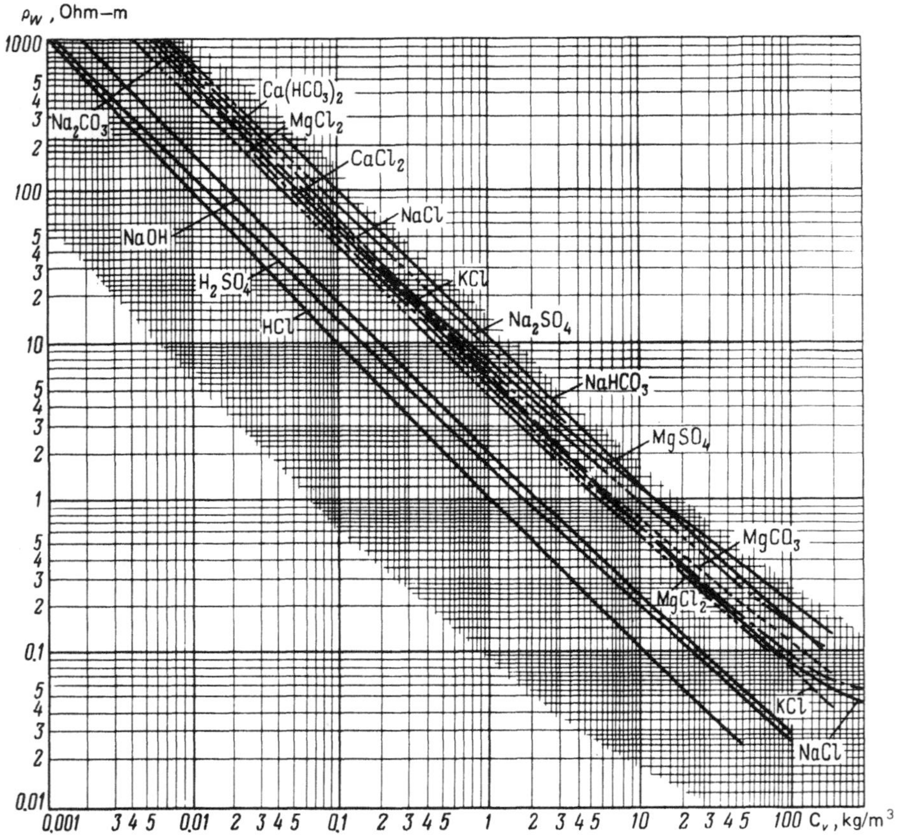

FIG. 74. Dependences of the specific resistance ϱ_w of electrolyte aqueous solutions on their volumetric concentration C_V at 20 °C

Since for certain salt types of stratal waters the difference in the specific resistance due to varying mobility of their ions is inappreciable (see Fig. 74), ϱ_w can be calculated from the total mineralization of solutions and mobility of ions of the predominant electrolyte (in natural waters this is generally NaCl).

Mineralization of natural waters varies from <0.1 to ≈ 500 g/l, and their temperature from <10 to >200 °C, ϱ_w varies accordingly from 10^{-2} to 10^3 Ohm·m. The minimum ϱ_w about hundredth fractions of the ohm-metre refer to highly mineralized waters of oil and gas deposits and some types of ore deposits. The specific resistance of the pore waters of oil- and gas-bearing and other deeply occurring deposits often diminishes with age, since their mineralization generally increases in this direction and the chemical composition changes. Maximum resistance from 8 to 10^3 Ohm·m is demonstrated by subsurficial ground, exclusively pure river and

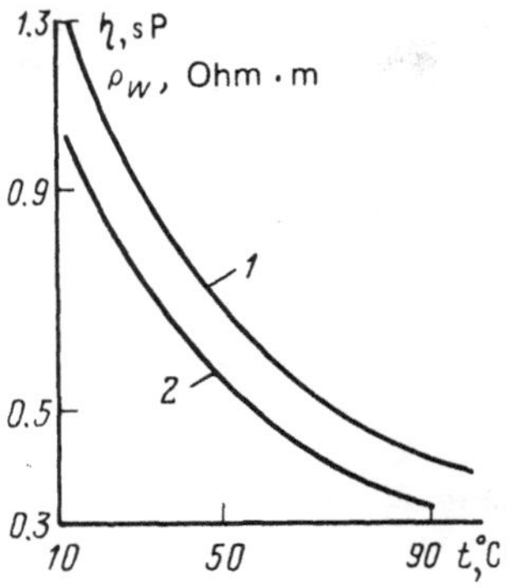

FIG. 75. Dependences of the specific resistance ϱ_w (*1*) of NaCl solution ($C_{NaCl} \approx 5$ g/l) and viscosity η (*2*) on temperature t

rain waters. The specific resistance of sea water alternatively known as salt water is about 0.2 Ohm·m.

Natural-water and oil (or gas) mixtures, or mixtures of all the three components have a lesser specific resistance, the greater is the fraction of their volume occupied by the natural water and the higher is the mineralization and temperature of the mixture.

The dielectrical permeability of water is governed by the concentration and composition of the salts dissolved in the water. According to H. Falkenhagen, for binary electrolytes

$$\varepsilon = \varepsilon_0 + 3.79\sqrt{C}$$

where ε is the dielectrical permeability of the solution; ε_0 is the dielectrical permeability of pure water; C is the concentration of the solution in moles per 1 dm^3.

The dielectrical permeability of water and oil mixtures is lower compared with that of these components taken separately (water has $\varepsilon_w \approx 80$ and oil has $\varepsilon_0 \approx 2$) and is governed by their volumetric ratio. The induced electrochemical activities of water and oil are close to zero since polarization types typical of these substances (electron and heat dipole for water and electron for oil) drop before they can be recorded by the developed methods. There is no way for determining these materials.

Sec. 35. **Minerals**

The specific resistance ϱ_m of most investigated minerals varies over a very wide range from 10^{-6} to 10^{15} Ohm·m. Minerals with higher or lower ϱ_m occur rarely. The dielectrical permeability of minerals varies from some 3 to >150 ε_m units. The wide range of variation of ϱ_m and ε_m values is accounted for by the appreciable differences in their composition and structure. The induced and diffusion and adsorption electrochemical mineral activities have been insufficiently understood. Some investigators label by a plus sign minerals with enhanced polarizability (their mixtures appreciably increases A_i of rocks), and by a minus sign their poorly polarizing varieties. Enhanced-polarizability minerals include native metals, graphite, many sulphides and some of the oxides.

In terms of ϱ_m and ε_m values, the character of processes in which they show themselves and the conditions of recording these quantities minerals can be divided into three groups. The first includes native metals, their natural paragenetic sequences, graphite; the second, a number of sulphides, their analogs, some oxides; the third, native nonmetals (sulphur, selenium, diamond) excepting graphite, silicates, carbonates, chlorides, phosphates and minerals of a number of other classes we have not specified except some of their representatives. All first-group minerals have ϱ_m values usually very small ($n \cdot 10^{-8}$ to 10^{-5} Ohm·m), and ε_m great (tending to infinity), a number of them are well polarized. These minerals provide natural conductors demonstrating an inappreciable specific resistance (a greater specific conductivity). The electric current in crystals of such minerals carries free electrons.

Free electrons are also contained in natural semiconductors: most sulphides (covellite, bornite, pyrrhotite, chalcopyrite, etc.), arsenides (nickeline or niccolite, skutterudite etc.), selenides and some of oxides (magnetite, hematite, cuprite etc.).

The specific resistance of second-group minerals is generally of the range $>10^{-6}$ to 10^{11} Ohm·m close to that of semiconductors ($>10^{-6}$ to $<10^{8}$ Ohm·m) manifesting electron and hole conductivity types. The dielectrical permeabilities of many of these (e.g. arsenopyrite, chalcosine alternatively known as chalcocite, galenite, molybdenite, pyrrhotite, ice, water, rutile, pyrolusite etc.) are more than 80 and sometimes much higher than ε_m of minerals of other classes. A number of them demonstrate an increased polarizability even though oxides include poorly polarized varieties.

High and increased values of ε of sulphides characterized by electron and ion displacement polarization are accounted for by a greater radius of the sulphur ion contributing to its appreciable polarization as well as by that their composition includes strongly polarized cations of metals (e.g., lead whose external shell has 18 electrons, copper, silver, iron etc.). The high ε of oxides appears to be associated with the presence of the oxygen ion. It has a sufficiently high polarizability and appreciable ratio to the ion radius which determines their packing density in the crystal with growth of which ε increases. The enhanced ε values are generally characteristic of oxides whose composition, apart from oxygen, includes such metals as Pb, Fe, Pt. The high values have also been established for braunite, pyrolusite, rutile, perovskite, anatase and brookite having Mg or Ti. For rutile and perovskite it is accounted for by a specific structure of the aforementioned minerals, and for water by that its molecules representing rigid dipoles are readily aligned with the field. The appreciable orientational polarization of water drops drastically with growth of the temperature and frequency of the field.

However, among sulphides and oxides one can encounter varieties with decreased ε values (e.g., sphalerite, cinnabar, antimonite, cuprite etc.) which is accounted for by the presence for some of them of solely electron displacement polarization (cuprite, euxenite) and high ϱ values due to the small number of free electrons.

The specific resistance of the third-group minerals has been poorly studied. Most investigated minerals (quartz, calcite, anhydrite, fluorite, sulphur, diamond) refer to dielectrics (insulators), their ϱ_m varies from 5×10^7 to 2.7×10^{16} Ohm·m and is generally greater than 10^{11} Ohm·m. Insulators are also provided by some halite, sylvine and biotite varieties. Lesser values of the specific resistance are shown by aegirine, serpentine, siderite and some other minerals.

At radio-frequencies, minerals of the group under study demonstrate electron polarization (e.g. diamond, selenium, sulphur) or electron and ion polarization; their ε are generally found in the range 4-6 or 6-8. It may vary, but to a lesser degree of likelihood, from 8 to 12. Strongly polarized minerals include diamond (ε = 16), sericite (hydromica) (ε = 19.6 to 25.4), cerussite (ε = 19.6 to 25.4) and anglesite (ε = 14). The composition of the latter two mineral species includes lead.

Thus, weakly polarized dielectrics include many rock-forming sedimentary rock minerals: quartz, feldspars (muscovite, biotite, phlogopite), halite, sylvine (sylvite), anhydrite, calcite, dolomite, sulpur, fluorite etc.

The mechanism of the electrical conductivity of dielectrics is of the ion type. Current carriers—light alkaline metal ions located either in the crystal lattice or in the mineral impurities—jump from the node to a local vacancy or internode ions move. The ion conductivity of ion crystals is associated with the imperfection of their crystal lattice and pattern of distribution of its defects.

All the three groups contain minerals showing a wide range of variation of the specific resistance of their individual varieties. E.g., the specific resistance of pyrite can vary in value from $<10^{-6}$ to 10^2 Ohm·m, that of magnetite from 10^{-4} to 10^4 Ohm·m. This is accounted for by impurities. They can disturb the structure of the minerals' crystal lattice, create in the polycrystal a network of paths with a conductivity that differs from that of the bulk of the mineral (owing to the nonuniform composition, structure and conductivity of the bulk of crystals and impurities), and coat individual crystal grains by a nonconducting or conducting shell. For example, the conductivity of chalcosine (chalcocite) increases by several orders of magnitude, given a minor excess of sulphur, due to the formation of current conducting paths of metallic conducting CuS.

According to the results presented by A.S. Semenov, the resistance of single crystals and polycrystalline mass of magnetite differs by several orders of magnitude due to the presence of very thin nonconducting intercalations between magnetite crystals. The poorly conducting impurities of sulphides and their analogs include quartz, sphalerite, carbonates and other minerals accompanying ore minerals. The higher the resistance compared with that of the bulk of the mineral the greater is the effect of the impurities.

Intercrystalline intercalations different from crystals in structure, composition, physical and chemical properties form during crystallization or following it under the action of chemical or physical factors and may differ in thickness from macroscopic to molecular values. Intercalations also form during the deformation of crystals. Then at the disrupted sites products of decay of solid solutions or the crystal lattice structure is disturbed.

An inappreciable number of impurities is sufficient for continuous films to form about crystals and electrically connected threadlike penetrations.

Single crystals of minerals are anisotropic. Due to this, their specific resistance is characterized by three values ϱ_{11}, ϱ_{22}, ϱ_{33}, determined in the direction of the three principal crystallographic axes with which the *X*, *Y*, *Z* coordinate system axes are caused to agree. If the coordinate axes fail to be aligned with the three principal crystallographic ones, the specific resistance of the single crystal is characterized by its nine components. Polycrystalline minerals have no electrical anisotropy owing to the different orientation of crystalline grains.

Dielectrical losses have been insufficiently studied. The values of tan δ vary from 2×10^{-4} (gypsum) to 1.5×10^{-1} (microcline). They have been obtained at a frequency 10^4-10^5 Hz and temperature 20-30 °C. The smallest tan δ values (periclase and spinel) correspond to dense ion packing in the crystal and its high symmetry. The imperfections of crystal structure and impurities cause increased losses. The dielectrical losses in single crystals are less than in their polycrystal varieties. The values of tan δ generally increase for minerals containing crystallization water. A particularly high tan δ must be shown by most native metals, graphite, sulphides and many oxides due to their high conductivity.

As a result of the heating of natural conductors electron concentration remains unchanged. It affects only inappreciably the speed of their thermal notion. However, oscillation amplitudes of the crystal lattice ions increase. This prevents a directed flow of electrons in the metal which results in a decrease of their average free path length λ and decreased mobility u and, consequently, electrical conductivity. Given a temperature t, the specific resistance of metal is

$$\varrho_t = \varrho_0(1 + \alpha\Delta t + \beta\Delta t^2)$$

where ϱ_0 is the magnitude of the specific resistance at the beginning of the temperature increment interval Δt; α and β are temperature coefficients of resistance.

The conductivity of minerals, semiconductors and dielectrics increases with temperature: that of semiconductors due to increased electron concentration in the conduction band, holes in the valence band and, in a lesser degree, due to the increased mobility of current carriers, that of dielectrics due to the growth of ion mobility and their kinetic energy. This latter contributes to a more intensive removal of ions from the crystal lattice. The increased conductivity σ_t of semiconductors and dielectrics with growth of temperature is found, respectively, from the relations:

$$\sigma_t = \sigma_0 \exp(-E_0/2kT)$$

and

$$\sigma_t' = \sigma_0 \exp(-E_0/kT)$$

where E_0, σ_0 are activation energy and conductivity coefficient for $1/T = 0$ (T is the absolute temperature); k is the Boltzmann constant.

The pattern of the drop in the specific resistance with temperature is governed by the chemical composition and structure of minerals. The conductivity of minerals is enhanced by the presence in their composition of Fe^{3+}, Al^{3+}, Mg^{2+} and Be^{2+} cations. Experimental data have been obtained upon varying the temperature from 200 to 1 000 °C and in the presence of a mineral in the air.

Dehydration, decarbonization and degraphitization facilitate the appearance of various anomalies in the log $\varrho = f(1/T)$ dependence graphs. The dependence of σ_m and ϱ_m of minerals on pressure has been studied to a lesser degree. However, according to the experimental data, the pressure in the range 0.05-2 GPa causes σ_m and ϱ_m to a much lesser extent than temperature variation over the range 100-1 000 °C.

For minerals where current carriers are provided by monovalent ions showing an appreciable ion radius a growth of resistance ϱ_m with pressure has been established. The presence in the composition of minerals or as impurities of iron cations leads to a decreased ϱ_m with growth of pressure. Few data are available on the specific resistance of minerals involving different oxides under conditions of the concurrent action of temperature and pressure. With growth of temperature the dielectrical permeability of minerals first remains constant or increases somewhat followed by its intensive increase. The greater the range of temperature variation where ε_m remains practically unchanged the higher are the frequences.

The dielectrical permeability of minerals generally increases with growth of pressure since in this case a unit volume will contain more polarized particles.

No data are as yet available on how tan δ, A_i and A_{da} of minerals vary with temperature and pressure.

Sec. 36. **The Solid Phase**

The solid phase of rocks is generally polymineral and is most commonly represented by natural dielectrics (quartz, feldspars, calcite, dolomite, olivine, pyroxenes etc.) mixtures. Note that its specific resistance is very great (may exceed 10^{11} Ohm·m), and the dielectrical permeability (at radio-frequences) is relatively small owing to high and close in value specific resistances and low dielectrical permeabilities of principal and secondary minerals of this rock component. Less often (much as for ores) the solid phase of rocks is represented by dielectrical and conductor (or semiconductor) minerals. In this case its average specific resistance $\varrho_{s1,2}$ and dielectrical permeability $\varepsilon_{s1,2}$ are governed by specific resistances ϱ_{s1} and ϱ_{s2} and dielectrical permeabilities ε_{s1} and ε_{s2} of conducting and nonconducting components, their volumetric fractional content w_{s1} and $w_{s2} = 1 - w_{s1}$, the shapes and distribution of dielectrics (or semiconductors) in the solid phase.

The dependences $\varrho_{s1,2}$ on ϱ_{s2} and ϱ_{s1}, $\varepsilon_{s1,2}$ and ε_{s1} as well as of these two quantities on the w_{s1} and w_{s2} content and component distributions in the mixture have been studied in sufficient detail theoretically whereas available experimental data are few.

In theoretical studies with the objective to derive relationships permitting one to evaluate $\varrho_{s1,2}$ and $\varepsilon_{s1,2}$ of a heterogeneous two-component solid phase, given specified volumetric components ratios, shape and volumetric distribution, the solid phase has been represented as simplified models.

Matrix models represent packings of uniformly distributed in space spheres, cubes, ellipsoids of revolution of the same or different size, extremely flattened and extended ellipsoids of revolution, cuboctahedra (Fig. 76), rectangular plates and other geometrical figures. Their pore space is completely filled by another component of the medium, alternatively these figures are separated by a second component. A two-component solid phase has also been represented by statistical mixtures with two components.

The material of inclusions of this medium has been assigned a resistance ϱ_{s2} or its dielectrical permeability ε_{s2}, and the filling component or the second compo-

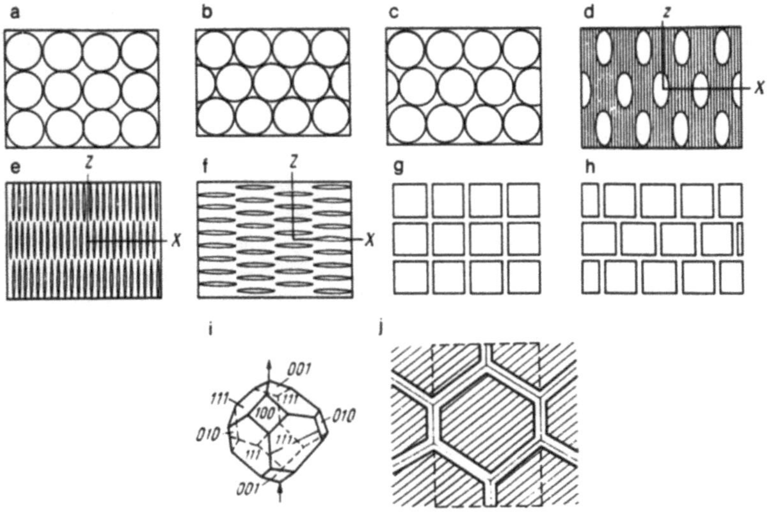

FIG. 76. Diagrams of perfect structures of a two-component solid phase of rocks.

Solid phase: *a-c*—with spherical inclusions of identical size, respectively, given loose, dense and intermediate packing; *d*—with ellipsoidal inclusions uniformly distributed over the volume; *e*—with needle-like inclusions; *f*—with oblate disks; *g*—with cubic inclusions; *h*—with cubic inclusions packed after a masonry pattern; *i*—with equal-edged cuboctahedron inclusions; *j*—section by a diagonal plane of a unit cell of the solid phase of rock (its outline being shown in the diagram by a dashed line) with cuboctahedral grains (after A.S. Semenov)

nent of the statistical mixture has been assigned, respectively, ϱ_{s1} and ε_{s1}. The values of ϱ_{s2} have first been specified high ($\sim 10^4$ Ohm·m) or very high, close to infinity, and ϱ_{s1} relatively small. Then, vice versa, ϱ_{s2} was inappreciable, and ϱ_{s1} very significant. The same procedure was followed in deriving equations to calculate average dielectrical permeabilities. Dielectrical permeabilities of components of statistical mixtures also differed very much.

The equations making it possible to calculate $\varrho_{s1,2}$ and $\varepsilon_{s1,2}$ using matrix models have been obtained by different methods. Let us consider, for example, the derivation of relations determining $\varrho_{s1,2}$ of the medium if one of its components has a cubical packing of inclusions (Fig. 77) or packing of n sphere series and the other is found in their pore space. Assume that faces of cubic inclusions are mutually parallel and the cubes are separated by a second component with a minor specific resistance. Suppose that sides of a model are $l = 1$ in length and the latter is divided into structural cubical elements a on the side. Each of these elements contains a cubical grain surrounded by a layer of the conducting component $\delta/2$ in thickness with the specific resistance ϱ_{s1} (see Fig. 77). Then the resistance of the model $R_{s1,2}$ equals its specific resistance $\varrho_{s1,2}$ and, consequently,

$$\varrho_{s1,2} = R_{s1,2} = l/F\varrho_{s1} = \frac{\varrho_{s1}}{(2\delta/a) - (\delta^2/a^2)} \qquad (26)$$

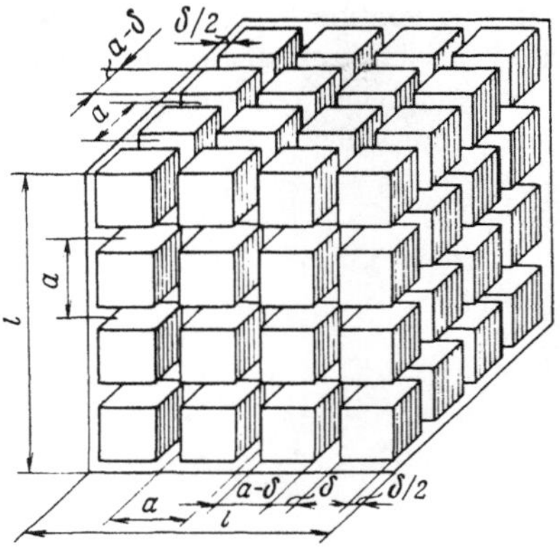

FIG. 77. A model of a two-mineral solid phase of a rock with cubic inclusions used to derive an equation determining the specific electric resistance

here the area of the conducting component between the cubical inclusions

$F = a^2 - (a - \delta)^2 N$

$N = l/a^2$ is the number of cubic inclusions accommodated in the cross section of the model passing the current.

Let us determine the relationship between δ/a and the specific volumetric content of w_{s1} of the filling component of the model:

$$w_{s1} = \frac{a^3 - (a - \delta)^3}{a^3} = 3\frac{\delta}{a} - 3\frac{\delta^2}{a^2} + \frac{\delta^3}{a^3}$$

This cubic equation is solved by substituting $Y = (\delta/a) - 1$:

$$w_{s1} = 3(1 + Y) - 3(1 + Y)^2 + (1 + Y)^3 = 1 + Y^3$$

Hence $Y = \sqrt[3]{w_{s1} - 1}$ and $\delta/a = 1 + \sqrt[3]{w_{s1} - 1}$.

By substituting δ/a into Eq. (26) we get

$$\varrho_{s1,2} = \varrho_{s1}/(1 - \sqrt[3]{(w_{s1} - 1)^2})$$

Suppose that a two-component model of a rock's solid phase after filling it by a series of the largest inclusions has a volume w_{s0} containing smaller spherical inclusions and the second component of the model having a total resistance ϱ_3'. Given $\varrho_2 \to \infty$, the specific resistance of such a model is

$$\varrho_{s1,2} = \left(\frac{3 - w_{s0}}{2w_{s0}}\right) \varrho_3'$$

which follows from the relationship

$$\varrho_{1,2} = \frac{w\varrho_1 + (3 - w)\varrho_2}{2\varrho_2 w + (3 - w)\varrho_1}\,\varrho_1 \qquad (27)$$

obtained by Maxwell for $\varrho_{1,2}$ of a packing of spheres of the same size having a specific resistivity ϱ_2 which is much greater than that ϱ_1 of its second component filling the space between the spheres and occupying volume w. Since the volume w_{s0} is, in its turn, filled to attain a 1-w_{s0} fraction of inclusions of the next smaller size and the remaining space contains its second component consisting of a third and remaining series of inclusions and the model's component between the spheres having a resistance ϱ_3'', then

$$\varrho_3' = \frac{3 - w_{s0}}{2w_{s0}} \varrho_3''$$

Then, proceeding along the same lines, we get

$$\varrho_3'' = \frac{3 - w_{s0}}{2w_{s0}} \varrho_3'''$$

. . . .

$$\varrho_3^{n-2} = \frac{3 - w_{s0}}{2w_{s0}} \varrho_3^{n-1}$$

and, finally,

$$\varrho_3^{n-1} = \frac{3 - w_{s0}}{2w_{s0}} \varrho_{s1}$$

where ϱ_{s1} is the specific resistance of the filling component between the series of spheres.

By removal from the equation set obtained of ϱ_3', ϱ_3'', ..., ϱ_3^{n-1}, we have

$$\varrho_{s1,2} = \left(\frac{3 - w_{s0}}{2w_{s0}}\right)^n \varrho_{s1}$$

The value of the index n is found from the following apparent relationship

$$w_{s0}^n = w_{s1}$$

where w_{s1} is the specific volumetric content of the filling component in the solid phase with n spherical inclusion series. Thus

$$n = \log w_{s1}/\log w_{s0}$$

and

$$\varrho_{s1,2} = \left(\frac{3 - w_{s0}}{2w_{s0}}\right)^{\log w_{s1}/\log w_{s0}} \varrho_{s1}$$

In like manner the equations to determine the specific resistance of the solid phase of rocks with n series of ellipsoidal inclusions are derived. In so doing the basis is provided by equations obtained by I. K. Ovchinnikov for the solid phase of a rock with identical-size ellipsoidal inclusions arranged according to a scheme shown in Fig. 76*d*.

A different method was used to get a number of other formulae permitting the evaluation $\varrho_{s1,2}$ of the solid phase, given its different composition and structure. Let us compare and analyse some of them (Table 9). By comparing the relationships presented in Table 9 we can draw the following conclusions: (1) the average specific

TABLE 9. Equations Permitting the Calculation of the Average Specific Resistance $\varrho_{sph1,2}$ of the Two-component Solid Phase, Given Its Different Structure and Composition ($\varrho^{*}_{sph2\to\infty}$, ϱ_{sph1} is inappreciable)

Solid phase model	Equation governing $\varrho_{sph1,2}$	Author
Packing of unidimensional spheres**	$\varrho_{sph1,2} = \frac{3 - w_{s1}}{2w_{s1}}\varrho_{sph1} = P_{sph1,2}\varrho_{sph1}$	M.A. Maxwell
Packing of unidimensional cubes	$\varrho_{shp1,2} = \frac{1}{1 - \sqrt[3]{(w_{s1} - 1)^2}}\varrho_{sph1} = P_{sph1,2}\varrho_{sph1}$	V.N. Dakhnov
Packing of unidimensional ellipsoids of revolution	$\varrho_{sph1,2Z} = \frac{[k - (1 - w_{s1})]}{kw_{s1}}\varrho_{sph1} = P_{sph1,2}\varrho_{sph1}$ $\varrho_{sph1,2X} = \varrho_{sph1,2Y} = \frac{[l - (1 - w_{s1})]}{lw_{s1}}\varrho_{sph1}$ $= P_{sph1,2X}\varrho_{sph1} = P_{sph1,2Y}\varrho_{sph1}$, where k and l are coefficients of geometry varying from $k \to 0$ and $l = -\infty$ (very much flattened ellipsoids of revolution) to $k \to \infty$ and $l \to -1$ (very much elongated ellipsoids of revolution). For spherical grains $k = l = -2$	I.K. Ovchinnikov
Packing of n series of spherical inclusions	$\varrho_{sph1,2} = \left(\frac{3 - w_{s1}}{2w_{s1}}\right)^{\log w_{s1}/\log w_{s0}}\varrho_{sph1}$ $= P_{sph1,2}\varrho_{sph1}$	A.S. Semenov

* The designation ϱ_{s2} is also possible (see list of symbols).
** The second component of the solid phase fills the voids space of packings.

resistance $\varrho_{s1,2}$ of a two-component solid phase, given $\varrho_{s2} \to \infty$ and $\varrho_{s1} \to 0$, is in direct proportion to the specific resistance ϱ_{s1} of the component filling the space between inclusions of the model; (2) the values of the proportionality coefficient $P_{s1,2}$ which can be called a parameter of composition and structure and governed by the fractional volume content w_{s1} of the filling component and sometimes by the coefficient k and l of the geometry of the inclusions. However, given specified w_{s1}, k and l, they are different for different rock solid phase models which is suggested by the different form of the dependence of the composition and structure parameter on w_{s1}. Given $\varrho_{s2} \to \infty$ and $\varrho_{s1} \to 0$, this conclusion holds good also for many other investigated models of the solid phase of rocks. This cannot be done if specific resistances both of the one and the other component are finite (for example, given $\varrho_{s2} = 10^4$ Ohm·m and $\varrho_{s1} = 10^{-3}$ Ohm·m). Under conditions of this and similar proportions of specific resistances of the components of the model the relation for $P_{s1,2}$ includes also such quantities as ϱ_{s2} and ϱ_{s1}. An example is provided by Eq. (27).

By evaluating the pattern of variation of $P_{s1,2}$ values depending on the content w_{s1} of the filling component in the solid phase of rocks and distribution of its

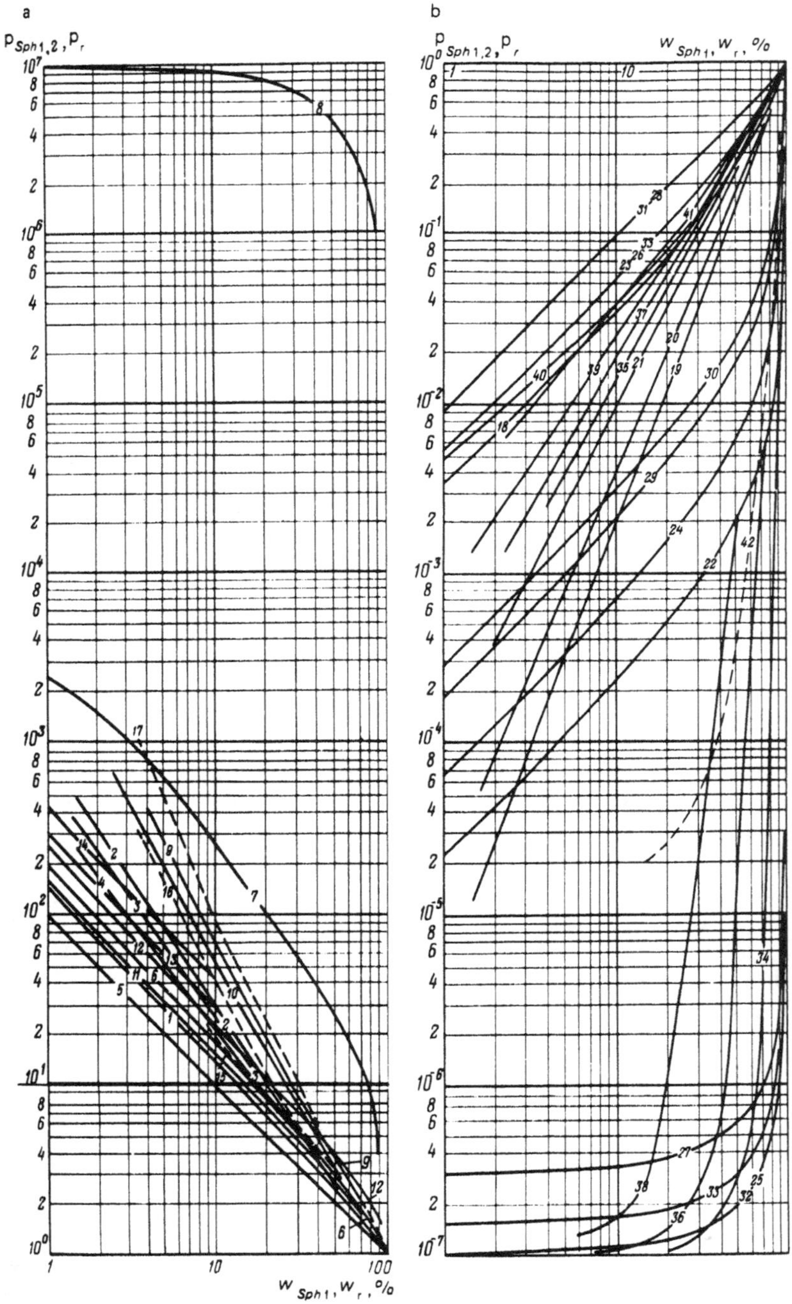
a
b
$p_{Sph1,2}, p_r$
w_{Sph1}, w_r, %

components throughout the model's volume, let us represent graphically and compare the results of the calculation of $P_{s1,2}$ using relationships presented in Table 9 and elsewhere (Fig. 78). A comparison of graphs in Fig. 78 shows that, given $\varrho_{s2} \to \infty$, $\varrho_{s1} \to 0$ and w_{s1} growing from 0 to 1, the quantity $P_{s1,2}$ varies from values close to infinitely great to unity. The pattern of decreasing $P_{s1,2}$ with increasing the content w_{s1} of the filling component is dramatically different for models with isometrically equal and different-dimension inclusions (spheres, cubes, cubooctahedron) and nonisometric inclusions (ellipsoids of revolution, extremely flattened or extended ellipsoids of revolution converted to disks or needles one of the principal dimension of which appreciably exceeds the others). Compared with the original increase of w_{s1} from 0 to 0.01 (in fractions of the volume of the solid phase), when $P_{s1,2}$ of models with isometric inclusion packings changes very rapidly from ∞ to $\sim 1.5 \times 10^2$, then, with changing w_{s1} from 0.01 to 1 the decrease of its values slows down and a direct relationship is observed between $P_{s1,2}$ and w_{s1} values.

The pattern of the quantity $P_{s1,2av}$ in models with nonisometric inclusion packings differs. The average geometric values of $P_{s1,2av} = \sqrt{P_{s1,2Z}P_{s1,2X(Y)}}$, where $P_{s1,2Z}$ and $P_{s1,2X(Y)}$ are parameters of composition and structure calculated from $\varrho_{1,2Z}$, $\varrho_{1,2X}$ and $\varrho_{1,2Y}$ in the direction of the Z and X (or Y) axes of models, are long preserved very great and drop rapidly to attain $w_{s1} = 1$ solely when w_{s1}

FIG. 78. Dependences of parameters $P_{sph1,2}$ and P_p of composition and structure on the content w_{sph1} (or w_{s1}) and w_t of the filling component for a perfect two-component solid phase and rocks proper having different structure and various directions of measurements.
a—$\varrho_{sph1} \ll \varrho_{sph2}$; $P_{sph1,2} = f(w_{sph1})$ relationships for solid phase: *1*—with equal-diameter spherical inclusions; *2-4*—with different-diameter spherical inclusions, given w_0 = 90, 75 and 50%; *5*—with inclusions in the shape of oblong ellipsoids of revolution having a 1:40 axis ratio, resistance measured in the direction of the Z axis; *6*—the same, but with resistance measured in the direction of the X axis; *7*—with inclusions in the shape of oblate ellipsoids of revolution, having a 1:40 axis ratio, resistance measured in the direction of the Z axis; *8*—with inclusions in the shape of extremely oblate ellipsoids of revolution that degraded to form flat disks, resistance measured in the direction of the Z axis; *9*—with inclusions in the shape of various-size ellipsoids of revolution for w_0 = 90%, resistance measured in the direction of the X axis; *10*—the same, yet for w_0 = 75%; *11*—with equal-size cubic inclusions (faces of two adjacent cubes lying in the same planes); *12*—with equal size-cubic inclusions (the latter being arranged after the masonry pattern); *13*—with the filling component as three systems of parallel pipes going in three mutually perpendicular directions; *14*—with the filling component as three systems of parallel pipes going in three mutually perpendicular directions (pipe length being 1.42 times longer than the cube's side); *15-17*—principal experimental $P_p = f(w_t)$ relationships for sandy-clayey rocks; *b*—$\varrho_{sph1} \gg \varrho_{sph2}$; $P_{sph1,2} = f(w_{sph1})$ relationships for the solid phase: *18*—with equal-diameter inclusions; *19-21*—the same as in *2-4*; *22*—the same as in *5*; *23*—the same as in *6*; *24*—the same, but the resistance is mean geometric; *25*—with inclusions as extremely elongated ellipsoids of revolution that degraded to form needles, the resistance is measured in the direction of Z axis; *26*—the same, but the resistance is measured in the direction of the X axis; *27*—the same, but the resistance is mean geometric; *28*—the same as in *7*; *29*—the same but the resistance is measured in the direction of the X axis; *30*—the same but the resistance is mean geometric; *31*—the same as in *8*; *32*—the same but the resistance is measured in the direction of the X axis; *33*—the same but the resistance is mean geometric; *34-36*—with inclusions as ellipsoids of different size, respectively, for w_0 = 90, 75 and 50%, the resistance measured in the direction of the Z axis; *37-39*—the same, but with the resistance measured in the direction of the X axis; *40*—the same as in *11*; *41*—with inclusions as equal-edged cuboctahedra of equal size; *42*—experimental curve for quartz porphyry with iron lustre inclusions

values attain 0.8-0.9. In this case the solid constituent is almost entirely represented by the filling component (conductor or semiconductor).

$P_{s1,2}$ values for packings of n series of spheres or ellipsoids of revolution decrease with growth of w_{s1} less intensively compared with packings with identical-size inclusions and over the range of w_{s1} values from 0 to 1 will always remain higher than $P_{s1,2}$ values of models with the inclusions of the same size. In this case, graphs of the $P_{s1,2} = f(w_{s1})$ dependence corresponding to them are found higher than those for packings of identical-size spheres, cubes etc., the higher is the content w_{s0} in the model of the first fictitious filling component* (compare graphs *1*, *11*, *12* and *2-4* or *9-10* in Fig. 78).

In the case of models with $\varrho_{s2} \to 0$ and $\varrho_{s1} \to \infty$, $P_{s1,2}$ varies from values approaching zero to the ones tending to unity with growth in them of the content w_{s1} of the filling component. However, much as for models where nonconducting element was represented by inclusions, and conducting ones by the component filling their pore space, the variation of the quantity $P_{s1,2}$ with growth of w_{s1} is nonuniform for models with identical-, different-size isometric and nonisometric elements of packings.

For models with packings of isometric inclusions with w_{s1} increasing to attain 0.01, $P_{s1,2}$ rose very rapidly to 1.25×10^{-3}, whereupon its growth slowed down and a definite proportion between values of $P_{s1,2}$ and w_{s1} was observed. If models included packings with n series of different-size inclusions, the growth of $P_{s1,2}$ with increasing w_{s1} was less intensive, particularly over the range of $w_{s1} = 1$ to 0.01. As to $P_{s1,2}$ values of solid-phase models with strongly nonisometric inclusions (needle- and plate-shaped) long preserved very small values of the parameter of composition and structure until the content of the filling mineral $w_{s1} \approx 0.8$ to 0.9. Then they increased very rapidly to attain 1 (100%) (see Fig. 78).

Thus, given specified ϱ_{s1} and ϱ_{s2}, volumetric ratio of the conducting and nonconducting components of the solid phase, one can obtain strongly different values of the quantities $P_{s1,2}$ (and $\varrho_{s1,2}$) solely on the account of the structural differences of the models. Consequently, according to theoretical studies, in the general case $\varrho_{s1,2} = P_{s1,2}\varrho_{s1}$ essentially depends not only on ϱ_{s1} and ϱ_{s2} and volumetric ratio of components, but also on their distribution pattern in the medium under consideration.

The results of theoretical studies of the dielectrical permeability of a two-component solid phase through matrix and statistical models also reduce to a number of relationships (Table 10) that permit the determination of the average dielectrical permeability $\varepsilon_{s1,2}$ of the medium being studied if the values of dielectrical permeabilities ε_{s1} and ε_{s2} of its components, fractional content of w_{s1} and w_{s2} and their distribution in the two-component solid phase are known (Fig. 79).

* The fictitious filling component in the method of successive filling is a mixture consisting of an actual filling component and $(n - x)$ series of one-dimension inclusions. The first filling component consists of an actual filling component and $(n - 1)$ series of the largest inclusions.

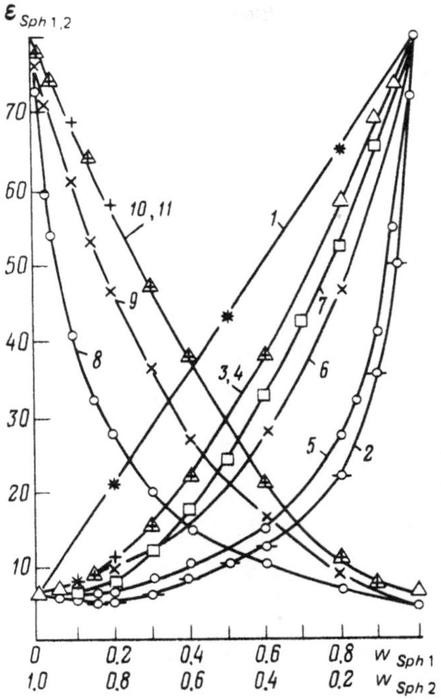

FIG. 79. Dependences of the average dielectric permeability of the solid phase $\varepsilon_{sph\,1,2}$ on specific volumetric contents, $w_{sph\,1}$ and $w_{sph\,2}$, of components having a low and a high dielectrical permeability.

$\varepsilon_{sph\,1,2} = f(w_{sph\,2})$ (given $\varepsilon_{sph\,1} = 6$ and $\varepsilon_{sph\,2} = 80$) relationships plotted by: *1, 2*—V.V. Rzhevskii and G.Ya. Novik (layers located, respectively, parallel and perpendicular to the field's lines of force); *3, 4*—B.I. Odelevskii and Buttcher; *5*—L.V. Lorentz-Lorentz; *6*—K.L. Lichtennecker; *7*—V.N. Dakhnov ($m = 2$); $\varepsilon_{sph\,1,2} = f(w_{sph\,1})$ relationships (given $\varepsilon_{sph\,1} = 80$ and $\varepsilon_{sph\,2} = 6$), plotted by: *8*—L.V. Lorentz-Lorentz; *9*—K. Lichtennecker; *10-11*—V.I. Odelevskii and K. Boetscher

All theoretical relationships hold good solely in the absence of the migration and other types of slow polarization, i.e. given such composition of the components and frequency of the field that prevent development of these polarization types. Some of them also fail to allow for the mutual polarization of the components. We see from a comparison of graphs $\varepsilon_{s1,2} = f(w_{s2})$ and $\varepsilon_{s1,2} = f(w_{s1})$ shown in Fig. 79, given that $\varepsilon_{s2} = 80$ and $\varepsilon_{s1} = 6$, and also given $\varepsilon_{s2} = 6$ and $\varepsilon_{s1} = 80$, the values of $\varepsilon_{s1,2}$ are greatly affected by the fractional volume proportion and distribution of the components in a two-component medium. The greatest value of $\varepsilon_{s1,2}$, given fixed ε_{s1}, ε_{s2}, w_{s1}, can be obtained for a laminar model, given that the polarizing field is oriented parallel to layers where $\varepsilon_{s2} \ll \varepsilon_{s1}$, and the least ones given that the field is at right angles to the layers or when calculating by the use of Maxwell's equation (see Fig. 79, Table 10).

Graphs represented in Fig. 79 also suggest that, given $\varepsilon_{s2} \gg \varepsilon_{s1}$ and decreasing concentration of w_{s2} of inclusions with ε_{s2}, $\varepsilon_{s1,2}$ first decreases rapidly, then the decrease of the values of the average dielectrical permeability of the model slows down followed by gradually attaining values of ε_{s1} of the filling component. However, the pattern of the drop of $\varepsilon_{s1,2}$ values is affected not only by the ratio of components with ε_{s2} and ε_{s1}, but also by their distribution pattern in the model.

Given $\varepsilon_{s2} \ll \varepsilon_{s1}$, the $\varepsilon_{s1,2}$ values first grow slowly with the content w_{s1} of the component showing a great ε_{s1} value, then the growth of $\varepsilon_{s1,2}$ is appreciably accelerated. Note that in this case as well the $\varepsilon_{s1,2}$ values are also governed not only by

TABLE 10. Equations for Calculating $\varepsilon^*_{s1,2}$ of a Two-component Solid Phase, Given Its Different Composition and Structure

Solid phase model	Equation determining $\varepsilon_{s1,2}$	Author
	Matrix models	
Similar-size spheres with ε_{s2} uniformly distributed in space. Between them is the filling component with ε_{s1} ($\varepsilon_{s2} \gg \varepsilon_{s1}$ or $\varepsilon_{s2} \ll \varepsilon_{s1}$)	$\varepsilon_{s1,2} = \frac{2\varepsilon_{s1} + \varepsilon_{s2} + 2w_{s2}(\varepsilon_{s1} - \varepsilon_{s2})}{2\varepsilon_{s1} + \varepsilon_{s2} + w_{s2}(\varepsilon_{s1} - \varepsilon_{s2})} w_{s1}$	M.A. Maxwell
Densely packed spheres of two sorts with ε_{s2}. Between them is the filling component with ε_{s1}.	$\varepsilon_{s1,2} = \frac{\varepsilon_{s1,2} - \varepsilon_{s1}}{3\varepsilon_{s1,2}} = \frac{\varepsilon_{s2} - \varepsilon_{s1}}{\varepsilon_{s2} + \varepsilon_{s1,2}} w_{s2}$	K. Boetscher
Layers of components with ε_{s2} and ε_{s1} located along the lines of force of the field $\varepsilon_{s2} \gg \varepsilon_{s1}$ or $\varepsilon_{s2} \ll \varepsilon_{s1}$.	$\varepsilon_{s1,2\parallel} = \sum \varepsilon_{si} w_{si}$	V.V. Rzhevskii, G.Ya. Novik
Ditto, layers are at right angles to lines of force (ε_{si} is dielectrical permeability of component with a relative volumetric content w_{si})	$\frac{1}{\varepsilon_{s1,2\perp}} = \sum \frac{w_{si}}{\varepsilon_{si}}$	Ditto
	Statistical models	
Component mixtures with ε_{s2} and ε_{s1} are randomly distributed in space ($w_{s2} + w_{s1} = 1$) for non-interacting components	$\log \varepsilon_{s1,2} = w_{s1} \log \varepsilon_{s1} + w_{s2} \log \varepsilon_{s2}$ where $\log \varepsilon_{s1,2} = \sum w_{s1} \log \varepsilon_{si}$	K. Lichtennecker
Unordered mixture of two components with $\varepsilon_{s2} \gg \varepsilon_{s1}$ ($w_{s2} + w_{s1} = 1$) for non-interacting components	$\varepsilon_{s1,2} = a + \sqrt{a^2 + \frac{\varepsilon_{s1}\varepsilon_{s2}}{2}}$ where $a = \frac{(3w_{s1} - 1)\varepsilon_{s1} + (3w_{s2} - 1)\varepsilon_{s2}}{4}$	V.N. Odelevskii
Unordered mixture or two components with ε_{s2} and ε_{s1} for non-interacting components	$\varepsilon_{s1,2} = \varepsilon_{s1} + \varepsilon_{s1} w_{s2}(\varepsilon_{s2} - \varepsilon_{s2}) \times \left[\varepsilon_{s1} + \frac{1 - w_{s2}}{3}(\varepsilon_{s2} - \varepsilon_{s1})\right]^{-1}$	L.W. Lorentz-Lorentz
A mixture of two components with ε_{s2} and ε_{s1}	$\frac{\varepsilon_{s2} - \varepsilon_{s1,2}}{\varepsilon_{s2} - \varepsilon_{s1}} \left(\frac{\varepsilon_{s1}}{\varepsilon_{s1,2}}\right)^{1/3} = 1 - w_{s2}$	D.A. Bruggeman
A mixture of two components with ε_{s2} and ε_{s1}	$\varepsilon_{s1,2} = \varepsilon_{s1} + (\varepsilon_{s2} - \varepsilon_{s1}) w_{s2}^m$ given $\varepsilon_{s2} \gg \varepsilon_{s1}$, m may vary from 1 to 2	V.N. Dakhnov

* The dependences of average dielectrical permeability in solid phase $\varepsilon_{s1,2}$.

the volumetric ratio of the components but also by their distribution throughout the model.

The εs values of sandy-clayey rocks vary from 3.8 to 8 given that the frequency of the polarizing field is 28×10^6 Hz. The maximum variation of ε_s values is likely at relatively low frequencies.

Sec. 37. **Rocks**

The values of the most important quantities (Table 11) (specific resistance, dielectrical permeability, tan angle of the dielectrical losses, induced and diffusion and adsorption electrochemical activities) of maximum moisture rock types and groups are governed by the sum total of values of characteristics: (a) composition and volumetric ratio of their constituents (chemical and mineral composition of the phases, voids ratio and moisture capacity coefficient); (b) volume components distribution throughout a rock (granulometric, voids composition, meandering, coefficient of clayiness, specific surface etc.); (c) phase interactions (ion exchange characteristics, coefficients of residual water saturation, maximum adsorptional moisture capacity).

Moreover, the values of the electrical quantities of the rocks saturated by gas and water, oil and water, and gas, water and oil are governed by those of the gas, oil, their volumetric content in the filling component of the rock voids (i.e. coefficients of water-, gas- and oil saturation) as well as by the distribution of these components in the pore space.

The values of the most important electrical quantities of all rocks are also governed by the thermobaric conditions of their occurrence (temperature and pressure of formation and transformation of rocks).

Maximum Moist Ion-Conducting Rocks

Model Investigations

The specific resistance and dielectrical permeability. If a rock does not practically contain well conducting ore inclusions or a clayey component and its solid phase is mainly represented by an aggregate of dielectrical minerals, and the liquid phase by aqueous solutions of natural electrolytes, such a rock can be considered as being a two-component one. To a definite degree of approximation, by grouping in this manner solid phase minerals, let us think of low-voids ratio electron conducting ores as being also two-component media. Clearly, when studying the specific resistance ϱ_{mr} or dielectrical permeability ε_{mr} of such rocks it is permissible to use the results of studies of the average values of $\varrho_{s1,2}$ and $\varepsilon_{s1,2}$ obtained by using matrix and statistical models of two-component solid media, given dramatically different specific resistances ϱ_{s1} and ϱ_{s2} or dielectrical permeabilities ε_{s1} and ε_{s2}, content w_{s1} and w_{s2} and distributions of the components in the volume of a rock's model.

Hence, we could assert that ϱ_{mr} of maximum moist clean (weakly clayey) sedimentary, igneous and many metamorphic rocks is directly related to ϱ_w of

TABLE 11. The Specific Resistance, Dielectrical Permeability, Losses, Induced and Diffusion-adsorption Activity of Rocks (after different authors)

Rock	ϱ, Ohm·m, given: moisture content, hygroscopic	moisture content, total	of ore inclusions	ε
			Sedimentary, metamorphic rocks	
Sand	-	$>0.1\text{-}2.5 \times 10^3$	-	4->30
Sandstone	$10^5\text{-}10^6$	$0.5\text{-}5 \times 10^3$	$1\text{-}10^2$	3->30
Quartzite	$10^6\text{-}10^8$	$10^3\text{-}10^5$	$50\text{-}10^3$	4.3-7
Quartz aleurite	-	-	-	25->80
Aleurolite	$10^4\text{-}10^6$	$<1\text{-}10^3$	$1\text{-}10^2$	-
Clay	$10^3\text{-}10^5$ (sometimes to 10^7)	$1\text{-}10^2$	-	>20
Argillite	$10^3\text{-}10^5$	$1\text{-}2 \times 10^2$	-	8-12
Slates:				
argillaceous	$10^3\text{-}10^5$	$1\text{-}5 \times 10^2$	$1\text{-}10^2$	-
argillaceous and coaly siliceous	-	Up to 10^3	-	-
micous	-	-	11-14	-
Phyllite	$10^4\text{-}10^5$	$10^3\text{-}10^4$	10-50	13
Para- and orthogneiss	$10^2\text{-}10^5$	$10^3\text{-}10^4$	10-50	9-10
Limestone	$10^4\text{-}10^6$ (sometimes to 10^8)	$<1\text{-}2 \times 10^5$	$1\text{-}10^2$	8->15
Marble	$10^6\text{-}10^7$	$10^5\text{-}10^6$	-	8.3-13.5
Dolomite	$10^5\text{-}10^6$	$<1\text{-}10^5$	$1\text{-}10^2$	7.7-12
Anhydrite	$10^4\text{-}10^5$	$10^2\text{-}10^5$	-	6.7
Gypsum	$10^5\text{-}10^7$	$10^5\text{-}10^6$	-	-
Rock salt	-	$10^4\text{-}10^8$	-	6
Ferruginous quartzite	-	-	-	-
Ore				
magnetite	-	$10\text{-}10^5$	-	30-35
hematite	-	-	-	-
martite	-	$10^4\text{-}10^5$	-	15-18
Cupreous sandstone	-	$3 \times 10^2\text{-}10^3$	-	-
Bauxite	-	$5 \times 10^{-2}\text{-}4.5 \times 10^3$	-	-
Graphite	-	$0.5 \times 10^{-6}\text{-}10$	-	-
Graphitized slate	-	-	-	-
Coal:				
brown	-	$10\text{-}2 \times 10^2$	-	-
black	-	$10^2\text{-}10^6$	-	2.5-15
anthracite	-	$10^{-3}\text{-}10$	-	-
			Igneous rocks, and metamorphic rocks	
Dunite	-	-	-	6.0-10
Pyroxenite	$10^6\text{-}10^7$	$10^5\text{-}10^6$	$10\text{-}10^3$	6.2-9.5
Peridotite	$10^6\text{-}10^7$	$10^5\text{-}10^6$	$10\text{-}10^3$	5-16

f, Hz; w, %; used in determining ε	tan δ	f, Hz in determining tan δ	A_B, %	A_{da} × mV
and ores derived from the former				
$2 \times (10^5\text{-}10^2)$; 0-44	0.1-1 and more	$10^2\text{-}10^6$	$2.5 \times 10^{-2}\text{-}1.6$*	(−5)-15
$10^5\text{-}10^2$; 0-40	0.1-1 and more	$10^2\text{-}10^6$	$3 \times 10^{-2}\text{-}4.5$	(−5)-50
$2 \times (10^5\text{-}10^2)$; dry	-	-	0.25-2.3	-
$2 \times (10^5\text{-}10^2)$; 100	-	-	-	-
-	-	-	0.7-10	(−2.5)-45
-	0.1-1 and more	$10^2\text{-}10^6$	0.11-1.6	7.5->70
$10^7\text{-}10^2$;	-	-	0.5-3.2	-
-	-	-	0.22-2.3 (sometimes to 4.5)	-
-	-	-	0.1-20	-
-	-	-	-	-
-	-	-	-	-
$10^7\text{-}10^2$; dry	-	-	1.3-4.6	-
$10^7\text{-}10^2$; dry	-	-	0.1-2 (sometimes to >5)	(−25)-55
$10^7\text{-}10^5$; -	-	-	0.1-1.5	-
$10^7\text{-}10^5$; dry	-	-	0.14-1.85 (sometimes to 9)	(−10)-50
$10^7\text{-}10^5$; dry	-	-	-	-
-	-	-	-	-
10^2; dry	-	-	-	-
-	-	-	0.5-10	-
10^5; dry	-	-	1.7-20	-
-	-	-	<1.7-21 (sometimes to 51)	-
10^5; dry	-	-	1.7-10	-
-	-	-	3-8	-
-	-	-	1.8-37	-
-	-	-	-	-
-	-	-	0.22-54	-
-	-	-	-	-
−; dry	-	-	-	-
-	-	-	95	-
and ores derived from them				
$10^7\text{-}10^2$; dry	-	-	-	-
$10^7\text{-}10^5$; -	-	-	0.2-2	-
$10^7\text{-}10^2$; -	-	-	-	-

Rock	ρ, Ohm·m, given moisture content: hygroscopic	ρ, Ohm·m, given moisture content: total	ρ, Ohm·m, given of ore inclusions	ε
Gabbro	10^6-3×10^7	$2 \times (10^3$-$10^5)$	50-5×10^3	8.8-15
Serpentinite	-	10^3-10^4	-	6.2-10
Diorite	10^6-3×10^7	$2 \times (10^3$-$10^5)$	50-5×10^3	8.6-10.8
Syenite	10^6-3×10^7	$2 \times (10^3$-$10^5)$	50-5×10^3	7-14
Granite	10^6-8×10^7	10^3-10^4	10-5×10^3	4.5-11.1
Basalt	5×10^3-10^5	5×10^2-10^5	-	6-18.8
Andesite	5×10^3-10^5	5×10^2-10^4	10^2-10^3	-
Liparite	-	10^1-10^3	-	-
Diabase	$5 \times (10^5$-$10^6)$	10^5-10^6	-	7.8-28
Porphyrite	$5 \times (10^4$-$10^5)$	10^3-5×10^4	-	-
Quartz porphyry	5×10^4-10^6	$10^3 10^4$	50-10^3	-
Schist:				
crystalline	10^3-10^5	5×10^2-10^5	50-10^3	-
hornblende	-	-	-	8.9-10.3
chlorite	-	-	-	6-8
talk	-	-	-	7.6-32
Amphibolite	10^6-10^7	10^5-10^6	-	7.9-8.9
Skarn	10^6-10^7	10^3-10^6	50-10^3	4-13
Hornfels	10^6-10^7	10^3-10^6	50-10^3	-
Ore:				
magnetite	-	10-10^5	-	30-35
chromite	-	10^3	-	10-16
nonferrous and rare metal	-	$n \times (10^{-2}$-$10^2)$	-	-
sulphide and copper nickel	-	10^{-5}-10^2	-	-
cassiterite-sericite-quartz	-	10^3-10^6	-	-
sulphide-cassiterite	-	1-10^2	-	-
quartz-molybdenum-chalcopyrite-sericite	-	10^2-8×10^2	-	-
polymetallic	-	10^{-5}-10^4	-	-
copper-zinc	-	5×10^{-2}-10^3	-	-
zinc oxidized	-	10^2-3×10^3	-	-
pyrite-chalcopyrite-sphalerite (massive)	-	10^{-7}-10^{-3}	-	-
chalcopyrite	-	-	-	-
pyrite	-	10^{-5}-10^{-1} (sometimes to 10^2)	-	25-28
lead	-	-	-	-
gold-sulphide	-	0.3-4×10^3	-	-

* The last value generally corresponds to rocks with ore inclusions unless they are ores; n is equal to several unities.

f, Hz; *w*, %; used in determining ε	tan δ	*f*, Hz in determining tan δ	A_B, %	A_{da} × mV
10^7-10^2; dry	-	-	0.5-1.8	-
10^7-10^2; -	-	-	<0.1-15 (sometimes to 20)	-
10^7-10^4; -	-	-	0.27-4	-
10^7-10^2; -	-	-	1-3	-
10^5-10^2; dry	-	-	0.1-4; (moist)	-
10^7-10^2; moist	-	-	0.2-4.2	-
-	-	-	0.4-3.8	-
-	-	-	0.5-3.9	-
10^7-10^2; dry	-	-	0.8-4.2	-
-	-	-	0.25-7.6	-
-	0.015-0.2	5×10^6	-	-
-	-	-	-	-
10^7-10^2; dry	-	-	-	-
-	-	-	0.13-5	-
10^7-10^2; dry	-	-	-	-
10^7-10^5; -	-	-	0.5-2.6	-
10^7; dry	-	-	0.4-26	-
-	-	-	-	-
10^5; -	-	-	1.7-20	-
10^5; -	-	-	<10-30 (sometimes to 32)	-
-	-	-	20-70	-
-	-	-	Up to 25	-
-	-	-	1-6	-
-	-	-	5-18	-
-	-	-	10-25	-
-	-	-	3-70	-
-	-	-	10-60	-
-	-	-	5×10^{-1}-2	-
-	-	-	10-20	-
-	-	-	7.6-70 (moist)	-
10^5; -	-	-	3-18	-
-	-	-	< 0.1-10	-
-	-	-	10-24	-

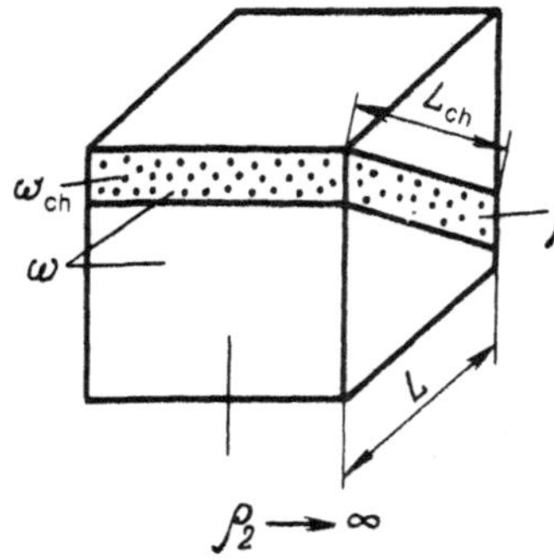

FIG. 80. Cubic rock sample with a conducting inclusion

saturating them natural water (filling component), i.e.

$$\varrho_{mr} = P_p \varrho_w$$

where $P_p = \varrho_{mr}/\varrho_w$ is a proportionality coefficient proposed by V. N. Dakhnov to be called a *parameter of porosity.*

The P_p values of rocks under study are mainly governed by the filling component (water) content which is characterized by the voids ratio k_p or the total moisture capacity coefficient w_t. With growth of k_p (or w_t) they vary roughly much as $P_{s1,2}$ values with content of the filling component w_{s1} in solid state models with n series of spherical or ellipsoidal grains where $\varrho_{s2} \to \infty$ and ϱ_{s1} is insignificant. This model best approximates actual non-clayey detrital materials.

Consequently, with growth of k_p (or w_t) the P_p values must change following the hyperbolic law, first dropping dramatically and then, given $k_p = (w_t) \approx 0.01$, the rate of their decrease slows down until the P_p values attain unity.

It has also been established that the parameter of porosity is a function of meandering of voids channel and the direction of the electrical field. In an analytical study of the $P_p = f(T_e)$ relationship where $T_e = L_{ch.av}/L$ is *electrical meandering* or the ratio of the average statistical length $L_{ch.av}$ of the ion path in a rock sample to its length L, the rock sample represents a cube L on the side and ω in cross section, penetrated by a conducting inclusion which is ω_{ch} in cross section and L_{ch} in length with a low ϱ_1 (Fig. 80); the remaining part of the sample has a specific resistance $\varrho_2 \to \infty$. Clearly, the electrical resistance of the model of an ion conducting rock is

$$R = \varrho_{1,2}L/\omega = \varrho_1 L_{ch}/\omega_{ch}$$

where $\varrho_{1,2}$ is the average specific resistance of the model.

Hence,

$$P_{1,2} = \varrho_{1,2}/\varrho_1 = P_p = L_{ch}\omega/L\omega_{ch}$$

Since for the adopted model the voids ratio

$$k_p = \omega_{ch}L_{ch}/\omega L, \quad \text{then} \quad \omega/\omega_{ch} = L_{ch}/k_p L$$

and

$$P_p = L_{ch}^2/k_p L^2 = T_e^2/k_p$$

Thus, the parameter of porosity of a maximum moist clean (with a low clay content) rock is in direct proportion to the quadratic electrical meandering and is inversely proportional to its voids ratio.

It was assumed that for $f > 10^6$ Hz the dielectrical permeability ε_{mr} of maximum moist rocks must decrease with k_p (or w_t) since much as it does in the case of a two-component solid phase with diminishing the content w_{s1} of the component with $\varepsilon_{s1} \approx 80$ (given an inappreciable $\varepsilon_{s2} \approx 6$) since the ε_s values of the solid component of these rocks generally approaches 6 and for the voids filling component (water) $\varepsilon_w \approx 80$. Thus the ε_{mr} values of maximum moist non-clayey rocks must, with increasing k_p (or w_t), first slowly grow (see Fig. 79, curves *1-7*) and then much more rapidly. Consequently, the $\varepsilon_{mr} = f(w_t)$ or $\varepsilon_{mr} = f(k_p)$ graph is similar to $\varepsilon_{s1,2} = f(w_{s1})$ curves (see Fig. 79).

To a lesser degree, but sometimes also sufficiently strongly the ϱ_{mr}, P_p and ε_{mr} values of various rocks are dependent on the temperature and pressure under which they occur. Clayey rocks and porous ores are represented as being three-component aggregates. The third component of such rocks is represented by moist clay, and that of highly porous ores by mineralized water. When considering moist clay as a third component of sedimentary rocks we must take into account the dependence of its specific resistance on mineralization of the water. Hence follows a fluctuation in the values of the parameter of porosity $P_{p.cl} = \varrho_{mr.cl}/\varrho_w$ of clayey rocks, yet the proportionality between the $\varrho_{mr.cl}$ and ϱ_w is preserved.

Let us consider the model studies determining the dependence of $P_{p.cl}$ on k_{cl}, ϱ_w, T_e and anisotropy of a clayey rock. In one of such studies the main component of the model of a strongly clayey rock was thought to be represented by a psammite-aleurite fraction with an infinitely high specific resistance whose voids space contained a clayey material with a much lower and variable specific resistance, and the central regions of voids contained free electrolyte solution with a still lower specific resistance (Fig. 81). The average specific resistance of such a model is calculated as follows.

Calculate the specific resistance of clayey cement on an assumption that the specific conductivity of the layer of molecularly bound water including an electrical double layer is equal to σ_{lr} and the fraction of the volume of the voids space taken by it equals $(1 - \xi)$ where ξ is the fraction of the volume of voids filled by free electrolyte. Let us denote the specific resistance of free electrolyte σ_w. Then the com-

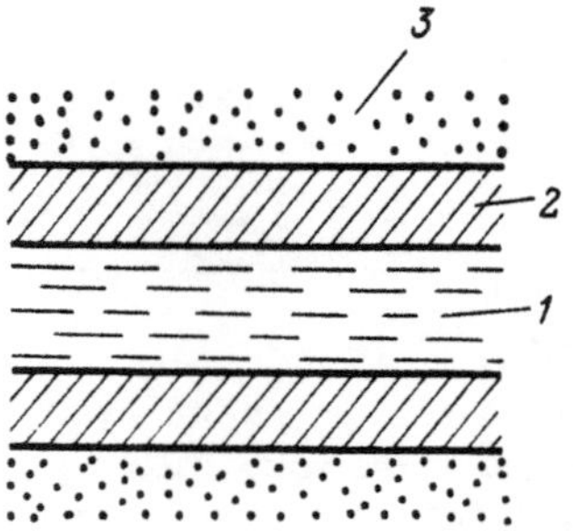

FIG. 81. Diagram of location of various materials in the pore space of strongly clayey rocks.

1—electrolyte of rock; *2*—clayey cement in pore space of rocks; *3*—psammite-aleurite fraction ponent filling the voids space of clayey cement is free electrolyte with σ_w and an electric double layer with σ_{lr} (see Fig. 81) will have a specific conductivity

$$\sigma_{f.cl} = (1 - \xi)\sigma_{lr} + \sigma_w \xi \tag{28}$$

By expressing the specific conductivities of electrolytes of the free and electric double layer through their concentrations and ion mobilities we obtain

$$\sigma_w = C_w(u_w + v_w) = \lambda C_w \tag{29}$$

$$\sigma_{lr} = u'q \tag{30}$$

where C_w is the concentration of free electrolyte; q is the content of adsorbed cations in a unit of volume occupied by an electrical double layer; u_w, v_w and λ are, respectively, mobilities of cations and anions and the equivalent specific conductivity; u' is the average mobility of adsorbed cations in the electrical double layer.

A substitution of u and q, Eqs. (29) and (30), into Eq. (28) will yield

$$\sigma_{f.cl} = (1 - \xi)u'q + C_w(u_w + v_w)\xi$$

Having found the specific conductivity of the component filling the voids space of clays we can calculate the specific resistance of the clayey portion of the rock:

$$\varrho_{cl} = \frac{P_{p.cl}}{\sigma_{f.cl}} = \frac{P_{p.cl}}{u'q(1 - \xi) + (u_w + v_w)C_w \zeta}$$

where $P_{p.cl}$ is evaluated from the voids ratio $k_{p.cl}$ of clays and the $P_{p.cl} = f(k_{p.cl})$ dependence.

The next step is to calculate the specific resistance of the component filling the pore space between sand and aleurite grains.

By defining as k_{clV} the clay content in a unit of volume of a rock with the voids ratio k_p we arrive at an equation determining the specific conductivity of the conducting component filling the voids space:

$$\sigma_{fil} = \sigma_{cl} \frac{k_{p.clV}}{k_{p\ \mathrm{Ps+Av}}} + \sigma_w \frac{k_{p\ \mathrm{Ps+Av}} - k_{clV}}{k_{p\ \mathrm{Ps+Av}}}$$

$$= \sigma_w + (\sigma_{cl} - \sigma_w) \frac{k_{clV}}{k_{p\ \mathrm{Ps+Av}}}$$

Then we calculate the specific resistance of a moist clayey rock:

$$\varrho_{mr.cl} = P_{p\ \mathrm{Ps+Av}} \varrho_{fil} = \frac{P_{p\ \mathrm{Ps+Av}}}{\varrho_w + (\sigma_{cl} + \sigma_w) k_{clV}/k_{p\ \mathrm{Ps+Av}}}$$

$$\varrho_{mr.cl} = \frac{P_{p\ \mathrm{Ps+Av}} \varrho_w}{1 + (\varrho_w/\varrho_{cl} - 1)(k_{clV}/k_{p\ \mathrm{Ps+Av}})}$$

$$\frac{\varrho_{mr.cl}}{\varrho_w} = P_{p\ \mathrm{Ps+Av+Pl}} = \frac{P_{p\ \mathrm{Ps+Av}}}{1 + (\varrho_w/\varrho_{\mathrm{cl}} - 1)(k_{clV}/k_{p\ \mathrm{Ps+Av}})} \tag{31}$$

If the structure of a sandy and aleurite clayey rock is more involved and is such that the clayey fraction is randomly arranged in the voids space of the sandy and aleurite portion, then the electric conductivity of the rock is comprised by the electric conductivity of the electrolyte filling the voids and moist clay. Let us assume that the specific volumetric content of the voids space electrolyte of the rock equals k_p and that of the moist clayey component is k_{clV} and their specific conductivities are, respectively, ϱ_w and ϱ_{cl}. The electric conductivity of such a clayey rock, it appears, will not much differ from the electric conductivity of a clean rock if this latter has a specific volume of the water filling the voids $k_p + \Delta k_p$ with a specific resistance ϱ_w.

Here Δk_p is the volume of the voids space electrolyte equivalent in terms of electric conductivity to the volume of clayey inclusions, $\Delta k_p = k_{cl}(\varrho_w/\varrho_{cl})$, i.e. its volume is as many times less than the specific volume k_{clV}, as many times the specific resistance of the voids space electrolyte is less than that of the moist clay filling the voids.

Since $P_p = a_p/k_p^m$ [see Eq. (35)], then

$$P_{p.cl} = \frac{a_p}{(k_p + k_{clV}\varrho_w/\varrho_{cl})^m} = \frac{a_p}{k_p^m} = \frac{1}{(1 + k_{clV}\varrho/k_p\varrho_{cl})^m} = P_p\Pi_p$$

where Π_p is *a parameter of surface conduction*. If rock clay inclusions show a laminated structure, one obtains different types of relation between $P_{p.cl}$ and other petrophysical quantities.

Hence follows that the parameter $P_{p.cl}$ of a clayey rock is a function not only of the voids ratio but also from the specific volumetric moist clay content k_{clV} and specific resistances ϱ_w and ϱ_{cl}.

Macroscopically heterogeneous laminated rocks, e.g. thin-intercalation sand and clay varieties are characterized by the specific resistance whose magnitude depends on the direction in which its measurement is made (phenomenon of anisotropy).

Suppose that a rock contains a layer having a specific resistance ϱ_p and ϱ_s (Fig. 82*a*). At this, the total thickness of the layers with resistance ϱ_p is ν times greater than that of layers with resistance ϱ_s. A cube of unit volume is cut out of this rock. In measuring the specific resistance at right angles to the layers of

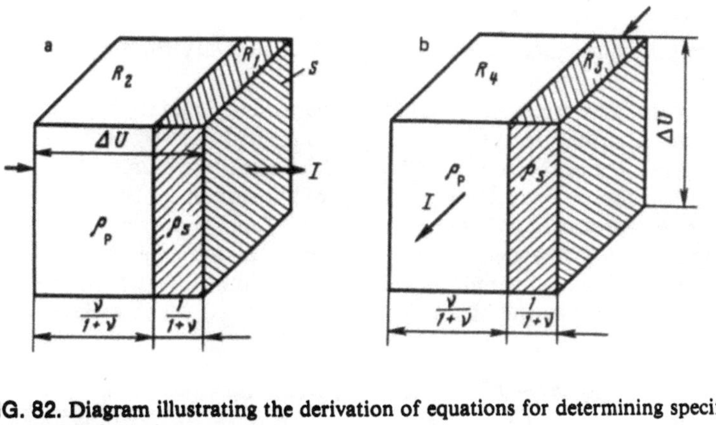

FIG. 82. Diagram illustrating the derivation of equations for determining specific resistances of anisotropic rocks

the rock

$$\varrho_{ws.p\ n} = R_1 + R_2 = \frac{\varrho_s + \nu\varrho_p}{1 + \nu}$$

where the total resistances of layers of rocks, respectively, with specific resistances ϱ_s and ϱ_p are

$R_1 = l_1\varrho_s/S_1 = \varrho_s/(1 + \nu)$

$R_2 = l_2\varrho_p/S_2 = \varrho_p\nu/(1 + \nu)$

and thicknesses and areas of the sample layers at hand (see Fig. 82*a*) are, respectively, $l_1 = 1/(1 + \nu)$, $l_2 = \nu/(1 + \nu)$ and $S_1 = S_2 = 1$.

Given that the current (Fig. 82*b*) is oriented parallel to the layers, the resistance is

$$\varrho_{ws.p\ t} = \frac{\varrho_s\varrho_p(1 + \nu)}{\varrho_p + \nu\varrho_s}$$

since

$$\frac{1}{\varrho_{ws.p\ t}} = \frac{1}{R_3} + \frac{1}{R_4}$$

where the total resistances of layers with specific resistances ϱ_s and ϱ_p are, respectively,

$R_3 = \varrho_s(1 + \nu)$ and $R_4 = \varrho_p(1 + \nu)/\nu$

From the ratio

$\varrho_{ws.p\ n}/\varrho_{ws.p\ t} = 1 + \nu(\varrho_p - \varrho_s)^2/(1 + \nu)^2\varrho_p\varrho_s$

it follows that the resistance of the rock and its parameter $P_{p,cl}$ in the direction of the stratification is less than in the direction at right angles to it. The difference is greater the more is the difference of the resistance of the layers. These conclusions are validated by measurements on samples. *The anisotropy coefficient*

$\lambda = \sqrt{\varrho_{ws.p\ n}/\varrho_{ws.p\ t}} = \sqrt{1 + \nu(\varrho_p - \varrho_s)^2/(1 + \nu)^2\varrho_p\varrho_s}$

given clay layers are alternated by sandstone the λ values vary from ones close to unity, to 2.2.

By knowing the anisotropy, it is possible, e.g. to characterize rock structure (lamination, schistosity and fracturing) and, consequently, identify sedimentation conditions, degree of metamorphism and tectonic alterations of rocks.

When studying the dielectrical permeability of maximum moist clayey rocks these must also be represented as being composed of a nonconducting solid phase, a clayey component and free electrolyte. These constituent elements appreciably differ in dielectrical permeability and the following relationship has been proposed by E. I. Leont'ev to calculate the dielectrical permeability $\varepsilon_{ws.p\ cl}$ of a clayey rocks as a whole:

$$\frac{\varepsilon_s - \varepsilon_{ws.p\ cl}}{\varepsilon_s + 2\varepsilon_{ws.p\ cl}}(1 - k_p) + \frac{\varepsilon_w - \varepsilon_{ws.p\ cl}}{\varepsilon_w + 2\varepsilon_{ws.p\ cl}}k_p(1 - k_{clV}) + \frac{\varepsilon_{cl} - \varepsilon_{ws.p\ cl}}{\varepsilon_{cl} + 2\varepsilon_{ws.p\ cl}}k_pk_{clV} = 0$$

where ε_s, ε_w and ε_{cl} are dielectrical permeabilities of the unswelling solid phase (represented by a sandy-aleurite, carbonate and other similar components), free water and a clayey fraction, respectively. This rock model has the clayey fraction found in the pore space of an unswelling component of the rock.

The equation, proposed by E. I. Leont'ev, bases on the generalized equation of V. I. Odelevskii for calculating the average specific resistance, dielectical permeability and thermal conductivity coefficient of mixtures which appears as

$$\sum \frac{\lambda_i - \lambda}{\lambda_i - 2\lambda} V_i = 0 \tag{32}$$

where λ_i and λ are the generalized conductivity of the i-th phase of a rock model and disperse medium: V_i is the volume of the i-th phase in a statistical mixture.

Consequently, in accord with the results obtained by using models, the dielectrical permeability of the highest moisture content of clayey rocks is governed not only by their total voids ratio k_p but also by their volumetric clayiness coefficient k_{clV}. Note that for rocks with the same porosity $\varepsilon_{ws.p\ cl}$ is greater the higher is their clayiness.

Diffusion-adsorption and induced electrochemical activities. The diffusion-adsorption of a maximum moisture rock is

$$A_{da} = \frac{k_{sw.clV}k_{clV}}{k_{sw.clV}k_{clV} + k^*_{eff.p}} A_{da.cl} = \eta_{cl}A_{da.cl} = k_{sw.clV}\eta_{cl.dr}A_{da.cl} \tag{33}$$

where $k_{sw.clV}$ is a clayey component swelling coefficient, i.e. the ratio of the volume of swelled clay to its volume in a dry state; k_{clV} and $k_{eff.p}$ are the dry rock volumetric clayiness coefficient and the effective voids ratio, respectively; $A_{da.cl}$ is the diffusion-adsorption activity of clean clays; the relative clayiness is

$$\eta_{cl} = k_{clV}/(k_{clV} + k^*_{eff.p}),$$

$k^*_{eff.p} = k_{p.sk} - K_{clV}k_{sw.clV}$ is the fractional content in the rock voids between the unswelling solid phase; the relative clayiness calculated from the value of the clayiness in an air-dry state is

$$\eta_{cl.dr} = \frac{k_{clV}}{k_{clV} + k_{o.p}} = \frac{k_{clV}}{k_{clV} + k^*_{eff.p} + (k_{sw.clV} - 1)k_{clV}} = \frac{k_{clV}}{k_{sw.clV}k_{clV} + k^*_{eff.p}}$$

As can be seen from Eq. (33), the diffusion and adsorption activity of maximum moisture rocks is in inverse proportion to their effective voids ratio and is governed by the content and diffusion and adsorption activity of their clayey fraction. The greater the relative clayiness of dry clay and the higher its swelling coefficient and diffusion and adsorption activity, the higher this activity is.

The induced electrochemical activity A_i of ion-conducting rocks having a maximum moisture content can be described, according to D. A. Friedrichsberg and M. P. Sidorova, by this relationship:

$$A_i = \frac{2\Delta nRT(C_1 - C_2)}{(A_1l_1 + A_2l_2)F}$$

where Δn is the transfer ratios in the narrow and wide void channels of the rock; R is a gas constant; T is absolute temperature; F is the Faraday constant; C_1 and C_2 are concentrations of the voids space solution at the ends of the capillaries (voids channels) (in planes *1* and *2*, see Fig. 63):

$A_i = I/\alpha_1\lambda S_1$ and $A_2 = I/\alpha_2\lambda S_2$

Here I is the current in the capillaries; S_1 and S_2 cross sections of narrow and wide capillaries; λ is the equivalent electric conductivity of the solution; l_1 and l_2 are lengths of capillaries; α_1 and α_2 are activity coefficients taking into account the conductivity of ions of the electrical double layer.

This relationship is converted to assume this form:

$$A_i = \frac{4\alpha_1\alpha_2\varphi(\Delta n)^2}{(1 + \varphi)(\alpha_1 + \alpha_2\varphi)} = \frac{4\gamma\varphi(\Delta n)^2}{(1 + \varphi)(\gamma + \varphi)} \tag{34}$$

where $\gamma = \alpha_1/\alpha_2 = \sigma_1/\sigma_2$; $\varphi = S_2l_1/S_1l_2$ is a structure parameter; σ_1 and σ_2 are the specific conductivities of the solution in a narrow and a wide channel.

One can see from Eq. (34) that A_i is governed by the structure of the rock and composition of its electrolytes.

Experimental Studies

In addition to the theoretical model study of ϱ_{mr}, ε_{mr}, A_{da} and A_i of maximum moisture rocks, a complex laboratory investigation and comparison of these petrophysical quantities and ones that determine them have been conducted.

In the laboratory studies of the electrical and other petrophysical quantities representative sample collections (200-300 specimens) have been selected from sediments and rock massifs whose origin, alteration, substance composition and structure of deposits have been determined. For each of these electrical and other essential petrophysical quantities, such as coefficients k_p, $k_{o.p}$, w_t, i.e. the total and open voids ratios, the total moisture capacity coefficient, meandering T_e of voids channels of samples, their mass, $k_{cl.m}$, or volumetric, k_{clV}, clayiness, specific surface S_V, exchange capacity Q_{100} (or Q_V or q), permeability coefficient k_{perm}, density δ_r, velocity of propagation of elastic longitudinal waves v_P or other petrophysical quantities. Electrical and some of the aforementioned petrophysical quantities have been determined under conditions of full saturation of samples by a natural electrolyte solution, most commonly at t = 20 °C and pressure 0.1 MPa. Less frequently these experiments were conducted at high temperatures and pressures. The quantities measured were then compared and the effect of the temperature and pressure on the quantities was also established.

Let us analyse the results of the experimental studies.

The dependence of electrical quantities on the voids ratio and moisture capacity coefficient. Let us first note a drop in the values of ϱ_{mr}, P_p, A_{da}, A_i with growth of the coefficient of volumetric moisture (or voids ratio) of individual varieties of rocks of the same type referred to definite sediments or rock massifs (Figs. 83-85) and an associated increase of ε_{mr} and tan δ (Fig. 86). This is in good agreement

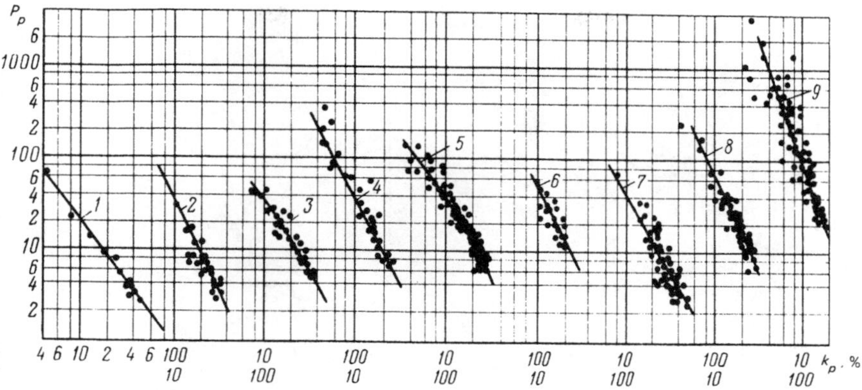

FIG. 83. Dependences of the porosity parameter P_p on the voids ratio k_p for sandy-clayey sediments of various regions of the USSR.

1—sands (after L.A. Gorbenko, Azerbaijan Research Institute, and L.P. Dolina, All-Union Research Institute of Geophysics); *2*—sandstones of IX, X and XI productive horizons of the Gazli oil field; *3*—sandstones of XII and XIII horizons of the same oil field; *4*—sandstones of the Cretaceous and Jurassic sediments of Trans-Terek Region (after A.M. Nechai); *5*—sandstones of the Devonian sediments of the Tuimaza oil field (after L.I. Orlov); *6*—Devonian sandstones of the Kuibyshev Region (after N.Ya. Kachurina); *7*—Paleogene sediment aleurolites of the Krasnodar Province (after materials submitted by the petrophysics laboratory of the Moscow Institute of Oil and Gas Engineering); *8*—Devonian sandstones of the Saratov Volga Region (after I.E. Eidman); *9*—sandstones and aleurolites of the Pre-Jivetian deposits of Bashkiria and Tataria (after materials of the petrophysics laboratory of the Moscow Institute of Oil and Gas Engineering)

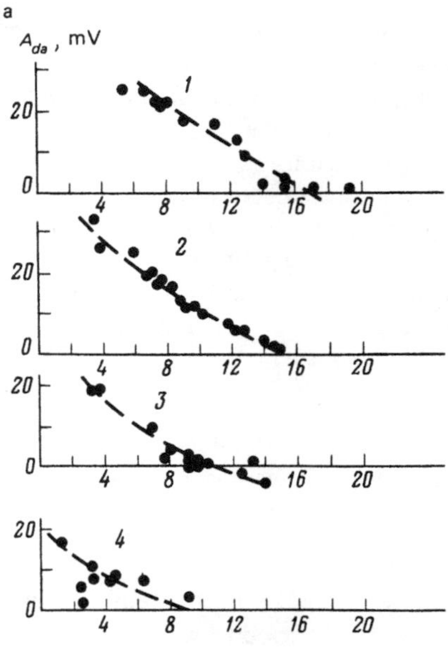

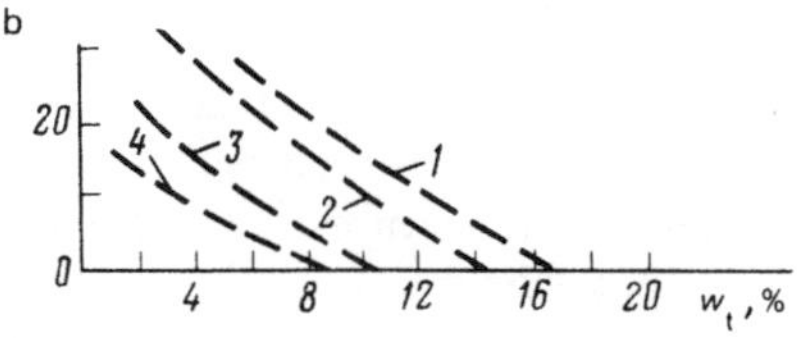

FIG. 84. Dependences of the diffusion-adsorption activity A_{da} on the volumetric moisture content w_t of polymictic rocks of sandy-clayey sediments of areas of Bashkiria and Tataria.

a—rocks with cement; *1*—ferruginous film about grains, pyrite and siderite; *2*—clayey, clayey-sericite and contact clayey; *3*—opal chalcedony; *4*—dolomite; *b*—comparison of median lines *1-4* expressing an $A_{d.a} = f(w_t)$ relationship for polymictic sandy-clayey rocks with different cement types (after V.N. Kobranova)

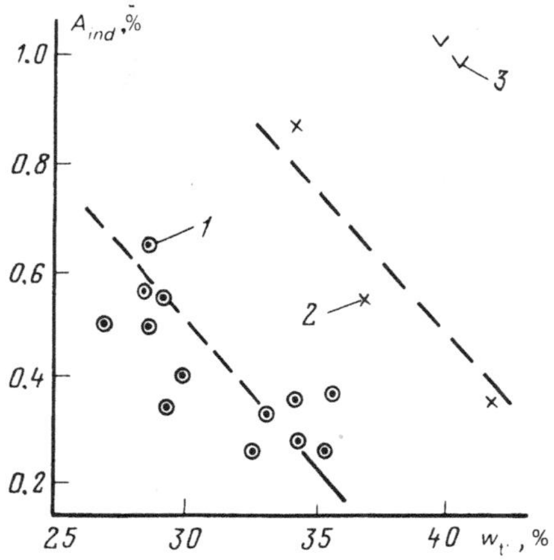

FIG. 85. Connection of the induced electrochemical activity A_{ind} with the coefficient of volumetric moisture content (after A.A. Vedenin).

Chalk-like limestones with an average Fe_2O_3 content (in %): *1*—0.65; *2*—1.32; *3*—1.72

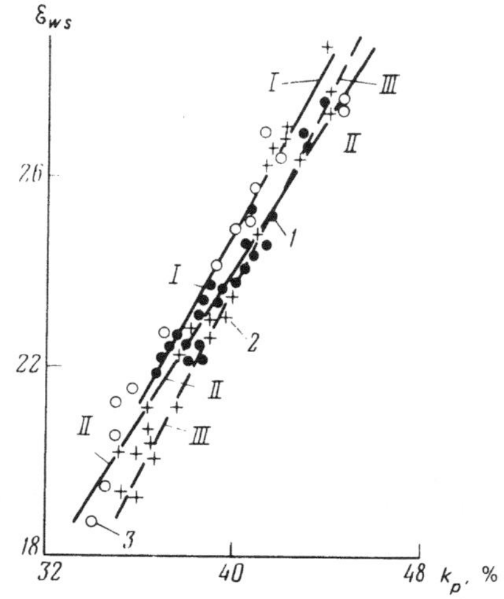

FIG. 86. Dependence of the dielectric permeability ε_{mr} on the voids ratio k_p for maximum moisture quartz sands (after G.Ya. Chernyak).

Sands: *I*—large-grained; *II*—fine-grained; *III*—medium-grained; grain diameter in mm: *1*—0.05-0.25; *2*—0.25-0.5; *3*—0.5-1

with the results of model studies. A reverse intensive (particularly given minor k_p or w_t values) and sufficiently close connection between ϱ_{mr} and w_t (k_p) or between P_p and w_t (or k_p) is accounted for by a great difference in values of the specific resistances ϱ_w of the solid and ϱ_w of the liquid phase of rocks. In case ϱ_s tends to infinity in the absence of ore electron conducting inclusions, ϱ_w is relatively small.

The dependences $\varrho_{mr} = f(k_p)$ or $P_p = f[k_p(w_t)]$ have also been obtained as averaged curves for weakly-, medium- and strongly cemented sandy and carbonate rocks. On a double logarithmic scale graphs averaging the experimental $P_p = f(k_p)$ dependences represent straight lines or curves somewhat concave with respect to the axis of the voids ratio (volumetric moisture capacity). On an arithmetic scale the dependence is averaged by hyperbolas satisfying the equation

$$P_p = a_p/k_p^m \quad (\text{or} \quad P_p = a_p/w_t^m) \tag{35}$$

where a_p is a coefficient constant for the given sediments whose values change from 0.4 to >1; m is a structure power index evaluating the effect on the magnitude of P_p of the geometry of the voids channels. Given $a_p = 1$,

$$m = -\log(\varrho_{mr}/\varrho_w)/\log k_p = -\log P_p/\log k_p$$

For various sedimentary rocks m is in the range from 1.3 to 2.2 and seldom attains greater values. Averaged dependence graphs show a good fit with theoretical dependences $P_{s1,2} = f(w_{s1})$ for models with inclusions in n series of ellipsoids of revolution or spheres (see Fig. 78). A great practical importance of $P_p = f(k_p)$ dependences must be noted. They often permit the voids ratio to be determined from data of the method of resistances.

By virtue of a different nature reverse connections $A_{da} = f(w_t)$ $[A_{da} = f(k_p)]$ (see Fig. 84) and $A_i = f(w_t)$ $[A_i = f(k_p)]$ (see Fig. 85) are less intense and close.

Since the values of the dielectrical permeability of the liquid phase of rocks having a maximum-moisture content in the radio-frequency region are much higher than the solid phase but differ lesser than ϱ_s and ϱ_w. Then the connections of $\varepsilon_{mr} = f(w_t)$ $[\varepsilon_{mr} = f(k_p)]$ are direct ones and are less intense than $\varrho_{mr} = f(w_t)$ or $P_p = f(w_t)$ connections. With increasing the water content in the radio-frequency region thermal losses in rocks, their tan δ, also increase.

However, given a specified moisture content, different samples of definite sediments or rock massifs generally exhibit different values of the aforementioned electrical and electrochemical quantities which is due to the variations in the composition, distribution and interaction of the solid and liquid phase throughout the volume of media under study.

The dependence between the values of electrical and electrochemical quantities on the characteristics of the composition, distribution and interaction of the phases in a rock, given their definite volumetric ratio. *The specific resistance and porosity parameter.* If rocks of definite sediments do not contain ore inclusions, then the larger their interphase surface, the higher the values of the aforementioned quantities are for the given moisture content and high mineralization of the electrolyte filling the pore space. The values of the latter are governed, for instance, by such quantities as the effective medium and median radius $r_{ef.m}$ (diameter $d_{ef.m}$) of voids or grains, the specific volumetric surface S_V, volumetric Q_V, and reduced q_V, capacity of exchange, specific volumetric clayiness k_{clV}, relative clayiness η_{cl}, specific volumetric cementation k_{cemV}, permeability coefficient k_{perm}, etc. The increase of ϱ_{mr} and P_p values at the given k_p (w_t) with decreasing the radius (diameter) of voids or particles and increasing Q, k_{clV}, k_{cemV}, η_{cl} etc. of samples is connected, under given conditions, with complications their ion conducting channels.

In this case the channels in rocks become more meandering, heterogeneous and small in cross section, and some of them are practically unipolar (passing a current in one direction). This changes differently and decreases, on the average, the transference numbers of ions in the voids channels compared with their transference numbers in free electrolytes and, consequently, so does the electric conductivity of samples decreases, given a particular moisture content, phase composition and mineralization of the electrolyte.

Since any sediments and rock massifs are heterogeneous in terms of composition and structure of the rock varieties composing them, it is possible to obtain several values of ϱ_{mr} or P_p for a number of different volumetric ratios between the solid and liquid phases in samples, given their different structure characteristics (S_V, q_V, k_{clV}, $w_{mh.m}$, k_{perm}, A_{da} etc.) for the sediments under study. Then a series of dependences $\varrho_{mr} = f(w_t)$ $[\varrho_{mr} = f(k_p)]$, each of which corresponds to a definite value of some of the structure characteristics, should be plotted. Examples of families of such dependences are connections $P_p = f(k_p)$, given q_V = const (Fig. 87), P_p =

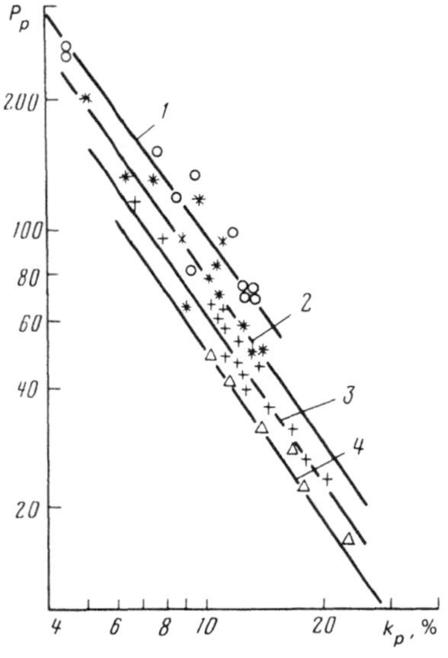

FIG. 87. Comparison of the porosity parameter P_p and the voids ratio k_p of sandy-clayey rocks of the Dnieper-Donetsk depression (after M.M. Ellanskii).

Reduced exchange capacity q_V (in mg-eqv/cm^3): *1*—0.6 and more; *2*—0.4-0.6; *3*—0.2-0.4; *4*—0-0.2

$f(k_p)$, given $k_{cl.m}$ = const (Fig. 88). Each of such dependences exhibits a greater crowdedness and sometimes intensity than $P_p = f(k_p)$ dependences (see Fig. 83) and therefore with a greater degree of reliability can serve to evaluate k_p, if P_p is known and q_V and $k_{cl.m}$ or k_{clV} are known. It is known that in a number of cases P_p is reliably evaluated by geophysical methods which cannot as yet be said about q_V and k_{clV}.

In this respect $P_p = f(w_t)$ relations are more promising, where graphs are differentiated in terms of quantities determined by methods of geophysics (diffusion- or γ-activity).

If rocks are saturated by a weakly mineralized electrolyte solution, then, given k_p = const, with growth of their interphase surface owing to the decreased size of grains or voids, increased q_V, k_{clV}, $w_{mh.m}$ and others a drop in ϱ_{mr} is observed. The minimum specific resistance, given the specified k_p and ϱ_w = 15 Ohm·m are shown, for instance, by clayey rocks, clays and argillites with the grain diameter d_{gr} = 0.005-0.01 mm, a higher resistance is shown by aleurolites (d_{gr} = 0.01-0.1 mm), and ϱ_{mr} of sandstones ($d_{gr} > 0.1$ mm) is still higher. The same is true of limestones; given k_p = const and C_w = 0.01 and their P_p is lower the higher is their clayiness (see Fig. 88*a*).

It has also been established that with decreasing the mineralization (increasing the specific resistance) of the liquid phase of rock samples of definite sediments their specific resistance ϱ_{mr} increases less, the greater their clayiness, given a definite voids ratio. As a result, a drop in the value of P_p with growth of ϱ_w is observed (see Fig. 88*b*). This leads to that the $P_p = f(k_p)$ dependence graphs obtained for definite sediments for different mineralization of waters filling the voids have different shapes. Under conditions of appreciable mineralization of voids pore waters, these graphs on a double logarithmic scale are generally close to straight lines, there-

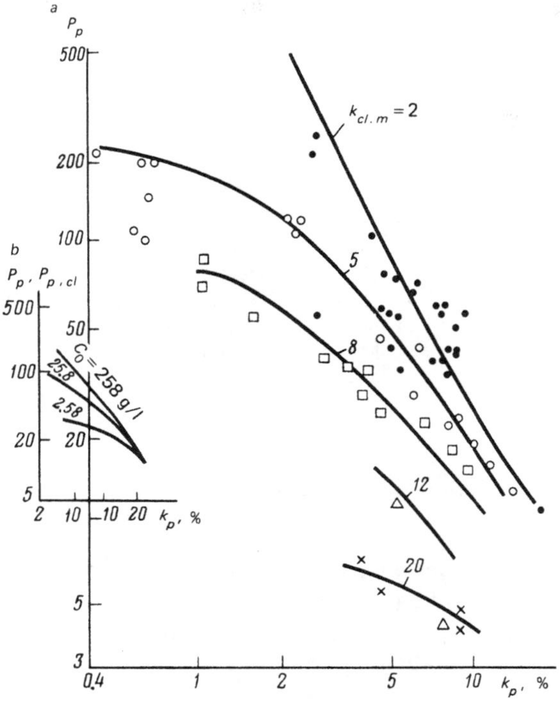

FIG. 88. A comparison of P_p with k_p of carbonate clayey rocks of the Upper Triassic of the area of Dolni Dybnik upon mineralization of the pore water C_0 = 0.01 n (*a*) (after I.B. Nikolova) and lines averaging the dependences of these quantities obtained for unequiargillaceous Devonian sandstone and aleurolite samples of Tataria (*b*) (after V.M. Dobrynin)

fore, with decreasing mineralization of the electrolyte filling the voids channels they become more curved and the greater the clayiness of the latter, the more the values of the porosity parameters of individual samples decrease.

Particularly dramatic is the drop in P_p values of low voids ratio samples with increasing ϱ_w since their voids ratio is appreciably decreased owing to voids being cemented by clays. This variation of ϱ_{mr} and $P_{p.cl}$ values (see Fig. 88*b*) of individual samples are accounted for by additional conductivity due to the electrical double layer and the surface conduction involved. This electric conductivity grows with the thickness of the electrical double layer and this becomes greater the lesser is the mineralization of free electrolyte filling the voids.

So does the surface conduction increase with increasing the interphase surface which is greater the higher is the clayiness of rocks. The effect of the surface conduction and specific resistance of clayey rocks is allowed for by the surface conduction parameter Π. This quantity is accounted for differently and we will explain why. As evidenced by a number of workers, the effect of the surface conduction

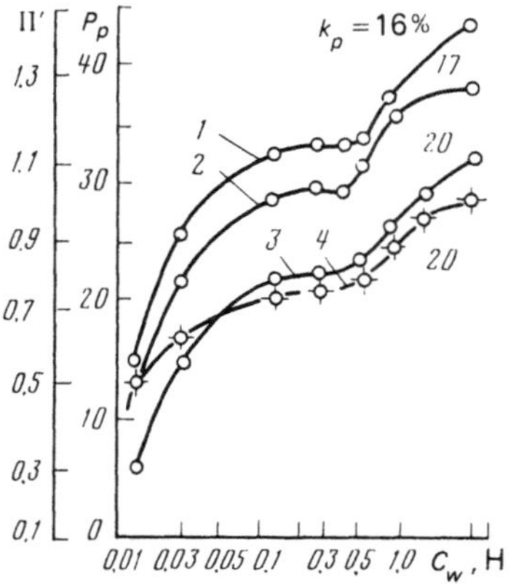

FIG. 89. Dependences of P_p (curves *1*, *3*, *4*) and surface conduction Π′ (curve *2* for $k_p = 17\%$) on mineralization C_w of pore waters for beds $BS_{1\text{-}15}$ of the Malobalyk deposit (after E.I. Leont'ev and A.Ya. Malykhin)

on the specific resistance of individual samples drops continuously as the specific resistance diminishes.

Other workers consider the effect of the surface conduction is observed only to a definite mineralization of the electrolyte filling the voids (e.g. for mineralization of a NaCl solution filling the voids 0.3-0.4 n (Fig. 89), at which the surface conduction of clayey material can be ignored since it does not differ from that of the free electrolyte. Given a greater concentration of the electrolyte saturating a rock the electrical double layer is ever narrower, the ion and water molecule density in it increases and the surface conduction, dropping continuously, becomes less compared with the free electrolyte. In this connection clayey particles present an ever greater obstacle to the passage of a current through a rock which leads to increased values of the specific resistance and porosity parameter of a definite clayey rock.

In accord with the studies concerned it is suggested to take as a parameter of surface conduction two quantities Π and Π′. The quantity Π is a ratio of the porosity parameter, P_{px}, values, given mineralization x, to the porosity parameter $P_{p\ max}$ for maximum mineralization of the voids space electrolyte (4 n for a NaCl solution), or, alternatively, a reciprocal quantity, $\Pi' = P_{px}/P_{0.3\text{-}0.4\ n}$ (where $P_{p\ 0.3\text{-}0.4\ n}$ is a porosity parameter, given voids space mineralization, providing the equality of the specific conductivity of the electrical double layer and the free electrolyte which seems to occur for electrolyte concentration 0.3-0.4 n).

If $\Pi = P_{p\ x}/P_{p\ max}$, its values for clayey rocks are less than unity, given that $\Pi = P_{p\,max}/P_{p\,x}$, its values are more than unity. When the surface conduction parameter is defined as a ratio $P_{p\ x}/P_{p\ 0.3\text{-}0.4\ n}$, its values are less than unity (for voids space electrolyte solution mineralization less than 0.3-0.4 n) and more than unity (for higher mineralizations of the same electrolyte). The surface conduction parameter $\Pi = P_{p\ x}/P_{p\ max}$ decreases continuously with growth of capacity of rock exchange, their diffusion and adsorption activity. An inverse and sufficiently close connection of Π with A_{da} is accounted for by that A_{da} is a function of the exchange capacity of rocks.

The higher the clayiness of the sandstone samples having different clayiness, the less are the Π' values for a given resistance ϱ_w of the pore electrolyte. Moreover, the Π values are first (given $C_w > 0.3$-0.5 n) greater than and later (for $C_w < 0.3$-0.5 n) are less than unity. The Π values of sandy-clayey rocks decrease over a greater range the less is the mineralization of the voids space electrolyte. For these rocks, given $\varrho_w = 2.6$ Ohm·m the Π values changes from 0.9 to about 0.25, and given that $\varrho_w = 0.275$ Ohm·m, from 0.97 to 0.7.

A comparison of the porosity parameter values of sandy-clayey rocks with an average $r_{ch.av}$ and median $r_{ch.m}$ effective radii of voids channels shows that P_p grows both with $r_{ch.av}$ and $r_{ch.m}$. This is attributed to the appreciable clayiness of rock varieties with minor $r_{ch.av}$ (or $r_{ch.m}$), given which, the meandering of voids channels increasing, as it does, yet an additional surficial conductivity appears owing to which ϱ_{mr} and, consequently, P_p values drop.

The dielectrical permeability and tan angle of dielectric losses. Even though the main effect on the dielectrical permeability and tan angle of dielectrical losses is made by the moisture content of rocks, the values of these quantities seem to be also governed by structural petrophysical quantities. This follows, for example, from model theories. These conclusions, however, have as yet been insufficiently validated by experimental studies.

The diffusion and adsorption activity. The A_{da} values are determined not only by the specific volumetric moisture content but also strongly depend on the phase composition, values of quantities characterizing the development of the interphase surface. A_{da} increases with growth of the content in the liquid phase of sandy-clayey rocks of aluminum and ferric oxides directly related to the clayiness of these rocks. The A_{da} values are also governed by the composition of cement for the same rock types and given moisture content of rocks. For example, A_{da} differs for k_p (w_t) = const of sandy and clayey rocks with ferrous, clayey, opal and chalcedony and dolomite cement (see Fig. 84) or of the same rocks with predominantly kaolinite and hydromicaceous and calcite cement. The greater the amount of crystallization water in their solid phase, the higher are the A_{da} values of effusive rocks. The diffusion and adsorption activity of sandy-clayey and calcareous-magnesial rocks increases under conditions of: (1) lessening of the effective diameter $d_{gr.eff}$ and average diameter $d_{gr.av}$ of their grains, the average $r_{ch.av}$ and median $r_{ch.m}$ radii of the voids channels, absolute permeability coefficient k_{perm} (Fig. 90), increasing their specific surface (Fig. 91), exchange capacity (Fig. 92), relative clayiness (Fig. 93), mass clayiness, volumetric clayiness k_{clV}, content in a sandy-clayey rock of clayey soluble portion, hygroscopic w_h, maximum mass hygroscopic $w_{mh.m}$, residual $k_{w.r}$ and reduced hygroscopic $q_{w.h}$ moisture capacity.

All the aforementioned direct and inverse relationships do not show much closeness and their intensity is different in all regions of values of quantities determining the interphase surface or volumetric moisture content. Most closeness is exhibited by connections between A_{da} and such relative or reduced quantities as the relative specific surface $S(1 - w_t)/w_t$ (see Fig. 91), the reduced exchange capacity q_V (see Fig. 92), relative hygroscopic moisture capacity as well as connections of A_{da} with structural characteristics of rocks, given definite moisture content values of samples

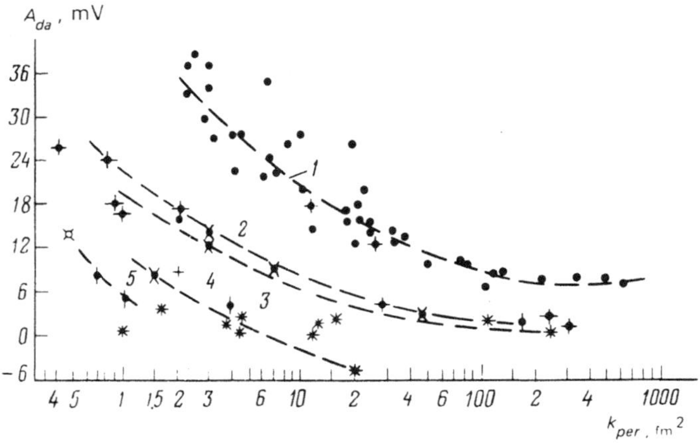

FIG. 90. Dependences of the diffusion-adsorption activity A_{da} on the coefficient of permeability k_{perm} of sandy-clayey rocks.

1—for quartz-glauconite aleurolites with clayey cement from the Paleogene sediments of the Akhtyr area; *2*—for quartz-feldspar-micaceous sandstones and aleurites with ferruginocement, ferrugineous film about grains, pyrite and siderite cement from Paleozoic Pre-Zhivetian sediments of exploratory areas in Bashkiria and Tataria; *3*—for quartz-feldspar-micaceous sandstones and aleurolites having a clayey cement from the same deposits; *4*—for quartz-feldspar-micaceous sandstones and aleurites with opal-chalcedony cement from the same sediments; *5*—for quartz-feldspar-micaceous sandstones and aleurolites with dolomite cement from the same sediments

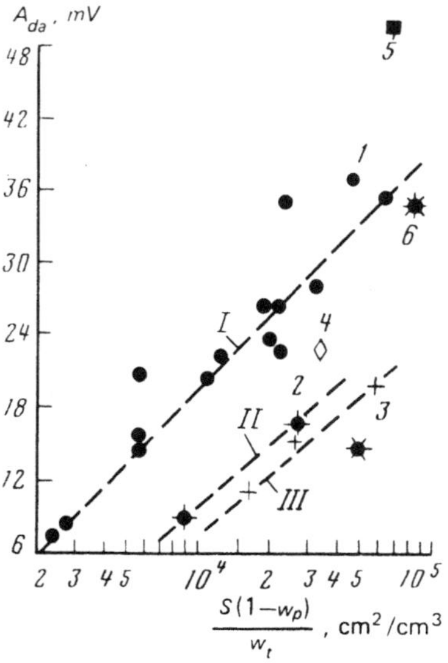

FIG. 91. Dependences of A_{da} on the reduced specific surface $S(1 - w_t/w_t)$ for sandy-clayey rocks of the Tertiary sediments of the Krasnodar region.

Rocks containing cement types: *I*—clayey; *II*—clayey-opal; *III*—calcite; rocks containing cement types (after F.Bak): *1*—clayey; *2*—clayey-opal; *3*—calcite; *4*—dolomite; *5*—siderite; *6*—mixed

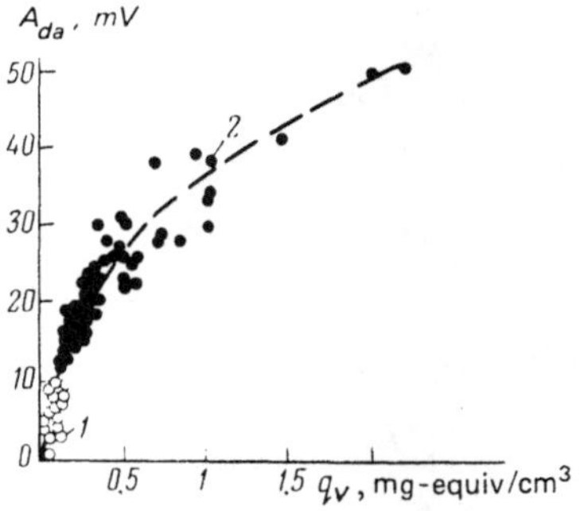

FIG. 92. Depencence of A_{da} on the reduced exchange capacity q_V for terrigenic rocks of the productive sediments of the oil fields in the Tyumen region (after B.Yu. Wendelstein).
1—reservoir bed; *2*—other than reservoir bed

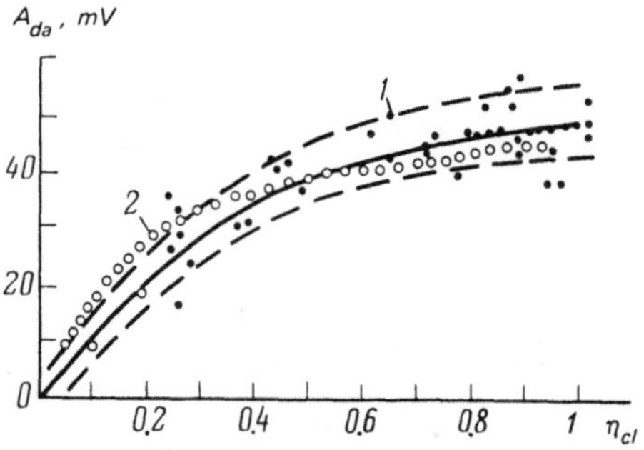

FIG. 93. Dependence of A_{da} on the relative clayiness η_{cl} for Triassic carbonate rocks of the Beli Isvor, Glavaci and Dolni Dybnik areas (*1*) (after I.B. Nikolova) and the Crimea (*2*) (after V.S. Kudryavtsev)

(e.g. $A_{da} = f(k_{cl.m})$ relationships for w_t = const or $A_{da} = f(\Pi)$ relationships and at k_{perm} = const, or relationships $A_{da} = f(k_{perm})$ and $A_{da} = f(k_{perm})$ relationships, given a constant composition of cement of sandy-clayey rocks (see Figs. 84 and 90).

More intimate connections must also be shown by $A_{da} = f(k_p)$ relationships for $k_{cl.m}$ = const and $A_{da} = f(k_p)$ for Π = const. As can be seen from a consideration of the above closer relations, they compare A_{da} values of rock samples showing the specified moisture content with values of any of structural characteristics or compare A_{da} values of rock samples determined from structural characteristics with any of their capacity characteristics. In these cases connections are closer since the quantities being compared depend differently on capacity and structural characteristics.

The induced electrochemical activity. The highest A_i values of maximum moisture pure quartz sands are possible for pore space electrolytes with alkaline reaction (e.g. Na_2CO_3 and NaOH solutions) with bivalent and, particularly, trivalent cations. The polarizability of the same rocks is directly related with the zeta potential for its negative values. According to experimental data, the induced electrochemical activity of detrital rocks is in direct proportion to the specific resistance ϱ_w of waters saturating them

$$A_i = U_{mr}F/IlP_p\varrho_w = A_{mr_1}\varrho_w$$

where F and l are the cross section area and length of a rock sample; P_p is a porosity parameter; I is current; A_{i1} is the value of the electrochemical activity reduced to the specific resistance of waters in 1 Ohm·m.

The induced electrochemical activity of detrital rocks increases with their specific surface (Fig. 94) to its definite values and then drops with increasing clayiness

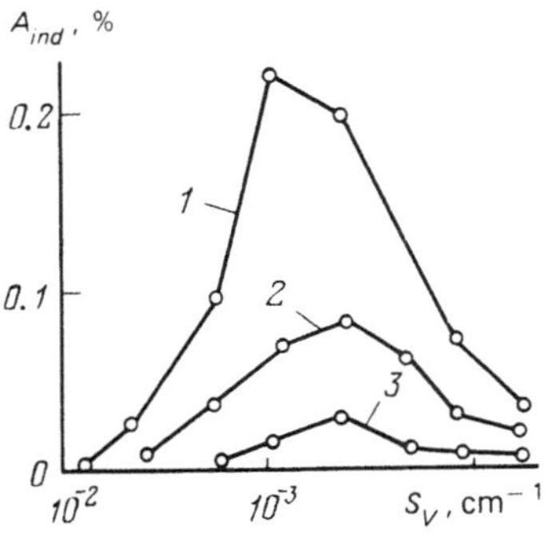

FIG. 94. Dependence of induced electrochemical activity A_{ind} on the specific volumetric surface S_V of sandstone samples, given different specific resistance of KCl solution saturating rock samples (after S.G. Komarov and P.T. Kotov)

ϱ_w (in Ohm·m): *1*—700; *2*—270; *3*—63

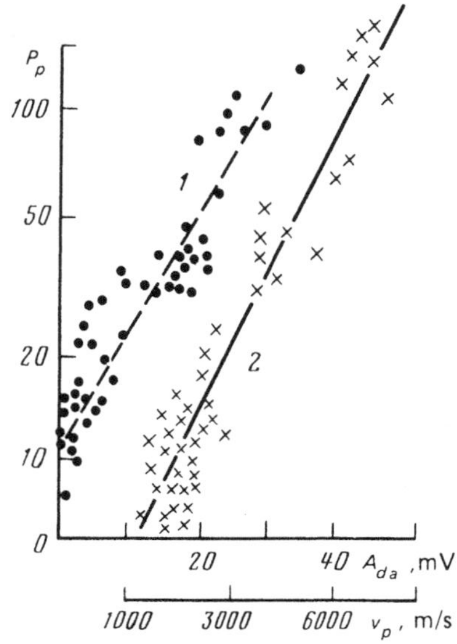

FIG. 95. Connections of the porosity parameter P_p with A_{da} (*1*) and speed of propagation of elastic waves v_P (*2*)

and exchange capacity. A growth in A_i values of detrital rocks with a drop in k_{perm} is also observed. A sufficient closeness between these connections is possible, given that the voids ratio of samples is constant. The rate of the drop of the value of $U_{mr\ 5}/U_{mr\ 20}$ decreases with increasing the diameter of grains of sandy rocks (here the drop times are, respectively, 5 and 20 s).

It should be noted that the electrical quantities are intimately connected with one another and with nonelectrical quantities also through the voids ratio and other structural characteristics. There have been discovered connections, e.g. $P_{p\parallel} = f(\delta_{dr})$ and $P_{p\perp} = f(\delta_{dr})$, $P_p = f(A_{da})$ and $P_p = f(v_P)$ (Fig. 95), $P_p = f(k_{perm})$, $\varepsilon = f(\sigma_{mr})$, $A_{da} = f(q_p)$, $\Pi = f(A_{da})$, $A_i = f(\varrho_{mr})$, $A_i = f(P_p)$, $A_i = f(v_P)$.

These relationships do not generally manifest much closeness since each of the quantities being compared is dependent on the voids ratio or structural characteristics in its own fashion. These connections are closer if they have been obtained for fixed values of structural characteristics or the voids ratio.

Ores and Rocks with Ore Inclusions

The specific resistance. Dry ores and dense rocks are two-phase or single-phase and two-component media. These rocks are represented as such because the specific resistance of the air and minerals, which enclose ore materials or are found among them, is practically identical and very great ($>10^{10}$ Ohm·m), and the well conducting ore component of rocks has a small specific resistance. Moist dense ores and rocks with ore inclusions, given isolated voids, can also be considered close to two-component ones if conductivity of a negligible volume of almost unbound water in minor voids volume of these media is taken into account. That is why $\varrho_{dr} = P_o \varrho_o$ and $\varrho_{dr} = P_{n\text{-}o.c} \varrho_{n\text{-}o.c}$ where ϱ_{dr} is the specific resistance of ore or rock with ore inclusions in a dry or moist state; ϱ_{dr} and $\varrho_{n\text{-}o.c}$ is that of the ore and non-ore components; P_o, $P_{n\text{-}o.c}$ are parameters whose values are governed by the volumetric ratio of the ore- and non-ore component and distribution pattern of the conducting component in the volume of the rock. The dependence of P_o values on the volumetric content w_{s1} of the non-ore component and its distribution in the volume of the ore (or volume of a dense rock with ore inclusions) is presented in Fig. 78. It is shown here that the values of the parameter $P_{1,2}$ and, consequently, those of $P_{n\text{-}o.c}$ of various models of a two-component solid phase with growth of the volumetric fraction w_{s1} of the filling component with a very high specific resistance ϱ_1 (i.e. $\varrho_{n\text{-}o.c}$), given a small specific resistance ϱ_2 (i.e. ϱ_o) of conducting inclusions very rapidly increase to the content of the filling non-conducting component equal to 0.01 in the case of isometric conducting inclusions; then the growth of the proportionality coefficient $P_{1,2}$ ($P_{n\text{-}o.c}$) slows down, its value attaining unity. Given non-isometric needle-like, plate-like and other inclusions $P_{1,2}$ ($P_{n\text{-}o.c}$) until the volumetric content of non-conducting inclusions equal to 0.8-0.9 remains very low and close to constant and rapidly grows tending to unity once the non-conducting component content becomes very great.

The results of theoretical studies of the specific resistance of a two-component solid phase for dry ores and rocks with ore inclusions is well validated by experiments on artificial dry ore samples (see Fig. 78, curve *42*). The laboratory curve obtained for quartz porphyry mixtures with iron lustre is located between theoretical ones for different models identical in structure of a two-component solid phase of rocks.

Experimental connections between the specific conductivity of water-saturated ores $\sigma_{ws\ r}$ and content ξ of ore inclusions revealed during a study and comparison of the aforementioned quantities for sample collections from definite ore deposits (magnetite, sulphide etc.) are insufficiently intimate. Point spread at these dependences will be diminished if each of this is plotted for a definite moisture content. In so doing, we will obtain for the particular ore deposit a family of dependences $\sigma_{ws\ r} = f(\xi)$, given w_t = const. They can be used for a geophysical determination of the ore mineral content in the ore.

Consequently, experimental studies of the electric resistance of ores validate the results of its calculation on rock models. An appreciable anisotropy of the electric conductivity of banded sulphide ores is pointed out.

Three components dramatically different in specific electric conductivity are distinguished in coals, viz. the organic mass of coals, specific resistance ϱ_{coal} which attains more than 10^6 Ohm·m for all coals excepting anthracites and semianthracites (their ϱ_{coal} drops to 10^{-3} Ohm·m), admixtures of non-burning minerals (with which a definite ash content is associated) having much smaller electric resistances than does the organic mass for all coals except anthracites and water with a relatively low ϱ_w (it is particularly abundant for brown coals). As a result correlations have been found between the specific resistance and ash content of dry coals as well as between ϱ_{mr} and moisture content of coals.

The dielectrical permeability and dielectrical losses. When studying the dielectrical permeability and tan angle of the dielectrical losses of ores and rocks with ore inclusions one can think of them as being two-phase three-component (dry ores and rocks with ore inclusions), three-phase four-component (the same rocks completely saturated by water) and three-phase five-component (the latter rocks partially water-saturated) media. Note that one of the components of moist rocks is an electrical double layer.

If we imagine ores and rocks with ore inclusions as being statistical mixtures of their components, then, by using Odelevskii's equation [see Eq. (32)] (as does E. I. Leont'ev for dry completely and partially water-saturated rocks), it is possible to calculate ε_{av} or tan δ, given definite values of these quantities for each of the components and their volumetric ratio. In the case of dry ores and rocks with ore inclusions the dielectrical permeability and tan angle of dielectrical losses, given a definite content of ore inclusions, lower with increasing their voids ratio since ε_{air} and tan δ_{air} of the air are much less than the same quantities referred to non-ore and, particularly, ore components.

An increased moisture content for a given content of ore inclusions contributes to the growth both of ε and tan δ of rocks under study since ε_w and tan δ_w for water are much greater than for non-ore minerals. With increasing the ore mineral content with $\varepsilon_{mr} > 80$, given a definite moisture content of rocks, the values of ε and tan δ increase.

The diffusion and adsorption activity. This quantity for maximum moisture-content ores and rocks containing ore inclusions seems to differ from A_{da} of rocks without ore inclusions yet having the same specific moisture content (porosity) and interphase surface. The diffusion and adsorption activity of ores and rocks with ore inclusions has not been practically studied.

The induced electrochemical activity. For rocks and ores this quantity increases with increasing a content in them of electron conducting minerals (sulphides, oxides etc.) (see Fig. 96). When deriving equations defining the values of the induced electrochemical activity Yu. I. Bulashevich proceeded from:

(1) a solution of a problem concerning the potential distribution for inclusion of a sphere with ϱ_i into a homogeneous medium with ϱ_e where a uniform electrical field with the intensity E_0 exists;

(2) an assumption that the emf of $\mathscr{E}_s$ of polarization appearing at the surface of the sphere as a result of passage of a current is proportional to its normal density: $\mathscr{E}_s = -k\vec{j}_n$ where k is a proportional coefficient;

(3) boundary conditions at the surface of the sphere

$$U_e - U_i = k\vec{j}_n;$$

$$\frac{1}{\varrho_e}\left(\frac{\partial U_e}{\partial r}\right)_{r=a} = \frac{1}{\varrho_i}\left(\frac{\partial U_i}{\partial r}\right)_{r=a}$$

where a is the radius of the sphere; r is a distance from the sphere centre to the

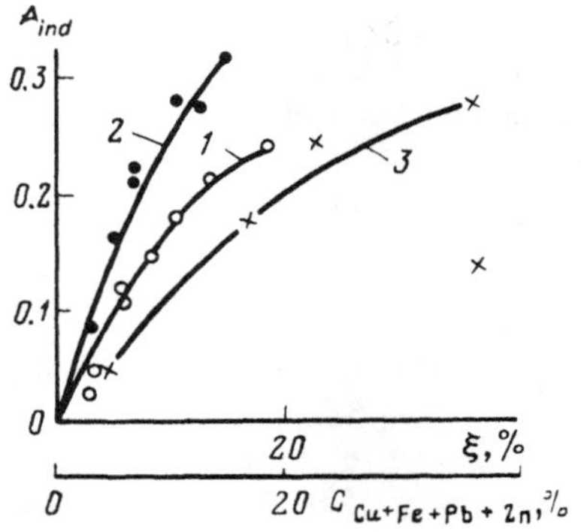

FIG. 96. Dependences of the induced electrochemical activity A_{ind} on mass content of ore minerals ξ in quartz sand (*1*, *2*) (after V.A. Komarov) and ore metals $C_{Cu+Fe+Pb+Zn}$ in Dzhezkazgan copper ore (after B.M. Urazaev).

1—polymetallic ore inclusions (52% sphalerite, 20% galenite, 10% chalcopyrite, 15% pyrite, 3% quartz) with grain size 2-4 mm; *2*—sulphide-nickel ore (90% pyrrhotite, 10% petlandite) with ore particle sizes from 5 to 10 mm; *3*—copper ore

point under consideration; U_e and U_i are potentials in the enclosing medium and inside the sphere.

The solution of the problem leads to equations determining the potential functions U_i inside the sphere and U_e in the space surrounding it as well a normal component of current density $\vec{j}_d$ at its surface which makes it possible to find the value of the emf of the induced polarization of the sphere and calculate the dipole electrical moment of the polarized sphere:

$$\vec{P}_{ip} = \mathscr{E}_s a^2 \frac{\varrho_e}{\varrho_e + 2\varrho_i} = -\frac{3k/a}{[1 + (2\varrho_i/\varrho_e)][\varrho_e + 2\varrho_i + (2k/a)]} a^3 \varrho_e \vec{j}_0$$

where $\vec{j}_0$ is the current density in a homogeneous medium: $\vec{P}_{ip}$ is the polarization vector of the rock.

Then one can find relationships determining A_i of a medium with conducting inclusions. Numerous operations involved in the derivation of this equation are given below:

$$A_i = \frac{U_{ip}}{U_{ap}} = \frac{E_{ip}}{E_{ap}} = \frac{3c\xi}{a^3 + c\xi} = \frac{9\xi\beta k/(a + 2k\nu)}{1 + 3\xi\beta k/(a + 2k\nu)} \tag{36}$$

where ξ is the volumetric content in the rock of the solid phase with electron conductivity; $\beta = \varrho_e/(2\varrho_i + \varrho_e)^2$; $\nu = 1/(2\varrho_i + \varrho_e)$;

$$c = \frac{3\alpha^3 k/a}{(1 + 2\varrho_i/\varrho_e)(\varrho_e + 2\varrho_i + 2k/a)}$$

As can be seen from Eq. (36), A_i of rocks with an electron-ionic conduction is in the general case dependent on the volumetric content of the electron conducting phase, diameter of rock grains, specific resistance of both phases and constant k determined, in particular, by the mineral composition of the latter. The above relationship has a number of constraints. In the limiting case, given $\varrho_i \ll \varrho_e$, $a \ll 2k\nu$ and inappreciable ξ, the induced electrochemical activity is $A_i = 9\xi/2$. It will be noted here that by contrast to specific resistance affected solely by the content of bound conducting components, A_i is governed both by isolated and bound conducting inclusions.

Dry Oil and Water, Gas and Water- and Gas-Oil and Water Saturated Rocks

The specific resistance. For weakly clayey rocks the decrease in the volume of confined water on oil and gas traps under conditions of their oil-, gas saturation or the presence of a mixture of these components gives rise to an increase of the specific resistance

$$\varrho_{os.r} = P_s \varrho_{ws.r} \tag{37}$$

It can be seen from this relationship that it is in direct proportion to the specific resistance of the same rock for a 100% water saturation of its voids.

The proportionality coefficient — *saturation parameter* P_s — is governed by the distribution and content of water k_w in a unit of volume of the rock pore space. It shows that, given a definite $\varrho_{ws.r}$, it is possible to obtain different $\varrho_{os.r}$ depending on the volumetric ratio in the sample between water and oil, water and gas (or water, oil and gas), composition of its enclosing and cementing solid components, meandering rate of the voids channels for partial saturation of the voids by water.

The parameter

$$P_s = \varrho_{os.r}/\varrho_{ws.r} = P_{k_w}\varrho_w/P_p\varrho_w = P_{k_w}/P_p \tag{38}$$

where $P_{k_w} = \varrho_{os.r}/\varrho_w$ is the water saturation parameter.

Unlike the parameter $P_{os.r}$ governed by the voids ratio and structure of the solid phase of rocks, the parameter P_{k_w} is also conditioned by the extent of water saturation of the rock and distribution of water, oil and gas in the voids.

Similarly to the relationship $P_p = T_e^2/k_p$ for maximum saturation rocks the following relation is valid for rock varieties partially saturated by confined waters:

$$P_{k_w} = T_{es}^2/w = T_{es}^2/k_p k_w$$

where k_w is the water saturation coefficient of the pore space; T_{es} is the electrical meandering rate of the voids channels of a partially saturated rock conditioned both by meandering of the voids channels and the distribution of the water, oil and gas in the pore space of rocks; w is the specific volumetric moisture content of a partially water saturated rock.

The distribution of the water, oil and gas in the pore space of a rock depends on the selective wettability of their solid phase. A substitution into Eq. (38) of values of P_p and P_{k_w} will yield

$$P_s = P_{k_w}/P_p = (T_{es}/T_e)^2/k_w \tag{39}$$

Analysis of Eqs. (37) and (39) shows that the resistance of a partially water-saturated rock differ more from that of a maximum moisture content rock the lesser is its water saturation. Given two rocks with the same water saturation, a higher $\varrho_{os.r}$ is manifested by the rock for which the ratio of meanders under conditions of partial and complete saturation is greater.

The dependence of P_s on k_w and distribution of the fluid and gas in the voids of rocks has been studied on pure samples of various sandy and clayey and carbonate sediments.

To determine P_s, the $\varrho_{os.r}$ of samples with known k_w and ϱ_{mr} of the same samples, given their total water saturation, has been measured. From known $\varrho_{os.r}$ and $\varrho_{ws.r}$ the P_s values have been calculated followed by plotting $P_s = f(k_w)$ relationships for particular sandy and clayey or carbonate deposits (Fig. 97). Studies have shown that P_s weakly depends on the rock type and heterogeneous fluids saturating them if water wets the solid phase. This enabled a number of authors to obtain analytical relationships for plotting curves of dependences of the average values of P_s on k_w.

According to V. N. Dakhnov,

$$P_s = a_s/k_w^n$$

at this, for sandy and clayey rocks (for $k_w < 40\%$) $a_s = 0.6$ and $n = 2.25$, and for carbonate rocks (given $k_w < 25\%$) $a_s = 0.4$ and $n = 2.1$;

according to G. Guiot,

$$P_s = 1/k_w^{1.98} \tag{40}$$

according to H. Archie,

$$P_s = 1/k_w^{2.0} \tag{41}$$

As can be seen, the index n allowing for the structure and the distribution of fluids in the pore space of rocks approaches two in value. However, some $P_s = f(k_w)$ graphs correspond to Eqs. (40) and (41) solely provided that the index n has higher or lower values.

According to G. W. Keller, the relationship between the quantities P_s and k_w is expressed by two equations for high and low saturations. The equation for high saturation values

$$P_s = k_w^{-n_1}$$

is valid for values of k_w that are greater than the critical water saturation $k_{w.cr}$.

For hydrophilic rocks, given $k_w < k_{w.cr}$, the water on the walls of the voids appears as drops (see Fig. 32*A*, *d*) and for hydrophobic rocks $k_{w.cr}$ is attained upon removing of practically all water from small voids and disturbing of the electrical bonding between individual parts of the conducting fluid (see Fig. 32*B*, *d*) and

$$P_s = bk_w^{-n_2}$$

where n_2 is a power index for low saturations k_w; b is a constant coefficient.

The absolute values of n_1, n_2 differ very much and are governed by the extent of saturation and distribution of the fluid and gases in the voids of rocks. The latter is conditioned by the structure of the pore space and on whether their solid phase is hydrophilic or hydrophobic.

The $P_s = f(k_w)$ curves averaging experimental data for a definite type of sandy and clayey and calcareo-magnesial sediments are of much practical value; by using them and from the specific resistance of the rock, given a maximum moisture content and an oil- or gas-saturated state, it is possible to find its oil- or gas-saturation coefficient.

In the case of clayey nonfoliated oil-saturated oil and gas traps, given $n = m$

$$P_s = \frac{\varrho_{os.r.cl}}{\varrho_{ws.cl}} = \left[\frac{1 + (k_{clV}\varrho_w/k_p\varrho_{cl})}{k_w + (k_{clV}\varrho_w/k_p\varrho_{cl})}\right]^m = \left(\frac{1 + A}{k_w + A}\right)^m \tag{42}$$

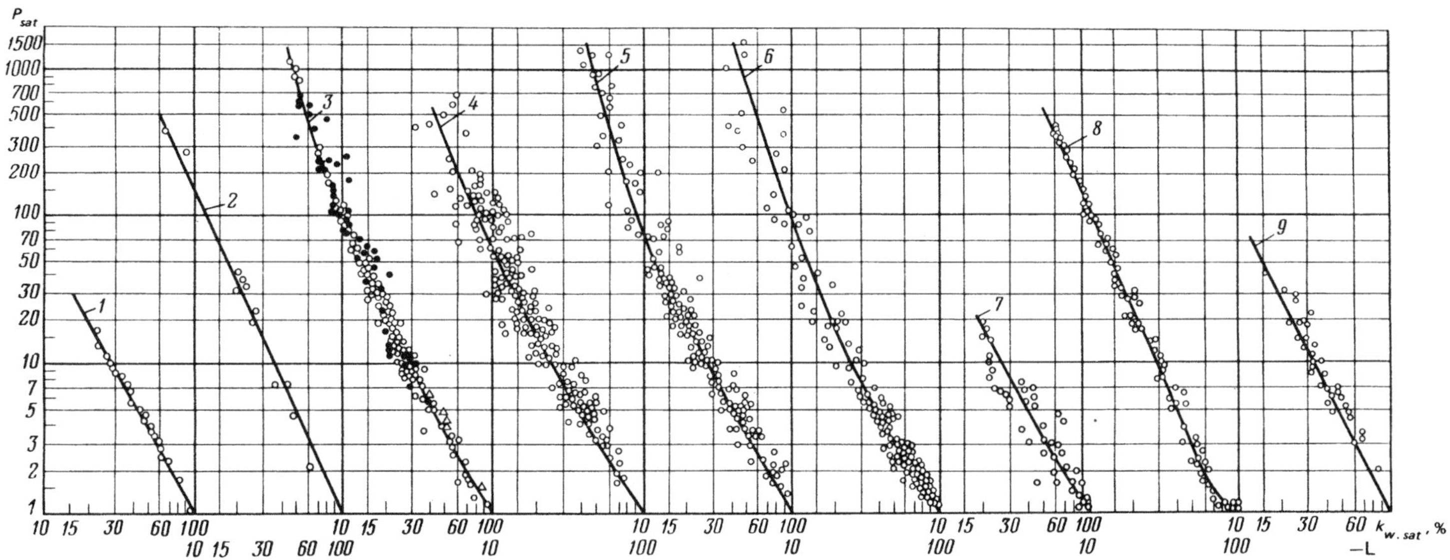

FIG. 97. Dependences of the saturation parameter P_s on the coefficient of water saturation k_w for various sandy-clayey sediments in the USSR.

1—sandstones of Tertiary sediments of Azerbaijan (after N.G. Kogan, Azerbaijan Research Institute); *2*—sandstones and aleurolites of the Devonian sediments of the Tuimaza oil field (petrophysics laboratory of the Moscow Institute of Oil and Gas Engineering); *3*—sands and sandstones of the Devonian sediments of the Tuimaza oil field (after G.S. Morozov, All-Union Geophysics Research Institute); *4*—sandstones of the coal bearing suite of Kuibyshev Volga region deposits (after N.Ya. Kachurina, Kuibyshev Geophysical Trust); *5*—Devonian sandstones of a Kuibyshev Volga region deposit (after N.Ya. Kachurina, Kuibyshev Geophysical Trust); *6*—Devonian sandstones of the Romashkinskoe deposit (after L.P. Dolina and G.I. Skoblikova, Ull-Union Geophysics Research Institute); *7*—Tertiary sandstones of the Grozny oil fields (after L.P. Dolina, All-Union Geophysics Research Institute); *8*—Devonian sandstones of the Tuimaza oil field (after A.V. Zolotov); *9*—sandstones of different age (after F.I. Kotyakhov)

where $\varrho_{os.r.cl} = [a_s/(k_p k_w + k_{clV}\varrho_w/\varrho_{cl})^m]\cdot\varrho_w$ and $\varrho_{ws.cl} = [a_p/(k_p + k_{clV}\varrho_w/\varrho_{cl})^m]\varrho_w$ are the specific resistances of an oil- and a water saturated clayey rock; $a_s \approx a_p$ and $A = (k_{clV}/k_p)\cdot(\varrho_w/\varrho_{cl})$.
$m = n = 2$, different values of the quantity $A = k_{clV}\varrho_w/k_p\varrho_{cl\ dis}$ and dispersed inclusion of clayey components. The same reference considers a theory referred to the layered position of the clayey component.

V. N. Dakhnov describes a family of $P_s = f(k_w) = \varphi(k_s)$ curves, given $m = n = 2$, different values of the quantity $A = k_{clV}\varrho_w/k_p\varrho_{cl\ dis}$ and dispersed inclusion of clayey components. The same reference considers a theory referred to the layered position of the clayey component.

Other theories are also available based on definite physical models. For example, in conformity with one of them the following relationship has been derived:

$$P_s = 1/k_w[(\varrho_{wb.av}/\varrho_w)(1 - k_{w.b}) + k_{w.b}] \tag{43}$$

where $\varrho_{wb.av}$ is the average effective specific resistance of a bound solution; $k_{w.b}$ is a water saturation of a rock by bound water; k_{ws} is a current-water saturation coefficient reflecting the total amount of available water (bound and free).

Given that $k_{ws} = k_{w.b}$, Eq. (43) takes on this form:

$$P_s = (\varrho_{wb.av.}/\varrho_w)[(1 - k_{w.b})/k_{w.b} + 1]$$

For clean rocks, given $\varrho_{wb.av} = \varrho_w$ and $k_w = k_{w.b}P_s = 1/k_w$.

Two $P_s = f(k_w)$ graphs, plotted from Eq. (43) for $\varrho_w/\varrho_{wb.av} = 0.1$ and $k_{w.b} = 17.4\%$, are given in some works.

Studies conducted on rock samples made it possible to derive an equation.

$$P_s = a_{ws}(k_w \mp C)^{-n} \tag{44}$$

where C is the total volume of bound, blind and drop-like water. Eq. (44) is valid for $k_w > C$. Given that $C \to 0$, both for weakly clayey, well sorted, weakly cemented and hydrophilic rocks the equation assumes a form: $P_s = a_{ws}k_w^{-n}$, and the graph of this dependence (on a bilogarithmical scale) represents a straight line. If with decreasing k_w the $P_s = f(k_w)$ relationship gradually departs upwards from the straight line, this means that a volume C of water, which conducts practically no electrical current given the particular water saturation, must be subtracted from the volume of water defined by the coefficient k_w. This relationship corresponds to low porosity oil and gas traps with deadlock voids and decreased hydrophily. A gradual flattening of $P_s = f(k_w)$ relationships with decreasing k_w implies that the volume of water defined by k_w must be increased by C of water bound with the clayey component. The content of this water fails to be determined by using volumetric laboratory techniques (centrifugation, capillary displacement). The flattened curve corresponds to clayey low-permeability hydrophilic oil and gas traps.

Let us assume that $T_g^2 \approx T_e^2 = P_p k_p$ and $k_{wr} = (1/P_s)^{1/n}$, $\tau = k_{w.b}k_p/S_{filV}$ and substitute these quantities into Eq. (11), then

$$k_{perm} = \frac{[1 - (1/P_s)^{1/n}]^3\tau^2}{P_p f} P_s^{2/n} \tag{45}$$

It can be seen from Eq. (45) that k_{perm} is closely connected with P_p and P_s and the thickness of the film of residual water.

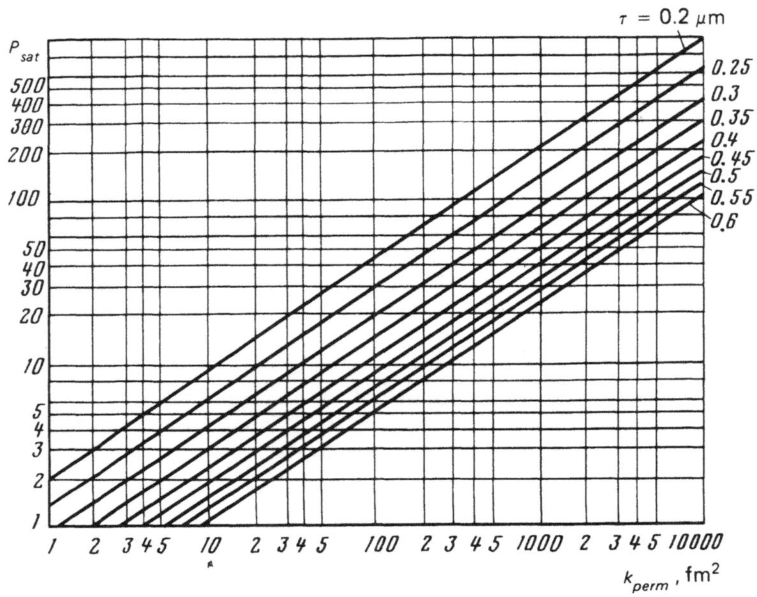

FIG. 98. Relationship between the saturation parameter P_s and coefficient of permeability k_{perm} for sand oil reservoir beds of Tertiary sediments of the productive rock mass of Azerbaijan (after L.A. Buryakovskii)

Figure 98 shows a set of $P_s = f(k_{perm})$ curves, given the thickness of the film of bound water τ = const (modules of curves) the use of which enables one to determine the permeability of oil and gas traps from values of τ and P_s. Design curves show a satisfactory fit with similar experimental dependences (see Fig. 98) used in practice to estimate the permeability of oil-saturated oil and gas traps.

The dielectrical permeability. Dry single-mineral rocks containing no ore inclusions are two-component and two-phase in electrical terms, so their average relative dielectrical permeability $\varepsilon_{s.air}$ can be appraised by using any of the tables presented in Table 10 or others. Such an estimate has been made by referring to equations of Lorentz-Lorentz. Lichtennecker, Odelevskii and others and the results are represented as graphs (Fig. 99) a consideration of which shows that with increasing the volumetric content of air the values of $\varepsilon_{s.air}$ decrease. They do so first more intensively compared with when the air content becomes appreciable. This is in good agreement with values of the dielectrical permeabilities of the components of a rock, viz. the solid phase ($\varepsilon_s = 6$) and air ($\varepsilon_{air} = 1$).

However absolutely dry rocks do not occur in nature, so E. I. Leont'ev proposes the following relationship for calculating the dielectrical permeability of rocks with residual water, $\varepsilon_{s.air.wr}$

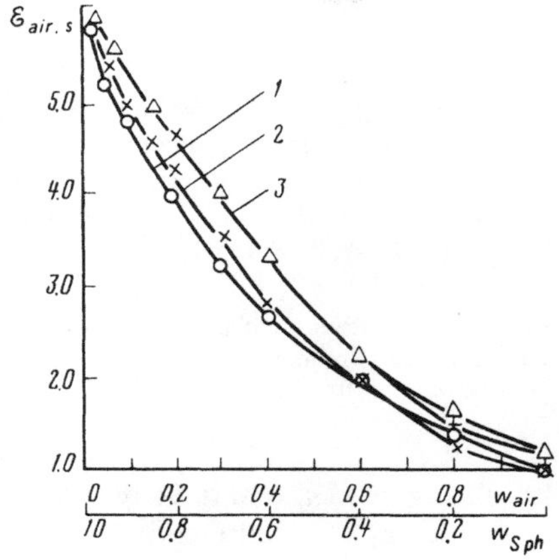

FIG. 99. Analytical relationships of the average dielectrical permeability $\varepsilon_{s.air}$ [$\varepsilon_{sph1,2}$ (see Fig. 79)] on the specific volumetric content of dry air $w_{air}(w_{sph1})$ in rock ($\varepsilon_{air} = \varepsilon_{sph1} \approx 1$) or $w_{sph}(w_{sph2})$ of solid phase ($\varepsilon_{sph2} \approx 6$).

Calculations made use of equations proposed by: *1*—Lorentz-Lorentz; *2*—Lichtennecker; *3*—Odelevskii or Boetscher

$$\frac{\varepsilon_s - \varepsilon_{s.air.wr}}{\varepsilon_s + 2\varepsilon_{s.air.wr}}(1 - k_g) + \frac{\varepsilon_{air} - \varepsilon_{s.air.wr}}{\varepsilon_{air} + 2\varepsilon_{s.air.wr}} k_p k_w$$

$$+ \frac{\varepsilon_{o(g)} - \varepsilon_{s.air.wr}}{\varepsilon_{o(g)} + \varepsilon_{s.air.wr}}(1 - k_{wr})k_p + \frac{\varepsilon_{w.b.av} - \varepsilon_{s.air.wr}}{\varepsilon_{w.b.av} + 2\varepsilon_{s.air.wr}} k_p k_{w.b} = 0$$

where ε_s, ε_w, $\varepsilon_{o(g)}$, $\varepsilon_{w.b.av}$ are dielectrical permeabilities, respectively, of the solid phase, aqueous solution of natural electrolyte, oil or gas, the average effective of the layer of bound water; $k_{w.r} = (1/n)k_w + k_{w.b}$ is a residual water saturation coefficient (free water does not differ in electrical properties from water taken in a great volume, $1/n$ is the fraction of the free water).

The dependence of $\varepsilon_{s.air.wr}$ on k_{wr} is also validated experimentally (Fig. 100).

The dielectrical saturation parameter $P_{s_\varepsilon} = \varepsilon_{osr}/\varepsilon_{w.sr}$ introduced by E. I. Leont'ev (where ε_{osr} and $\varepsilon_{w.sr}$ are dielectrical permeabilities of, respectively, partially and completely water saturated rock) is also closely connected with the voids ratio k_p of the residual water saturation coefficient $k_{w.r}$ and the specific volumetric gas saturation $w_{gV} = V_g/V_{dr}$ (V_g and V_{dr} are volumes of gas and dry rock). The quantity

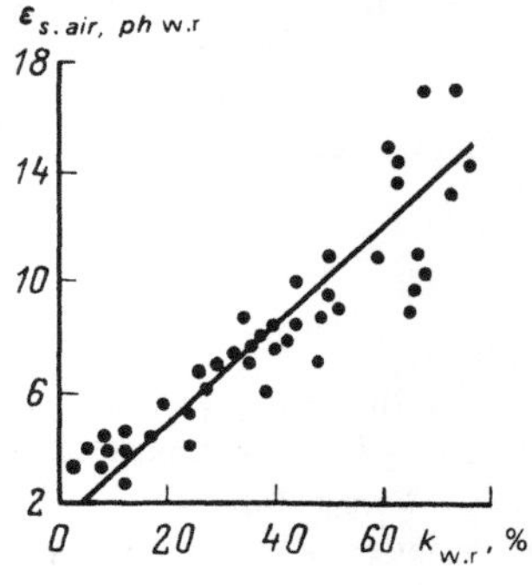

FIG. 100. Dependence of dielectric permeability $\varepsilon_{s.air,phw.s}$ of gas-saturated polymictic Mesozoic reservoir beds of West Siberia on the coefficient of residual water saturation $k_{w.r}$ (frequency 10^6 Hz) (after Yu.A. Brylkin, L.N. Dubman, E.I. Leont'ev)

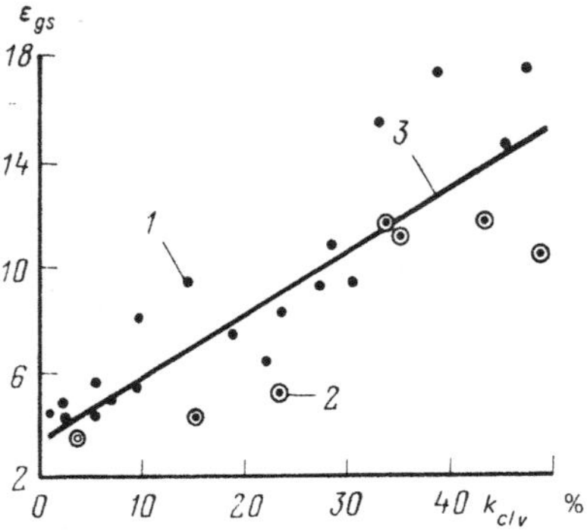

FIG. 101. Dependence of dielectrical permeability of gas saturated polymictic rocks $\varepsilon_{gs} = \varepsilon_{s.air,ph w.s}$ on the coefficient of volumetric clayiness for Mesozoic sandstones and aleurolites of West Siberia (after Yu.A. Brylkin, L.I. Dubman, E.I. Leont'ev)

1—polymictic sandstones and aleurolites; *2*—rocks with increased hydromica cement content; *3*—regression line; frequency $f = 10^6$ Hz.

P_{s_ε} is inversely related with the coefficients $k_{w.r}$ and $w_g = k_p k_g$ and directly related with $k_{w.r}$. E. I. Leont'ev suggests also a quantity $P_{0_\varepsilon} = \varepsilon_{o.sr}/\varepsilon_{dr}$ representing a ratio of the dielectrical permeabilities of a partially saturated and a dry rock (the voids of the latter being filled with air). This quantity is inversely related to k_p and directly to $k_{w.r}$.

The dielectrical permeability of partially water-saturated rocks is governed not only by the solid to the liquid or the solid to the gaseous phase ratio but also by their composition. The values of ε_{sr} of sandy-silt-rocks decrease, e.g. according to E. I. Leont'ev, with an increase in their composition of the quartz solid phase, and increase as the clayey cement content in them increases. This is accounted for by abnormally low values of the dielectrical permeability of quartz and the presence of a film of bound water in clayey particles, ion and adsorption types of an electrical double layer. The spread of points in Fig. 101 is associated with a change of the clayey cement. The decreased values of the dielectrical permeability referred to a definite clayiness are manifested by samples where the active hydromica cement content in the clayey component exceeds 45%.

Tan angle of dielectrical losses. For a dry rock, given a random distribution of voids of various configuration, the average value is

$$\tan \delta_{av} = \tan \delta_s^{1 + a k_p} \qquad (46)$$

where a is a coefficient evaluating the dependence of $\tan \delta_{av}$ on the configuration of voids (a = 1-6); δ_s is the angle of dielectrical losses of the solid component of rocks.

As follows from Eq. (46), with increasing the voids ratio $\tan \delta_{dr}$ of a dry rock decreases.

Types and Groups of Different Rocks

The different role played by different varieties of different rock types and their definite groups in the electrical and electrochemical processes is evidenced by the wide range of variation of ε_{dr}, ϱ_{mr}, $\tan \delta$, A_i and A_{da}. This is accounted for by the different composition of phases of the principal varieties of separate types of various rock types, specific content, their distribution and interaction in the volume of these media, temperature and pressure variations, conditions of laboratory and field determinations of the quantities of interest to us.

When describing experimental data referred to the limiting values of electrical quantities of rock groups (see Tables 11, 12), it is necessary to point out the following.

1. The variation range of tan δ and A_{da} of metamorphic rocks, ores, rocks with ore inclusions and coals has not been practically established; nor have the limiting values of A_{da} of igneous rocks; the quantities characterizing the electrical properties of sedimentary rocks have been studied better than others; however, this group includes rock types for which it has not proved possible to evaluate the limiting values of definite electrical quantities.

2. The variation range of values of electrical quantities of sedimentary rocks are wider compared with others (see Tables 11 and 12) which is due to great intervals of variation of their volumetric moisture content, specific surface, mineralization of waters in voids space. Maximum values of ϱ, A_i, A_{da} are shown by low-porosity, hygroscopically moist sedimentary rocks as well as different voids size, highly air-saturated, gaseous hydrocarbons, oil and mixtures of these materials of rocks. The values of ε and tan δ of the aforementioned rocks are minimum (see Tables 11 and 12).

3. The values of $\varrho_{ws.r} \cdot \varepsilon_{dr}$ and A_i of igneous and metamorphic rocks vary over a narrower range compared with those of sedimentary rocks (see Tables 11 and 12). This is conditioned by a smaller variation range of their voids ratio and moisture content coefficient as well as by that laboratory determinations of ε_{dr} have been mainly made on dry rock samples and at a frequency 10^2 and 10^7 Hz, rather than over the entire frequency range (0-10^{14} Hz).

4. The greatest variation range of the specific resistance is characteristic of ores and coals (see Tables 11 and 12). The specific resistance of certain varieties of ores, rocks with ore inclusions, graphites and coals can attain much lower and the induced electrochemical activity of high values (see Tables 11 and 12) compared with other rocks owing to the presence in their composition of electron-conducting inclusions even though some ores and rocks with ore inclusions may exhibit low A_i values.

5. The dielectrical permeability of ores and rocks with ore inclusions changes over a relatively narrow range. Coal may have both low (given a small content of the electron conducting component) and sufficiently high values of ε_{dr} provided that they are characterized by an increased content of the electron conducting component.

6. As suggested by the available evidence, tan δ values vary over the narrowest range ($\sim$0.015 to 1), this range is somewhat wide for ε_{dr} (2.5-35), a still greater variation range is shown by A_i values (2.5×10^{-2}-70-95) and A_{da} values (-25 to 70 mV), and the greatest variation range is shown by ϱ_{mr} (10^{-7}-10^8 Ohm·m) of rocks.

7. A comparison of values of dielectrical quantities of various types of maximum moisture and dry rocks points to the following.

TABLE 12. Ranges of Variation of Electrical Quantities of Various Rock Groups

Rock group	Moisture content	Specific resistance ϱ, Ohm·m	Moisture content	Frequency, Hz	Dielectrical permeability ε
Igneous	Maximum	$10\text{-}10^6$	Air-dried	10^2 10^7	7.2-23.5 5-11.9
Metamorphic	Ditto	$10\text{-}10^6$	Ditto	10^2 10^7	4.3-31.5 4-13
Sedimentary	Ditto	$10^{-1}\text{-}10^8$	From air-dried to maximum moisture rocks	10^2 10^7	>30->80 3->4
Ores and rocks with ore inclusions	-	$10^{-7}\text{-}10^6$	-	-	10-35
Graphite	-	$10^{-6}\text{-}10$	-	-	-
Coal	-	$10^{-3}\text{-}10^6$	Air-dried	-	2.5-15

* For rocks with ore inclusions.

Igneous rocks. The values of ϱ_{mr} of maximum moisture intrusive rocks increase from their acidic to intermediate, basic and ultrabasic types. The same is also true of ϱ_{mr} of unaltered effusive rock types. This regularity in the variation of the specific electric resistance is connected with a general growth of the water content from basic to acidic rock varieties.

The values of the dielectrical permeability of dry acidic rocks are lower than other igneous rock types since they contain more hardly polarized quartz, air, and ore inclusions are less likely to occur in them compared with basic and ultrabasic rocks (gabbro, diabases, basalts, peridotites).

The greater the content of Fe^{2+}, Fe^{3+} cations, magnetite or other electron conducting minerals and the more the dispersion of ore minerals, the higher the values of tan δ of individual types of intrusive rocks: tan δ also grows if the ore component is found along the boundaries of rock-forming mineral grains.

The induced electrochemical activity of individual types of massive intrusive igneous rocks is generally 1-2%, upon their decomposition it drops to 0.5-1%; A_i of igneous rocks with ore mineral inclusions increases to 6% and more.

Sedimentary rocks. Given maximum moisture content, the highest values (from 10^4 to 10^8 Ohm·m) of the specific resistance are demonstrated by rock salt; they are less for anhydrite (from 10^2 to 10^6 Ohm·m). The values of ϱ (from 1 to 10^5 Ohm·m for dolomites and 2×10^5 Ohm·m for limestones) of calcareous and magnesial rocks vary over a wide range. Still smaller ϱ values (0.1 to 0.5 Ohm·m) are manifested by sands, sandstones and aleurolites. However, ϱ of fresh-water saturated tight sands and sandstones may attain as much as several thousand ohmmeters whereas for clays, argillites and schists it is often $\varrho = 1$ Ohm·m and may attain one (for clays) or several (for argillites and argillaceous slates) hundred ohmmeters.

Moisture content	Frequency, Hz	Tan angle of dielectric losses, tan δ	Moisture content	Induced electrochemical activity, A_i, %	Moisture content	Diffusion-adsorption activity, A_{da}, mV	Notes
From air-dried to maximum moisture rocks	5×10^6	0.015-0.04 0.12-0.2	Maximum	0.1-7.6*	-	-	Rocks with ore inclusions
-	-	-	-	<0.1-26*	-	-	-
From air-dried to maximum moisture rocks	10^2-10^6	0.1-1 and more	-	2.5×10^{-2}-10*	Maximum	-25->70	-
-	-	-	-	0.1-70	-	-	-
-	-	-	-	-	-	-	-
-	-	-	-	95	-	-	-

The highest ε values (>80 for $f = 10^2$ Hz) have been found for high moisture sands and sandstones. With an increase in the voids ratio, oil saturation coefficient and growth of frequency of gas- and oil-saturated rocks, the dielectrical permeability of these rocks decreases since their composition shows an increased content of hardly polarized gas or oil. The values of the dielectrical permeability of dry limestones and dolomites are higher compared with dry sandstones, given the same k_p, since their rock-forming minerals calcite and dolomite show greater ε values than quartz which often is the main rock-forming mineral of sandy aleurite clayey rocks.

The dielectrical permeability of hydrochemical sediments, anhydrites, gypsums, rock salt insignificantly differ from ε of their principal rock-forming minerals due to inappreciable and sufficiently stable porosity of these rocks. Relatively high ε values have been found for sedimentary magnetite and martite ores (see Table 11).

The dielectrical losses of tan $\delta = 0.01$ are referred to low-porosity weakly moist rocks and the highest ones (up to 1) to high-porosity moist sedimentary rocks.

The induced electrochemical activity of sands and sandstones varies over the greatest range from several hundredths of a per cent to over 2%. The lowest A_i values are shown by well sorted rounded quartz sands and other rocks with a homogeneous granulometric composition. By contrast, the highest A_i values are characteristic of clayey varieties. As to gas-bearing representatives of these rock types, A_i can increase to attain 3-3.7%. The maximum A_i values correspond to the moisture content (mass) 3-5%. Sandstones showing A_i = 15-30% contain iron sulphide impregnations.

The values of A_i of aleurolites varies over a rather wide range (0.7-10% and more) attaining sufficiently high values which is accounted for by their greater fine-grained content, clayiness, possibly, sulphide content and content of some oxides.

A smaller range of variation is manifested by A_i of clays and argillites. Low A_i values refer to fat, dense, mineralized-water saturated clays, and high ones to their varieties with ore inclusions. The value of $A_i = 0.1\text{-}2$ is typical for limestones and dolomites if these rocks have ore inclusions, their number exceeds 2. Minimum (as low as 15 mV) and sometimes negative A_{da} values (up to -25 mV for limestones) are shown by pure sand, sandstone, aleurolite, limestone and dolomite varieties. High positive (as great as 45 mV) A_{da} values are demonstrated by clayey aleurolites, sandstones, limestones and dolomites. Maximum A_{da} values (up to 70 mV) are characteristic of clays. Appreciable values of A_{da} (up to 50-55 mV) have been also found for low-porosity, strongly cemented, strongly clayey sandstones, marls, limestones and dolomites.

Metamorphic rocks. The specific resistance of individual types of these rocks, even given their complete water saturation, is generally inappreciable ($>10^3$ Ohm·m). An exception is argillaceous slates whose ϱ is within $1\text{-}5 \times 10^2$ Ohm·m and is even less in case when these rocks contain ore inclusions. Particularly high values ($10^5\text{-}10^6$ Ohm·m) are shown by the specific resistance of marbles, amphibolites and hornfelses. Somewhat lower ϱ values ($5 \times 10^2\text{-}10^6$ Ohm·m) have been found for individual crystalline schist, quartzite and skarn varieties. A great number of other rocks of the group (phyllites, gneisses, serpentines) frequently have lower specific resistances ($10^3\text{-}10^4$ Ohm·m). Should these rocks contain electron conducting inclusions, their specific resistances appreciably decrease, but, given poor electrical bonding and a small content of electron conducting grains, ϱ of metamorphic rocks may attain 10^3 Ohm·m.

Quartzites are characterized by the least values of dielectrical permeability. The ε values of marbles are higher which is in agreement with dielectrical permeabilities of their principal rock-forming minerals (quartz and calcite) and low k_p of these rocks. The ε values of gneisses and serpentinite are inappreciable. The dielectrical permeabilities of other metamorphic rock types have been little studied. The A_i coefficient of metamorphic rocks varies from 0.1-0.5 to 2.5% in the absence of ore inclusions. In the presence of these latter A_i is in the range from 5 to 25%.

Ores. A specific position, in terms of ϱ, ε, tan ε and A_i values, is occupied by ores and sedimentary and metamorphic rocks enclosing them with an increased content of ore inclusions. The specific resistance of ores attains 10^3 and even $10^5\text{-}10^6$ Ohm·m as is the case, e.g. of magnetite, martite, cassiterite and cericite quartz and other rocks, and may drop to $10^{-5}\text{-}10^{-7}$ Ohm·m. A high ϱ value is shown by sparsely impregnated ores with electrically non-bound electron conducting inclusions, and low ϱ values by massive ores and ores with electrically bound electron conducting inclusions. For well conducting ore varieties the values of ε and A_i are high ($\varepsilon = 25\text{-}35$ and A_i up to 20-70%) according to a relatively small amount of evidence. In the case of poorly conducting ores these characteristics drop appreciably (ε to 10-18) and A_i varies from $<0.5\text{-}1$ to 2-10%.

Graphites. The specific resistance of graphites is small (from 0.5×10^{-6} to 10), and ε, A_i and tan δ values have been insufficiently determined.

Coals. Coals are poor conductors of electric current: their $\varrho_{coal} = 10^2$-10^6 Ohm·m, brown coals conduct current somewhat better and have $\varrho_{coal} = 10$-2×10^2 Ohm·m owing to their higher porosity. Both coal varieties represent natural semiconductors. Anthracites with an electron ion conductivity frequently provide good conductors of current, their ϱ varying from 10^{-3} to 10 Ohm·m. The dielectrical permeability of these minerals varies over the range from 2.5 to 15, and A_i of anthracites attains 95%.

Electrical Quantities, Given High Temperatures and Pressures and Various Field Frequencies

The geostatic (overall) pressure at depths currently reached by prospecting boreholes attains (1.5-2)·10^2 MPa and temperature 200 °C. Viewed in this context, it is important to know the way electrical quantities vary depending on the effective pressure and temperature as well as on the field frequency. Few data are as yet available on variations of tan δ, A_i and A_{da} values with increasing the pressure, temperature and current frequency, that is why we will mainly dwell on the variation of ϱ and ε depending on these factors. Unless rocks are decomposed, due to the action of the effective pressure $P_{efg} = p$-p_p (where p, p_p are an overall and a pore pressure, respectively), the ϱ and ε values change mainly due to volume deformations (compressibility of their voids space and solid phase).

The following facts have been established experimentally. With growth of an overall (geostatic) pressure, given p_p = const, the specific resistance $\varrho_{ws.r}$ of maximum-moisture, mainly sedimentary rocks (sands, sandstones, limestones, dolomites etc.) first rapidly increases then ever more slowly, attaining the limiting values, given high p; at this, the highest $\varrho_{ws.r}$ are attained for the lowest p_p values (Fig. 102, $\varrho = f(p)$ curve family, given p_p = 15, 30, 60 and 90 MPa). This is accounted for by the fact that the pore pressure p_p counteracts the overall pressure p due to which the latter deforms a rock less the greater is the pore pressure. With increasing p, given p_{eff} = const, the specific resistance drops and its values for specified p prove higher the greater are p_{eff} values, (see Fig. 102, $\varrho_{ws.r} = f(p)$ curve family for p_{eff} = 15, 30, 45 MPa). In this case the growth of the overall pressure leads to an increase in the effective pressure p_{eff}, and in order that it should remain constant, the pore pressure p_p must be increased such as to facilitate a lessening of the rock deformation owing to the compression of its solid phase and a decrease in $\varrho_{ws.r}$.

Relative changes of the specific resistance of rocks induced by the pressure are mainly governed by their structural changes (decreased porosity, increased meander-

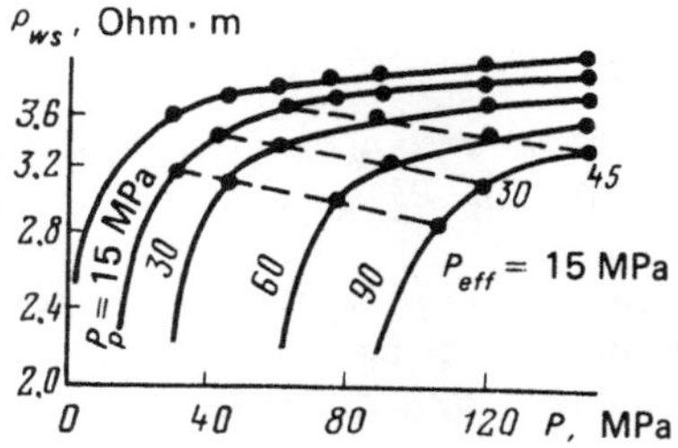

FIG. 102. Dependence of the specific resistance of highly porous sandstone $\varrho_{ws.r}$ on the overall pore pressure of electrolyte p_{op} for t = const

ing rate and specific surface, narrowed voids channels, lessened size and complicated grain configuration and sorting). In the case of sandstones these changes are affected by their cement type. The least structural changes are typical of sandstones with opal cement, and the maximum ones by the same rocks with clayey cement. The quantity $\varrho_{ws.r}$ is intricately related to temperature, too. The latter affects the specific resistance of rocks due to increased conductivity of their voids space electrolyte, under the effect of temperature, however, the structure of rocks also changes. The voids ratio values of weakly clayey sandstones increase with temperature. Given clayey sandstones, the form of this relationship is influenced by the increased electric conductivity of the clayey component.

The electrical double layer changes in structure and conductivity, with growth of temperature this leads to an increase in p_p values and a decrease in $\varrho_{ws.r}$. Given $t = 250$ °C, $\varrho_{ws.r}$ of sedimentary rocks is 20 times and more less compared with $t = 20$ °C. Maximum changes are observed at temperatures from 20 to 50-60 °C. The rate of variation of the specific resistance is then lowered. For most investigated rock types ϱ_{op} decreases with depth (see Table 46), since temperature affects this quantity more than does pressure. The decrease of $\varrho_{ws.r}$ with depth is particularly pronounced for high porosity sandstones; at a depth 5 to 6 km it is on the average 60% of the original one corresponding to $t = 20$ °C and $p = 0.1$ MPa. If solely pressure is to be taken into account, $\varrho_{ws.r}$ proves to be greater by 16-20%. In the case of low porosity rocks ($k_p \approx 5\%$) due to the concurrent and opposite action of p and t the resistance $\varrho_{ws.r}$ decreases inappreciably. The specific resistances of sedimentary rocks are generally diminished upon giving corrections for temperature and pressure since the decreasing effect of temperature is greater than the increasing effect of pressure. Sometimes, however, they do not practically change. The total relative variation of the specific resistance of sedimentary rocks with growth of pressure and temperature, according to V. M. Dobrynin, is as follows:

$$\frac{\varrho_{p,p_s,t}}{\varrho} \approx \frac{\varrho_{p-p_{s1}}}{\varrho} \frac{\varrho_{p_s}}{\varrho_{p_{s1}}} \frac{\varrho_t}{\varrho}$$

where $\varrho_{p,p_s,t}$ is the specific resistance of a rock, given overall p, stratal p_s pressure and temperature t; ϱ is the same in laboratory (atmospheric) conditions $\varrho_{p-p_{s1}}/\varrho$, $\varrho_{p_{s1}}/\varrho_{p_{s1}}$ and ϱ_t/ϱ are relative changes of the specific resistances of a rock due to an increase, respectively, of the effective pressure, stratal pore pressure and temperature; $\varrho_{p_{s1}}$, ϱ_{p_s} are specific resistances at the initial and current pressure of the electrolyte filling the channels.

The relative variation of rock resistance is

$$\varrho_t/\varrho = (P_p)_t(\varrho_w)_t/P_p\varrho_w$$

where ϱ_t, $(P_p)_t$, $(\varrho_w)_t$ are, respectively, the specific rock resistances, voids ratio and specific resistance of the pore water at temperature t; ϱ, P_p, ϱ_w are the same at a specified temperature (e.g., 20 °C).

The dielectrical permeability of rocks ε_r generally grows with pressure (Fig. 103) connected with an increase in the volume concentration of polarized particles. ε_r first, as a rule, remains practically unchanged followed by a dramatic increase in the rate of its variation (Fig. 104).

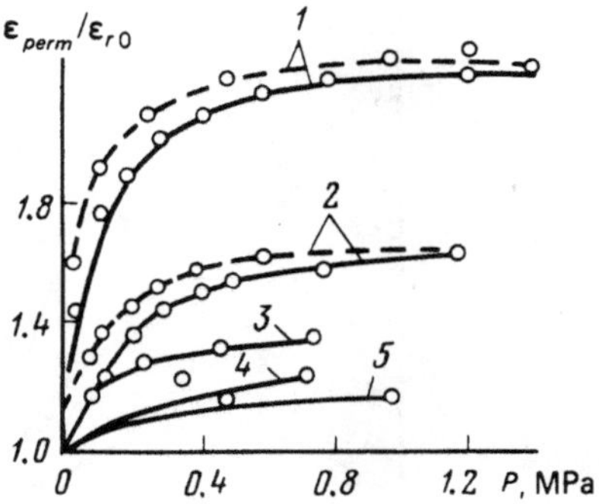

FIG. 103. Dependences of the relative variation of dielectric permeability $\varepsilon_{perm}/\varepsilon_{r0}$ on overall pressure p for rocks (after E.I. Parkhomenko)

1—sandstone; *2*—syenite; *3-5*—various granite samples

The growth of the frequency of the electrical field contributes to an expansion of ε_r values over which they change only inappreciably. Given a frequency $f = 10^5$ Hz, ε_r varies from 5 to 10, given $f = 10^2$ Hz and increasing t from 200 to 1000 °C, over a much greater range (see Fig. 104). Data on the joint effect of pressure and temperature on ε_r are unavailable. They are, however, needed, primarily, for determining the distribution pattern of ε_r with depth.

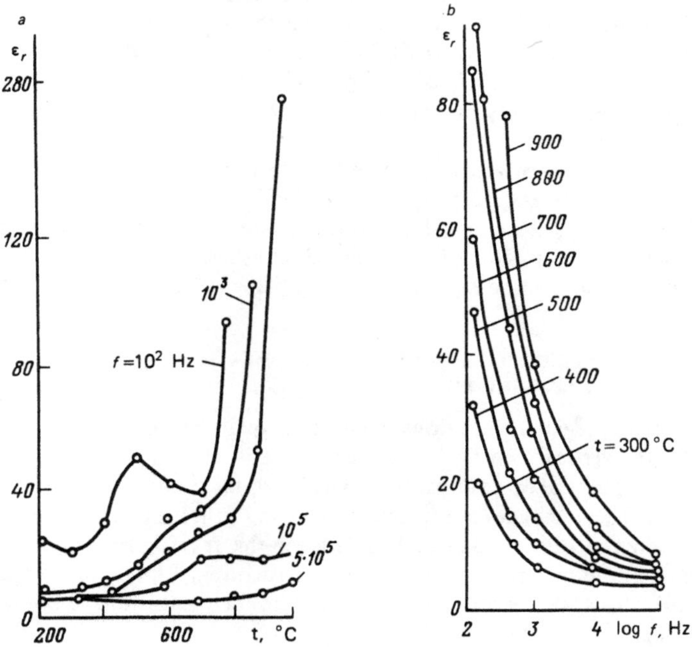

FIG. 104. Dependences of the dielectric permeability of rocks ε_r on frequency f and temperature t (after E.I. Parkhomenko).

a—granosyenite; *b*—granite

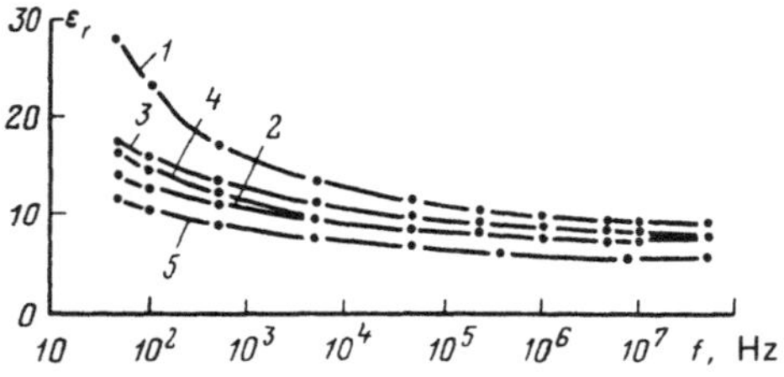

FIG. 105. Dependence of the dielectric permeability ε_r on frequency f for definite magmatic rocks (after B.F. Howell and P.H. Leecastre).

1 and *2*—diabases; *3* and *4*—gabbros; *5*—peridotite

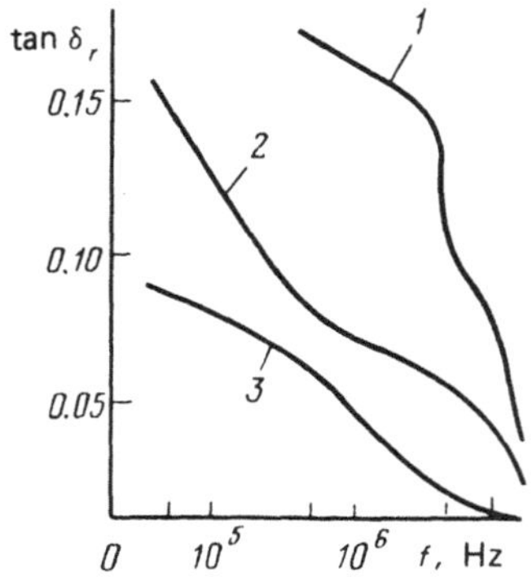

FIG. 106. Dependences of tan δ_r on frequency f for syenite.

Samples: *1*—maximum-moisture; *2*—air dry; *3*—dried

Whenever a frequency increases and given $t = 20$ °C, ε and tan δ of all igneous rock types diminish. For peridotite, gabbro, diabase, diorite a decrease has been established in ε by a factor of 2 or 3 with increasing the frequency from 50 to 5×10^7. The character of dependences $\varepsilon = F(f)$ and tan $\delta_r = F(f)$ mainly governed by the availability of moisture is illustrated in Figs. 105 and 106. Owing to a great value of the dielectrical permeability of water ($\varepsilon_w = 80$) even its inappreciable content may noticeably increase ε_r.

Given water in the pore space of rocks and a sufficiently high specific surface value, volumetric polarization is also possible whose effect on ε_r decreases with frequency. The dielectrical permeability of sedimentary and metamorphic rocks at $t \approx 20$ °C and f increasing from 10 to 10^7 Hz also diminishes, while in the rocks with a high moisture content more significantly (Fig. 92).

Local and Regional Variations of Electrical Quantities

A knowledge of local and regional variations of electrical quantities of strata, suites and rock masses in the vertical and horizontal directions is of paramount importance for successful geophysical operations. It, in particular, facilitates petrophysical zoning of large areas. These changes are traced by referring to joint electrical cross sections and maps prepared on a small scale solely for the specific resistance. A layer-by-layer zoning of the specific resistance at local structures of individual areas differs very much. For instance, for local structures of dense sediments of the Lower Paleogene of the Tajik depression (according to works of S. I. Dembitskii) an increase is observed in the values of specific resistance toward the arch of structures whereas at periclines and in the direction of depression ϱ_{op} values diminish. For plastic sediments of the Upper Paleogene a drop in ϱ_{op} is observed toward the arch of structures due to the increased porosity of rocks. The layer-by-layer zoning is

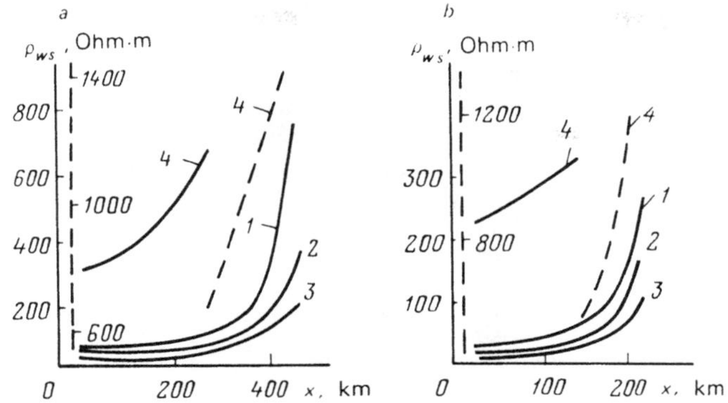

FIG. 107. Regional variations of the specific resistance of rocks of the Rudnitskaya subsuite of the Vorkuta series of the Pechora basin (after V.V. Grechukhin).

1—sandstone, *2*—aleurolite; *3*—argillite; *4*—coal; studies conducted in the following directions: *a*—from SW to NE; *b*—from NW to SE. To plot graphs, use was made of average ϱ_{ws} values referred to individual deposits obtained on samples and layers

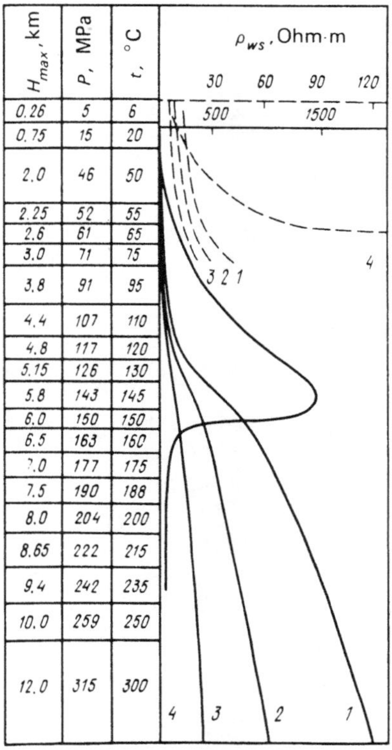

FIG. 108. Typical electric cross section of sediments of geosyncline group coal basins (after V.V. Grechukhin).

1—sandstone; *2*—aleurolites; *3*—argillite; *4*—coal (ash content of coal A_c is 20%)

smoothed with increasing the thickness of the sedimented cover and total drop of the voids ratio.

A comparison of electrical cross sections of definite stratigraphic horizons or sedimented masses referred to different boreholes within areas and regions (e.g. the Pechera, Kuznetsk, Donets coal basins of a geosyncline type) has enabled V. V. Grechukhin to establish characteristic variations in ϱ_{op} of rock types over the areas of the basins (Fig. 107*a*, *b*) and a regular growth of the specific resistance with depth (Fig. 108).

Chapter IX

Thermal Conductivity, Thermal (or Heat) Capacity, Thermal Diffusivity

Sec. 38. The Processes and Laws Governing the Heat Distribution in Rocks

The distribution of temperatures at the Earth's surface and in its inner reaches—the natural thermal field of the earth—is dependent on: (1) the spatial distribution and the power of heat sources (the sun, atmospheric precipitation, radioactive elements, chemical reactions, crystallization, polymorphic transformations, compaction and other processes); (2) the ability of rocks for heat exchange—heat energy transfer; (3) spatial rock distribution having different thermal conductivity. We distinguish also such types of heat transfer as thermal conductivity, convective and radiant heat exchange.

Thermal conductivity is a directed transfer of internal heat energy, a process of heat distribution from more to less heated volumes of a nonuniformly heated substance (without convection and radiation) contributing to the equalization of the temperature of this medium.

The internal heat energy of dielectrics is due to thermal vibrations of their crystal lattices, and that of conductors and semiconductors, besides that, due to the thermal movement of electrons. The nodes of the lattices of dielectrics accommodate interacting atoms, molecules or ions in a thermal motion and, since the vibrations of particles of crystal structure are not isolated, thermal oscillation waves propagate in the substance. The transfer of energy of the connected vibrations of the lattice nodes are thought to be the propagation in a substance of harmonic elastic acoustic waves of different frequency ν_i.

Acoustic waves of high frequencies ν_{max}, according to quantum theory, are a gas of acoustic quanta called phonons. The anharmonicity of the lattice node vibrations leads to phonon distribution and impairs heat transfer via thermal conductivity. This effect increases with temperature so the lattice thermal conductivity of dielectrics decreases in the particular case. Thus heat transfer in dielectrics is considered as being a transfer by microscopic waves of kinetic energy toward its decrease.

Heat energy transfer of conductors and semiconductors is mainly implemented by the diffusion of free electrons—the transfer of heat energy by the conductivity of electrons, the lattice thermal conductivity being much less than electron thermal conductivity.

In a stationary field the amount of heat Q transferred through a flat layer x in thickness is proportional to the temperature gradient $(t_1 - t_2)/x$, area of the lay-

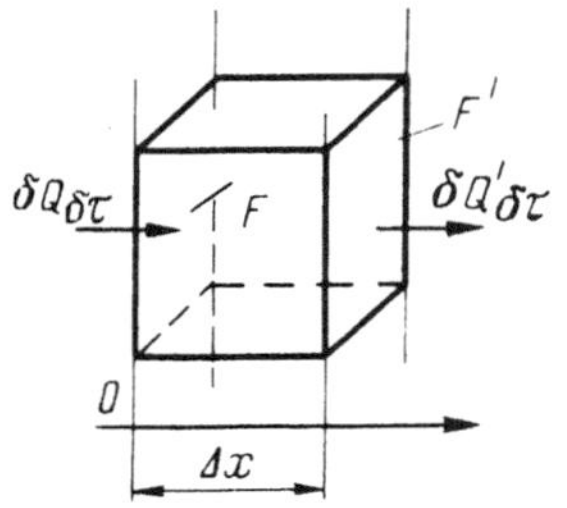

FIG. 109. Diagram illustrating the law of conservation of thermal energy

er's surface F and time τ:

$$Q = \lambda \frac{t_1 - t_2}{2} \tau F \tag{47}$$

$$q = \frac{Q}{F} = \lambda \frac{t_1 - t_2}{x} \tau$$

where λ is a coefficient of thermal conductivity depending on the specific features of the thermal properties of the layer's material, J/(c·m·K) or Wt/(m·K); q is the thermal flux density, J/m^2 is the amount of heat passing in a unit of time, τ, s, across the area of the surface F, m^2, which is at right angles to the X axis.

If we neglect transformations of heat energy to other types then, as follows from the law of conservation of energy, in the case of a unidimensional problem (Fig. 109)

$$\delta Q_{\delta\tau} + \Delta V w \delta\tau = \delta Q'_{\delta\tau} + c_{pm}\delta_r \Delta V \delta t \tag{48}$$

where $\delta Q_{\delta\tau}$ and $\delta Q'_{\delta\tau}$ are, respectively, minor (elementary) amounts of heat entering the volume ΔV across the cross section area F and leaving it across the cross section area F' in the time interval $\delta\tau$ (see Fig. 109); $\Delta V w \delta\tau$ is the amount of heat released in a volume ΔV in the time interval $\delta\tau$ (w is power of heat sources in per unit of volume); $c_{pm}\delta_r\Delta V\delta t$ is heat amount in ΔV volume in the time interval τ (here δt is the temperature change which is a function of the x and τ coordinates); c_{pm} is the specific isobaric mass thermal capacity; δ_r is the density of the medium under study.

Since $\delta Q_{\delta\tau} = -\delta q F \delta\tau = -\lambda F(\partial t/\partial x)_x \delta\tau$, $\delta Q'_{\delta\tau} = \delta q' F \delta\tau = -\lambda F(\partial t/\partial x)_{x+\Delta x}\delta\tau$ and $\Delta V = F\Delta x$, then Eq. (48), after transformations, takes on this form:

$$-\lambda F(\partial t/\partial x)_x \delta\tau + F\Delta x w \delta\tau = -\lambda F(\partial t/\partial x)_{x+\Delta x}\delta\tau + c_{pm} d_r F \Delta x \delta t \tag{49}$$

The minus sign preceding the quantity λ in the latter relationships and in Eq. (49) indicates that an increase in temperature proceeds in the direction opposite to that of the thermal flux δq.

By taking into account the dependence of t on x and τ, λ on t we can consider the product $\lambda(\partial t/\partial x)$ also a function of x and τ: $\lambda(\partial t/\partial x) = \Phi(x, \tau)$.

If we divide Eq. (49) by $F\Delta x\delta\tau$ and carry the first term of its right-hand side to the left-hand side we get

$$\frac{\Phi(x + \Delta x, \tau) - \Phi(x, \tau)}{\Delta x} + w = c_{pm}\delta_r \frac{\delta t}{\delta\tau}$$

or, in the limiting case, given $\Delta x \to 0$,

$$\frac{\partial \Phi(x, \tau)}{\partial x} = \frac{\partial}{\partial x}\left(\lambda \frac{\partial t}{\partial x}\right) + w = c_{pm}\delta_r \frac{\partial t(x, \tau)}{\partial \tau} \tag{50}$$

which, following transformations, takes on this form

$$\frac{\partial t(x, \tau)}{\partial \tau} = \frac{\partial}{\partial x}\left(a \frac{\partial t}{\partial x}\right) + \frac{w}{c_{pm}\delta_r}$$

and, given that a = const and $w = 0$ (in the absence of a heat source in the medium)

$$\frac{\partial t(x, \tau)}{\partial \tau} = a \frac{\partial^2 t(x, \tau)}{\partial x^2} \tag{51}$$

where the coefficient of thermal conductivity (in m^2/s) is

$$a = \lambda / c_{pm}\delta_r \tag{52}$$

For heat distribution in a three-dimensional space, according to the above relationships, we have

$$\frac{\partial}{\partial x}\left(\lambda_x \frac{\partial t}{\partial x}\right) + \frac{\partial}{\partial y}\left(\lambda_y \frac{\partial t}{\partial y}\right) + \frac{\partial}{\partial z}\left(\lambda_z \frac{\partial t}{\partial z}\right) + w = c_{pm}\delta_r \frac{\partial t}{\partial \tau} \tag{53}$$

given λ = const,

$$\lambda \nabla^2 t + w = c_{pm}\delta_r \frac{\partial t}{\partial \tau} \tag{54}$$

or

$$\frac{\partial t}{\partial \tau} = \frac{\lambda}{c_{pm}\delta_r} \nabla^2 t + \frac{w}{c_{pm}\delta_r}$$

where

$$\nabla^2 t = (\partial^2 t/\partial x^2) + (\partial^2 t/\partial y^2) + (\partial^2 t/\partial z^2)$$

If heat sources are absent in the volume of a homogeneous medium under study ($w = 0$), then $\lambda \nabla^2 t = c_{pm}\delta_r\,(\partial t/\partial \tau)$ or $\partial t/\partial \tau = a\nabla^2 t$.

Consequently, under conditions of conductive heat exchange thermal properties of homogeneous isotropic and anisotropic substances are characterized by *the thermal conductivity and thermal diffusivity coefficients* λ and a and *the specific mass heat* c_m. The coefficient λ determines the different capacity of single-phase substances for heat transfer via thermal conductivity.

The heat capacity C (in J/K) of a sample of any substance is the amount of heat needed to increase its temperature by 1° under conditions of a specified thermodynamic process (p = idem, V = idem etc.):

$$C = (\lim \Delta Q/\Delta t)_{\Delta t \to 0} = \delta Q/\delta t$$

where δQ is an infinitesimal amount of heat transferred to the specimen; δt is an infinitesimal variation of its temperature.

The average heat capacity of a sample following a variation of its temperature from t_1 to t_2 is

$$C_{av} = Q_{1,2}/(t_2 - t_1)$$

where $Q_{1,2}$ is the amount of heat spent for the variation of a sample's temperature from t_1 to t_2. Since samples of various materials generally differ in mass or volume, for a comparative evaluation of their properties for differently absorbing heat energy during heat transfer the mass, volumetric and molar specific heat capacities are introduced.

The specific mass heat c_m [in J/(kg·K)] is the amount of heat necessary for changing by 1 degree a unit mass m of the sample in a particular thermodynamic process (p = idem, V = idem etc.):

$$c_m = \delta q/\delta t = C/m \qquad (55)$$

where δq is the infinitesimal amount of heat imparted to a unit mass of the substance; δt is an infinitesimal variation of its temperature; m is its mass. The average specific mass heat c_{mav} [in J/(kg·K)] is the amount of heat $q_{1,2}$ (in J/kg) to be imparted to a unit mass of the substance for causing its temperature to vary by 1 degree, given a definite thermodynamic process and temperature interval from t_1 to t_2:

$$c_{mav} = q_{1,2}/(t_2 - t_1)$$

The specific volumetric heat c_V [in J/(m^3·K)] is the amount of heat to be imparted to a unit mass of a substance for causing its temperature to change by 1 degree in a definite thermodynamic process (p = idem, V = idem):

$$c_V = \delta Q/V_n\delta t = \delta_n c_m$$

where δQ is an elementary heat amount, J; δt is an elementary temperature increment, °C; V_n and δ_n are, respectively, volume, m^3, and density, kg/m^3, of a sample of a substance at normal physical conditions (t = 0 and p = 0.101325 MPa or 760 mm of column of mercury).

The average specific volumetric heat c_{Vav} [in J/(m^3·K)] is

$$c_{Vav} = Q_{1,2}/V_n(t_2 - t_1) = \delta_n c_{mav}$$

where $Q_{1,2}$ is the amount of heat imparted to a sample for causing its temperature to change from t_1 to t_2.

The specific molar heat c_ν (in J/(kmole·K)) is the amount of heat to be imparted to a unit molar amount of substance for causing its temperature to change by 1 degree in a definite thermodynamic process (p = idem, V = idem etc.):

$$c_\nu = \delta Q/\nu\delta t = Mc_m$$

Here δQ and δt are, respectively, an infinitesimal amount of heat, J, and temperature variation, °C; ν is the molar amount of the substance, kmole; M is a relative molecular mass of the substance, kmole/kg. The molar heat capacity is constant for all materials and equals $25\cdot10^{-3}$ J/(kmole·K).

The average specific molar heat is

$$c_{\nu av} = Q_{1,2}/\nu(t_2 - t_1) = Mc_{mav}$$

where $Q_{1,2}$ is the amount of heat imparted to a substance sample upon varying its temperature from t_1 to t_2; ν is the molar amount of a substance.

Since under specified conditions of heat transfer the heat capacity is expressed by means of $W = E + pV$ of the sample state, i.e. the specific mass enthalpy alternatively known as heat content, sensible heat or total heat (where E is the internal energy of the sample; V is its volume; p is its pressure), then we also distinguish heat capacity C_{Vm} for a constant volume and c_{pm} for constant pressure acting on the sample during the heating. The specific mass heat capacity for V = const of the sample being heated $c_{Vm} = (\partial E/\partial t)_{V=const}$ characterizes the heat capacity when in the period of heating (cooling) its volume remains constant and all heat $\delta Q = \delta E$ is spent for incrementing the internal energy.

Given constant pressure, the specific mass heat is determined by relationship

$$c_{pm} = (\partial W/\partial t)_{p=const} \tag{56}$$

The character p in Eq. (56) means that the derivative is partial for p = const.

Given p = const, the heat $\partial Q = \delta E + \delta A = \delta E + p\delta V = \delta(E + pV)$ is spent on the variation of its internal heat energy and work $\delta A = p\delta V$ of sample expansion since sample expansion uses a fraction of thermal energy during the heating.

Far from the melting temperature the crystal structure is stable, i.e. particle vibrations are small and harmonical. A substance containing N particles has a system of three classical unidimensional harmonical oscillators (and particles vibrate in three individual directions). According to the law of distribution of energy with respect to degrees of freedom, an average energy kT (where T is the absolute temperature of the crystal) refers to each classical oscillator. Consequently, the internal energy conditioned by vibrations of the mineral lattice is

$$E = 3NkT$$

therefore $c_{Vm} = (\partial E/\partial t)V = 3Nk$.

For one mole of a substance ($N = N_A$) $c_{Vm} = 3N_A k = 3R = 25 \times 10^3$ J/(kmole·K).

The heat capacity of substances in laboratory conditions is evaluated as being the amount of heat needed for heating 1 kg or 1 m^3 of a dry substance by 1 °C under standard constant or varying thermobaric conditions available here.

For low-porosity substances heat exchange is implemented solely by thermal conductivity, and for porous and permeable substances thermal energy is also transferred by convection, so, when calculating the temperature diffusivity a from Eq. (52), use must be made of the quantity c_{pm}. Since in this case the components of a substance that are mobile at a pressure gradient are at rest, then, given a temperature gradient, they do not transfer additional amounts of heat.

The coefficient a characterizes the speed of propagation of the isothermic surface during temperature equalizing in a nonuniformly heated mineral or rock.

Sec. 39. Minerals

Thermal conductivity and thermal resistance. The conductive heat exchange is characteristic of minerals. The thermal conductivity of minerals λ_m has been studied rather insufficiently. Very few λ_m values have been obtained for sulphides, not have been adequately determined for oxides, sulphates, and carbonates. Data are available, though few, on wolframates, molybdates, phosphates and borates. The thermal conductivity of silicates has been investigated best. Sufficient data are also available on the thermal conductivity of native metals; their λ values little differ from those of relevant pure elements that have been sufficiently studied.

The coefficient of thermal conductivity of minerals varies from ~ 0.3 (for sulphur) to >420 W/(m·K) (for Ag). A high thermal conductivity [up to 300 W/(m·K)] is demonstrated by gold, copper, some other native nonmetallic elements, such as graphite and diamond, and a number of metals that do not occur in a free state but enter mineral compounds (Al, K, Na, Mg, Ca etc.), their λ varying from 100 to 200 Wt/(m·K). Some native metals, however, as well as other elements which do and do not occur in a free state enter the composition of the aforementioned minerals and show average [from 10 to 50 Wt/(m·K)] for Pb, Sb, Mn, Th, U, Zr etc.), decreased [from 1.5 to 10 W/(m·K) for Hg, Bi, Cd etc.], low [from 0.5 to 1.5 W/(m·K) for B] and very low [<0.5 W/(m·K) for H, N, F, Cl_2, O_2, S, Se etc.] values of the coefficient of thermal conductivity.

Mechanisms of variation of λ of metals are connected with the difference of the electron structure of their atoms, nonuniform concentration of electrons of conductivity that transfer the bulk of thermal energy. It is well validated by the close connection between the coefficients of electric, σ, and thermal, λ, conductivity of metals (Fig. 110), close to linear in the range of very high to decreased λ values, so the λ/σ ratio is considered as being approximately constant. For example, for Cu it is $6.4 \cdot 10^{-8}$, for Pb $7 \cdot 10^{-6}$ J·Ohm·m/(s·K). The λ/σ varies in proportion to the absolute temperature according to the Wiedemann-Franz law (also known as the Lorentz relation); $\lambda/\sigma = 3(k/e)^2 T = 2.23 \cdot 10^{-8}$ T.

The presence in the mineral composition of elements having a high thermal conductivity [λ_m = 50-300 W/(m·K)], such as Fe in pyrite, Zn in sphalerite, Al in corundum, Mg in periclase etc. often causes their λ_m to increase. At the same time, the high concentration in minerals of elements having an average, decreased and very low thermal conductivity decreases the values of this quantity. Minerals having a relatively low thermal conductivity [λ_m in W/(m·K)], partly due to a high concentration of poorly conducting elements, seem to include: ice (H_2O-2.3), gypsum ($CaSO_4H_2O$-1.2), kaolinite (Al, MG)$_2$(OH)$_2$[Si_4O_{10}]H_2O-0.88), phlogopite (0.5), biotite (2).

The composition affects λ_m not only due to the difference in the thermal conductivity of elements entering minerals but also due to different packing density (different numbers of close neighbours and different interatomic distance r_0) of definite atoms or ions in minerals of different composition. All this provides for definite lattice (or mixed electron and lattice) thermal conductivity of minerals and concentration in them of phonons (or also electrons) responsible for their one λ_m or another. The greater the interatomic distances, the less is the thermal conductivity of minerals.

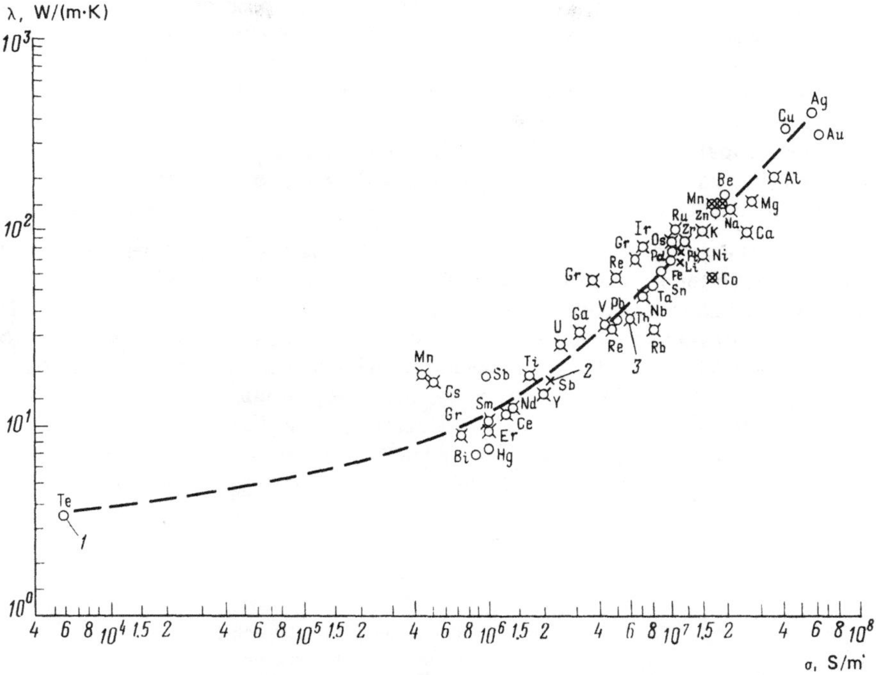

FIG. 110. A connection between the coefficient of thermal conductivity λ and specific resistance σ for metals.

Metals: *1*—native; *2*—pure; *3*—not occurring in a free state

The λ_m values generally decrease with a decrease in the density of their atoms and, consequently, the density of minerals, but the close relation between λ_m and δ_m is absent. Defects of the crystal structure of minerals also contribute to a decrease of their coefficient λ_m. The decrease of λ_m of central members of the plagioclase series is ascribed to local disturbances of their structure following a substitution of sodium by calcium and of aluminium by silicium.

Despite large fluctuations in λ_m values for individual varieties of a definite mineral class of their proximity for various class minerals due to specific features of their composition and structure, the principal investigated mineral classes in the order of decreasing average λ_{mav} values are arranged in the following series: native metals and such elements as graphite and diamond [$\lambda_{av} \sim 120$ W/(m·K)—sulphides (~19)—oxides (~11.8)—fluorides and chlorides (~6)—carbonates (~4.0)—silicates (~3.8)—sulphates (3.3)—nitrates (~2.1)—native elements—nonmetals (selenium, sulphur ~0.85)]. One of the last places in this sequence is occupied by silicates. Different mineral groups are far from being uniform in terms of $\lambda_{m.av}$. Comparatively large $\lambda_{m.av}$ values are observed for nesosilicates, less for inosilicates, still less for tectosilicates and laminated silicates. This sequence of variation of $\lambda_{m.av}$ of individual silicate groups demonstrating phonon thermal conductivity is governed by the specific features of their chemical composition and structure. In addition, one may point to a still lower thermal conductivity of aqueous sulphates compared with non-aqueous varieties and complex oxides compared with elementary ones except ice and cuprite. The latter demonstrate the lowest thermal conductivity among oxides.

The thermal conductivity of tourmaline and feldspars increases with temperature which is unusual for other minerals being the result of their certain thin-scale crystal imperfections.

Thermal capacity. In the case of minerals it has been studied rather poorly and particularly so for elementary and complex oxides, molybdates, wolframates, phosphates containing radioactive elements and borates. As follows from estimates made on samples of minerals, their isobar mass specific heat c_{pm} varies from 0.125 to >2-4 kJ/(kg·K) and is mainly dependent on their chemical composition and structure. Since density δ_m is also governed by their composition and structure, a fairly close connection between c_{pm} and δ_m is observed (Fig. 111).

In terms of the mass specific heat principal mineral classes can be arranged in this series: native metals < sulphides and their analogs < oxides < sulphates <

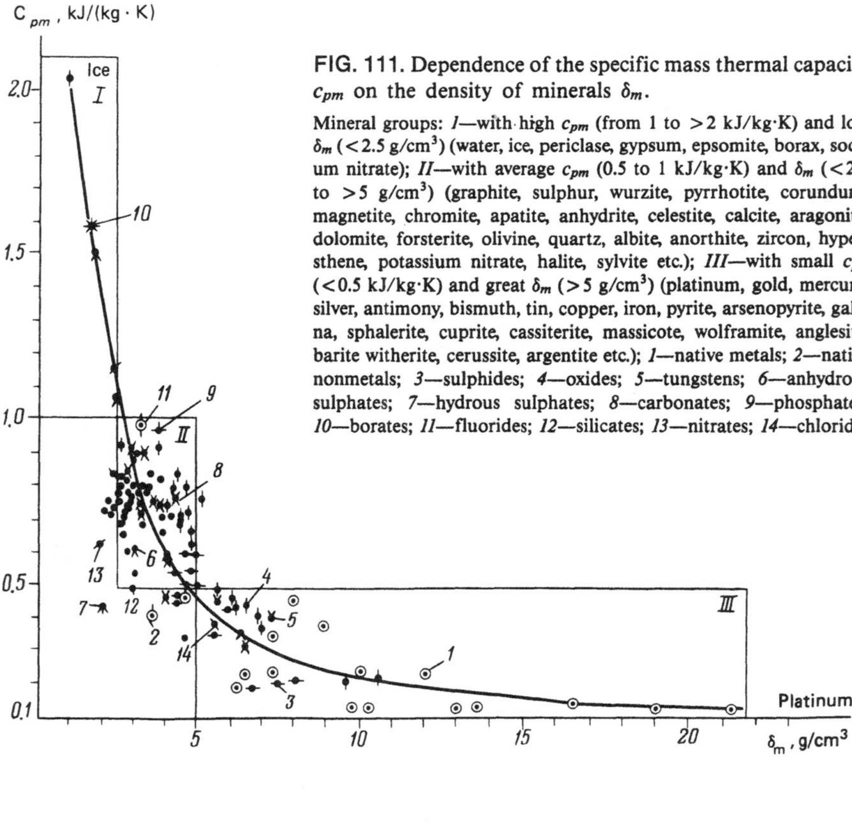

FIG. 111. Dependence of the specific mass thermal capacity c_{pm} on the density of minerals δ_m.

Mineral groups: *I*—with high c_{pm} (from 1 to >2 kJ/kg·K) and low δ_m (<2.5 g/cm^3) (water, ice, periclase, gypsum, epsomite, borax, sodium nitrate); *II*—with average c_{pm} (0.5 to 1 kJ/kg·K) and δ_m (<2.5 to >5 g/cm^3) (graphite, sulphur, wurzite, pyrrhotite, corundum, magnetite, chromite, apatite, anhydrite, celestite, calcite, aragonite, dolomite, forsterite, olivine, quartz, albite, anorthite, zircon, hypersthene, potassium nitrate, halite, sylvite etc.); *III*—with small c_{pm} (<0.5 kJ/kg·K) and great δ_m (>5 g/cm^3) (platinum, gold, mercury, silver, antimony, bismuth, tin, copper, iron, pyrite, arsenopyrite, galena, sphalerite, cuprite, cassiterite, massicote, wolframite, anglesite, barite witherite, cerussite, argentite etc.); *1*—native metals; *2*—native nonmetals; *3*—sulphides; *4*—oxides; *5*—tungstens; *6*—anhydrous sulphates; *7*—hydrous sulphates; *8*—carbonates; *9*—phosphates; *10*—borates; *11*—fluorides; *12*—silicates; *13*—nitrates; *14*—chlorides carbonates < silicates. The c_{pm} values [in kJ/(kg·K)] of native metals vary from 0.13 to 0.2 (for Pt, Au, Bi, Pb etc.) to 0.35-0.45 (for Cu, Fe, Zn), sulphides and their analogs from 0.21-0.22 (for halenite, cinnabar) to 0.5-0.6 (for coveline, wurzite etc.), oxides from 0.22-0.24 (for massicot, thorianite, uraninite etc.) to 0.9-1 (for limonite, pyrolusite, periclase) and even to 2-4 (for ice and water), sulphates from ~0.35 (for anglesite) to 1-1.5 (for gypsum, epsomite), silicates from 0.5-0.6 (for tourmaline) to 0.9-0.98 (for spodumene, zurcon). Extremely low c_{pm} values for each of the aforementioned, except native, metals are accounted for by a high concentration in them of elements having a high density and low specific thermal capacity [up to 0.1 kJ/(kg·K)], such as Bi, Hg, Pb, Th, U and others and limiting high values by a high concentration in them of low-density elements [B, C (graphite), Mg, Na, Si, O etc.] having a comparatively high mass thermal capacity [0.2-0.3 kJ/(kg·K)] which is particularly true of hydrogen.

The average thermal capacity of the latter element in the gaseous state, in the temperature range 0-200 °C, attains 14.235 kJ/(kg·K) and much exceeds the specific thermal conductivity of other gases and elements. Hydrogen may enter the composition of minerals (as for H_2O and ice) and may be contained in their crystallization water (as for gypsum, epsomite, borax). The high thermal capacity of borax may also be associated with that its composition may contain B having a comparatively high thermal capacity [1.0467 kJ/(kg·K)].

The highest values of the average mass specific heat are shown by nesosilicates [$c_{pm.av}$ = 0.783 kJ/(kg·K)], lower values are shown by inosilicates (0.774) and still lower by laminated silicates (0.742) and tectosilicates (0.743). The c_{pm} value of phyllosilicates has not been practically studied.

Sec. 40. **The Solid Phase**

Thermal conductivity. The λ_s value of the solid phase of rock is governed by its mineral composition, shape, size and spatial orientation of crystals or grains, temperature and pressure. The character of dependence of λ_s on all the above specific features and conditions of its existence has not been practically studied.

According to F. Birch and H. Clark, the thermal conductivity of a nonporous mineral aggregate is approximately estimated, in conformity with the notion of a successive inclusion of relative amounts of constituent minerals, i.e. λ_s of practically a nonporous rock (igneous, metamorphic or densely cemented sedimentary) can be roughly found from the relationship

$$\frac{100}{\delta_s\lambda_s} = \frac{m_{sm1}}{\delta_{sm1}\lambda_{sm1}} + \frac{m_{sm2}}{\delta_{sm2}\lambda_{sm2}} + \ldots + \frac{m_{smn}}{\delta_{smn}\lambda_{smn}} = \sum_{i=1}^{n} \frac{m_{smi}}{\delta_{smi}\lambda_{smi}}$$

where δ_s, δ_{smi} are the densities of the rock solid phase and i-th of minerals composing it; m_{smi} is the mass of the solid phase of the i-th mineral (mass per cent); λ_s and λ_{smi} are coefficients of thermal conductivity of the solid phase of the rock and the i-th mineral composing this phase.

Since quartz and plagioclases have one of the highest thermal conductivities of rock-forming minerals [$\lambda_{s.qrz}$ = 9 W/(m·K)], the λ_s values of igneous rocks must decrease from basic to intermediate (due to plagioclases) and increase from intermediate to acidic ones. The quartz content in the rocks solid phase increases in this direction. The solid phase of various sandy rock types contains from 24 to >70% (mass) quartz. It has been found that impurities in crystals leading to increased nonuniformity of distribution in them of the atomic mass facilitate the spread of phonons and, due to this, to decreased thermal conductivity of crystals and, in particular, SiO_2 crystals.

Any specific feature of a crystal decreasing the free path of waves of thermal vibrations diminishes the thermal conductivity. The greatest average free path of waves is observed in simple and symmetrical lattices (for example in the crystal lattice of diamond). The thermal conductivity of crystalline quartz is several times greater than that of an extremely disordered structure of quartz glass. Quartz crystals in the solid phase of various sediments can include pure ones to varieties having a great number of impurities and the latter can be nonuniformly distributed in a crystal. It appears that $\lambda_{s.qrz}$ of pure quartz is higher several times than quartz that contains different impurities. The $\lambda_{s.qrz}$ meaning which must be used for calculations can be obtained by averaging $\lambda_{s.qrz}$ for several quartz crystal types or extrapolating λ_s for low-porosity quartzites toward zero values of k_p.

Thermal capacity. The isobar mass specific heat of the solid phase of rocks is governed by its mineral composition and is found from the relationship

$$c_{pm} = \sum_{i=1}^{n} c_{pmi} m_i \tag{57}$$

where m_i is the relative mass content of the i-th mineral having a mass specific heat c_{pmi}.

The solid phase of rocks having in their composition sulphides and oxides have on the average a smaller thermal capacity compared with the same rock not containing minerals of these classes. A comparatively great λ_s value may be manifested by igneous, metamorphic, sedimentary rocks due to the fact that the main components of the solid phase are various silicates having a relatively high thermal capacity. Close λ_s values are also shown by the solid phase of carbonate rocks except the differences in the composition of the solid phase which contains such minerals as witherite and cerussite involving relatively low-thermal capacity barite and lead. A high thermal conductivity is manifested by the solid phase whose composition includes minerals (gypsum, epsomite, borax etc.) with much crystallization water.

Sec. 41. The Liquid Phase

Thermal conductivity. The coefficient of thermal conductivity λ_l of a liquid is proportional to its heat capacity c_{pm}, density δ_l, average interatomic distance L and speed v_s, of shifting molecules of this substance from a hot layer to one that is less heated. The latter is identical to the speed of sound in this medium and exceeds the speed of the thermal molecular motion of particles. These assumptions make it possible to write

$$\lambda_l = \delta_l c_{pm} v_s L \tag{58}$$

where $L = \Delta - d$ (Δ is the distance between the centres of molecules of a diameter d).

The difference between λ_l values obtained from Eq. (58) and experimental ones is generally in the range 5-15% but may sometimes attain as much as 50%. B. Wahrgaft has proposed the following experimental formula to calculate the coefficient of thermal conductivity:

$$\Delta_l = \dot{A} c_{p\nu} \delta_l^{4/3} / M^{1/3} \tag{59}$$

where $c_{p\nu}$ is the specific isobaric molar thermal capacity of a fluid at a constant pressure p; A is a proportionality coefficient governed solely by the temperature ($Ac_{p\nu} \approx$ const); δ_l is the density of the fluid; M is the molar mass of the fluid.

The coefficient of thermal conductivity of distilled water, λ_w, at zero temperature and atmospheric pressure equals 0.582 W/(m·K). With increasing the temperature its values first increase [$\lambda_w = 0.589$ at 20 °C and 0.67 W/(m·K) at 75 °C), then starting at approximately 120 °C, diminish since the attraction between the molecules of the fluid decreases. The λ_w values increase with pressure: at $p = 1\,200$ MPa the coefficient has a value half as great as the atmospheric pressure (0.1 MPa). At $p < 3$-4 MPa the effect of pressure on λ_w is very small.

At t and $p =$ const, λ_w increases in a minor way with increasing the salinity of water. Given $t =$ const and growth of pressure p, the λ_w values for the salinity of water $s = 35$‰, also increase from 0.571 (at 1 MPa) to 0.619 W/(m·K) (at 140 MPa).

Crude oil at 20 °C has $\lambda_o = 0.13$-0.14 W/(m·K) which is less than λ_w by a factor of 4 and more. The presence of oil in rocks decreases their thermal conductivity more the greater is their oil saturation.

The most important liquid components of oil (penthane, *n*-hexane, *n*-octane, *n*-decane) little differ in thermal conductivity and at 20 °C and 0.1 MPa have a much lower thermal conductivity compared with that of distilled and mineralized water; their λ_o varies from 0.135 for penthane to 0.147 W/(m·K) for *n*-decane. The λ_o values also increase with pressure; for example, for penthane, given $p = 1\,200$ MPa, it exceeds by 70% the value obtained at atmospheric pressure.

Heat capacity. For water, c_{pm} values are in the range 4.1816-3.9688 kJ/(kg·K) at 20 °C and variation of its salinity from 0 to 40‰.

At all other temperatures for which values of this quantity have been obtained, c_{pm} values decrease with growth of mineralization. Given a specified mineralization, with increasing temperature c_{pm} decreases rather inappreciably, for example, from 4.1920 kJ/(kg·K) at $t = 0$ to 4.117 at 40 °C and water salinity 40‰.

The isobaric specific heat can be determined by using L. Atkinson's and F. Richard's relationship: $c_{pm} = 1.005\text{-}0.0004226t + 0.000006321\,t^2$. The c_{pm} value calculated by a number of researchers for water of a definite mineralization and temperature decreases with growth of pressure: e.g., from 3.886 to 3.820 kJ/(kg·K) upon variation of temperature from 10 to 30 MPa and for water mineralization ~35‰.

The heat capacity of oil upon temperature variation from 40 to 80 °C and pressure variation from 0.1 to 30 MPa in the range 1.884-2.763 kJ/(kg·K).

Sec. 42. The Gaseous Phase

Thermal conductivity. Energy transfer by gas molecules is closely connected with their average free path (average distance travelled by a molecule in gas before its collision with another molecule). The coefficient of thermal conductivity of gaseous mixtures, $\lambda_g = 0.5\,(\lambda_{sm} + \lambda_{rm})$ where λ_{sm} and λ_{rm} are average coefficients of thermal conductivity calculated from relationships: $\lambda_{sm} = x_1\lambda_1 + x_2\lambda_2 + x_3\lambda_3 + \ldots$; $1/\lambda_{rm}) = (x_1/\lambda_1) + (x_2\lambda_2) + (x_3\lambda_3) + \ldots$.

Here $\lambda_1, \lambda_2, \ldots$ are coefficients of thermal conductivity of the mixture's components; $x_1, x_2, \ldots$ are their molar fractions. According to these data, λ_{air} equals 0.2441 W/(m·K) at $p = 0.1$ MPa and $t = 0$. With increasing the temperature, the values [in W/(m·K)] increase from 0.02571 at 20 °C to 0.03056 at 100 °C and to 0.03684 at 200 °C. It must be noted that nitrogen which is the main constituent element of air has a specific thermal conductivity [in W/(m·K)] equal to 0.02596 at 27 °C and 0.02721 at ~50 °C. Oxygen and carbon dioxide have a lower specific heat [in W/(m·K)]: 0.02386 at $t = 0$ and 0.01675 at 26.7 °C. The thermal conductivity of air is a function of pressure. Given 20 °C and variation of p from 0.1 to 40 MPa, its coefficient λ_{air} increases almost twofold. At higher temperatures λ_{air} increases with pressure ever less intensively.

Methane and ethane show specific heat values increasing with temperature, being greater for methane [0.03391 W/(cm·K)] and smaller for ethane [0.02135 W/(cm·K)]; given $t = 26.7$ °C and $p = 0.1$ MPa, these values are similar to that of air at the same pressure and close temperatures.

Thermal capacity. Polyatomic gas molecules, apart from a forward motion, may have a rotation motion, and the constituent atoms vibrating at about an equilibrium state at appreciable temperatures.

In the case of natural, not very high temperatures reached during the drilling work (somewhat above 200 °C) there occurs a forward and rotation motion of gas molecules. This is explained by the independence of thermal capacity from temperature at comparatively small values of the latter and the conditions of gas molecule structures. The thermal capacity of air at $p = 0.1$ MPa and $t = 0$ is 1.007 kJ/(kg·K) and at 50 °C it attains 1.011 kJ/(kg·K).

The thermal capacity of air is much less than that of methane and ethane, that is why for trapped gas it is much higher than for air varying from 2.612 to 3.584 kJ/(kg·K) upon varying t from 40 to 80 °C and p from 0.1 to 30 MPa. For single-atom gases (argon, helium etc.) $c_{pvav} = 20.934$ kJ/(kmole·K), average isobaric mass specific heat c_{pmav} of gases varies in kJ/(kg·K) from 0.5 (argon) to ~5 (helium). The isobaric mass specific heat of air regularly increases with temperature. For example, at $t = 0$ and $p = 0.1$ MPa it is 1.007, and at 100 °C, 1.015, and at 250 °C, 1.027 kJ/(kg·K). With increasing pressure at a constant temperature c_{pm} of air also grows. Given $t = 0$ and $p = 0.1$, 10 and 22 MPa, it is, respectively, 1.007, 1.188 and 1.333 kJ/(kg·K). The same is true of the thermal capacity of formation gases.

Thus, $\lambda_s > \lambda_l > \lambda_g$. The highest thermal capacity values are demonstrated by the liquid phase of rocks, lower ones by the gaseous phase and still lower ones by the solid phase unless an appreciable fraction of the latter includes ice or minerals having a great crystallization or constitutional water content.

Sec. 43. **Rocks**

Thermal conductivity. The heat transfer process in multiphase rocks is much more involved as compared with their individual phases since heat energy is transferred from a more to a less heated region of the rock, not only as a result of thermal conductivity processes—conduction (differing in mechanism in their solid, liquid and gaseous phase)—but also as radiation at high temperatures. In addition, rocks have an electric double layer where the heat exchange mechanism undoubtedly differs from interphase ones and has been insufficiently studied. The heat transfer mechanism across the boundaries and relatively homogeneous phases adjacent to it have also not been sufficiently studied.

Convective heat exchange in rocks is a process of transfer of heat energy by a fluid seeping through them or gas to their solid phase (heat transfer proper) or other regions of the fluid (gas) with a smaller energy of thermal motion of particles. The character of convective heat exchange is due not only to thermal and other properties of the moving phase and ones in contact with it but is dependent on the specific features of the motion of the liquid or gas through the pore space of the rock which occurs because of the effect of applied (forced convection) or natural (natural convection) forces.

Under conditions of forced convection the displacement of a liquid or gas in the voids of oil and gas trap rocks occurs under the effect of an external excitation agent. The intensity of heat exchange is governed by the speed of forced motion. Under conditions of natural convection the material filling the voids infiltrates due to the temperature difference, and, consequently, the densities of the voids filling at separate regions of a nonuniformly heated rock. The intensity of natural convection increases with the temperature gradient, speed of heat transfer from the liquid (or gas) to the solid phase and drops with increasing the viscosity of the substance filling the voids. The intensity of heat transfer of the moving component to the immovable one is characterized by the coefficient of heat transfer $\alpha = q/(T_1 - T_2)$ where q is the heat flux density; $T_1 - T_2$ is the difference of absolute temperatures. Since at the surface of the solid phase the dense or immovable part of the electric double layer is found where the speed of the motion of the liquid (gas) is zero, transfer of heat of liquid is here due solely to thermal conductivity. At this,

$$q = \lambda_l \,|\text{grad } T|_{st}$$

where λ_l is the coefficient of thermal conductivity of the liquid, $|\text{grad } T|_{st}$ is the module of the temperature gradient of the liquid at the interface with the solid phase.

A definite relationship has been established between α and λ_l.

An appreciable role may be played in porous gas-saturated rocks by radiant heat exchange. In the given case the radiation in associated with the excitation of the atoms and molecules of the solid phase by thermal motion. The intensity and spectral composition of this radiation are conditioned by the temperature and state of the surface of the luminiscent pore walls. At a not very high temperature of rocks ($\sim$200 °C) radiant heat exchange is practically absent.

If convective heat transfer α_{cv} is followed by radiant one α_r, then $\alpha = \alpha_{cv} + \alpha_r$. The intensity of heat transfer and heat exchange between the surface bounding the pore space and the substance filling the voids taking place due to convective and radiant heat exchange characterize the coefficient of heat transfer

$$\alpha_{av} = dQ/dt_{av}dF$$

where dQ is a heat flux, the amount of heat being transferred; dF is the heat exchange surface, the area bounding the pore space of the rock; $dt_{av} = t_f - t_s$ is a temperature head; t_f and t_s are temperatures of the filling substance and of the rock's solid phase. Convection together with other exchange types facilitate equalization of temperatures in the filling substance and between the phases of the rock and, consequently, in the rock proper.

Thus the equalization of temperature of three-phase highly porous and non-uniformly heated rocks in the volume they occupy occurs due to thermal conductivity, convection and radiation. At this, the different capacity of various rocks for heat exchange is characterized by a quantity called the effective coefficient of thermal conductivity, $\lambda_{eff} = \lambda_c + \lambda_{cv} + \lambda_r$, where λ_c, λ_{cv} and λ_r are, respectively, coefficients of conductive, convective and radial thermal conductivity.

The values of λ_c, λ_{cv} and λ_r and, consequently, λ_{eff} are governed not only by the composition and ratio of the phases and cement of rocks, their interaction, structure of the solid phase but also by temperature and pressure.

For multiphase rocks the Fourier heat equation is also valid but in the particular case

$$dQ_{eff} = -\lambda_{eff} \frac{\partial t}{\partial x} dF d\tau$$

where $Q_{eff} = Q_c + Q_{cv} + Q_r$ is the effective amount of heat passing through an element of isometric surface of a multiphase rock in the time interval $d\tau$, given a temperature gradient $\partial t/\partial x$.

However, the principal role in the heat distribution in the volume of rocks is played by the thermal conductivity process. The λ_{cv} and λ_r values seem to be small although their estimate is rather approximate, particularly, given forced convection. Laboratory studies of the thermal conductivity of rocks are generally focussed solely on the main constituent of a thermal flux viz. conductive one (otherwise laboratory determinations of λ are much complicated). Then the density of a convective heat flux is

$$q_c = Q_c/dFd\tau = -\lambda_c(\partial t/\partial x)$$

and densities of a convective heat flux and radiation are $q_{cv} = q_r = 0$.

It has been previously proved [see Eq. (51)] that in the absence in a rock of internal heat sources ($w = 0$) and given a unidimensional thermal flux the following relationship is valid:

$$c_{pm}\delta_r(\partial t/\partial \tau) = \lambda_c(\partial^2 t/\partial x^2)$$

or

$$\partial t/\partial \tau = a(\partial^2 t/\partial x^2)$$

and, given a three-dimensional thermal flux,

$$c_{pm}\delta_r(\partial t/\partial \tau) = \lambda_{cx}(\partial^2 t/\partial x^2) + \lambda_{cy}(\partial^2 t/\partial y^2) + \lambda_{cz}(\partial^2 t/\partial z^2)$$

Since λ determinations on rock samples do not generally take into account convective and radiant thermal conductivity, all data that follow on λ and a refer to conductive heat exchange.

The coefficient of thermal conductivity, specific heat and thermal diffusivity of rocks cannot only be determined in laboratory but also evaluated analytically. In the latter case use can be made of very much simplified models of a multiphase rock. They outline the phase composition, structure, values of the thermal conductivity of the solid, liquid and gaseous components and sometimes the interphase layers (electric double layers) of these heterogeneous media.

To calculate the coefficient of thermal conductivity, the rocks can be thought of as being a packing of unidimensional spheres having a coefficient of thermal conductivity λ_s between which a liquid or gaseous phase is found together with a coefficient of thermal conductivity corresponding to λ_l and λ_g. The calculation of the coefficient λ_{sl} of this model used an equation similar to Maxwell's equation for calculating the average electric conductivity of the identical packing of spheres. Upon filling the space between the spheres by a fluid

$$\lambda_{sl} = \frac{3\lambda_s - 2w_t(\lambda_s - \lambda_l)}{[3 + w_t(\lambda_r/\lambda_l - 1)]} \tag{60}$$

where w_t is the specific volume of the space between solid spheres (and in the case of unswelling rocks this quantity can be substituted by k_p).

Equation (60) defines a regular decrease of the coefficient λ_{sl} of a maximum moisture rock with the growth of the coefficient w_t or k_p (Fig. 112). Graphs *I-III* in Fig. 112 have been obtained for $\lambda_l = 0.57$ W/(m·K) (a value close to λ_w of natural waters) and λ_s equal to, respectively, 8.8, 2.0 and 1.4 W/(m·K) characteristic of principal rock-forming minerals (quartz and feldspar sandstones, clayey rocks). A close character of variation of the coefficient of thermal conductivity is also found for maximum moisture limestones with growth of k_p (Fig. 113). The λ_{sl} values determined from Eq. (60) are in sufficiently good agreement with experimental data (see Figs. 112 and 113).

However, the solid phase of sandstones, limestones and other rocks is usually polymineral which appreciably tells on λ_{sl} values. We will illustrate this. Since Eq. (60) is valid for any two-component aggregate we will use it for a calculation of λ_{sl} of a polymineral rock. If λ_l of a maximum moisture sandstone is still equal to 0.57 W/(m·K) and the solid phase includes quartz with $\lambda_{quar} = 8.84$ W/(m·K), feldspar with $\lambda_{fs} = 2.0$ W/(m·K) and kaolinite having $\lambda_{cl} = 0.88$ W/(m·K), then, according to B. G. Kokshenev, given definite assumptions, we can calculate λ_s as follows. Let us find λ_s' of the solid phase consisting of quartz and feldspar by using Eq. (60), assuming that between quartz grains whose fraction is the greatest (e.g. 0.7 of a unit mass of the solid phase) there are feldspar grains as a result of which we have

$$\lambda_s' = \frac{3\lambda_{quar} + 2w_{fsm}(\lambda_{quar} - \lambda_{fs})}{3 + w_{fsm}(\lambda_{quar}/\lambda_{fs} - 1)} = \frac{26.52 - 13.68\,w_{fsm}}{3 + 3.42\,w_{fsm}}$$

where w_{fsm} is the mass feldspar content.

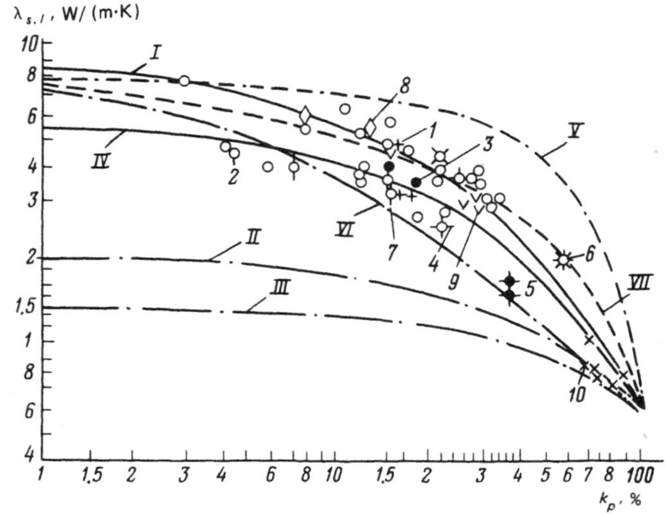

FIG. 112. Analytical and experimental dependences of the coefficient of thermal conductivity λ_{sl} on the voids ratio of maximum moisture rocks and abyssal sediments in the composition of the solid phase of which quartz, feldspars, clayey minerals are predominant or it is polymineral.

I—curve for a rock whose solid phase is represented by quartz calculated from the equation

$$\lambda_{sl} = \frac{3\lambda_s - 2k_r(\lambda_s - \lambda_l)}{3 + k_r(\lambda_s/\lambda_l - 1)}$$

given $\lambda_s = 8.84$ W/(m·K) and $\lambda_l = 0.57$ W/(m·K); *II*—the same with the solid phase represented by feldspar, $\lambda_s = 2$ W/(m·K); *III*—the same for a rock whose solid phase is represented by clay; *IV*—curve calculated from the Maxwell-Kokshenev equation for a rock whose solid phase is represented by quartz, feldspar and clay (their content being, respectively, 0.7, 0.2 and 0.1 of a unit volume of the solid phase, and coefficients of thermal conductivity being 8.84, 2 and 0.88 W/(m·K), $\lambda_l = 0.57$ W/(m·K); *V*—analytical curve obtained from the equation $\lambda_{sl\perp} = (1 - k_p)\lambda_s + k_p\lambda_l$, given $\lambda_s = 8$ W/(m·K) and $\lambda_l = 0.57$ W/(m·K) where the solid and liquid phase are connected in parallel; *VI*—the same but obtained from the equation

$$\lambda_{sl\parallel} = \frac{1}{k_p/\lambda_l + (1 - k_p)/\lambda_s}$$

(here the solid and liquid phase are connected in series); *VII*—curve of $\lambda_{sl.av} = f(k_p)$, where $\lambda_{sl.avi} = \sqrt{\lambda_{sl\perp i}\lambda_{sl\parallel i}}$. Water saturated rocks: *1*—sand; *2*—quartz sandstone; *3*—quartz sandstone with ferruginous cement; *4*—feldspar sandstone; *5*—feldspar sandstone with ferruginous cement; *6*—tripolite; *7*—subgraywackes; *8*—quartzite; *9*—sandstones without their characteristics; *10*—abyssal red-clay and silt

Suppose that it is 0.2 of a unit mass of a rock solid phase. By assuming that kaolinite particles which are the fewest in the solid phase (0.1 of a unit mass of the solid phase) are found in the pore space of the quartz and feldspar component of the solid phase and using Eq. (60) we get a relationship for calculating the coefficient of thermal conductivity of a three-component solid phase. In our case

$$\lambda_s = \frac{79.56 - 47.76(w_{clm} - 0.859\,w_{fsm}) + 33.38\,w_{fsm}w_{clm}}{9 + 27.14(w_{fsm} + 0.36\,w_{clm}) - 18.97\,w_{fsm}w_{clm}}$$

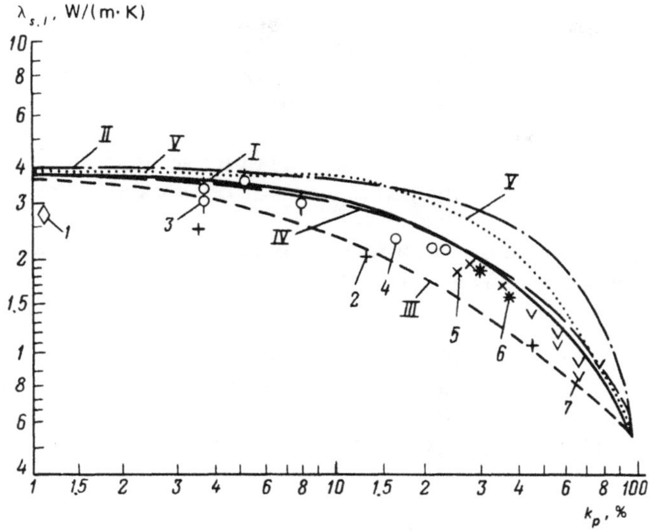

FIG. 113. Analytical and experimental dependences of the coefficient of thermal conductivity λ_{sl} on the voids ratio k_p for maximum moisture rocks and abyssal sediments where calcite is predominant in the composition of the solid phase.

I—curve calculated from the equation

$$\lambda_{sl} = \frac{3\lambda_s - 2k_p(\lambda_s - \lambda_l)}{3 + k_p(\lambda_s/\lambda_l - 1)}$$

given $\lambda_s = 4.0$ W/(m·K) and $\lambda_l = 0.57$ W/(m·K); *II*—curve obtained from the equation $\lambda_{sl\parallel} = (1 - k_p)\lambda_s + k_p\lambda_l$ at $\lambda_s = 4$ W/(m·K) and $\lambda_l = 0.57$ W/(m·K), where the solid and liquid phases are connected in parallel; *III*—the same but found from the equation

$$\lambda_{sl\perp} = \frac{1}{k_p/\lambda_l + (1 - k_p)/\lambda_s}$$

(the solid and liquid phase being connected in series); *IV*—curve of $\lambda_{sl.av} = f(k_p)$, $\lambda_{sl.avi} = \sqrt{\lambda_{sl\parallel i}\lambda_{sl\perp i}}$; *V*—curve calculated from the equation

$$\lambda_{sl} = X + \sqrt{X^2 + \lambda_s\lambda_l/2} \quad \text{where}$$

$$X = [(3V_s - 1)\lambda_s - (3V_s - 2)\lambda_l/4],$$

given $\lambda_s = 4$ W/(m·K), $\lambda_l = 0.57$ W/(m·K).

1—marble; *2*—limestone; *3*—very fine-grained calcareous subgraywackes (40-50% being dissolved in HCl); *4*—subgraywackes (85% being dissolved in HCl); *5*—very fine-grained sandy limestones; *6*—organogenic and oölite limestones; *7*—globigerine ooze

where w_{clm} is the clayey component content. If λ_s has been obtained which in our case proved to be 5.6 W/(m·K) and $\lambda_l = 0.57$ W/(m·K), we can, by using Eq. (60) again, calculate λ_{sl} values for specified k_p values and plot a graph of the dependence $\lambda_{sl} = f(k_p)$ (see Fig. 112, curve *IV*). As can be seen from Fig. 112, the curve for quartz and feldspar sandstone is located much lower than for quartz sandstone, and the experimental λ_{sl} values for quartz and feldspar clayey sandstones are located closer to curve *IV* than to curve *I*.

In case the pore space is filled by oil and water or water and gaseous hydrocarbon mixtures, then λ_f is determined on an assumption that the thermal conductivity of the filling substance equals the weighted mean (in terms of volumetric content) value of the conductivities of its components. If a rock has a great number of combined component, a major role in heat transfer is also played by water films in rock particles. The latter join together rock particles (due to surface tension) thus increasing their thermal contact and the thermal conductivity of a rock as a whole. In this case averaging is impossible.

The calculation of the thermal resistance of filling substance representing a mixture of air with $\lambda_{air} = 0.024$ W/(m·K) and water with $\lambda_w = 0.57$ W/(m·K) uses the equation:

$$\lambda_f = \lambda_w k_w + \lambda_{air}(1 - k_w) = 0.57k_w + 0.024(1 - k_w)$$

where k_w is a coefficient of water saturation of the pore space.

For rocks the pore space of which is filled by a mixture of water and oil with the coefficient $\lambda_o = 0.125$ W/(m·K) and gaseous hydrocarbons with $\lambda_g = 0.012$ W/(m·K)

$$\lambda_f = 0.012k_g + 0.13k_o + 0.57k_w$$

where k_o is the coefficient of oil saturation of the pore space.

By assuming elements of a rock liquid and solid phase to be series-connected and the heat flux to be normal to this structure, then an average value can be calculated for such a model:

$$\lambda_{sl\parallel} = (1 - k_p)\lambda_s + \lambda_l k_p \tag{61}$$

If the results of this calculation (for $\lambda_s = 8.0$ W/(m·K), $\lambda_l = 0.57$ W/(m·K), $k_p = 0.01$-1) are represented graphically (see Fig. 112, curve *V*), it can be easily seen that λ_{sl1} values for specified k_p values are higher or much higher than ones determined from Maxwell's equation [see Eq. (60) and compare curves *1* and *5* in Fig. 112]. At the same time for a structure showing a shunt-connection of solid and liquid elements, given a normal direction of a heat flux to this aggregate,

$$\lambda_{sl\perp} = 1/[k_p/\lambda_l + (1 - k_p)/\lambda_s] \tag{62}$$

The $\lambda_{sl\perp}$ values for specified k_p calculated by using Eq. (62), given the same λ_s, λ_l values as previously, prove to be less or much less than ones determined from Eq. (60) (see Fig. 112, curves *I* and *VI*).

It is more likely that liquid and solid elements of rocks are both series- and shunt-connected with one another, so, given a normal direction of a thermal flux to a sample of such a rock, we must get some average values λ_{slav} of those obtained by using Eqs. (61) and (62). Suppose these are geometrical mean values $\lambda_{slav} = \sqrt{\lambda_{sl1}\lambda_{sl2}}$. Let us calculate them from values λ_{sl1} and λ_{sl2}, plot a graph of the dependence $\lambda_{slav} = f(k_p)$ (Fig. 112, curve *VII*) and see that it is practically agrees with a $\lambda_{sl} = f(k_p)$ curve obtained by calculating λ_{sl} by using Maxwell's equation [see Eq. (60) and compare curves *I* and *VII* in Fig. 112]. So do curves of $\lambda_{sl} = f(k_p)$ and $\lambda_{slav} = f(k_p)$ agree that have been obtained for limestones [$\lambda_s = 4.0$ W/(m·K), $\lambda_l = 0.57$ W/(m·K)] from Maxwell's equation and as a geometric mean for media demonstrating series- and shunt-connected structure elements (see Fig. 113, curves *I* and *IV*).

We could refer to a number of other equations derived by using matrix models and represent the results graphically. They are, however, unlikely to show a better fit with the experimental data than the analytical correlations we have considered.

Of equations corresponding to statistical models of rocks we will only describe analytical correlations established by E. I. Leont'ev based on Eq. (32) which holds not only for two-phase but also for rocks with any number of phases.

According to E. I. Leont'ev, for a two-phase (maximum moisture) rock having a low content of a highly dispersed (clayey) component Odelevskii's equation has this form:

$$\frac{\lambda_s - \lambda_{sl}}{\lambda_s + 2\lambda_{sl}}(1 - k_p) + \frac{\lambda_w - \lambda_{sl}}{\lambda_w + 2\lambda_{sl}} k_p = 0$$

for a dry rock this equation takes on a form:

$$\frac{\lambda_s - \lambda_{sg}}{\lambda_s + 2\lambda_{sg}}(1 - k_p) + \frac{\lambda_g - \lambda_{sg}}{\lambda_g + 2\lambda_{sg}} k_p = 0$$

and then, for a weakly clayey water and oil-saturated rock it can be represented as follows:

$$\frac{\lambda_s - \lambda_{sw.o}}{\lambda_s + 2\lambda_{sw.o}}(1 - k_p) + \frac{\lambda_w - \lambda_{sw.o}}{\lambda_w + 2\lambda_{sw.o}} k_p k_w + \frac{\lambda_o - \lambda_{sw.o}}{\lambda_o + 2\lambda_{sw.o}} k_p(1 - k_p) = 0$$

and for a water-saturated clayey rock

$$\frac{\lambda_s - \lambda_{swcl}}{\lambda_s + 2\lambda_{swcl}}(1 - k_p) + \frac{\lambda_w - \lambda_{swcl}}{\lambda_w + 2\lambda_{swcl}} k_p(1 - k_{cl}) + \frac{\lambda_{cl} - 2\lambda_{swcl}}{\lambda_{cl} + 2\lambda_{swcl}}(k_p k_{cl}) = 0$$

etc.

A comparison of $\lambda_{sl} = f(k_p)$ dependence graphs obtained by using V. I. Odelevskii's equation*

$$\lambda_{sl} = x + \sqrt{x^2 \lambda_l \lambda_s / 2} \tag{63}$$

where

$$x = [(3V_s - 1)\lambda_s - (3V_s - 2)\lambda_l]/4$$

$V_s = 1 - V_{por} = 1 - k_p$, since $k_p = V_{por}/V_p = V_{por}/1$ is the volume of the solid phase in a unit volume of the rock, and other relationships (see Fig. 112) suggest the following. The results of calculations using Eq. (63) disagree the most with the experimental data. They are close to ones obtained by using Eq. (61) for a model with a successive packing of structure elements and normal direction of the thermal flux to its faces (see Fig. 113, curves *II* and *V*). This model yields an equation determining the highest λ_{sl} values for specified k_p, because here it is seemingly the thermal energy must expand through both phases and at the same time the thermal

* The equation is presented in a form convenient for calculations.

energy must spread for parallel packing of structure elements of the model as well as for parallel direction of the thermal flux the solid phase that has a higher thermal conductivity bypassing the liquid phase. One can also find λ_{sl} from the relationship

$$\log \lambda_{slav} = \sum_{i=1}^{i=n} V_i \log \lambda_i$$

where λ_i is the coefficient of thermal conductivity of components with a relative volumetric content V_i.

Thus the use of structure models makes it possible not only to evaluate the order of λ_{sl} values but also to determine their variations depending on the volumetric ratio of phases, distribution in the volume and interaction.

Thermal capacity. The thermal capacity of rocks can be determined by laboratory analysis and evaluated analytically. The isobaric mass specific heat of a low-porosity rock is roughly estimated by using Eq. (57). For porous and moist rocks having a mass moisture content w_m not exceeding 15-20%

$$c_{pmmr} \approx c_{pmdr}(1 - w_m) + w_m c_{pmw}$$

where c_{pmmr}, c_{pmdr} and c_{pmw} are isobaric mass specific heats of a moist, dry rock and water, respectively.

The isobaric volumetric specific heat of a moist porous rock is

$$c_{pV} = (c_{pm}\delta_r) = c_{pms}\delta_s(1 - k_p) + c_{pmw}\delta_w k_p$$

where c_{pV}, c_{pm}, c_{pms}, c_{pmw} are isobaric specific heats (volumetric and mass) of a rock and mass ones of its solid phase and water; δ_r, δ_s, δ_w are densities of the rock, its solid phase and water, respectively.

Thermal diffusivity. This quantity is evaluated by using Eq. (52) into which the relevant values of the quantities λ_{sl}, c_{pm} and δ_r or δ_{sl} and c_{pV} are introduced if determinations are conducted for a maximum moisture rocks.

Sec. 44. Various Rock Types and Groups

Thermal conductivity. The values of the coefficient of thermal conductivity of abyssal marine sediments, muds, vary over a very small range 0.66-1.78 W/(m·K) (Table 13) owing to a small concentration in them of a thermal conductive solid phase, particularly, if this latter is represented by weakly thermal conductive clayey minerals. In the case of sedimentary, igneous and metamorphic rocks λ varies, respectively, from 0.14 to 7.4, from 0.25 to 5.0 and from 0.25 to 7.6 W/(m·K) (see Fig. 114, Table 13). Consequently, the widest range of its values is manifested by sedimentary rocks. The limiting λ values of metamorphic rocks are fairly close to them. Those of igneous rocks vary over a narrower range. This is due to their mineral and phase composition, in particular, the mineral composition of the cement and granulometric composition of clastic rocks, degree of crystallinity of rocks having a crystalline, amorphous or crystalloamorphous structure.

Analysis of data on the ranges of variation of λ of main sedimentary rock types suggests that the thermal conductivity increases in a series clays → argillites →

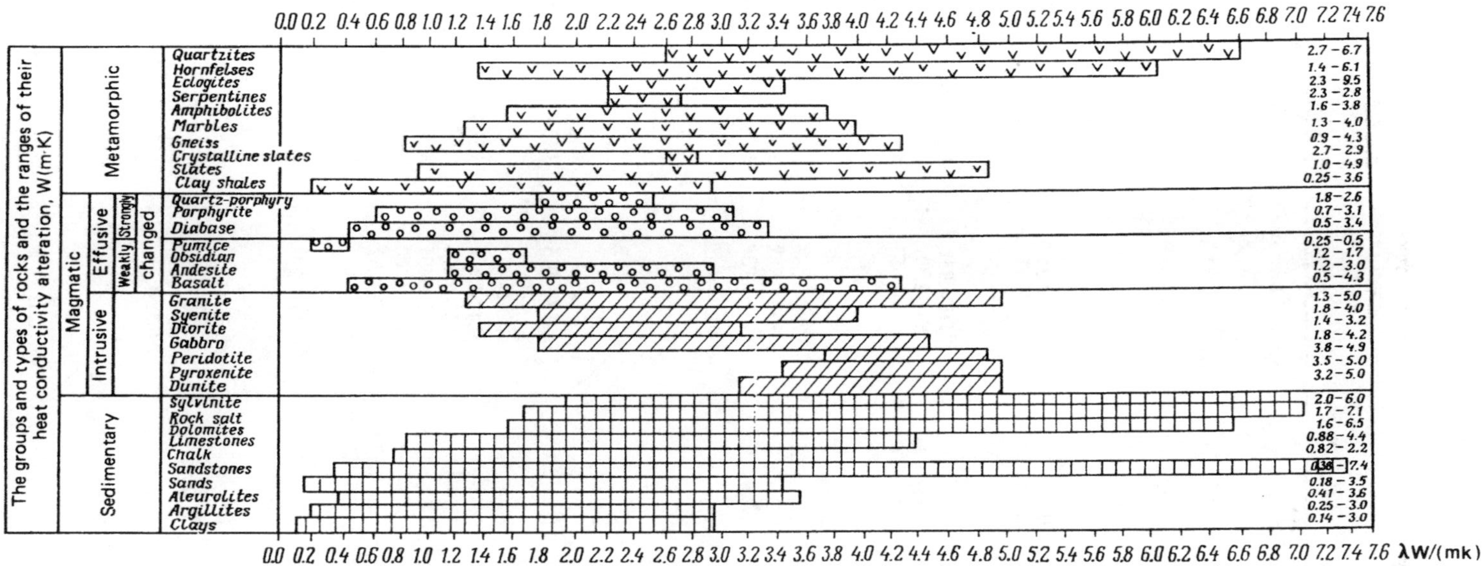

FIG. 114. Values of the coefficient of thermal conductivity λ of various rock groups and types (from air-dry to maximum moisture rocks including their altered varieties)

TABLE 13. Thermal Conductivity, Thermal Capacity and Diffusivity of Primary Sediments and Rocks

Primary sediment or rock	Thermal conductivity coefficient λ, W(m·K)	Thermal capacity c_{pm}, kJ/(kg·K)	Diffusivity coefficient a, 10^{-7} m^2/s
Sediments and sedimentary, metamorphic rocks and ores derived from them			
Breccia	0.92-3.3	-	-
Conglomerate	1.05-5.0	0.75-0.84	6.3-11.5
Dry gravel	0.38	-	-
Sandy mud	0.98-1.78	1.04-2.06	-
Sand:			
dry	0.18-3.49	-	-
moist	1.63-4.75	-	-
oil-saturated	2.3-4.28	-	-
Sandstone:	0.38-5.17	0.67-3.35	2.54-20.43
dry	0.67-6.49	-	-
moist	1.1-7.41	-	-
oil-saturated	0.84-4.24	-	-
Quartzite	2.68-7.6	0.718-1.331	13.6-20.90
Aleurite-clayey mud	0.88-1.13	2.42-2.49	-
Aleurolite	0.41-3.58	0.75-1.65	5.36-15.42
Clayey mud	0.66-0.86	-	-
Clay:			
dry	0.14-0.24	-	-
moist	0.38-3.02	0.753-3.596	2.54-11.56
Argillite	0.25-3.01	0.737-0.988	5.94-15.28
Clayey slate	0.25-3.01	0.737-0.988	5.94-15.28
Phyllite	2.9	-	-
Slate	2.51	-	-
Tripolite (diatomite and diatomaceous tripoli):			
dry	0.53	-	-
moist	2.03	-	-
Globigerina ooze	0.87-1.07	-	-
Chalk	0.82-2.2	0.837-3.915	3.13-6.20
Limestone:			
dry	0.69-2.51	-	-
moist	0.92-4.4	0.42-1.712	3.91-16.94
Marble	1.59-4.0	0.753-0.879	7.8-12.0
Dolomite	1.63-6.5	0.648-1.465	8.26-16.8
Marl	0.5-3.61	0.586-3.100	3.14-13.89
Gypsum	0.6-1.67	0.9-<1.1	-
Anhydrite	2.5-5.8	0.58-0.61	-
Pure rock salt	6.1-7.1	-	-
Rock salt	1.67-5.50	1.447-4.651	11.2-17.7
Sylvinite	2.6-6.0	-	-
Ore:			
martite and magnetite	5.03	-	-
martite and jaspilite	4.94	0.536	25.3
magnetite-martite hornfels	4.17	0.601	22.5
martite hornfels	4.3-4.8	0.58-1.04	-
Peat:			
dry	0.11	-	-
moist	0.46	-	-

TABLE 13 (*concluded*)

Primary sediment or rock	Thermal conductivity coefficient λ, W(m·K)	Thermal capacity c_{pm}, kJ/(kg·K)	Diffusivity coefficient a, 10^{-7} m²/s
Coal	0.13-2.24	0.863-1.528	0.7-7.02
Graphite	1.16-17.4	0.67	
Magmatic and metamorphic rocks and ores derived from them			
Dunite	1.11-1.85	0.586-0.795	7.16-8.48
Peridotite	3.78-4.85	0.921-1.088	11.97-14.10
Pyroxenite	3.48-5.02	0.879-1.214	9.44-14.86
Serpentinite	2.31-2.87	0.963-1.130	-
Gabbro	1.59-2.98	0.897-1.130	9.32-12.17
Diorite	1.38-2.89	1.118-1.168	3.34-8.64
Syenite	1.80-2.97	-	5.4-7.9
Granite	1.12-3.85	0.257-1.548	3.33-16.5
Basalt	0.44-3.49	0.544-2.135	3.44-13.45
Andesite	1.42-2.79	0.808-0.823	6.17-6.44
Trachyte	1.7-2.6	-	-
Obsidian	1.38-1.56	0.963-1.214	5.47-5.87
Pumice:			
dry	0.25	-	-
moist	0.50	-	-
Diabase	1.71-3.25	0.791-0.929	5.04-12.0
Porphyrite	0.72-3.52	0.716-1.557	2.64-12.8
Quartz-porphyry	1.76-2.60	-	5.00-6.94
Pegmatite	2.85-3.3	-	-
Tuff	1.30-3.95	0.795-1.411	9.99-12.36
Lava	0.25-0.73	0.670-1.382	2.35-4.13
Schist	0.65-4.76	0.699-1.643	2.87-22.5
Crystalline schist	2.66-2.94	-	-
Gneiss	0.94-4.86	0.754-1.176	6.30-8.26
Amphibolite	1.57-2.89	1.063-1.201	5.25-8.14
Eclogite	2.31-3.5	-	-
Hornfels	2.12-6.10	1.476-1.482	13.44-15.64
Skarn	1.48-2.97	-	-
Charnockite	1.06-1.5	-	-
Ore (deposit):			
pyrite (Belorechensk, Degtyarsk)	4.19-4.17	0.836-1.327	9.9-10
copper pyrite (Degtyarsk, Novolevinsk)	4.21	0.862-0.858	10.3-10.4
thick pyrite impregnation in quartz	3.36	0.607	13.6
staffelite-magnetite (Kovdor)	1.6-1.7	-	5.0-5.4
apatite-forsterite-magnetite (Kovdor)	3.0-4.7	-	8.6-14.1
magnetite (Krasnoyarsk)	2.0	-	-
magnetite (Sheregesh, Tashtagol)	2.7	-	-

sands → aleurolites → limestones → dolomites → rock salt. The sandstones do not enter this series. The lowest thermal conductivity is shown by dry and highly porous sandstones and the highest one by their low-porosity silicified or mineralized varieties having a very high concentration of relatively high thermal-conductivity quartz or ore minerals (pyrite, magnetite etc.). Clays and argillites usually demonstrate a relatively narrow range of variation of λ, and their dry and highly porous varieties show low λ values (see Fig. 114). The range of variation of λ values of aleurolites and sands is somewhat greater due to a large range of variation of their k_p and smaller clayey mineral content. The range of variation of λ limestones, rock salt is still wider (except sandstones) and is shifted toward higher λ values as a result of decreasing their k_p and increasing λ of rock-forming minerals (calcite, dolomite, halite).

Let us note a regular increase in extremely high and low λ values as we pass from clayey rocks to salts (see Fig. 144) (except sandstone and chalk).

In the case of intrusive igneous rocks one can observe a decrease in the highest and lowest λ values in a series of ultrabasic → basic → intermediate rocks, growth of extremely high and widening of limiting values for syenite and granite. For syenite this is due to an increased concentration in the rock of higher heat-conducting minerals (potassium feldspars and hornblende) and for granite, quartz.

Weakly altered effusive rocks typically demonstrate a narrowed range of variation of the coefficient of thermal conductivity, a decrease of extremely high λ values owing to the presence of pyroxene in the composition of these rocks (see Fig. 114, Table 13). Much lower λ values are exhibited by obsidian and pumice containing much silici acid which values are due to amorphous structure of these materials. The limiting small λ values of highly porous basalts are much lower than those of andesite and obsidian. It is of interest to point out that despite a high silica content, obsidians and pumices do not show high λ values not only because of a high porosity as is the case of pumice but also due to a low thermal conductivity of glassy quartz whose λ value, according to data of recent determinations does not exceed 1.34 W/(m·K).

Limiting high λ values of strongly altered effusive rocks decrease from diabase to quartz porphyry and limiting low values increase in the same direction. Low λ values of diabase and porphyrite (see Fig. 114) are associated with the crystallo-amorphous structure of these rocks. For porphyries they seem to be due to the presence of glassy varieties in the collection of samples investigated. Relatively small λ values of quartz porphyry, given a fairly high quartz content (34%) do not seem strange in the view of the fact that the given rock contains as much as 66% minerals (orthoclase, oligoclase, biotite) whose λ values are much lower than those of quartz.

Many metamorphic rock types (see Fig. 114) show a wide range of variation of λ values. What is more, this range is particularly great in the case of hornfelses and quartzites. An exception are some crystalline schists, seprentinites and eclogite. Very low λ values of clay shales and very high ones of quartzites are sufficiently easily accounted for by the specific features of their mineral composition. Clay shales are characterized by a high concentration of clayey minerals with low λ values, and quartzites by a high silica content. It must be noted that they show a different ratio of quartz, opal and chalcedony that seem to have very much differ-

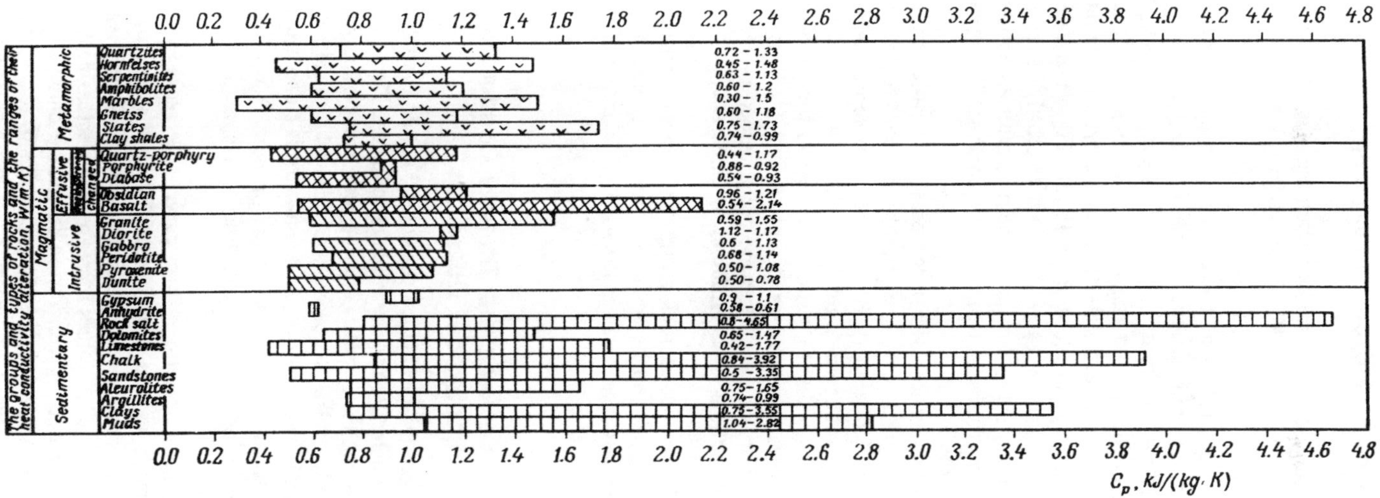

FIG. 115. Values of the thermal capacity c_{pm} of various rock groups and types

ent λ values. The voids ratio of these rocks sometimes has very low values. The λ values of ores vary from 1.6 to 4.9 W/(m·K), these values being close to λ of low-porosity rocks.

Thermal capacity. The c_{pm} values of rocks vary from 0.42 (limestone) to 4.65 (rock salt) kJ/(kg·K). For some rock groups it varies [in kJ/(kg·K)] as follows: from 0.42 to 4.65 (sedimentary rocks), from 0.45 to 2.13 (igneous rocks), from 0.3 to 1.72 (metamorphic rocks). The greatest range of c_{pm} among sedimentary rocks is shown by rock salt, sandstones, chalk, limestones and clays and the narrowest one by anhydrites, gypsums and argillites (see Table 13 and Fig. 115). The wide range of c_{pm} variation of rock salt has not as yet been satisfactorily accounted for, and for most of other sedimentary rocks they are associated with an appreciable variation of their voids ratio and coefficient of moisture content. The greater c_{pm}, the higher are the values of the latter.

The value c_{pm} of basalts varies over a comparatively wide range since their moisture content changes appreciably; c_{pm} of all other igneous rocks, except granite, is in the range from 0.5 to 1.2 kJ/(kg·K). For granite c_{pm} is somewhat higher owing to a greater variation of its moisture content. A relatively small variation of c_{pm} values of metamorphic rocks is due to a minor change in their moisture content. The thermal capacity of ores is much less [from 0.6 to 1.3 kJ/(kg·K)] than that of rocks. The thermal capacity of these latter are not governed by their granularity, lamination and state (amorphous or crystalline) of minerals.

Thermal diffusivity. For individual rock groups and types it varies over the range $(2\text{-}20)\cdot 10^{-7}$ m^2/s. The a values can be evaluated from λ, c_p and δ_r. The thermal diffusivity of ores is much less than a values $[(0.5\text{-}2.5)\cdot 10^{-7}$ m^2/s] of the aforementioned rocks.

Sec. 45. Relationship Between the Thermal Conductivity and Other Petrophysical Quantities

Inverse correlations between the product of the coefficient of thermal conductivity λ_{mr} with the porosity parameter P_p and the voids ratio k_p (see Fig. 116) and between the coefficient of thermal conductivity λ_{mr} and the coefficient of clayiness k_{clm} have been established for sandy and argillaceous rocks. A relationship represented in Fig. 116 for a specified sediment according to its authors, is closer than correlations λ_{sl} $(\lambda_{mr}) = f(k_p)$ (see Figs. 112 and 113) since the product $P_p\lambda_{mr}$ better allows for the structural heterogeneity of rocks contributing to a spread of points in Figs. 112 and 113.

A $\lambda_{mr} = f(k_{clm})$ graph is a hyperbolic form curve; a decrease in λ_{mr} values with growth of the clayey cement content is accounted for by a sharp increase in loose contacts between highly disperse clayey particles of rocks which leads to growth of their thermal resistance. The λ_{mr} values increase with decreasing k_{clm}, it appears, due to an increased role played by convection in heat exchange.

The role of convection and radiation in heat transfer in rocks has been insufficiently studied. There are different views relating to the contribution to the thermal

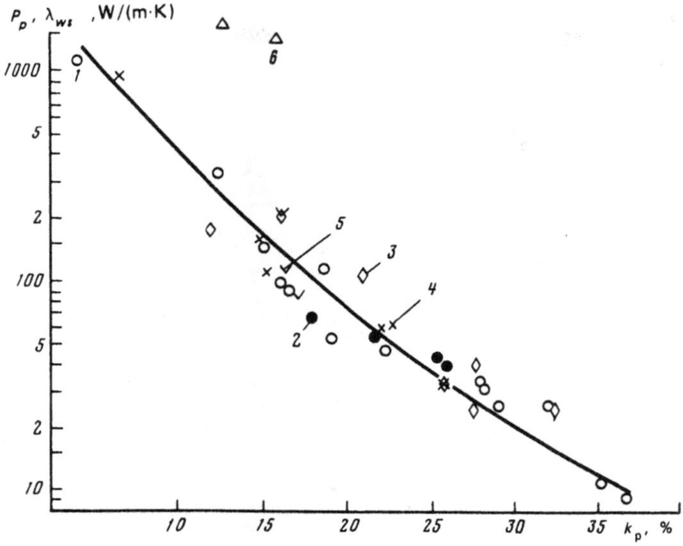

FIG. 116. Connection between the product of the porosity parameter P_p with the coefficient of thermal conductivity $\lambda_{w.s}$ and the voids ratio k_p for sandstones and graywackes (after experimental data of H. Zierfuß and G. Van der Vliet).

Sandstones: *1*—quartz; *2*—feldspar; *3*—with unknown mineral composition; *4*—subgraywacke; *5*—migrating sands; *6*—quartzite

conductivity of its individual constituents. It is determined by the mineral composition of the rock, its voids ratio and coefficient of saturation, size and geometry of the voids and grains, temperature and pressure. With increasing k_p and k_w, $k_l\lambda$, as has been demonstrated, decreases which is accounted for by a lower thermal conductivity of a fluid and, particularly, gas in the voids of rocks compared with that of their solid phase.

Convection is possible if voids and channels of different effective diameters are communicating with one another. Some investigators believe that convection can be ignored in channels of an effective diameter less than 3 μm whereas others believe it to be already negligible in the air space 6 μm in width. Convection is ruled out in channel regions filled by bound water.

The size of grains governing the effective diameter of the channels must certainly affect thermal conductivity. The low thermal conductivity of clays is associated not only with the small thermal conductivity but also with the appreciable dispersion of its particles.

For sandy-argillaceous and other rocks from definite sediments direct correlations have also been found, those between λ and δ_{dr} (Fig. 117), λ and c_V (Fig. 118) and λ and E etc. These three relationships are accounted for by that each of the quantities being compared taken separately is definitely connected with k_p. By using equations established enabling one to obtain, for example, the dependences of the

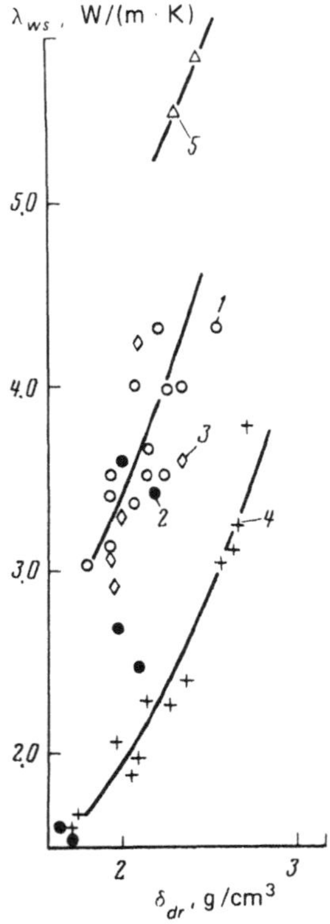

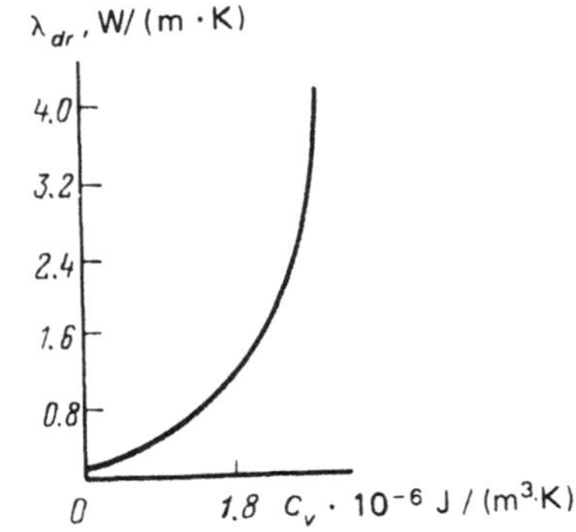

FIG. 117. Connections between the coefficient of thermal conductivity $\lambda_{w.s}$ and density δ_{dr} for sandstones and limestones (after experimental data of H. Zierfuß and G. Van der Vliet).

Sandstones: *1*—quartz; *2*—feldspar; *3*—of unknown mineral composition; *4*—limestones; *5*—quartzite; λ_{mr}—determined on maximum moisture samples, δ_{dr}—on dry samples

FIG. 118. Connection between the coefficient of thermal conductivity λ_{dr} and volumetric thermal capacity c_V (after V.V. Rzhevskii and Ya.G. Novik)

coefficient of thermal conductivity of a maximum moisture rock and density of a dry rock on the voids ratio, it is possible, by jointly solving them, to find a correlational equation of the dependence of λ on δ_{dr} (see Fig. 117). In like manner one can also obtain for dependences $\lambda = f(c_V)$ and $\lambda = f(E)$.

Dirrect correlations $\lambda = f(E)$ and $\lambda = f(\delta)$ have been also established for ferruginous quartzites (Fig. 119) and hematite ore (Fig. 120). Their presence, however,

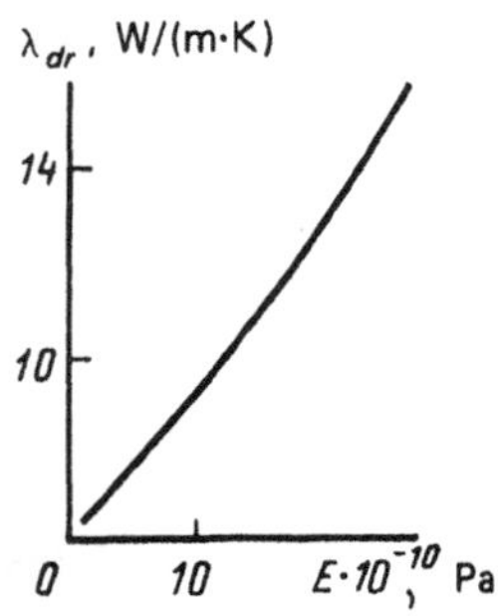

FIG. 119. Connection between the coefficient of thermal conductivity λ_{dr} and Young's modulus E for ferruginous quartzites (λ_{dr} measured at right angles to the lamination) (after V.V. Rzhevskii and G.Ya. Novik)

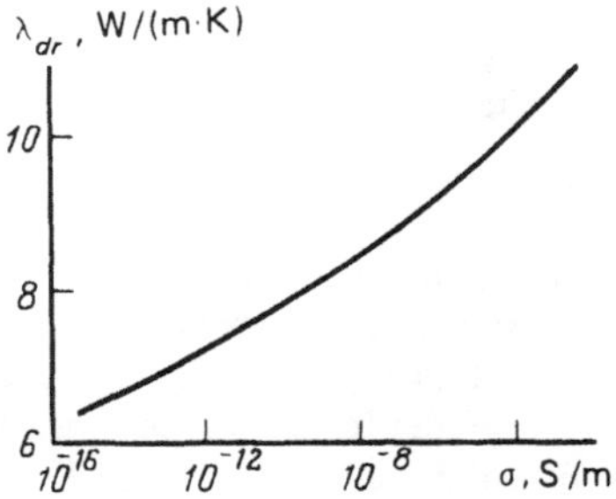

FIG. 120. Connection between the coefficient of thermal conductivity λ and specific electric conductivity σ for hematite ore (after V.V. Rzhevskii and G.Ya. Novik)

is accounted for differently, viz. by that each of the quantities being correlated is in a definite fashion governed by the content in an ore of an electron-conducting component (iron oxides and hematite). By referring to these correlations for each of the quantities involved it is possible to find also dependences $\lambda = f(E\perp)$ and $\lambda = f(\delta)$ in the same way as has been described above.

Sec. 46. Dependence of Thermal Conductivity and Capacity of Rocks on Temperature and Pressure

The effect of temperature. The thermal conductivity of rocks decreases with increasing the temperature and particularly so until 200-427 °C. The λ values of some rocks (olivinite, granite, diorite) upon attaining minimal values with increasing the temperature somewhat enhances. The minimum value of thermal conductivity generally agrees with the beginning of melting rocks. The different behaviour, during the heating, for example, of such close in composition rocks as granite and obsidian is accounted for by the difference in their structure. In terms of their behaviour during heating rocks are divided into three groups: crystalline (granite, diorite, eclogite etc.), amorphous (obsidian) and rocks having crystalloamorphous structure (diabase, porphyrite etc.). The thermal conductivity of rocks having a crystalline structure is conditioned by the scattering of phonons on crystalline grains and on one another. The latter process accounts for the relationship $\lambda = f(\lambda_0/t)$ where λ_0 is the λ value at 20 °C.

Heat transfer of amorphous unordered-structure rocks refers to random processes and $\lambda = f(t)$.

Rocks characterized by a crystalloamorphous structure show a heat transfer mechanism usual both for crystalline and amorphous bodies. Due to this their thermal conductivity is not practically or is inappreciably affected by temperature.

The temperature conductivity drops with growth of t. This process is common for rocks having a crystalline structure and to a lesser degree for rocks with a crystalloamorphous structure, it is not practically observed for purely amorphous varieties. The volumetric specific heat of rocks increases upon their heating to a temperature 850 °C.

The effect of pressure. The thermal conductivity increases with pressure, its maximum variations referring to pressures from 0.1 to 10 MPa. Further the coefficient λ little changes or remains practically unchanged. The growth of λ is supposed to be due to closer contacts between the grains since after the pressure is removed λ becomes higher than the original one.

Temperature conductivity increases with pressure.

The spatial variation of the coefficient of thermal conductivity. Little is as yet known about local and regional regularities of variations of the thermal quantity values of rocks in the earth's crust. Design data are available that give rough idea of the coefficient of the thermal conductivity of structural and formation complexes of various layers of the earth's crust. A consideration of these data shows that the least average thermal conductivity 1.2 W/(m·K) is displayed by the sedimentary cover (regolith) of the earth's crust composed by weakly lithified sandy and argillaceous sediments of recent-origin platforms. The thermal conductivity of lithified calcareous and magnesial and sandy and argillaceous sediments of ancient-origin platforms, edge depression and sediments of folded regions is twice that of the average thermal conductivity. The λ values, given normal p and t, for the granite metamorphic and diorite layers of the crust remain practically unchanged to again increase appreciably to 3.5 W/(m·K) in the basalt belt.

Chapter X
Magnetism

Magnetism is the property of rocks to be magnetized by the magnetic field, vary it and sometimes retain the magnetized state after the action of the magnetic field discontinues. Upon magnetization, any rock volume acquires a magnetic moment. Magnetism shows itself during the interaction of two magnetized rock samples or a rock sample and a conductor through which an electrical current flows. It is determined by concentration and distribution in the rock of dia-, para-, antiferro-, ferro- or, more commonly, ferrimagnetic components, their chemical composition, minerals' crystal lattice structure, in particular, the type of bounding of atoms or ions in it. Magnetism is governed by the origin and living conditions of rocks. It varies in magnitude if the pressure and temperature affecting the sinking mass grow with time; in so doing, not only the rock's structure but also its mineral composition changes.

Sec. 47. Magnetization Theory

Most minerals are dia- and paramagnetic; antiferromagnetic minerals are much less in number. Very rarely occurring meteorite iron, α-hematite and some hemoilmenite species may be ferromagnetic; ferrimagnetic minerals are about fifteen in number, yet it is their content that principally conditions magnetization of rocks, since magnetism of other minerals (excepting native iron) is appreciably less.

The property of dia- and paramagnetic minerals to be magnetized in field $\vec{H}$ is characterized by *specific volumetric* $\vec{J}_V$ and *mass magnetization* $\vec{J}_m$

$$\vec{J}_V = \varkappa\vec{H} \tag{64}$$

$$\vec{J}_m = \chi\vec{H} \tag{65}$$

The proportionality coefficients in relationships (64) and (65) are known as *specific volumetric susceptibility* $\varkappa$ and *specific mass susceptibility* χ; these are values reflecting unequal ability of minerals and rocks to be magnetized.

The value $\varkappa$ is dimensionless. Specific mass magnetization (magnetic moment of a rock's unit mass $\vec{J}_m = \vec{J}_V/\delta_{dr} = \chi\vec{H}$; hence $\vec{J}_V = \delta_{dr}\chi\vec{H}$ and, taking into account Eq. (64) $\delta_{dr}\chi = \varkappa$ and $\chi = \varkappa/\delta_{dr}$ [m^3/kg (or cm^3/g)].

As follows from Eqs. (64) and (65), given that $\vec{H} = 0$, $\vec{J}_V = 0$ and $\vec{J}_m = 0$, i.e. dia- and paramagnetic minerals are not magnetized unless a magnetic field is applied to them. This is accounted for differently. Each of the moving elementary atomic particles (molecules or ions) of dia- and paramagnetic minerals has its own (or spin) and orbital magnetic moment. Neutron and proton moments and the resulting magnetic moment of atomic nuclei are very small. In this connection one can consider solely the total magnetic moments of atomic electron shells of dia- and paramagnetic minerals. Each of the electrons of these shells has its own (spin) magnetic moment whose projection on the direction of the magnetic field is

$$\mu = \pm he/4\pi m_e = \pm |e|\hbar/2m_e = 9.27 \times 10^{-24} \text{ A} \times \text{m}^2$$

$\mu_B = |e|\hbar/2m_e$—Bohr magneton—natural atomic unit of magnetic moment, and the orbital moment of circular current is μ_{orb}. For the atom with one electron (hydrogen)

$$\mu_{orb} \approx \frac{ve\pi r^2}{2\pi r} = \omega e\pi r^2$$

where v is the speed of the electron moving along the circular orbit; e, m_e are the charge and mass of electron; r is the orbit's radius; h is Planck's constant; $\hbar = h/2\pi$, ω is the frequency of rotation of electrons about the nucleus. But diamagnetic mineral atoms and these minerals on the whole are not magnetized unless they are affected by the field since the structure of their electron shells is symmetrical, so the spin orbital moments of the electrons are compensated. By contrast, paramagnetic mineral atoms possess the resultant magnetic moment due to the fact that their electron shells have inner ($3d$ and $4f$) energetic sublevels incompletely constructed and spin orbital moments of electrons are not balanced.

However, neither are paramagnetic minerals magnetized in the absence of a magnetic field since their resultant atomic magnetic moments having whatever spatial orientation are mutually compensated.

In the magnetic field electrons of any materials have additional speeds due to Larmor precession. The latter consists in that, as a result of the induction action of the magnetic field, the electrons of the atomic shell acquire an additional angular speed at which their orbits precess about the external field (Fig. 121) and, consequently, additional magnetic moment proportional to the field and oriented opposite to it in conformity with Lenz's law.

The additional speed of electrons

$$\Delta v_{lar} = erH/2m_e$$

and the additional magnetic moment for the atom with one electron (hydrogen) rotating about a circular orbit

$$\Delta\mu_{orb.lar} = -i_{lar}S_{orb} \approx -\frac{\Delta v_{lar} e\pi r^2}{2\pi r} \approx -\frac{er}{2m_e} H \frac{e\pi r^2}{2\pi r} \approx -\frac{e^2 r^2}{4m_e} H$$

where i_{lar} is current appearing during Larmor precession; S_{orb} is the area of the orbit of the electron.

Whereas the moment of the gramme-atom of this element is

$$\mu_{g\text{-}a} \approx -\frac{e^2 r^2}{4m_e} N_A H$$

where N_A is Avogadro constant.

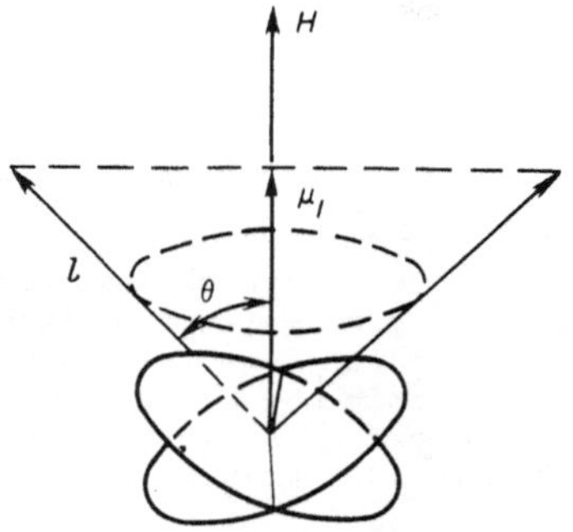

FIG. 121. Precession of electron orbit l about vector of magnetic field H (after S.V. Vonsovskii)

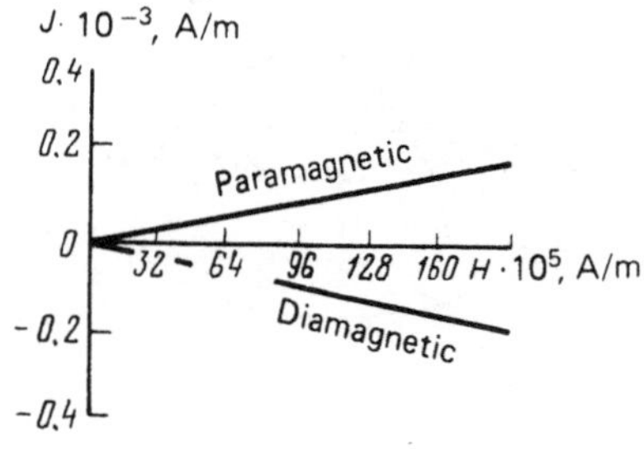

FIG. 122. Magnetization curves of weakly magnetic materials

For practically single-element minerals, such as sulphur, graphite etc., whose electron orbits are at different angles to the direction of the magnetic field and the electrons rotate in elliptical orbits, magnetization of the gramme-atom

$$\mu_{g\text{-}a.av} \approx -\frac{e^2 \sum_{n=1}^{n=z} r^2_{n.av}}{6m_e A} N_A H = -\chi_{g\text{-}a} H \tag{66}$$

where $\sum_{n=1}^{n=z} r^2_{n.av}$ is the sum of the average values of quadratic projections of radius vectors of orbits on a plane which is at right angles to the field.

$$\vec{J}_m = -\frac{e_2 \sum_{n=1}^{n=z} r^2_{n.av}}{6m_e A} N_A H = -\chi H \tag{67}$$

where A is the relative atomic mass;

$$\vec{J}_V = -J_m \delta_m = -\frac{e^2 N_A \delta_m \sum_{n=1}^{n=z} r^2_{n.av}}{6m_e A} H = -\varkappa H \tag{68}$$

where δ_m is the density of the mineral.

As can be seen from Eqs. (66)-(68), the magnetization of diamagnetic minerals is in direct proportion to the magnetic field strength H yet has an opposite sign (Fig. 122). The latter circumstance is the principal feature of diamagnetic minerals.

Statistically determined magnetization of paramagnetic minerals occurs under conditions of equilibrium between the action of the magnetizing field causing magnetic moments to be oriented and the action of the thermal molecular (atomic) motion causing their disorientation.

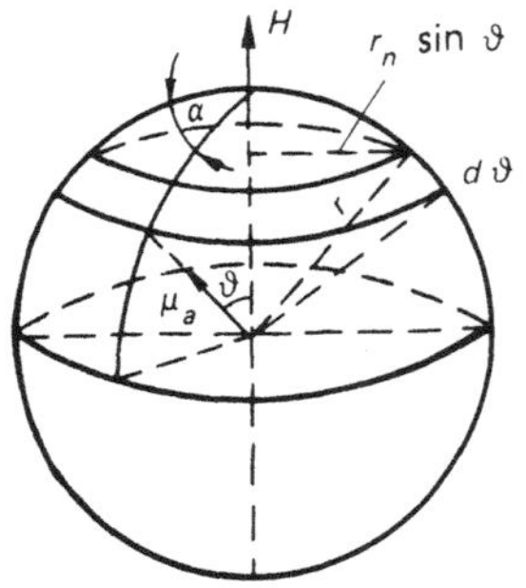

FIG. 123. Diagram illustrating the derivation of the Langevin equation.

A sphere of a unit radius where polar coordinates are adopted; ϑ is the polar angle and α is the longitude; the polar axis is parallel to the H axis; the direction of the moment μ_a of a random atom is characterized by the ϑ and α coordinates of the trace of the atom on the unit sphere

According to the Langevin theory, in the absence of a magnetic field the axes of the magnetic moments of unit dipoles are uniformly distributed in all directions due to thermal motion of molecules (atoms). Consequently, the number of dipoles (atoms) dN the traces of whose axes intersect the spherical belt on a sphere with a unit radius bounded by two parallel circles whose coordinates are ϑ and $\vartheta + d\vartheta$ is in direct proportion to the value of this element $2\pi \sin \vartheta d\vartheta$ (Fig. 123), i.e.

$$dN = C \sin \vartheta d\vartheta \qquad (69)$$

where C is constant.

Dipole magnetic moment orientation differs (according to Boltzmann) from that in the absence of a field [see Eq. (69)] by a factor $e^{-U/kT}$, where U is the potential energy of dipoles (atoms) in the field under consideration; k is the Boltzmann constant; T is the absolute temperature.

Since

$$U = -\mu_a H \cos \vartheta,$$

then

$$dN = Ce^{-\mu_a H \cos \vartheta / kT} \sin \vartheta d\vartheta$$

where μ_a is the dipole (atom) magnetic moment.

By defining $\mu_a H/kT = a$, we have

$$dN = Ce^{a \cos \vartheta} \sin \vartheta d\vartheta$$

If N dipoles with a magnetic moment μ_a are found in the sphere, then C can be found from the relationship

$$N = \int dN = C \int_0^\pi e^{a \cos \vartheta} \sin \vartheta d\vartheta = C/a(e^a - e^{-a})$$

whence

$$C = aN/e^a - e^{-a}$$

Since the magnetization vector $\vec{J_N}$ is parallel to the magnetic field, its value is determined as the sum of projections of magnetic moments of N dipoles (atoms, molecules) on the direction of the magnetic field.

The summed up dN dipole moment with axes in the range from ϑ to $\vartheta + d\vartheta$ is equal to $\mu_a dN$, and its projection on the direction of the field H is $\mu_a dN \cos \vartheta$. Then the magnetization vector of a sphere containing N dipoles is

$$J_N = \int \mu_a \cos \vartheta dN = C\mu_a \int_0^{\pi} e^{a \cos \vartheta} \sin \vartheta \cos \vartheta d\vartheta$$

$$= C\mu_a \left(\frac{e^a + e^{-a}}{a} - \frac{e^a - e^{-a}}{a^2} \right) = N\mu_a(\cot a - 1/a)$$

Since $a \ll 1$, then

$$\coth a = 1/a + a/3 - a^3/45 + \ldots \approx 1/a + a/3$$

and consequently, magnetization of one dipole (atom) is

$$\frac{J_N}{N} = \frac{\mu_a^2}{3kT} H = \frac{\mu_a^2 N_A}{3RT} H$$

where $R = kN_A$ is gas constant (known also as a gas-law constant).

As to the magnetic moment of one gramme-atom, it is

$$\mu_{g\text{-}a} = \frac{\mu_a^2 N_A^2}{3RT} H = \chi_{g\text{-}a} H \tag{70}$$

and

$$J_m = \frac{N_A^2 \mu_a^2}{3RTA} H = \frac{M_0^2}{3RTA} H = \chi H \tag{71}$$

where $M_0^2 = N_A^2 \mu_a^2$;

$$J_V = \frac{N_A^2 \mu_a^2 \delta_m}{3RTA} H = \frac{M_0^2 \delta_m}{3RTA} H = \varkappa H \tag{72}$$

It can be also easily shown that the mass paramagnetic magnetization of one mole is

$$\vec{J}_{mole} = \frac{\mu_m^2 N_A^2}{3RTM} \vec{H} = \frac{M_m^2}{3RTM} H = \chi_{mole} \vec{H} \tag{73}$$

where $M_m = \mu_m^2 N_A^2$; μ_m is the magnetic moment of a molecule; M is the mass of the molecule.

Thus, given that $a \ll 1$, i.e. in such cases where the energy $\mu_a^2 H$ of magnetics is small (H is inappreciable) compared with their thermal energy kT, the magnetization of single-element minerals much as of diamagnetic minerals [see Eqs. (70)-(72)] is directly proportional to the intensity $\vec{H}$ of the external magnetic field. Yet here it is always positive, in contradistinction to diamagnetic magnetization [see Eqs. (66), (68) and Fig. 122)] is governed by the temperature and has the same sign as the magnetic field strength vector. This is the principal feature of paramagnetic minerals.

Given appreciable H and minor T, i.e. given $a \to \infty$, $\coth a \to 1$, one observes the effect of saturation of paramagnetics, the proportionality $\vec{J}_V$ and $\vec{H}$, and J_V

tends to the limiting value $J_{V0} = N_{\mu_a}$. The relation

$$J_V/J_{V0} = L_j(a) \qquad (74)$$

(Langevin function) characterizes magnetization of a mineral (rock) with respect to the maximum possible value of this quantity.

The quantum theory has improved Langevin's concept relying on the discreteness of orientation of the atoms' magnetic moments. This made it possible to represent Langevin's formula in this form:

$$L_j(a) = \frac{2j+1}{2j} \coth \frac{(2j+1)a}{2j} - \frac{1}{2j} \coth \frac{a}{2j}$$

where j is the quantum number representing the sum of the orbital and spin quantum numbers and taking on values multiple to one half (0, 1/2, 1, 3/2 etc.).

Given that $j = 1/2$, i.e. the atomic magnetic moment is of spin origin and there are only two of its orientations relative to the magnetic field (parallel and antiparallel), i.e. $L_{1/2}(a) = \tan h\ a$.

Proportionality coefficients in the relationships (66)-(68) for diamagnetics and in (70)-(73) for paramagnetics, respectively, are provided by *atomic* $\chi_{g\text{-}a}$, *mass* χ, *volumetric* $\varkappa$, *molar* χ_{mole} *dia-* and *paramagnetic susceptibilities.*

The magnetic susceptibility of single-element dia- and paramagnetic minerals is governed by their chemical properties, viz. the electrical configuration of atoms (molecules) of the mineral [to be defined by the sequence number z of the mineral's element and be responsible for the sum $\sum_{n=1}^{n=z} r_{n.av}^2$, A and δ_m in Eqs. (66)-(68) as well as μ_a^2 and M_a in Eqs. (70)-(73)]. The density δ_m is conditioned, besides, by the difference in minerals' structure (diamond and graphite, crystalline and amorphous sulphur).

As the temperature rises the magnetic susceptibility of paramagnetic minerals decreases [see Eqs. (70), (71) and (73)], whereas for diamagnetic minerals it is generally unrelated to the temperature. For the given temperature this characteristic of both mineral varieties is a constant quantity (see Fig. 122).

Multi-element minerals may contain several purely dia- or paramagnetic elements, yet more often than not they are mixtures of these both; in the latter case their magnetic susceptibility is governed not only by that of individual atoms but also by their ratios in the multi-element mineral and its crystalline structure.

Magnetization and magnetic susceptibility associated with the Larmor precession of atomic electron orbits are characteristic of all minerals but they are taken into account solely in the case of pure dia- and paramagnetic minerals since J and $\varkappa$ of this process is much less than those resulting from ferri- and ferromagnetism.

The specific volumetric magnetic susceptibility of diamagnetic minerals is inappreciable and negative in value. The values of $\varkappa$ are in the range from $(-5.024) \times 10^{-4}$ SI units (grossular), $(-1.63) \times 10^{-4}$ SI units (bismuth) and zero.

Silicate rock-forming minerals are diamagnetic unless their composition includes Fe^{2+}, Fe^{3+}, Mn^{2+} and some other ions, otherwise they are paramagnetic.

Quite a number of important rock-forming sedimentary rock minerals (quartz, potassium feldspars, plagioclases, calcite, anhydrite, gypsum, halite, sylvite, graphite) refer to diamagnetic. So are also sulphur, bismuth, copper, gold, lead, mercury etc.

The anisotropy of the magnetic properties found in single crystals of natural diamagnetics is often unusual for their polycrystalline varieties as a result of the different crystal orientation.

For example, muscovite, spinels, rutile, dolomite, magnesite are paramagnetic; their magnetic susceptibility $\varkappa$ is also inappreciable; it varies from 12.56×10^{-6} SI units (dolomite), 37.68×10^{-6} SI units (magnesite) to 21.352×10^{-5} SI units (muscovite), not exceeding, it appears, 25.12×10^{-5} SI units. It is, however, positive in value. As to olivines, pyroxenes, garnets, biotites, amphiboles, cordierites, wolframites, cassiterites, chromites, whose magnetic susceptibility is of the order of several thousands of millionth fractions of an SI unit, according to recent studies, cannot be considered pure paramagnetics. Many varieties of these minerals contain microimpurities (10^{-5}-10^{-1}%) of ferrimagnetics to be found in the composition of relatively large crystals of iron-containing minerals. It is consistent with their residual magnetization $\vec{J}_{rs}$ of saturation and with important differences in values of $\varkappa$ obtained by measuring the same quantities for pure materials and natural minerals.

In minerals of igneous and less commonly metamorphic rocks magnetite is generally found to be a constant impurity. In addition, the amount of impurities is greater in magnesial and less in ferruginous varieties of dark-coloured minerals.

The appreciable anisotropy of $\varkappa$ is typical of paramagnetic mineral single crystals, is seldom observed for their polycrystalline varieties due to the same causes as for diamagnetic polycrystalline minerals.

Sec. 48. A Theory of Magnetization and Characteristics of Magnetic Properties of Ferro- and Ferrimagnetic Minerals

Isolated atoms of ferromagnetic minerals display appreciable resulting spin orbital magnetic moments due to the vacancies in their shell $3d$ and $4f$ sublevels. However, the resulting orbital atomic magnetic moments of electrons in single-element minerals (native meteorite iron) are mainly compensated and are unable to be oriented in a magnetic field owing to the great atom packing density of these materials. By contrast, electron spin moments in definite minor-size (about 10^{-9} cm^3) domains are oriented parallel and, consequently, ferromagnetics are here magnetized to saturation. This spin magnetic moment orientation is accounted for by the pronounced positive exchange interaction between electrons of unfilled shells of the adjacent atoms. This interaction causes electron distribution of atoms of ferromagnetics (Fe, Ni, Co, Cd) to be changed with respect to possible states and spin magnetic moments to be oriented parallel, even though thermal motion tends to disturb this structure. The spatial orientation of parallel spin magnetic moments is similar solely within definite, different in magnitude domains of a single-element ferromagnetic mineral because of the specific features of its crystalline structure and presence of crystal anisotropy energy. The resulting spin moments are also different in value and orientation, so the ferromagnetic mineral is on the whole not magnetized before being acted upon by a magnetic field. As a magnetic field is superposed, the magnetic moments of domains vary in magnitude and orientation to conform to the field which results in the magnetization of the ferromagnetic material. A definite fraction of magnetization is retained after the applied magnetic field (remanent magnetization) has been removed.

The Weiss theory (alternatively known as a molecular field theory) of magnetization of domains bases on the Langevin theory which suggests that magnetization of ferromagnetics* results from an established equilibrium of the aligning action of an external magnetic field H_{ex} and disorienting action of thermal motion and, besides, due to the magnetizing effect of an additional internal molecular field wJ_V (appearing during exchange interaction between electrons of the adjacent atoms) which is in direct proportion to magnetization of the substance concerned (w is a proportionality coefficient).

Consequently, dipoles are found in the field $H_{ef} = H_{ex} + wJ_V$.

Similarly to paramagnetics we have

$$J_V = \mu_a N(\coth a' - 1/a')$$

where

$$a' = \frac{\mu_a H_{ef}}{kT} = \frac{\mu_a(H_{ex} + wJ_V)}{kT}$$

In case $a' \to \infty$ (i.e. given an appreciable magnetic field strength and low absolute temperature) the magnetized state of the rock J_V approaches the limiting value $J_{V0} = N\mu_a$.

The ratio

$$\frac{J_V}{J_{V0}} = \coth a' - \frac{1}{a'} = \coth \frac{\mu_a H_{ef}}{kT} - \frac{kT}{H_{ef}\mu_a}$$

$$= \coth \frac{\mu_a(H_{ex} + wJ_V)}{kT} - \frac{kT}{(H_{ex} + wJ_V)\mu_a}$$

The latter relationship, when taking into account the discrete orientation of atomic magnetic moments in space is caused to take on the following form:

$$\frac{J_V}{J_{V0}} = \frac{2j+1}{2j} \coth \frac{2j+1}{2j} a' - \frac{1}{2j} \coth \frac{a}{2j}$$

In the simplest case, given $j = \pm(1/2)$

$$\frac{J_V}{J_{V0}} = \tanh \frac{\mu_a(H_{ex} + wJ_V)}{kT} \tag{75}$$

Eq. (75) is basic in the theory of ferromagnetism. If $H_{ex} \to 0$, then, by defining $\mu_a wJ_{V0}/k$ in terms of t_C (Curie point), we have

$$\frac{J_V}{J_{V0}} = \tanh \frac{J_V/J_{V0}}{T/t_C} \tag{76}$$

---

* Ferromagnetics are considered as being isotropic substances.

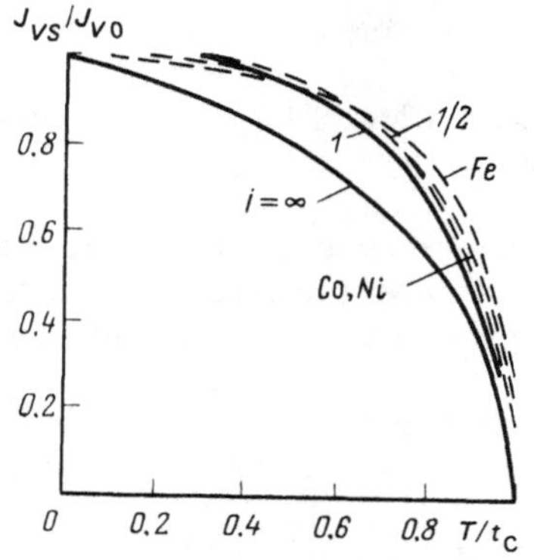

FIG. 124. Comparison of theoretical and experimental curves determining the dependence of the saturation magnetization on the temperature for iron, cobalt and nickel

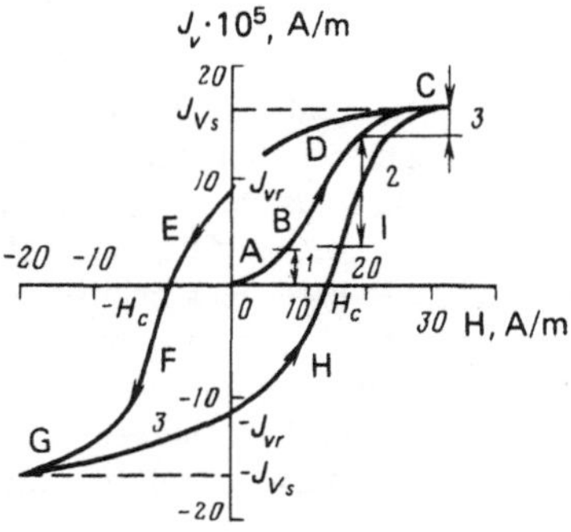

FIG. 125. Magnetization curve *ABC* and hysteresis loop *CDEFGHIJ* of ferromagnetic and its sections *1-3*

Eq. (76) connects the relative magnetization with relative temperature. By determining J_V on a sample under the conditions of technical saturation i.e. when all domains are aligned parallel to the magnetic field and $J_V = J_{Vs}$, we have

$$\frac{J_{Vs}}{J_{V0}} = \tanh \frac{J_{Vs}/J_{V0}}{T/t_C} \tag{77}$$

The pattern of connection between J_{Vs}/J_{V0} and T/t_C is validated by experimental results (Fig. 124).

At temperatures above Curie point ferromagnetics turn to paramagnetics. The Weiss theory describes the magnetization of internal portions of individual domains of a ferromagnetic demagnetized on the whole since it does not take into account the effect of other magnetic energy types. Magnetization in a magnetic field is characterized by a $\vec{J_V} = f(\vec{H})$ curve (Fig. 125) which differs much for ferromagnetics from similar curves for dia- and paramagnetics (see Fig. 122).

An increase in magnetization with increasing the field strength is not linear. The magnetization curve has characteristic regions *1-3* (see Fig. 125). The region *1* corresponds to a reversible process in which the connection between J_V and H is linear and magnetization disappears as the magnetizing field is removed. This is a process of reversible boundary displacement in which magnetization increases owing to the increase of the magnetic moment of individual domains due to the increase in their size without variation in the boundary orientation.

At the region *2* magnetization drastically increases as H increases due to the displacement of the boundaries: during this process the domains increase in size, the magnetization vectors orientation being identical and close to that of the magnetic field. The process is irreversible since an appreciable fraction of its energy is dissipated as heat. At the region *3*, J_V slowly grows with increasing H tending to a limiting value due to reversible rotation. At this, the volume of individual domains remains unchanged and their magnetization vectors are aligned with the direction of the field.

As the field strength drops to zero, magnetization lowers to a definite value J_{Vr} called residual isothermal magnetization. Zero magnetization is attained at a field strength equal to the coercive force H_c (see Fig. 125).

Ferromagnetics showing a relatively small hysteresis curve area and a coercive force (see Fig. 125) not exceeding thousands of amperes per 1 m are known as soft, by contrast to hard, ferromagnetics demonstrating a wide hysteresis loop and coercive force equal to several tens or even hundreds of amperes per 1 m.

Residual isothermal magnetization $\vec{J}_{Vr}$ and coercive force $\vec{H}_c$ together with $\vec{J}_V$, $\vec{J}_{Vs}$, $\varkappa$, χ, μ and others characterize the magnetic properties of ferromagnetics under conditions of normal magnetization.

Normal magnetization appears as a constant magnetic field is applied to a rock at a normal temperature (20°) and pressure (0.1 MPa). In so doing, magnetization vectors of domains that do not have this variation followed by the negotiation of appreciable energy barriers vary their orientation for one that is parallel to the magnetizing field. Normal magnetization is demonstrated, e.g. by a sedimentary rock in a modern geomagnetic field ($H_e \approx 40$ A/m).

Apart from normal, we still distinguish ideal and thermomagnetization. Ideal magnetization occurs during a joint effect of a constant and a variable magnetic field with an amplitude that varies from saturation fields to zero. Under conditions of ideal magnetization more magnetization vectors of domains are aligned in the direction of the field than in a constant field. In this case (see works by G.N. Petrova and T. Nagata) magnetization J_{Vri} shows a greater magnitude, so are the values of magnetic susceptibility $\varkappa_e$ appreciably much greater as are those of relative residual magnetization J_{Vri}/J_V, coercive force $H_{c.e}$ or relative coercive force $H_{c.e}/H_e$ compared with the same quantities for normal magnetization, given the identical strength of the magnetizing field H_e and residual magnetization.

Thermomagnetization is commonly manifested by rocks heated to temperatures above the Curie temperature (Curie point) and cooling to normal temperatures in a stationary magnetic field. This gives rise to the highest values of the residual magnetization of rocks J_{Vrt} since at elevated temperatures under conditions of minor mechanical stresses the most magnetization vectors of domains are aligned in the direction of the field. After the rocks cool the orientation of domain becomes fixed.

The volumetric magnetic susceptibility $\varkappa_t$, relative residual magnetization J_{Vrt}/J_V or J_{Vnt}/J_V and relative coercive force H_{cot}/H_e or its absolute values H_{cot} under conditions of identical residual magnetization are also maximum for this process.

The magnetic susceptibility of ferromagnetics for a field intensity approaching zero is constant. Then it increases attaining a maximum value, $\varkappa$ further decreases.

Most natural strongly magnetized materials, showing magnetic properties approaching those of ferromagnetics, much differ from these latter in magnetic structure, and with respect to this feature are referred to a special group of ferrimagnetics.

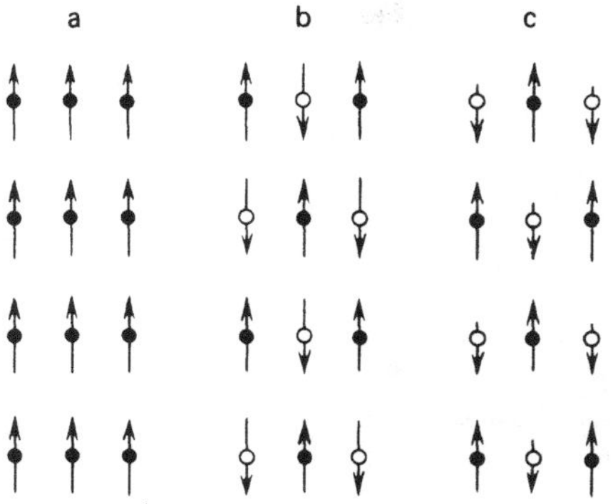

FIG. 126. Diagram of the distribution of magnetization vectors in crystals.
a—ferromagnetic; *b*—antiferromagnetic; *c*—ferrimagnetic

Magnetization vectors in crystals of ferrimagnetics form two groups; magnetization vectors in identical groups are parallel, in different groups antiparallel. Since the resulting magnetic moments are not equal (Fig. 126) appreciable intrinsic magnetization of ferrimagnetics appears.

This magnetic structure is conditioned by an inverse spinel structure of ferrimagnetics. These latter include most ferrites, in particular, magnetite.

Apart from the aforementioned minerals, the properties of ferromagnetics are demonstrated by ferromagnetic titanomagnetites, weakly ferromagnetic hematite, iron hydroxides (goethite, hydrogoethite and lepidocrocite), siderite, such ferrites as maghemite, trevolite, jacobsite, franklinite, magnesioferrite, hemoilmenites (rare antiferromagnetic species showing weak ferromagnetism) and sulphoferrites [pyrrhotite (some of its varieties), vallerite, cubanite]. Minerals representing titanomagnetites form titanomagnetite series I, and hemoilmenites form hemoilmenite series II. Thus, principal para-, ferro- and ferrimagnetic minerals mainly contain the Fe^{2+}, Fe^{3+}, Tio^{4+}, O^{2-} ions, and as impurities Mg^{2+}, Mn^{2+}, Al^{3+}, Cr^{3+}, V^{3+} ions.

The most susceptible minerals of those under consideration can be arranged in a series in the order of decreasing values of $\varkappa$ and Z_s: jacobsite, magnetite, maghemite, trevolite, magnesioferrite, titanomagnetites, hemoilmenites, pyrrhotite.

Magnetite demonstrates appreciable susceptibility (from 1.25 to >25 SI units) and saturation intensity of magnetization $>490 \times 10^3$ A/m. Magnetite is diagnosed from its Curie temperature (575-578 °C).

Cubical single crystals of magnetite commonly demonstrate magnetic anisotropy: magnetization occurs the easiest along the cube's diagonal axis and with most difficulty in the direction of its edges [see (Fig. 137*B*)]. Magnetization shows intermediate values in the direction of the diagonals of the cube faces. Almost similar values ($\sim 435 \times 10^3$ A/m) are attained by saturation intensity and over the same range (<4-25 SI units) the magnetic susceptibility of maghemite varies. The latter material, however, is unstable to heating, starting from $t \approx 200$ °C it converts to hematite.

Titanomagnetite series I is a family of minerals with a composition $x Fe_2TiO_4 (1-x) Fe_3O_4$, the values of x being in the range $0 < x < 1$ and showing

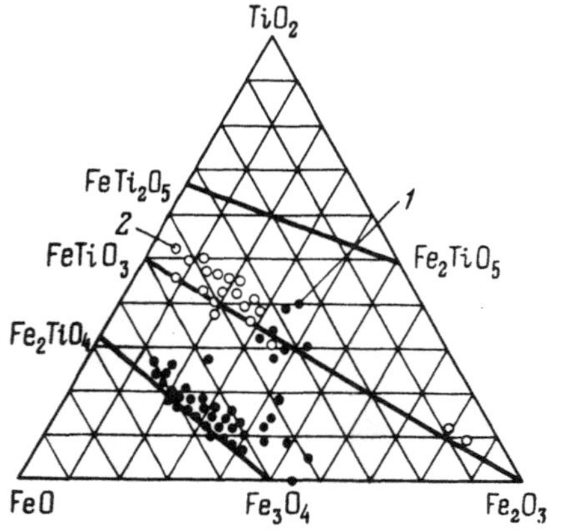

FIG. 127. FeO-Fe_2O_3-TiO_2 triple diagram and chemical composition of natural minerals (in molecular %) (after T. Nagata).
1—titanomagnetites; *2*—hemoilmenite

spinel structure; its parental materials are magnetite (Fe_3O_4) and ulvöspinel (Fe_2TiO_4) to which the Fe_2O_4-Fe_2TiO_4 line averaging laboratory data referred to the composition of synthesized titanomagnetites corresponds (Fig. 127). With increasing the ulvöspinel content in titanomagnetites their crystal lattice constant increases (from 0.839 to 0.853 nm) and Curie temperatures drop [from 578 to (−153) °C] as do the magnetic moment of saturation, saturation intensity of magnetization from 430×10^3 to 75×10^3 A/m and magnetic susceptibility from $>12.5 \times 10^{-1}$ to 12.5×10^{-6} SI units. The most probable values of x for these naturally occurring minerals are $x = 0.2$-0.8.

The properties of titanomagnetites as well as other ferrimagnetics are considered by taking into account the cation distribution of inverted spinel in their structure. As is known, the basis of spinel structure is an extremely dense oxygen ion packing with tetrahedral (A) and octahedral (B) interion spaces (sublattices). Eight tetrahedral and sixteen octahedral spaces are here taken by cations.

There may be two types of the distribution of different valency cations. Eight bivalent cations of normal structure spinel are located in tetrahedral, and 16 trivalent ions are located in octahedral structure nodes. This is manifested by a record $A_8^{2+}[B_{16}^{3+}]O_{32}^{2-}$. As to inverted-structure spinel, it has 8 bivalent and 8 trivalent ions located in octahedral nodes, and in tetrahedral nodes 8 remaining trivalent ions are located; this is written as follows: $B_8^{3+} \cdot [A_8^{2+} + B_8^{3+}]O_{32}^{2-}$.

Cations tend to a definite coordination environment. So, for example, lesser-size tetrahedral nodes of inverted-structure spinel are occupied by lesser-radius trivalent iron ions which is more advantageous in energy terms. As to titanum ions, they tend to be located in octahedral nodes. The resulting magnetic moment of saturation is conditioned by the difference of the resulting moments of A and B sublattices.

In spinel crystal lattice, appreciable interaction between A and B sublattice cations is assumed leading to the antiferromagnetic-antiparallel orientation of atomic magnetic moments of sublattices (see Fig. 126*b*) in strong magnetic fields and at low temperatures. Inverted-structure spinel magnetite has the resulting magnetic moments of trivalent cations of A and B sublattices compensated since these are located there in equal numbers and are antiparallel, whereas magnetic moments of bivalent cations of B sublattices prove to be uncompensated and their sum governs the resulting magnetic moment of magnetite. It is equal to $4\mu_B$ (μ_B is the Bohr magneton).

Ulvöspinel demonstrates a normal spinel structure. Cation distribution in titanomagnetites is in the general form characterized by this formula: $Ti^{4+}{}_{1-mA-nA}Fe^{3+}{}_{mA}Fe^{2+}{}_{nA} \times [Ti^{4+}{}_{2-mB-nB}Fe^{3+}{}_{mB}Fe^{2+}{}_{nB}]$ where bracketed ions are in octahedral positions, and unbracketed ions in tetrahedral positions; m_A, m_B, n_A, n_B concentration of, respectively, Fe^{3+} and Fe^{2+} ions in A and B positions. It can be shown, that cation distribution is responsible for the quantity J_{Vs} and Curie temperature.

The composition of titanomagnetites is governed by magma crystallization conditions. Decomposition products of titanomagnetites are hemoilmenite resembling ilmenite and titanomagnetite resembling magnetite in their properties; oxidation products are cation-deficit titanomagnetites undergoing further alterations. The chemical composition of cation-deficit titanomagnetites is conditioned not only by the paramater x determining cation distribution in octahedral and tetrahedral spaces, but also by the oxidation parameter p_{O_2}. It determines the fraction of the original number of the Fe^{2+} ions in a sample converted to Fe^{3+} by releasing the Fe^{2+} ion's external $6d$ shell electron to the absorbed oxygen atom. For the lattice neutrality to be preserved as Fe^{2+} ions convert to Fe^{3+} ions, structural vacancies $\square$ should be provided in the crystal of titanomagnetites. As a result, the chemical formula of single-phase cation-deficit titanomagnetites takes on this form: $Fe_aTi_b\square_cO_4$. Here $a + b + c = 3$. Cation distribution in oxidized titanomagnetites is unstable and is governed by the mechanism of low-temperature oxidation, chemical bounding type and Fe^{2+} ion mobility in spinel's two sublattices.

Curie temperatures of natural titanomagnetites are between 578 and 50 °C. Titanomagnetites appreciably change on heating.

Hemoilmenite series II is a mineral family having a composition xFeTiO$_3$(1 − x)Fe$_2$O$_3$ (x values being in the range $0 < x < 1$) showing a rhombohedral structure—solid solutions of α-hematite (α-Fe_2O_3) and ilmenite $FeTiO_3$. The Fe_2O_3-$TiFeO_3$ line averaging laboratory data referred to the composition of synthesized hemoilmenites corresponds to this series in Fig. 127.

Hematite of a rhombohedral structure is widely distributed: in the lithosphere J_{Vs} of hematite is by two orders of magnitude less than J_{Vs} of magnetite. Single crystals of hematite have a low $H'_{c.s} \approx 3200$-4800 A/m, whereas fine-grained polycrystalline hematite varieties exhibit high values of H'_c (tens of thousands of amperes per 1 m). The features of hematite are as follows: (1) high stability of J_{Vn} to a variable magnetic field; (2) lack of saturation in fields in excess of 6×10^5 A/m; (3) Curie temperature is about 675 °C; (4) the J_{mrs} curve shows a Morin point* corresponding to (−23) °C.

Ilmenite $Fe^{2+}Ti^{4+}O_3^{2-}$ demonstrates a rhombohedral structure and iron and titanum layers alternating in the crystal lattice. Fe^{2+} ions are arranged antiparallel

* Morin point is a temperature equal to (−23) °C providing a low-temperature point of hematite alteration.

in the adjacent layers. Below (−205) °C natural ilmenite is a paramagnetic and above this temperature it is an antiferromagnetic. The Curie temperature of hemoilmenites varies over the range from 675 °C (hematite) to (−205) °C (ilmenite). At room temperature hemoilmenites having a composition $1 > x \geqslant 0.45$ are ferrimagnetic, and having a composition $0.45 > x > 0$ are antiferromagnetic showing weak ferromagnetism. Solid solutions are ferrimagnetic at room temperature in the composition interval $0.75 > x \geqslant 0.45$.

There occur other mineral series with abnormally high values of magnetic parameters: pseudobrookite Fe_2TiO_5-$FeTi_2O_5$, Fe_3O_4-Mn_3O_4 series (it includes, e.g., jacobsite), substituted magnetite, hematite, maghemite.

Ferromagnetic pyrrhotite varieties generally demonstrate even a lesser magnetic susceptibility compared with the aforementioned minerals having ferromagnetic properties over the range $>12.5 \times 10^{-3}$-$>12.5 \times 10^{-2}$ SI units and saturation intensity of magnetization $J_{Vs} = (17\text{-}70) \times 10^3$ A/m.

Prismatic crystals of pyrrhotite in a magnetic field of whatever orientation and inappreciable strength are magnetized solely in a plane that is at right angles to the prism axis and to the greatest extent in the principal axial direction *OX*.

Jacobsite, trevolite, magnesioferrite are of less value in terms of magnetism since they occur less often.

Magnetite displays a greater residual isothermal magnetization compared with pyrrhotite, and hematite has that less by an order of magnitude compared with pyrrhotite. The greatest coercive force is demonstrated by hematites; magnetite, maghemite and pyrrhotites demonstrate a lesser coercive force and, for titanomagnetites it is still less.

Curie temperatures are the highest for maghemite and hematite (675 °C), they are somewhat lower for magnetite (~578 °C). Curie temperature of titanomagnetites are governed by the $TiFe_2O_4$ content in minerals and are in the range from 222-524 °C as the $TiFe_2O_4$ content varies from 42 to 8%. Curie temperatures of pyrrhotites vary from 300 to 325 °C.

Magnetic properties of ferromagnetic minerals are also governed by their grain size. For example, the magnetic susceptibility of hematite lowers with increasing grain dispersion. The latter fact is accounted for by a drop in the number of domains in grains as these diminish in size. Under such conditions the magnetization of a unit volume of a rock takes more energy since magnetization is mainly carried out due to the rotation of magnetization vectors of domains rather than due to the displacement of their boundaries which process requires less energy expenditures. In the limiting case (for grains with one domain) magnetization occurs solely due to the rotation of magnetization vectors of grains and requires maximum energy.

For the same reason with decreasing the grain size of minerals and the number of domains in each grain the coercive force of minerals grows.

The joint action of the magnetic field and mechanical stresses as well as of temperatures produces a great effect on the characteristics of magnetic properties of rocks. In a magnetic field ferri- and ferromagnetic mineral samples may exhibit both positive and negative magnetostriction, the dependence of dimensions of a sample on the direction and extent of its magnetization. With the growth of strains the extent of magnetization of magnetic minerals showing positive magnetostriction (e.g., magnetite) increases, and of ferromagnetic minerals showing negative magnetostriction decreases.

The magnetic susceptibility of minerals also varies under the effect of uniaxial compression and does not only decrease but also increase. Given a uniaxial compression 1.02×10^3 MPa and magnetizing field 800 A/m, titanomagnetites have $\varkappa$ which is 6 times less than under atmospheric pressure, and given $H \approx 1700$ A/m, 7 times. The $\varkappa$ values of magnetite under the same pressure and magnetizing fields decrease, respectively, by a factor of 4.5 and 6. Given appreciable stresses, $\varkappa$ of magnetite is stabilized and is independent of the field. The variation of $\varkappa$ under the effect of uniaxial pressure p is accounted for by the rotation of vectors J_{Vs} of domains.

Isothermal residual magnetization J_{mr} of minerals as a result of their being present in a stationary field also drops with loading them. The variation of J_{Vrs} with pressure for single crystals is characterized by hyperbolas showing different curvature rate. Practically all types of residual magnetization of hematite, magnetite, titanomagnetites and other minerals decrease under the action of total pressure.

The pattern of stability of characteristics J_{Vr}, I_{Vrt} and J_{Vri} of isothermal, thermal residual and ideal magnetization upon pressure variation is as follows: thermal residual magnetization is most stable followed by ideal and isothermal magnetization.

Since J_{Vr}, J_{Vrt} and J_{Vri} magnetizations show different stability, under effect of pressure the direction of total magnetization can vary if it is a component of the above magnetization types, a variation in the sign of residual magnetization is also possible. Thus pressure generally decreases the capacity of minerals to magnetization.

The effect of pressure diminishes when minerals attain Curie temperatures. At this temperature the saturation intensity of magnetization and coercive force also decrease.

If a mineral is stable to heating (magnetite), then $\chi = f(t)$ of the first heating of igneous rocks where it is located do not differ from the same curves during repeated heatings (Fig. 128*a*) which cannot be said about $\chi = f(t)$ curves for the same rocks with ferro- and ferrimagnetics (titanomagnetite, maghemite, pyrrhotite) (see Fig. 128*b*) unstable to heating. Maghemite is unstable if temperature increases to attain 275 °C (sometimes more); under these conditions it converts to a stable variety (α-hematite). On heating pyrrhotite to 270 °C there occurs an alteration of its crystal lattice.

A definite zoning has been established in the distribution of para-, ferro- and ferrimagnetic materials. Ferrimagnetic minerals do not occur in the upper mantle or, it appears, in the deeper reaches of the continental crust due to thermobaric and redox conditions unsuitable for their origination. We distinguish the following zones in the earth's crust and mantle: (1) a high-oxidation Fe^{3+} hematite zone near the earth's surface, minerals that have in their composition Fe^{2+} and $Fe^{2+} + Fe^{3+}$ are unstable here; (2) a magnetite zone where minerals with $Fe^{2+} + Fe^{3+}$ ions are mainly formed represented by Fe-Ti oxides; (3) a silicate zone where Fe^{3+} is practically non-existent but which contains ilmenite, ulvöspinel, wüstite and other oxides

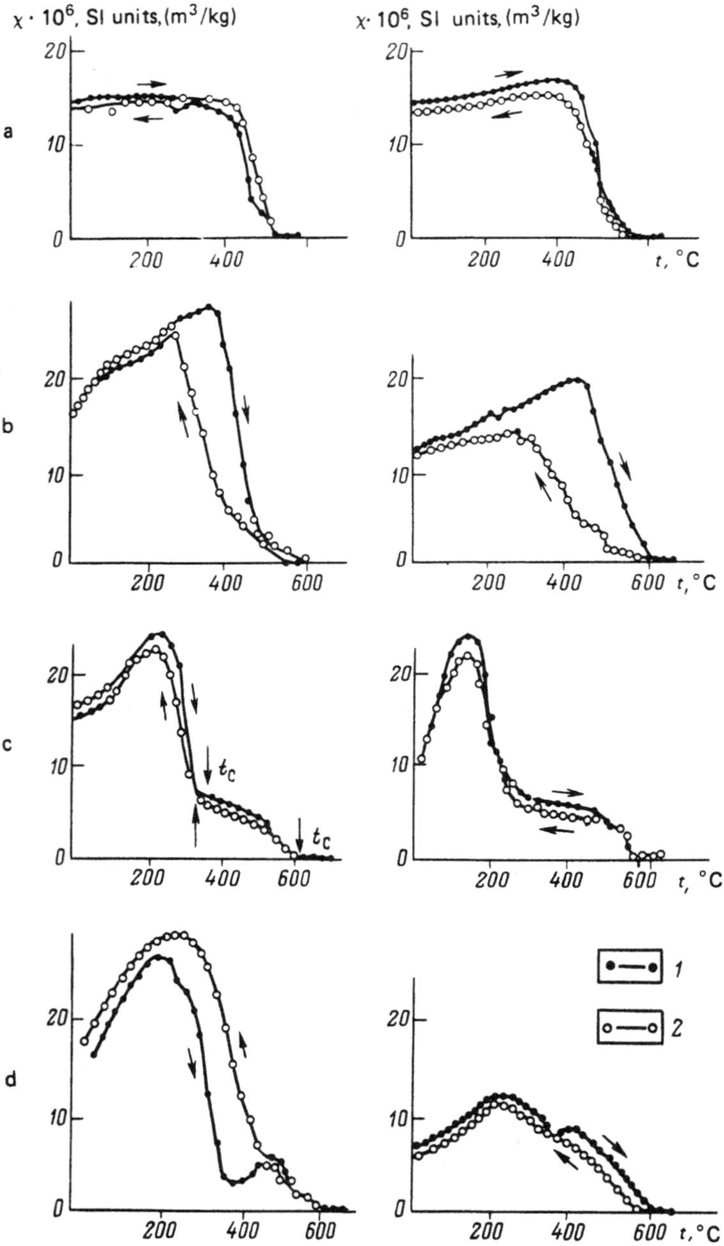

FIG. 128. Dependences of magnetic susceptibility χ on the temperature t.

Curve types: *a*—conventional reversible: *b*—conventional irreversible; *c*—nonconventional reversible; *d*—nonconventional irreversible; *1*—heating; *2*—cooling that have in their composition Fe^{2+} and Ti^{4+}, iron principally concentrating in silicates; (4) an iron metal (Fe^0 + Fe^{2+}) zone where, apart from silicates, metallic iron has been found.

The zones are characterized by thermobaric (T-p) and oxidational conditions. Consequently, ferri- and paramagnetic minerals may provide indicators of physicochemical conditions of the origination of magnetism in rocks.

Since some mineral species responsible for magnetic properties of rocks form concurrently with a rock, and others during the life of the latter, time period of formation of these minerals compared with that of a rock contains information on various events in the rock's life. Viewed in this context, we need information on the minerals responsible for the magnetization of rocks.

Let us consider antiferromagnetics. Given $H = 0$, they have no resulting magnetic moment, but magnetic moments of the adjacent atoms in their crystals are antiparallel. This arrangement of moment is governed by negative exchange interaction that occurs when atoms approach one another closer compared with ferromagnetics (see Fig. 128*b*).

The magnetic susceptibility of such antiferromagnetics as pyrolusite, siderite, albandite is about the same as that of paramagnetic minerals. However, with increasing temperature the magnetic susceptibility first grows attaining its maximum at a perfectly definite temperature for the particular mineral known as a λ point or Curie temperature (Curie point) and then decrease according to the Curie-Weiss law for paramagnetics (Fig. 144). Below the Curie temperature magnetic susceptibility is conditioned by the field strength. Because of practically common magnetic properties antiferromagnetics and paramagnetics are referred to paramagnetic materials.

Sec. 49. **The Solid, Liquid and Gaseous Phase**

The solid phase. It generally consists of a mixture of dia-, para-, ferri- and less commonly ferromagnetic minerals; it may sometimes be represented by one or two of these minerals. The magnetic properties of this component of rocks are mainly governed by the content and distribution throughout its volume of ferri- and rarely ferromagnetic minerals. The $\varkappa$ values of this phase of rocks vary from $< -0.4 \times 10^{-5}$ to 1 SI unit.

The liquid phase. The main components of this phase are water and oil which are diamagnetic (for water $\chi_w = \varkappa_w = 0.9 \times 10^{-5}$ SI units, and for oil $\chi_{oil} = -1.04 \times 10^{-5}$ SI units), i.e. these components of naturally occurring fluids are practically not magnetic, and $\varkappa$ of their mixtures is close to $\varkappa_w$.

The mineralization of waters little affects their magnetic properties since salts common for them ($NaCl$, $CaCl_2$, $MgCl_2$ etc.) are also diamagnetic, and their magnetic susceptibility is inappreciable.

The gaseous phase. This phase of rocks represented by air is magnetized to a still lesser degree and manifests less susceptibility compared with the liquid phase, and all its components, except oxygen, are diamagnetic,

Since the magnetic susceptibility of paramagnetic oxygen much exceeds $\varkappa$ of other gases, its volumetric content in the air is relatively great, and $\varkappa$ is very small (0.17×10^{-5} SI units), the air is also a paramagnetic, and its $\varkappa_{air} = 0.04 \times 10^{-5}$ SI units.

The diamagnetic susceptibility of dry hydrocarbon gases is likewise inappreciable. A certain idea of it can be inferred from the magnetic susceptibility values of methane ($\varkappa = -0.0008 \times 10^{-5}$ SI units), ethane ($\varkappa = -0.0015 \times 10^{-5}$ SI units), pentane ($\varkappa = -0.0003 \times 10^{-5}$ SI units) determined at room temperature. They enter in the composition of dry naturally occurring hydrocarbon gases and under definite conditions of bedding are gaseous.

Sec. 50. **Air-dry and Moist Rocks**

Magnetic properties of rocks are evaluated from the values of their: (1) specific mass J_m* [in (A·m^2)/kg)] and volumetric magnetization types $\vec{J_V}$**, $\vec{J}_{ms}$, $\vec{J}_{Vs}$, $\vec{J}_{mr}$, $\vec{J}_{Vr}$, $\vec{J}_{mn}$, $\vec{J}_{Vn}$ (in A/m). They are defined, respectively, as magnetization types: (a) current (values of $\vec{J}_m$ and $\vec{J}_V$ are governed by the field's intensity); (b) saturation magnetization ($\vec{J}_{ms}$ and $\vec{J}_{Vs}$ are attained in a field with greater intensity); (c) residual normal magnetization ($\vec{J}_{mr}$ and $\vec{J}_{Vr}$ are produced in a stationary field under normal (room) temperature and pressure and are retained as this latter is removed); (d) thermal residual ($\vec{J}_{mn}$ and $\vec{J}_{Vn}$ are produced if rocks are plutonic showing a temperature above the Curie temperature as a result of magma cooling in the earth's stationary magnetic field); (e) detrital (or sedimentary) residual magnetization ($\vec{J}_{mn.d}$ and $\vec{J}_{Vn.d}$ occurs for sedimentary rocks whose ferrimagnetic fragments are deposited in a magnetic field); (2) specific χ (in m^3/kg) and volumetric $\varkappa$ (in SI units) magnetization; (3) coercive forces H_c (in A/m) (under normal), and $H_{c.t}$ (in A/m) (under thermomagnetization); (4) Curie and λ (in °C) temperatures t_C and t_λ.

The determination of the values of magnetic parameters is conducted by using air-dried samples. The values of the above quantities differ (varying within definite limits) not only for individual rock groups but also for their definite types. This is accounted for by the dissimilar material composition of solid-state minerals, their different proportion and distribution in its volume, the multi-phase pattern of rocks and specific conditions of the origination and life of the latter. However, the principal role in the formation of definite magnetic properties of the solid phase of

---

* Specific mass magnetization $J_m = dM_{mn}/d_m$ where M_{mn} is the magnetic (in A·m^2) moment of a unit mass m_r (in kg) of a rock; dM_{mn} is the magnetic moment of the medium with a mass dm_r which is a vector sum of magnetic moments of all molecules, ions etc. contained in this mass, when m_n drops to infinitesimally small physical value yet appreciably greater than the atomic inhomogeneity of the medium.

** Specific volumetric magnetization $J_V = dM_V/dV$ where dM_V is the magnetic moment (in A·m^2) of a unit volume dV (in m^3) of a rock, dV is an infinitesimal physical quantity which is, however, much more than atomic inhomogeneity. Under definite conditions the definitions of other magnetization types are close to this one.

rocks is played by ferro- and ferrimagnetics, in particular, magnetite, maghemite and titanomagnetites.

Given equal values of J_V, J_{Vn}, $\varkappa$ etc. of their solid phase, rocks will show smaller values of these characteristics, the higher is their voids ratio and the less is their moisture content. However, the multi-phase pattern of rocks has a lesser effect than differences connected with dissimilar properties of their solid phase since the contribution of this latter to the magnetic properties of rocks is incomparably greater. The only characteristic unaffected by the multi-phase pattern of rocks is the Curie temperature.

Thus most of magnetic characteristics of dry rocks have smaller values compared with moist rocks. Their magnetism grows, although inappreciably, with increasing their moisture content.

Sec. 51. **Magmatic Rocks**

Magnetic Susceptibility

Magnetic properties of magmatic (or plutonic) rocks are governed by the magnetism of their solid phase. The high and elevated magnetic susceptibility of this component is accounted for by the presence in its composition of magnetite, hematite, titanomagnetite and hemoilmenite series minerals. Less commonly such values of $\varkappa$ can be ascribed to the maghemite, jacobsite, magnesioferrite and pyrrhotite content in the solid phase. These minerals appear both concurrently with the formation of a rock and during its lifetime. The most important of these are magnetite, titanomagnetites and hemoilmenites; they form practically in all plutonic rocks at different stages of magmatism.

Intrusive Rocks

The composition of titanomagnetites and hemoilmenites of intrusive rocks is governed by thermobaric and redox conditions, and consequently, by the melt crystallization depth; their composition is also affected by the rate of magma crystallization. Under conditions of slow crystallization of the melt at its later stages magnetite and hematite form due to the decomposition of titanomagnetites and hemoilmenites. Rapid magma cooling facilitates the preservation of titanomagnetites by a rock. Magma cooling gives rise to: (1) decomposition of titanomagnetite solid solutions; (2) their high- and low-temperature oxidation; (3) transformation of cation-deficit titanomagnetites.

Following oxidation of hemoilmenites hematite and hemoilmenite of a different composition are formed. High-temperature oxidation of silicate minerals may also give rise to magnetite and hematite.

The concentration of readily magnetized minerals in an eruptive lava is greater, the less is its cooling rate. The concentration of readily magnetized minerals in plutonic rocks is little governed by the rock composition and iron content.

The formation of readily magnetized minerals in normal series ultrabasic rocks showing a high iron content is commonly associated with secondary late and post-magmatic processes, for example, serpentization.

A comparison of the properties of ferromagnetics in acidic, intermediate and basic rocks suggests that: (1) their specific magnetic susceptibility grows with increasing the size of grains; (2) the initial specific magnetic susceptibility $\varkappa_0^*$ reduced to identical grain size is the greater, the less is the titanum oxide content in these minerals.

For intrusive complexes the magnetic properties of all rocks are governed by ferrimagnetic or both ferri- and paramagnetic minerals. There exist acidic, intermediate, basic and ultrabasic rock varieties which include ones that vary from practically nonmagnetic to strongly magnetic (Table 14).

The magnetic susceptibility of magmatic rocks may vary from several units (n) to several tens of thousands ($n \times 10^5$) fractions of a SI unit. The wide range of variation of $\varkappa$ values is mainly governed by the composition of the original melts, thermobaric and redox conditions of the origin and life of rocks and is responsible for their appreciable differentiation in terms of induced and residual magnetization.

The average $\varkappa$ values increase from acidic to basic to ultrabasic rock groups (see Table 14).

Basic and intermediate rocks of early introduction phases are characterized in all intrusive complexes of various formations by a higher $\varkappa$ compared with subsequent phases. This is connected with the growth of rocks' acidity from phase to phase (weakly magnetic complexes), or with decreased content of ferrimagnetics. However, rocks having identical basicity (in particular, acidic rocks) may vary from strongly to weakly magnetic (see Table 14).

Acidic intrusive rocks and their transition varieties. The principal rock-forming minerals (quartz, orthoclase, plagioclases) of acidic magmatic rocks and transition rocks from acidic to intermediate species are diamagnetic. Biotite, muscovite, hornblendes, less commonly cordierite, tourmaline and pyroxenes (augite) whose content does not exceed 12% for acidic rocks and 25% for transition varieties are paraferrimagnetic.

Ferrimagnetic minerals are represented in magnetic granitoids and their erupted analogues by pure comparatively large-grained magnetite having an inappreciable isomorphous titanum oxide impurity and, it appears, maghemite. Small magnetite grains are incorporated in weakly magnetic granitoids.

The content of ferrimagnetic minerals in acidic and transition plutonic rocks varies from thousandths and hundredths of one per cent to about 2-3.5% and rarely more. This is what mainly determines the wide range of variation of the magnetic susceptibility of granitoids, viz. from $<6 \times 10^{-5}$ to $>7\,500 \times 10^{-5}$ SI units (see Table 14). At this, fairly often practically nonmagnetic and very weakly and weakly magnetic granitoids are encountered.

What is known about the magnetic susceptibility of granitoids is the following: (1) minor and intermediate-size massifs of plagiogranites in the tectonomagnetic cycle's early stage formations generally demonstrate relatively small magnetic susceptibility $(60\text{-}380) \times 10^{-5}$ SI units; (2) intermediate and later-stage granites and granodiorites, principal representatives of plutonic rocks in a nonuniform composition batholith formation, show, as a rule, high magnetic susceptibility; (3) in other formations of the intermediate and later stage their magnetic characteristic varies in value: (a) in granite-granodiorite formations of the later intermediate stage a

TABLE 14. Limiting Values of Quantities Characterizing Magnetic Susceptibility of Primary Sediments and Rocks

Primary sediment or rock	Specific volumetric magnetic susceptibility $\varkappa \times 10^5$ SI units		Ferrimagnetic and other minerals with a relatively high $\varkappa$ found in ores
	limiting	most probable	
Primary sediments, sedimentary and metamorphic rocks and ores derived from them			
Sand	0-3 800	-	-
Sandstone	−3->50 000	12.5->125	-
Quartzite	0-380	0-50	-
Aleurolite	−3-6 300	12.5-125	-
Argillaceous mud	15->50	-	-
Clay	2.5-3 800	12.5-125	-
Argillite	2.5-1 250	12.5-125	-
Phyllite	~0-125	-	-
Mud:			
calcareous	~6-~9	-	-
foraminiferal	1.3-~6	-	-
Limestone	0.15-~3 800	1.25->30	-
Marmorized limestone	−6-~0	-	-
Marble	−5-~6	2.5-~6	-
Dolomite	0.15->125	~0-12.5	-
Marl	0.15-~250	1.25->30	-
Gypsum	0.15-125	-	-
Anhydrite	0.4-12.5	-	-
Rock salt	up to >10	-	-
Ores of formations of:			
ferruginous quartzites	125 000-380 000	-	Magnetite, titanomagnetite, maghemite, magnesioferrite
bauxite-terrigenous	~0-4 400	-	Maghemite, magnetite, hematite, (the latter are less abundant)
bauxite-carbonate	~0-1 250	-	Magnetite
leptochlorite (iron ore)	<90->300	-	Martite, siderite, hydrogoethite
martite and brown iron ore	>125->250	-	Ditto
siderite rocks	125-625	-	-
cuprous sandstones	>6->125	-	-
Coals	−7.5-−9	-	-
Plutonic and metamorphic rocks and ores derived from them			
Intrusive:			
dunite	>1->50 000	-	-
peridotite	>1->50 000	-	-
pyroxenite	>1->50 000	-	-

TABLE 14 (*continued*)

Primary sediment or rock	Specific volumetric magnetic susceptibility $\varkappa \times 10^5$ SI units		Ferrimagnetic and other minerals with a relatively high $\varkappa$ found in ores
	limiting	most probable	
serpentinite	190->60 000	3 800->50 000	-
gabbro	<1->25 000	-	-
diorite	<1->15 000	-	-
quartz diorite	<1->7 500	-	-
syenite	<1->8 000	-	-
granite	<1->7 500	-	-
granodiorite	<1->7 500	-	-
Effusive:			
basalt	~0->25 000	-	-
andesite	~0-<3 800	-	-
liparite	>1-5 000	-	-
diabase	~0->15 000	-	-
quartz porphyry	~0-<900	-	-
Orthometamorphic rocks:			
Slate:			
quartz-sericite	~0-1 900	12.5-125	-
quartz-chlorite	~0->8 800	12.5-125	-
mica schist	~0-1 900	12.5-125	-
actinolite-chlorite	~0-3 800	-	-
Amphibolite	~0->1 500	(12.5-125)-(1 900-3 800)	-
Eclogite	-	38-<90	-
Gneiss:			
biotite	~0->7 500	12.5-200	-
pyroxene	~0-3 800	1 900-3 800	-
Granulite:			
acidic	~0-880	-	-
basic	380->7 500	-	-
Plagiomicrocline granite	~0-3 800	-	-
Migmatite	-	0->7 500	-
Contact charnockite	~0-1 900	380->1 250	-
Spotted slate	~0-380	-	-
Biotite hornfels epidote pyroxenic	~0->7 500	(0-880)-(125-1 250)	-
Skarn	Up to 3 800	12.5->15 000	-
Listvenite	0-380	-	-
Carbonatite	0->380	-	-
Greisen:			
quartz	0->125	-	-
micaceous topaz quartz	0->125	-	-

TABLE 14 (*concluded*)

Primary sediment or rock	Specific volumetric magnetic susceptibility $\varkappa \times 10^5$ SI units		Ferrimagnetic and other minerals with a relatively high $\varkappa$ found in ores
	limiting	most probable	
Ores of formations			
magnetite-skarn	> 75 000-1 000 000	-	Magnetite, titanomagnetite, maghemite, magnesioferrite
magnesioferrito-skarn	125 000-<570 000	-	Ditto
titanomagnetite	38 000-440 000	-	Ditto
stanniferous skarns	125->750 000	-	Magnetite
cassiterite-sulphide	8 800-10 000	-	Pyrrhotite
cassiterite-silicate	<125	-	Magnetite, hematite
cassiterite-quartz	<12.5	-	-
stanniferous pegmatites	>5	-	-
gabbro-peridotite (sulphide copper-nickel)	~125-44 000	-	Pyrrhotite, magnetite
peridotite-pyroxenite (sulphide copper-nickel)	3 800-125 000	-	Ditto
chalcopyrite	>5-25 000	-	Ditto
copper-porphyry	>6-12 500	-	Magnetite, hematite
copper-molybdenite	12.5	-	-
chromite (serpentinous dunite)	1 000-7 000	-	Magnetite
chromite	<90-<2 300	-	Ditto
rare metal quartz (auriferous)	Up to 1 900-2 500	-	Magnetite, pyrrhotite
gold-quartz (gold-sulphide)	630_{av}	-	Pyrrhotite
quartz-antimonite	$(<12.5)_{av}$	-	-
quartz-sericite metasomatites	$(6\text{-}12.5)_{av}$	-	-
gold-silver	$(<6)_{av}$	-	-
skarn-sulphide-scheelite (wolfram)	$(3.8\text{-}630)_{av}$	-	Pyrrhotite
wolfram-sulphide-skarnoid	<12.5-1 000	-	Ditto
quartz-vein-greisen (wolfram)	12.5	-	-
pyrite-polymetallic	>6-1 000	-	Pyrrhotite (at high $\varkappa$)
skarn (polymetallic ores)	n*-n1 000	-	Ditto
quartz-sulphide (polymetallic)	$(6\text{-}20)_{av}$	-	Ditto
lead-zinc (polymetallic)	$(<6)_{av}$	-	Ditto

* Several units.

subformation is distinguished where high $\varkappa$ values are determined by ferrimagnetic minerals as is another one, containing paramagnetic minerals where, as a result, all rock groups of various complexes starting from gabbro (or diorites) are very weakly magnetic; (b) in granite formations showing the same features where granites, however, are predominant, rocks manifest a minor content of ferrimagnetics and contain paramagnetic minerals (here $\varkappa$ of granodiorites is not more than 60×10^{-5} SI units and of granites from $(6\text{-}12.5) \times 10^{-5}$ to $(25\text{-}38) \times 10^{-5}$ SI units); (c) two subformations can also be distinguished in a later-stage granodiorite-granite formation: with para- or ferrimagnetic minerals predominant; (d) in a later, granite-leucocratic formation the magnetism of rocks is commonly due to ferrimagnetics, consequently, at a late intermediate to early late stage of tectonomagmatic cycles rocks are formed demonstrating a very weak magnetic susceptibility; (4) in the extrageosynclinal period granodiorites, biotite and leucocratic granites showing high magnetic susceptibility are mainly formed.

In platform regions rapakivi granites (known also as wibrogites) are formed with $\varkappa < 50 \times 10^{-5}$ SI units.

Intermediate intrusive rocks. The content of very weakly and weakly magnetic minerals (augite, biotite, hypersthene, muscovite, hornblende etc.) in intermediate plutonic rocks attains 35-45% (in diorites).

Fine-grained and higher TiO_2, compared with granitoids, content magnetite concentration has been estimated to be equal for quartz diorites to 0.2-2.1%, for dacites to 0.2-3.5%, and for andesites to 1.3-6.61%. Practically nonmagnetic varieties showing a smaller ferromagnetics content also occur. Their magnetic susceptibility varies over a wide range $[(<1\text{-}15\,000) \times 10^{-5}$ SI units]. Intermediate intrusive rocks may include species that are practically nonmagnetic to strongly magnetic (see Table 14).

The diorites of early formations demonstrate a high magnetic susceptibility. Nonuniform-composition batholith formation diorites of the intermediate stage are generally magnetic, yet, on the average, to a less extent than early formation diorites.

Diorites in formations of the later intermediate and early later origination stage are, as a rule, rather weakly magnetic, their magnetic properties are governed both by ferri- and paramagnetic minerals. The later stage of the tectonomagnetic cycle is characterized by diorite massifs showing a high magnetic susceptibility.

Basic intrusive rocks. Minerals that are predominant in basic rocks and their transition varieties (gabbro-diabases etc.) are very weakly (some of pyroxenes, olivine*, epidote, muscovite and chlorites) and weakly (hornblende, augite, hypersthene, biotite etc.) magnetic. Practically nonmagnetic minerals are represented by basic plagioclases. That is why the magnetism of basic rocks is to a certain degree due to weakly magnetic minerals.

The principal carriers of magnetism of unaltered basic rocks of the early and intermediate stage are primary magnetite and titanomagnetites whose content is up to 2-6%.

---

* Magnetism of this mineral is variable; it can also be weakly magnetic as well.

The composition of metamorphosed varieties includes, in addition, finely pulverized secondary magnetite and, seemingly, maghemite. The values of $\varkappa$ of unaltered basic rocks vary over the range (<1-25 000) × 10^{-5} SI units (see Table 14). For weakly altered basic rocks $\varkappa$ is high, its average values varying from massif to massif.

Gabbro and gabbro-diorites of the intermediate stage nonuniform batholith formation are on the average less magnetic than the same rock groups of the early formation. Gabbros and gabbro-diabases of the late intermediate and early late stage formations are relatively weakly magnetic. The magnetic susceptibility of gabbro massifs of the late stage of the tectonomagmatic cycle is appreciable.

Ultrabasic intrusive rocks. The principal ultrabasic rock minerals (paraferrimagnetic olivine, hypersthene, augite, less often hornblende) are very weakly or weakly magnetic. Ferrimagnetics in these rocks are provided by magnetite and pyrrhotite occurring in rocks separately or jointly. Mineralization of hypersthenites whose composition may sometimes not contain any magnetite are due to pyrrhotite.

Among ultrabasic rocks, no varieties of practically nonmagnetic due to a relatively high content in these rocks of very weakly and weakly magnetic minerals have been discovered. Most of other ultrabasic rocks demonstrate other magnetization degrees.

Unaltered ultrabasic rocks of the early stage formations show rather a small magnetic susceptibility $\varkappa$ = (20-300) × 10^{-5} SI units even if their iron content is greater as compared with other intrusive rocks.

However, different-age ultrabasic rocks are commonly serpentized as a result of disintegration of pyroxenes, which gives rise to magnetite formation and increase of magnetic susceptibility 0.1 n SI units (see Table 14).

Unaltered hyperbasites of alkaline massifs of the epoch of the platform development of the earth's crust may demonstrate appreciable magnetite and titanomagnetite concentrations. Their magnetic susceptibility then grows attaining n SI units.

Effusive Rocks

The rapid rate at which effusive rocks cooled is what is responsible for the presence in them of not only magnetite but also of titanomagnetites, particularly in basic rocks. Given $0 < x < 1$, $Fe_{3-x}Ti_xO_4$ titanomagnetites and oxidized cation-deficit $(FeTi_x\square)_3O_4$ titanomagnetites (titanomaghemites) are principal easily magnetized minerals of basic-composition effusive rocks (basalts, diabases etc.) both of continental and bathygenic oceanic origin.

Ferrimagnetics are small-grained varying from pulverized to 0.01-0.5 mm in diameter, so their $\varkappa$ is on the average less than $\varkappa$ of ferrimagnetics of their intrusive analogues. The different grain size, their nonuniform distribution in effusive rocks are what governs the large spread of magnetic characteristics of their individual varieties (see Table 14).

We will point out the difference in $\varkappa$ of the surficial, pyroclastic and extrusive facies forming under dissimilar conditions. Surficial facies have a stable magnetic susceptibility due to an adequately constant mineral composition of the constituent rocks which governs $\varkappa$ as well. For pyroclastic-facies rocks (conglomerates, breccias, agglomerate, tuffs, ashes) an appreciable inhomogeneity in $\varkappa$ values is characteristic due to the nonuniform concentration and distribution of various ferrimagnetics in structure elements. The magnetic susceptibility of extrusive bodies varies throughout their volume much more than that of surficial-facies rocks yet less compared with pyroclastic-facies rocks.

Effusive rocks of the spilite and diabase formation (the earliest geosyncline formations) contain mainly ferrimagnetic minerals and show different magnetic susceptibility values varying in the range (100-2000) $\times 10^{-5}$ SI units.

Rocks of the andesite and andesite-liparite formation of the late intermediate and early late stage of tectonic magmatic cycles include subformations where para- and ferrimagnetic minerals are predominant. Rocks of the liparite and granite formation are paramagnetic. Ferrimagnetic mineralization is common for rocks formed in the late stage of the period of folded region tectonic activization. Principal-composition platform effusive rocks generally demonstrate a high magnetic susceptibility.

The Coercive Force, Natural Residual Magnetization and Curie Temperatures

The coercive force of investigated plutonic effusive rocks varies from -400 to $-37\,000$ A/m. For most rocks it does not exceed $-8\,000$ A/m. Hard magnetic varieties have been mainly found in basalts, granites and syenite porphyry.

The residual magnetization is demonstrated by magnetites, pyrrhotites and such rocks as bauxites, peridotites etc.

Residual magnetization is governed by the nature, concentration and grain size of ferromagnetics of rocks. Thermoremanent dual magnetization grows with decreasing ferromagnetic grain size.

The natural specific volumetric residual magnetization of plutonic rocks J_{Vn} varies from 10^{-3} to $>10^2$ A/m and be greater or less than induced magnetization $J_{Vi} = \varkappa H_e$ determined using pulverized samples in a magnetic field about 40 A/m.

The natural residual magnetization of crystalline rocks with a minor content of large-grained ferrimagnetics, e.g. of leucocratic granites, is equal to decimal fractions of J_{Vi}. The J_{Vn} values that exceed J_{Vi} by a factor of several units or even tens are more commonly exhibited by basic rocks with a high ferrimagnetics content. Exceptionally high residual magnetization value [up to (1-2) $\times 10^3$ A/m] have been recorded, for pyrrhotite ores.

Specific features of natural residual magnetization are: (1) nonuniform values and orientation even within a single massif of magmatic rocks; (2) relatively high values for some rocks species; (3) stability to demagnetization. With increasing temperatures thermoremanent magnetization increases to attain zero at Curie temperatures.

Curie temperatures (or Curie points) of magmatic rocks do not depend on the grain size and ferromagnetics content and adequately determine their petrographic composition. Magmatic rocks manifest one—ordinary t_{C0}, or two—ordinary t_{C0} and extraordinary t_{C_e}—Curie points established, for example, from relationships $\chi = f(t)$ of the first heating of samples in fields with intensity about 160 A/m (see Fig. 128).

The Curie point t_{C0} is obtained from the familiar equations $\chi = f(t)$ (see Fig. 128*a* and *b*) whereas the two points t_{C0} and t_{Ce} are found from extraordinary, stepwise, relationships $\chi = f(t)$ (see Fig. 128*d*, *c*). In so doing, $\chi = f(t)$ curves (see Fig. 128*a*) are described as reversible if they can be obtained not only as a result of heating a rock sample but also during cooling or repeated heating. This type of curves corresponds to rocks where ferromagnetics are provided by magnetic or magnetite with a minor titanum oxide impurity.

Irreversible curves (see Fig. 128*b*) are $\chi = f/t$ curves that cannot be repeated upon cooling or subsequent heating samples with ferromagnetics, e.g. titanomagnetites that are unstable to heating.

Stepwise curves are obtained if rocks have ferromagnetic minerals of different chemical composition incorporated in them. Note that irreversible curves are shown by rocks containing magnetite and ferromagnetics (titanomagnetite, maghemite etc.) that are unstable to heating.

As suggested by findings of A.S. Semenov and F.S. Fainberg, the reversibility and irreversibility of $\chi = f(t)$ curves are governed by the environment where ferromagnetics are heated and cooled. Under conditions of continuous heating of samples of hematite and limonite ores, siderite and pyrite to 800 °C with access of air their magnetic properties prove to be invariable. Following heating in a reductional environment, practically nonmagnetic samples acquired elevated or high values of J_n and χ due to the appearance in their composition of magnetite and sometimes iron.

Under natural conditions heating the rocks to appreciable temperatures in the reductional environment is observed, for example, in contact zones in the vicinity of intrusive rock massifs. At depths corresponding to temperature variations from 250 to 750 °C there are particularly favourable conditions for the formation of magnetite, titanomagnetites, magnesioferrite and other magnetic materials due to other iron-containing minerals and for appearance in rocks of appreciable magnetic properties.

Plutonic rocks have the ordinary Curie point close to 585 °C, the extraordinary point much lower (about 300 °C).

The Curie point is connected with the depth of occurrence of the magmatic chamber and, consequently, composition of rocks that are formed. For the Czech massif basalts it has been found, for example, that their Curie points t_C generally decrease with increasing the olivine content or decreasing the ratio $\omega = MgO/(CaO + Na_2O + K_2O)$, t_C is governed by the olivine content in the rock since with increasing the olivine concentration the titanum content in titanomagnetites grows. At the same time the olivine concentration and composition of titanomagnetites are closely connected with the depth of formation of the magmatic chamber. For the Czech massif basalts the depth H is related in a linear way with:

(1) Curie point t_C

$$H = 82 - 0.14t_C \tag{78}$$

(2) modal olivine content (in %)

$$H = 36.61 + 0.96 \text{ ol}$$

(3) value of ω in % (mole)

$$H = 24.17 + 22.28\omega$$

(4) ratio in % (mole) $Fe_2O_3/FeO = (0.066H - 0.8)^{-1}$ (Kropačék, Pokorná).

Since the Fe_2O_3/FeO ratio, to a certain degree, characterizes the volatility of oxygen, then, by determining H from the Curie temperature and Eq. (78), it is possible from the last relation to determine Fe_2O_3/FeO and, consequently, the variation of redox conditions (p_{O_2}) with depth.

The relation $\omega = f(H)$ is valid for many regions of the globe. The relation $t_C = f(H)$ is even more universal. It is conditioned by thermobaric and redox conditions (p-T-p_{O_2}) in the magma.

Upon averaging data of large collections the regional effect is felt to a lesser degree, and a distinct connection is found between Curie temperatures and the ω ratio.

Sec. 52. **Sedimentary Rocks**

The principal rock-forming minerals of these rocks: quartz, feldspars, calcite, dolomite, gypsum, anhydrite, halite, sylvite, and others are dia- or paramagnetic; secondary: biotite, pyrite, ilmenite, siderite, chlorite, clayey minerals are para- or paraferrimagnetic; accessory minerals: magnetite, maghemite, hematite, goethite, hydrogoethite, hydrohematite and lepidocrocite are ferro- or ferrimagnetic. Thus, the magnetism of sedimentary rocks is mainly due to their accessor minerals.

Ferro- or ferrimagnetic minerals occur here as grains of magnetite, martite and hematite having an effective diameter 0.01 to 2 mm; magnetite inclusions are also revealed in quartz and biotite grains of the same size. In addition, such ferrimagnetics as maghemite, hematite, less commonly magnetite, are found in the clayey fraction of these rocks in a fine-dispersed state or as iron hydroxide accumulations and films. In this case their grains vary from fractions of the micrometre to several tens of micrometres. Finally, in sedimentary rocks ferri- or ferromagnetics may occur as post-diagenetic and epigenetic formations involving limonite, products of oxidation and replacement of siderites, pyrite and magnetite.

The magnetic susceptibility of rocks of the sedimentary cover (including ores) in this country is in the range $(-3) \times 10^{-5}$ to 50×10^{-2} SI units (see Table 14). So does $\varkappa$ of separate sedimentary rock types vary over a considerable range. What is more, in the case of clays, argillites, sandstones and aleurolites varieties with a relatively low magnetic susceptibility in the range $(12.5\text{-}125) \times 10^{-5}$ SI units are predominant; and in the case of limestones, dolomites and marls, varieties with $\varkappa = (1.25\text{-}30) \times 10^{-5}$ SI units (see Table 14).

Maximum $\varkappa$ values have been found in the case of sandstones and aleurolites in the neighbourhood of the ablation sources and they are conditioned by a relatively high magnetite concentration. Enhanced magnetic susceptibility $[(125\text{-}625) \times 10^{-5}$ SI units] is manifested by certain siderite detrital rocks.

The least magnetic susceptibility (generally $< 30 \times 10^{-5}$ SI units) is shown by limestones, dolomites, anhydrites, gypsums, rock salt and coals. Anhydrites, gypsums, rock salt, pure limestone varieties, coals mainly containing diamagnetic minerals may also be diamagnetic.

Sedimentary rocks have been found to demonstrate a very weak (from $a \times 10^{-4}$ to 10^{-1} A/m, where a is a factor equal to several units) but rather stable natural residual magnetization. It grows with increasing the content of ferromagnetic minerals, in particular, magnetite.

The orientation of natural residual magnetization of sedimentary rocks generally little differs from that of the modern geomagnetic field. However, rocks occur whose vector orientation $\vec{J}_{Vn}$ does not agree with the orientation of today's magnetic field of the earth and is, sometimes, of an opposite direction.

During coprecipitation in water basins of parent rock fragments showing thermoremanent magnetization and without it there appears, for example, detrital magnetization. The resulting magnetic moments of single- and multidomain fragments are aligned with the direction of the earth's magnetic field and the sediment which forms is magnetized. No direct connection between J_n and the mineral composition of sedimentary rocks has been found. High J_n values are typically shown by sedimentary bauxite and iron ores, and small values by limestones, dolomites, quartz sandstones, gypsums and dolomites. Thermoremanent magnetization has also been established during repeated heatings of these rocks to temperatures observed in the atmosphere and in the presence of the geomagnetic field, the value $J_{Vn} \ll \varkappa H_e$ (H_e is the intensity of the geomagnetic field) does not appear to exceed 10^{-2} A/m.

The $\varkappa$ values of the sedimentary cover of the Russian platform and West Siberian lowland are on the average less by an order of magnitude than the magnetic susceptibility of the identical rocks of the folded regions of the Urals, Transbaikalia and the Far East. The magnetic susceptibility values make it possible to determine the direction and length of the path of ablation and, to a certain degree, the composition of the sources of ablation.

Sedimentary rocks generally show anisotropy of magnetic susceptibility

$$\lambda_\varkappa = (\varkappa_{max} - \varkappa_{min})/\varkappa_{av}$$

It attains maximum values ($\lambda_\varkappa = 1.5$) for strongly metamorphosed sedimentary rocks.

The anisotropy of $\varkappa$ of sedimentary rocks is greater than that of platform region rocks.

Sec. 53. **Metamorphic Rocks**

The magnetic susceptibility of metamorphic rocks varies over about the same range as that of plutonic rocks. These rocks vary from practically nonmagnetic to strongly magnetic (see Table 14).

Minor magnetic susceptibility is characteristic of metamorphic rocks originating from practically nonmagnetic sedimentary (clay shales, phyllites, crystalline schists, paraamphibolites, quartzites, paragneisses, marble etc.) or from practically nonmagnetic plutonic (granitogneisses etc.) rocks.

Very high values of magnetic susceptibility (generally in excess of $7\,500 \times 10^{-5}$ SI units) are shown by ferruginous quartzites (jaspilite) abounding in small magnetite and hematite impregnations or, sometimes, intercalations of hematite, hornfelses, serpentinites, orthoamphibolites, skarns, magnetite schists (see Table 14).

The magnetic susceptibility of other metamorphic rocks, for example, chlorite and talc schists, diorite-composition gneisses and others is generally in the range $(<12.5\text{-}7\,500) \times 10^{-5}$ SI units (see Table 14). These rocks originate from plutonic rocks poorly enriched by ferromagnetics.

The values of rock magnetization characteristics are proportional to the content of ferrimagnetics which are represented by magnetite in regionally metamorphosed rocks, and in rocks altered due to hydrothermal metasomatic and hypergene processes, by magnetite, hematite, maghemite and other ferrimagnetics. Pyrrhotite is often encountered in sulphide deposits.

Given different types of metamorphism, the magnetic susceptibility and other rock magnetization characteristics transform differently.

Regional metamorphism of greenstone slate and epidotamphibole facies gives rise to microcrystalline schists with $\varkappa = (0\text{-}15\,000) \times 10^{-5}$ SI units. At this, rocks are developed showing mainly a high magnetic susceptibility yet much lower than for original rocks. Under conditions of regional metamorphism the amphibolite and granulite facies gneisses, amphibolites and other rocks generally more magnetic than slates develop.

In the course of ultrametamorphism rocks may increase their $\varkappa$ (for example, during their granitization and silica and alkaline metasomatism*; in this case iron-containing silicates decompose forming, in particular, magnetite). Maximum values of $\varkappa$ are attained during the intermediate granitization stage, and minimum ones during the final stage which is accounted for by removal of iron.

Dynamometamorphism and contact metamorphism give rise to rocks with an unstable magnetic susceptibility which is determined both by the specific features of the original rocks and thermobaric conditions.

Autometamorphism and hydrothermal and metasomatic processes are responsible for the most pronounced change of the magnetic properties of rocks, e.g. during serpentinization, amphibolization and other processes.

---

* Metasomatism is a process in which one mineral or a mineral assemblage is replaced by another due to a chemical reaction between its solid phase and the solution.

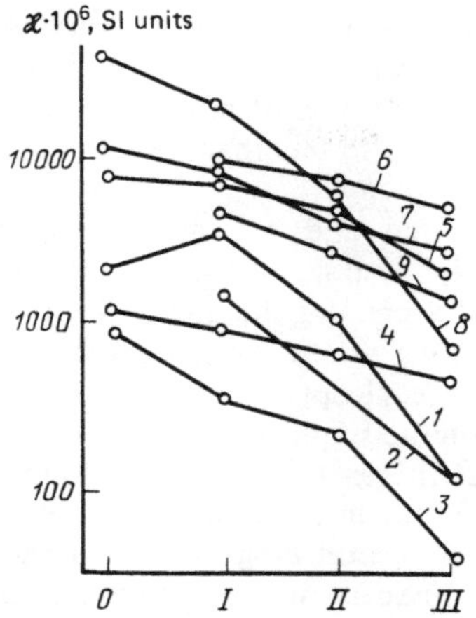

FIG. 129. The effect on the magnetic susceptibility ϰ of metasomatic rock alterations (after A.A. Smelov).

Alteration processes: *1*—silicification of porphyrites; *2*—the same, of tuff; *3*—quartz-sericite alterations of porphyrites and tuffs; *4*—the same of felsite-porphyries; *5*—sericitization of granodiorites; *6*, *7*—chloritization of granodiorites; *8*—epidotization of porphyrites; *9*—the same of quartz diorites; *I-III*—increase in degree of rock alteration. 100 to 500 samples have been taken to plot individual graphs

Greisenization, argillitization, seritization and chloritization of rocks result in decreased magnetic susceptibility owing to the recrystallization or removal of magnetite (titanomagnetite). Variations of ϰ due to a number of metasomatic processes are well illustrated in Fig. 129.

Supergene transformations cause the magnetic properties of rocks to be changed in a number of ways. They may result in a drop of magnetism of rocks because of the conversion of magnetite and titanomagnetite to hematite and other minerals or due to removal of ferrimagnetics.

Sec. 54. **Ores and Paleomagnetism**

Metallic ores. The values of the magnetic susceptibility of metallic ores vary over a wider range compared with other rocks (0-10 SI units). The highest values (from 38×10^{-2} to 10 SI units) are shown by the magnetite skarn, magnesioferrite-skarn, titanomagnetite and ferruginous-quartzite formation rocks (see Table 14). Appreciable values of ϰ are due to a relatively high, important for industrial processing, concentration of ore minerals (easily and strongly magnetized ferrimagnetics): magnetite, titanomagnetite, maghemite and magnesioferrite.

The values of ϰ from minor ($\sim 25 \times 10^{-5}$ SI units) to very high ($750\,000 \times 10^{-5}$ SI units) depending on the magnetite or pyrrhotite content are manifested by rocks of the stanniferous skarns, gabbro-peridotite and peridotite-pyroxenite (sulphide copper-nickel ores) formation rocks (see Table 14). Given increased contents of magnetite and pyrrhotite, magnetite and hematite or pyrrhotite, the high values of ϰ [up to $(10\,000\text{-}125\,000) \times 10^{-5}$ SI units] are shown by cassiterite-sulphide, copper-porphyry and chalcopyrite rock associations. However, the magnetic susceptibility of rocks of these formations, given minor ferrite or sulphoferrite concentrations, may drop to 5×10^{-5} SI units (see Table 14).

Bauxite-bearing, chromite, rare metal-quartz, wolfram-sulphide-scheelite, chalcopyrite-multimetal, skarn (polymetallic ores) formation rocks have $\varkappa$ varying from 0 to >7 000 × 10^{-5} SI units owing to the different content of respectively, magnetite, maghemite, hematite or pyrrhotite. The values of $\varkappa$ of other para- and ortho-ore rocks do not exceed several hundred and drop to less than several 10^{-5} fractions of SI units.

Analysis of the data on remanent magnetization and on the parameter $Q = J_{Vn}/J_{Vi}$ (J_{Vn} and J_{Vi} are natural remanent and induced magnetization of rock samples) permits us to conclude that in the order of decreasing J_{Vn} and Q the ores under consideration may be grouped into the following series: magmatogene and metamorphic-origin ores (Q = 0.1-20), polymetallic (chalcopyrite-metallic formation) ores (0.1-9.0), tin (cassiterite-silicate and cassiterite-sulphide formation) ores (0.5-7.0), wolfram (skarn-sulphide-scheelite formation) ores (0.5-7), gold (wolfram-sulphide-skarn formation) ores (0.3-0.7), sulphide copper-nickel (gabbro-peridotite formation) ores (0.3-3.0), polymetallic (skarn) (0.1-3.0), iron exogenic (0.2-1) and tin ores of all other formations (⩽0.2). Increased remanent magnetization is also characteristic of chromite ores and bauxites. Some of the aforementioned ores have the increased and high remanent magnetization linked with the presence of maghemite and magnetite in their composition, other ores with the presence of pyrrhotite.

Nonmetallic ores. Of nonmetallic natural materials, only asbestos ores include varieties from very slightly to well magnetized ones due to an enchanced magnetite content. A much less magnetic susceptibility is shown by micas, the values of this characteristic are not greater than 25 × 10^{-5} SI units for them, and for flaky graphite they are $\varkappa$ = (0-60) × 10^{-5} SI units. The magnetic susceptibility of other nonmetallic materials (sulphur, crystalline graphite) is about zero or amounts to (1.25-12.5) × 10^{-5} SI units (for fluorite), and that of common salt and sylvine may even have negative values varying over the range [(−4) − 8)] × 10^{-5} SI units. Fossil coals seem also to be diamagnetic, their magnetic susceptibility, as shown by the first measured results, attains (−9) × 10^{-5} SI units.

Paleomagnetism. It should be noted that modern residual magnetization results from residual rock magnetizations that originated and disappeared with time. Its fluctuations are due to the variations of the earth's magnetic field and a number of other factors. It is not uncommon, however, that the $\vec{J}_n$ vector can be represented by the geometric sum of the original natural remanent magnetization $\vec{J}_{n0}$ and secondary modern magnetization $\vec{J}_{nh}$ with the direction of today's field. This makes it possible to study the life history of the earth's magnetic field. There exist other aspects of using paleomagnetic investigations (regional geological investigations, geological mapping).

Sec. 55. Rock Classification According to Magnetic Susceptibility

By referring to the values of magnetic susceptibility, L.D. Bersudskii classifies rocks as very much weakly magnetic [$\varkappa$ < 40 × 10^{-5} SI units], very weakly magnetic [$\varkappa$ = (40-125) × 10^{-5} SI units], weakly magnetic [$\varkappa$ = (125-1 250) × 10^{-5} SI

TABLE 15. Classification of Rocks According to Their Magnetic Susceptibility Values

Group	Description of rocks according to their magnetic susceptibility	Group index	Value intervals for groups $\varkappa \times 10^{-5}$ SI units
I	Very weakly magnetized	A	0-100
II	Weakly magnetized	B	100-300
III	Intermediately magnetized	C	300-700
IV	Ditto	D	700-1500
V	Well magnetized	E	1500-3000
VI	Ditto	F	3000-6000
VII	Strongly magnetized	G	6000-12 000
VIII	Ditto	H	12 000-20 000
IX	Very strongly magnetized	I	20 000-40 000
X	Ditto	J	>40 000

units], magnetic [$\varkappa$ = (1 250-6 000) $\times 10^{-5}$ SI units] and strongly magnetic rocks ($\varkappa > 6\,000 \times 10^{-5}$ SI units).

The All-Union geological research institute suggests, to separate rocks as they are mapped into the following ten groups according to their $\varkappa$ values (Table 15).

Sec. 56. **Relationship Between Magnetic and Other Characteristics**

The following direct and inverse relations between the magnetic and other petrophysical or petrochemical characteristics have been established analytically and experimentally. Analysis has established a direct relationship between the magnetic susceptibility $\varkappa$ and the content k_{fm} of ferro- or ferrimagnetic minerals for homogeneous single-phase two-component rocks (e.g. rocks that contain practically only the solid phase and consist of ferro- and diamagnetic minerals).

The problem of defining the average magnetic permeability μ_{av}, and, consequently, magnetic susceptibility

$$\varkappa_{av} = \mu_{av} - 1 \tag{79}$$

of the particular rock, depending on the ferromagnetic mineral content, has, for example, this solution:

$$\mu_{av} = \frac{\mu_2(1 + 2V_0) + 2\mu_1(1 - V_0)}{\mu_2(1 - V_0) + \mu_1(2 + V_0)} \tag{80}$$

Here μ_2 is the magnetic permeability of spherical grains n of identical radius r_0 occupying volume $V_0 = 4\pi n r_0^3/3$ in a unit volume of the rock; μ_1 is the magnetic permeability of the second component found between the spheres.

The average magnetic susceptibility of rocks comprising minerals having different magnetic susceptibility and found in different ratios can also be found from this relationship:

$$\varkappa_{av} = f(\varkappa_{pm}, V_{pm}) + f(\varkappa_{fm1}, V_{fm1}) + f(\varkappa_{fm2}, V_{fm2}) \tag{81}$$

where $\varkappa_{pm}$ and V_{pm} are, respectively, the susceptibility and volumetric content of dia- and paramagnetic minerals; $\varkappa_{fm1}$, V_{fm1} are those of dispersed ferrimagnetic micrograins for concentration less than 0.01-0.1%; $\varkappa_{fm2}$, V_{fm2} are those of ferrimagnetic minerals, given concentrations more than 0.01-0.1%. By taking into account the weak magnetic interaction between dia-, para and dispersed micrograins of ferrimagnetic minerals, one can assume that their magnetism can be found from the product of susceptibility with concentration, viz.

$$f(\varkappa_{pm}, V_{pm}) = \sum_1^n \varkappa_{pm} V_{pm} f(\varkappa_{fm1} \times V_{fm1}) = \sum_1^n \varkappa_{fm1} V_{fm1} \tag{82}$$

The equation evaluating average magnetic susceptibility by taking into account magnetic interaction of ferrimagnetics, given concentration common for plutonic rocks, according to A.K. Veinberg, assumes this form:

$$\varkappa_{av} = \sum_1^n \varkappa_{pm} V_{pm} + \sum_1^n \varkappa_{fm1} V_{fm1} + \frac{\varkappa_{fm_2}}{1 + 4/3\pi\varkappa_{fm_2}} V_{fm_2} \tag{83}$$

The equation is valid for an idiomorphic* inclusion of ferrimagnetics common for plutonic rocks. If a ferrimagnetic inclusion is of a xenomorphic** pattern then the denominator of the last term of Eq. (83) takes on a form $\sqrt[3]{1 + 4\pi\varkappa_{fm2}}$.

For rocks that have ferri- and paramagnetic minerals in their composition Eq. (83) appears as:

$$\varkappa_{av} = \varkappa_{pm1} V_{pm1} + \varkappa_{pm2} V_{pm2} + \dots + \varkappa_{fm1} V_{fm1}$$

where $\varkappa_{pm1}$ and $\varkappa_{pm2}$ are the magnetic susceptibilities of different dia- and paramagnetic minerals; V_{pm1} and V_{pm2} are their volumetric content.

Since the last right-hand term of Eq. (83) is much greater than the other right-hand terms, then, given idiomorphic ferrimagnetic inclusions,

$$\varkappa_{av} \approx [\varkappa_{fm2}/(1 + 4/3\pi\varkappa_{fm2})]$$

Experiments conducted by K. Putzich and other investigators have established a close relationship between magnetic susceptibility and magnetite content for homogeneous structure rock models in high magnetic fields. This relationship, however, fails to be validated by studies of the magnetic susceptibility of many rocks.

The magnetic properties of rocks are governed not only by the content of magnetite and titanomagnetite grains but also by the size, shape, arrangement of the latter

---

* Idiomorphic (or, more correctly, automorphic) or euhedral. Of minerals in igneous rocks bounded by their own crystal faces independent of the adjacent crystalline grains.

** Xenomorphic (allotriomorphic). Of minerals in igneous rocks not bounded by their own faces and originating after the adjacent minerals that have their outlines impressed on them by these latter.

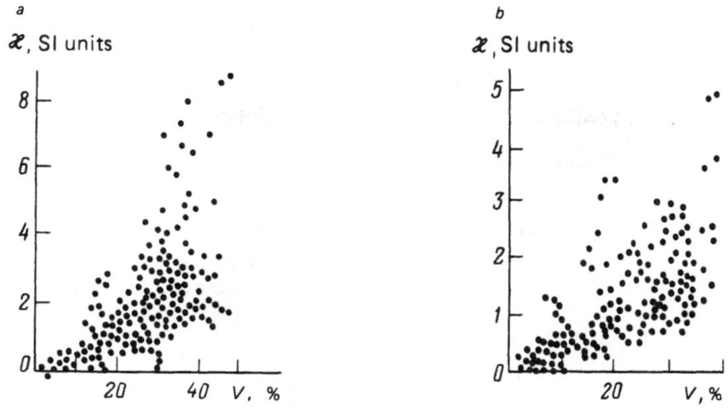

FIG. 130. Values of magnetic susceptibility of ferruginous quartzites of the Ukrainian iron ore province (*a*) and Kursk Magnetic Anomaly region (*b*) depending on the volumetric content of magnetite V (after V.N. Zavoiskii and Z.L. Krutikhovskaya)

in the particular medium. In addition, the magnetic properties of rocks are appreciably affected by the presence of other readily magnetized minerals, such as maghemite, pyrrhotite, and sometimes weakly magnetic varieties of augite, biotite, hornblende, chromite.

Moreover, the direct relationship between magnetic susceptibility and ferromagnetic mineral content is valid solely for a sample series taken from quite definite and inappreciable-volume regions of intrusive bodies, veins, and other geological assemblages that originated under particular conditions and containing solely one of the total number of ferrimagnetics. A convincing example of this generally loose connection is provided by the relation $\varkappa = f(V)$ for ferruginous quartzites (Fig. 130). Figures 131 and 132 present relationships between $\varkappa_{av}$ and V_0 plotted from the data submitted by A.V. Veshev who has calculated $\varkappa_{av}$ values for two probable types of structure-textural magnetite isolations compared with the experimental data.

Figures 131 and 132 show that the design values of $\varkappa_{av}$ are in fair agreement with the experimental results. However, given the same magnetite content, but different arrangement of its inclusions in a rock, we can find the values of magnetic susceptibility of rocks differing almost by an order of magnitude (see Fig. 132).

The spread of points over the experimental dependence is accounted for by the different nature of ferromagnetics as well. Rock samples showing pyrrhotite and titanomagnetite mineralizations show a lower $\varkappa$ compared with magnetite containing samples whose magnetic susceptibility also varies over a wide range. In addition, Eq. (80) is perfectly valid, solely given ferromagnetics content in excess of 5%. This is also corroborated by the relation $\varkappa = f(k_{fmm})$ obtained for plutonic rocks (Fig. 133).

Starting from the ferrimagnetics content k_{fmm} equal to 0.01% (mass) magnetic susceptibility of intrusive volcanic rocks is mainly governed by the value of this

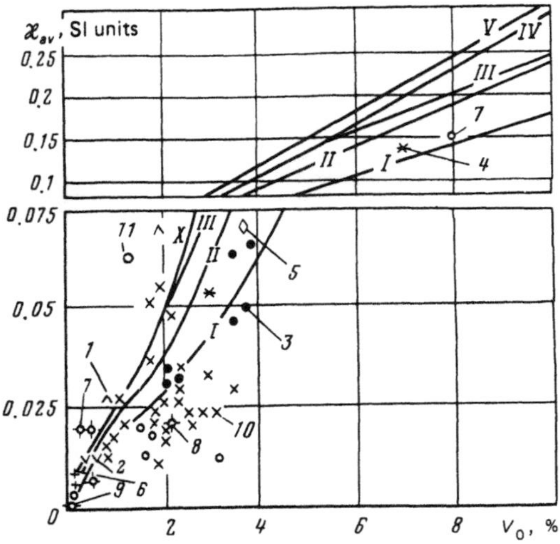

FIG. 131. Dependences of the average magnetic susceptibility $\varkappa_{av}$ on the content V_0 in ferromagnetic mineral rocks (plotted by V.N. Kobranova).

I-V—curves calculated from Eqs. (85) and (84), given the magnetic susceptibility of inclusions $\varkappa_2$ = 3.88; 8.8; 17.6; 47.7; 94.2 SI units and magnetic susceptibility of the filling mineral $\varkappa_1$ = 0; *1*—granodiorites; *2*—quartz diorites; *3*—andesites; *4*, *5*—gabbro-diabases (after data submitted by different workers); *6*—gabbro; *7*—metamorphized gabbro; *8*—gabbro-norite; *9*—hypersthene gabbro-amphibolites; *10*—basalts; *11*—pyroxenites

parameter; given smaller k_{fmm} values, there is no direct connection between them and $\varkappa$ (see Fig. 133). This is accounted for by the effect on the $\varkappa$ value, given k_{fmm} <0.01%, of the magnetic susceptibility of paraferrimagnetic minerals. Experimental values are averaged over the range of values of k_{fmm} = 0.01->10.

Fairly close and consistent direct relations have been found between: (1) $\varkappa$ and FeO_c; (2) magnetic susceptibility and Fe^{3+}/Fe^{2+} for rocks with ferrimagnetic

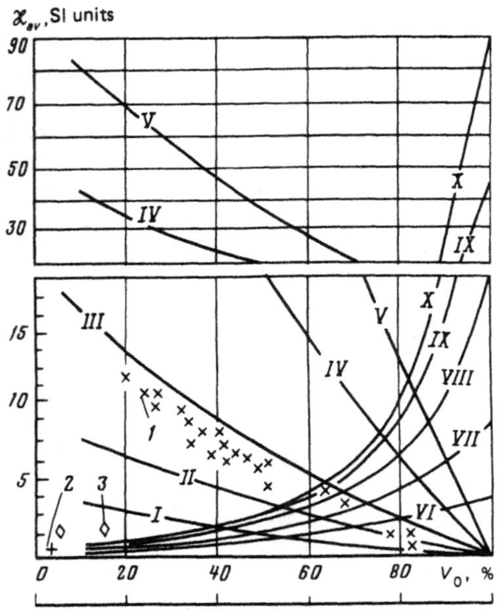

FIG. 132. Dependences of the average magnetic susceptibility $\varkappa_{av}$ on the content V_0 in rocks of mineral grains (compiled by V.N. Kobranova).

I-V—theoretical curves calculated from Eqs. (85) and (84), given $\varkappa_2$ = 0 and $\varkappa_1$ = 3.88; 8.8; 17.6; 47.7; 94.2 SI units; *VI-X*—the same, given $\varkappa_2$ = 3.98; 9.5; 18.7; 49.0 and 98.6 and $\varkappa_1$ = 0; *1*—core samples of the Persberg deposit (after A.V. Veshev); *2*—nepheline-basalts (after F. Birch, D. Shere, G. Spicer); *3*—diabases (after N.I. Spiridovich); V_0—grain content; $(1-V_0)$—content of the filling mineral

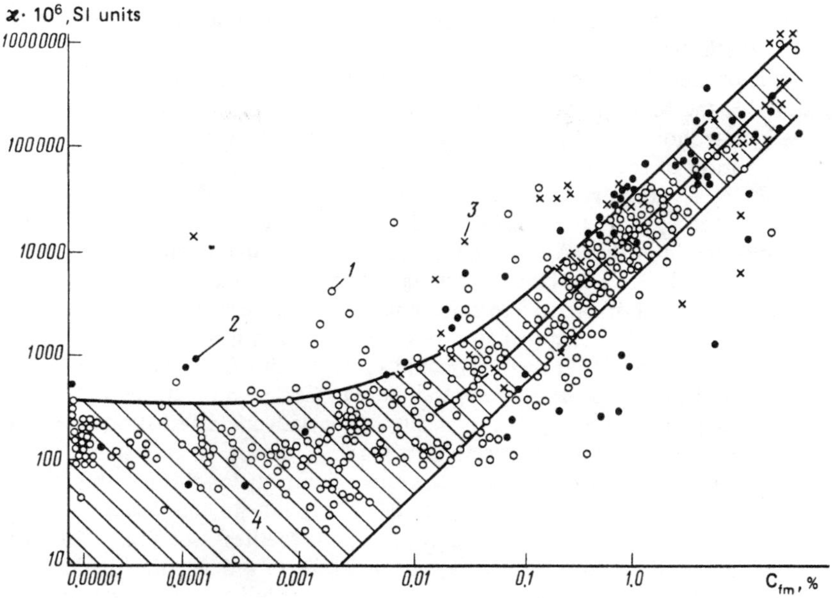

FIG. 133. Dependence of the magnetic susceptibility $\varkappa$ on the content of ferromagnetic inclusions C_{fm} in intrusive rocks (after laboratory data of N.B. Dortman and V.V. Lyakhovich and theoretical calculations of A.K. Weinberg).

1—granitoids; *2*—diorites and gabbros; *3*—hyperbasites; *4*—correlation field from theoretical calculations

minerals and rocks with ferri- and paramagnetic minerals; (3) magnetic susceptibility and sulphide content; (4) magnetic susceptibility and density; (5) J_{Vs}/J_{Vrs} ratio and coercive force H_c; (6) natural residual magnetization J_{Vn} and magnetite content, and inverse connections between (a) magnetic susceptibility and chromium oxide content; (b) magnetic susceptibility and voids ratio.

Sec. 57. Relationship of Magnetization Characteristics and Pressure, Temperature, Depth of Occurrence of Rocks

Magnetic susceptibility determined in a direction parallel to that of uniaxial pressure generally decreases with its growth. Susceptibility measured at right angles to the pressure first increases for some samples more dramatically than afterwards, and further compression produces no effect on the $\varkappa$ values. There exist samples that demonstrate an inappreciable decrease in $\varkappa$ upon compression. Negative variations of $\varkappa$ are attributable to the presence of magnetite in rocks.

The difference in the pattern of variation of the magnetic susceptibility of rocks following compression seems to be governed by the different minerals elasticity with ferromagnetic inclusions as well as the size, shape and mutual arrangement of the grains of the latter.

A study of the effect of overall pressure (800 MPa) on the stability of isothermal J_{Vr}, ideal J_{Vri} and thermoremanent J_{Vrt} magnetization of gabbro-diabases, alkaline granites, quartz diorites and plagiogneisses has shown that, whatever the rock type and composition of their ferrimagnetic minerals, values of J_{Vr} decrease with pressure almost by 85% under experimental conditions, about 60% of this variation being due to the pressure interval 0-200 MPa. As the field intensity increases the role of pressure diminishes. A relative variation of thermoremanent magnetization J_{Vrt} is not more than 15%. Stability to demagnetization J_{Vrt} has intermediate values. Thus J_{Vrt} is the most stable to pressure, and J_{Vr} is the least stable.

Following the heating of magmatic rocks containing a heat-resistant ferrimagnetic mineral (e.g. magnetite) their magnetic susceptibility first grows, then drops to values approaching zero. At temperatures close to Curie temperatures and following repeated heatings the $\varkappa = f(t)$ curves may agree with the original one (see Fig. 128). In case a ferrimagnetic is unstable to heating, then a similar shape of the $\varkappa = f(t)$ curve corresponds to the rock under identical conditions; this shape, however, is not preserved following repeated heating due to the variation in the composition of ferrimagnetics (for example, titanomagnetites) (see Fig. 128*b*). The rocks that have in their composition two or more ferri- and ferromagnetics (e.g., magnetite and titanomagnetite or magnetite and pyrrhotite) demonstrate more complex-shaped $\varkappa = f(t)$ curves (see Fig. 128*c*) accounted for by the presence of two or more Curie points for ferrimagnetics. These curves may also be reversible (if ferromagnetics are stable to heating) or irreversible (if one or both ferrimagnetics are unstable to heating) (see Fig. 128*d*).

An increase of temperature to 200-300 °C causes not only a drop of the magnetic susceptibility but also of isothermal, ideal, viscous and other magnetization type. As rocks are being heated to Curie temperatures, the thermoremanent magnetization that appeared during cooling of rocks in a field H is lost. The coercive force also diminishes with increasing temperatures.

Sec. 58. Regional Variation of Magnetic Characteristics of the Sedimentary Cover of Platforms and Depressions

The values of the magnetic characteristics of the sedimentary cover of platforms and depressions are very much affected by the chemical and thermobaric conditions of the origination and life of sedimentary rocks. A regular variation of the magnetic susceptibility values has been found, for example, for sedimentary deposits of the West Siberian plate. In some eastern areas of this region sandy clayey sections of the Ilek rock series of the Lower Cretaceous (or Kiyala rock series of the Hauterivian-Barremian), against the background values of $\varkappa = (25\text{-}30) \times 10^{-5}$ SI units, have shown values amounting to 200×10^{-5} SI units. This fact is attributed to ferrugination of the cross section where rock grains are enveloped by a thin iron hydroxide film. The Itat rock series of the Jurassic of the same sequence is charac-

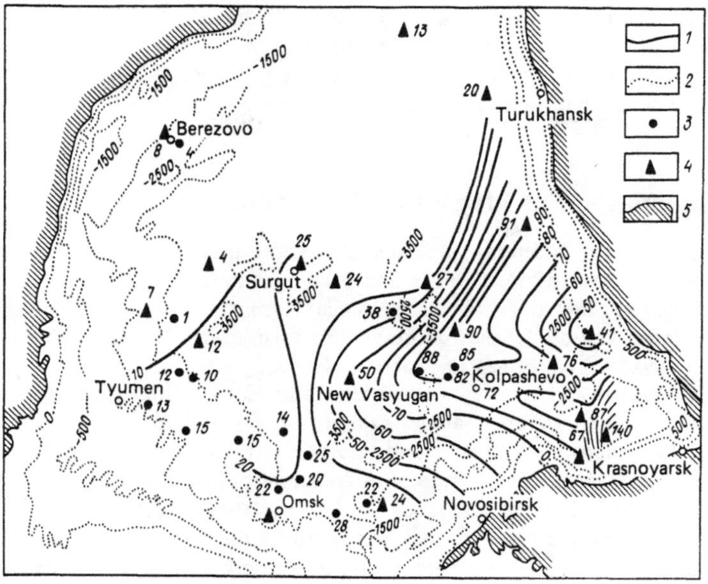

FIG. 134. Schematic map of isolines of equal values of the magnetic susceptibility of the Hauterivian-Barremian of the West Siberian plate (after N.A. Tuezova).

1—magnetic susceptibility isolines; *2*—isohypses on the roof of Meso-Cenozoic sediments; *3*—boreholes; *4*—reference boreholes; *5*—Paleozoic framework

terized by low values of $\varkappa = (0\text{-}6.5) \times 10^{-5}$ SI units owing to the presence of coal intercalations here. The sediment sequences of the central West Siberian plate most commonly demonstrate $\varkappa = (12.5\text{-}19) \times 10^{-5}$ SI units. The maximum values of this parameter can be singled out equal to $\varkappa = (30\text{-}38) \times 10^{-5}$ SI units corresponding to the Kiyala formation, in northern regions more appreciable values of $\varkappa$ are demonstrated by the rocks of the Tarskaya and Kulomzinskaya formation. The least $\varkappa$ values are shown by rocks of the Lyulinovorskaya, Ipatovskaya, Slavgorodskaya formations composed by opokas and opoka-like clays by the Maryanovskaya formation represented by bituminous argillites.

Meso-Cenozoic sediments of the West Siberian plate, according to $\varkappa$ values, can be referred to the regions: (1) with maximum $\varkappa$ values (Eastern region); (2) with lower variations of $\varkappa$ with depth (central region) and (3) with $\varkappa$ values generally not exceeding $100n$ SI units (Western region); Meso-Cenozoic sediments do not differentiate with respect to values of this characteristic.

Practically no magnetic anomalies are to be found in diagrammatical charts of equal $\varkappa$ values plotted for Aptian-Albian-Cenomanian and Jurassic sediments. Maximum $\varkappa$ values that little differ from other values refer to Narym and Parabel, and the minimum ones have been discovered in the western areas of the plate. Lower Cretaceous rock equal magnetic susceptibility charts (Fig. 134) show in the east two definitely oriented maxima with $\varkappa_{max} = 157 \times 10^{-5}$ and 113×10^{-5} SI units, and in the west with $\varkappa = (12.5\text{-}25) \times 10^{-5}$ SI units.

These regularities in the distribution of $\varkappa$ enable one to determine the source and orientation of ablation of detrital material as Lower Cretaceous sediments were formed.

Sec. 59. **Magnetization of Magmatic Formations**

Based on the data on $\varkappa$ of different-age folded structures an identical distribution of its values for a number of formations has been found.

The geosyncline stage includes these petrophysical formation groups: (1) ultrabasite: gabbro-peridotite and dunite-pyroxenite-gabbro formations; these are characteristic of major geosynclinal zones and depressions and are deep mantle strata assemblages; (2) anorthosite (referred to Archean and Early Proterozoic mantle magma assemblages); (3) gabbro-diabase (gabbro-diorite-diabase and gabbro-plagiogranite, spilite-diabase, basalt and andesite porphyry formations developed at the early geosynclinal stage and, referred to geosynclinal zones, depressions, abyssal faults, mantle magma assemblages at a depth 100-150 km); (4) diorite-plagiogranite and magmatite-plagiogranite (andesito-basalt composition mantle magma formations); (5) diorite-andesite (gabbro-diorite-granodiorite, andesite, andesite-dacite and andesite-porphyrite formations, product of andesite-basalt composition mantle magmas with a lesser depth of origination compared with gabbroid group magmas); the group is widely distributed in a number of folded regions; (6) granodiorite ferromagnetic (represented by a granite-granodiorite subformation, this group includes also andesite-dacite massifs and subcrustal magma formations); (7) migmatite-granite; (8) granite-liparite paramagnetic (granite, liparite formations and subformations of granitoids and effusive rocks mainly of acidic composition, composing large blocks, batholithic massifs, covers, a product of supracrustal magmas); (9) granite-liparite ferromagnetic and granosyenite (granite-leucogranite, granite-granosyenite formations and granitoid and acidic and moderately acidic composition effusive rock subformations, supercrustal magma assemblages, alkaline rocks); (10) basaltoid and basalt (basalt-liparite, andesite-basalt, trachy-basalt and basalt formations, mantle magma products).

Group 1, 2, 4 and 8 rock show weak magnetization. Group 1 rocks have $\varkappa = (25\text{-}400) \times 10^{-5}$ SI units, and practically all of their iron content is contained in the composition of dark-colour minerals. In case diaphorite masses are serpentinized, then $\varkappa$ values attain as much as 0.5 SI unit or more. Group 3, 5, 6, 7, 9 and 10 rocks generally are highly magnetized. The wide range over which $\varkappa$ values of gabbros (Group 3) vary is accounted for by the nonuniform content and composition of ferro- or ferrimagnetics of these rocks. Sometimes $\varkappa$ of gabbros does not exceed 125×10^{-5} SI units owing to the specific conditions of their origination. Amphibolization and other autometasomatic transformations aid in decreasing the magnetic susceptibility of rocks.

Seven formation groups have been identified for the extrageosynclinal stage: (1) granodiorite-dacite-andesite ferromagnetic; (2) granite-liparite paramagnetic

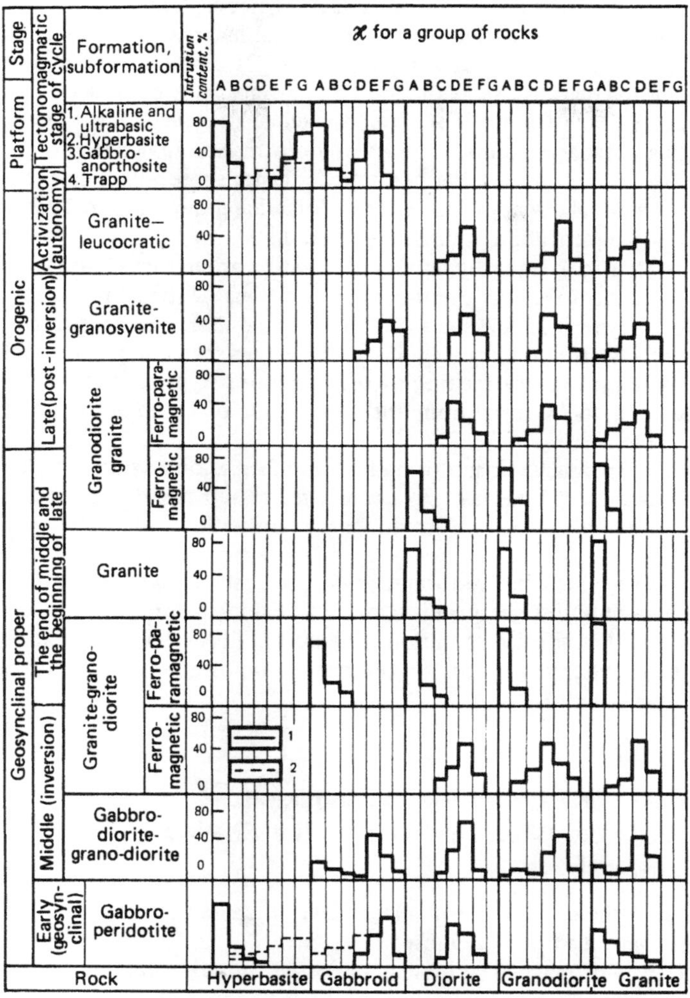

FIG. 135. The magnetic susceptibility $\varkappa$ of typical intrusive formations of various stages of the tectono-magmatic cycle (after N.B. Dortman).

The $\varkappa$ values (in 1.26×10^{-5} SI units) for groups: *A*—0-100; *B*—100-300; *C*—300-700; *D*—700-1500; *E*—1500-3000; *F*—3000-6000; *G*—600012 000; *1*—unaltered rocks; *2*—serpentinized and amphibolized rocks (after N.B. Dortman) (referred to the origination of supercrustal magmas); (3) leucogranite-liparite ferromagnetic; (4) granosyenite-trachytoid texture (subcrustal magma formation); (5) alkaline (product of subcrustal magmas); (6) basaltoid; (7) basaltic (products of mantle magmas). For the platform stage the following groups have been identified: (1) trap (or trappide), close to Group 10 of the geosynclinal stage due to its specific features; (2) alkaline-ultrabasic group (product of deep mantle magmas).

All of the above formation groups, except Group 2, demonstrate a high magnetization of rocks; it is particularly appreciable for the alkaline group. Group 2 rocks display a weak rock magnetization common for Group 8.

A periodical variation in the magnetization of formation group rocks is associated with the periodical change of the thermobaric conditions of their origination. High magnetization is typical of rocks that are products of deep mantle magmas, it decreases for mantle- and subcrustal magmas. The least magnetization is demonstrated by rocks resulting from the melting of the uppermost crustal strata. Then subcrustal magmas and rocks form from the former, all of them are strongly magnetized.

The appearance of ferromagnetic subformations and formations is due to higher pressures and temperatures, that is why they are referred to subcrustal, more basic, magmas, and that of paramagnetic subformations and formations, to supercrustal magmas.

The magnetization of the intrusive formations of early stages is very much affected by the processes of autometamorphism and regional metamorphism that occur due to a change in thermobaric conditions. Figure 135 illustrates the variation of the magnetic susceptibility of rocks of different composition formed under different conditions.

Sec. 60. Regional Variation of the Composition and Magnetic Characteristics of the Shields and Folded Regions

Definite regularities in the distribution of magnetization characteristics of shields and folded regions have also been disclosed.

The widest spectrum of magnetization characteristics of individual megablocks is displayed by the Baltic shield. The Ukrainian, Anabar and Aldan shields have magnetization characteristics that correspond to a definite megablock of the Baltic shield.

The southern, major part of the investigated area of the Baltic shield shows mainly weakly magnetic rocks. Things are different as regards its northern part, as well as the greater portion of the Ukrainian massif and Anabar shield. They have a variable but generally higher rock magnetization. In the eastern and western portions of the Aldan shield rocks have been found containing ferro- and paramagnetic minerals as in the central part of it the rocks with ferro- and ferrimagnetic minerals are predominantly developed. In addition, it has been established that weakly magnetized assemblages are typical of anticlinorium zones, and rocks showing nonuniform but usually ferrimagnetic mineralization are common for major synclinorium structures. This regularity is revealed fairly distinctly not on all occasions, since the magnetization characteristics of rocks are affected to a definite degree by the processes of metamorphism and granitization.

Rock magnetization characteristics of ancient series and formations of the same age are close for different shields and massifs.

The magnetization of rocks occurring in folded regions and zones of tectonic activity is connected with the strike of their principal structural elements and is governed by the confinement of the particular magmatic formations and facies of regional metamorphism.

Chapter XI

Radioactivity

Radioactivity is the spontaneous decay of the atoms of certain isotopes into new isotopes which may be stable or undergo further decay until a stable isotope is finally created. As a result, energy is liberated and more stable or new radioactive elements form. The nuclei of the latter decompose again and this process continues until a stable isotope appears.

Decay Types

Alpha decay involving emission by nuclei of α-particles ($_2He^4$) and formation of nuclei of new elements ($_{Z-2}X^{A-4}$) according to the scheme

$$_ZX^A \rightarrow {}_{Z-2}X^{A-4} + {}_2He^4$$

In this formula X is the element with a relative atomic mass A and atomic number (known alternatively as proton number) Z.

Beta decay involving radioactive transformation in a nucleus either of the neutron $_0n^1$ to a proton $_1H^1$ according to the scheme:

$$_0n' \rightarrow {}_1H^1 + {}_{-1}\beta^0 + {}_0\nu^0$$

the resulting electron $_{-1}\beta^0$ and neutrino $_0\nu^0$ are ejected from the nucleus giving rise to a new element whose atomic number is greater by unity compared with the initial element, or, alternatively, transformation of the $_1H^1$ proton to a neutron occurs according to the scheme:

$$_1H' \rightarrow {}_0n^1 + {}_1\beta^0 + {}_0\nu^0$$

in which case positron ($_{+1}\beta^0$) and neitrino emerge a new element appears with the atomic number less by unity compared with the initial. Beta decay also includes electron capture in which process the nucleus absorbs an electron from the K shell (more seldom from the L shell) to produce an element whose atomic number is unity smaller than the initial element.

Isomeric transition followed by gamma radiation causing the energy level of the nucleus to be changed.

Nuclear (or atomic) fission, the division of an atomic nucleus into parts.

Radioactive Particles and Their Interaction with Substances

Alpha particles, identical with the nuclei of the helium atoms have a charge equal to 2, a mass 6.6947×10^{-27} kg, an initial velocity $(1.42\text{-}2.054) \times 10^7$ m/s (depending on the type of the decaying element) and initial energy $(3.20\text{-}12.82) \times 10^{-13}$ J

(2 to 8 MeV). They exhibit a high ionizing capacity (an alpha particle produces in the air as many as 200 000 ion pairs and have an inappreciable rectilinear path $R_{a\upsilon}$ in a substance from 19.1 μm (uranitite) up to 42.7 μm (barite). During atom ionization the alpha particle loses its energy and at the end of the path absorbs two electrons to transform to a helium atom.

The interaction of alpha particles with a substance gives rise to secondary electrons showing a small path length (in the air up to 0.5 mm) and X-rays.

Apart from alpha particles recoil atoms are produced, i.e. a flux of atoms whose atoms have lost alpha particles. These atoms have a comparatively small ability for ionizing the environment.

Beta rays, a stream of electrons (positrons)* with a mass $m = 9.109 \times 10^{-31}$ kg or 0.00054 atomic mass unit, a charge equal to one elementary charge or 1.602×10^{-19} Cl.

A characteristic feature of beta radiation is the continuous energy and speed distribution from zero to maximum values approaching the speed of light (300 km/s). Owing to the variable energy the path length of individual particles differs. However, it is on the average much greater than that of alpha particles.

The probability of scattering β-particles by the atomic nucleus is proportional to the squared nuclear charge, and by the electron shell to its first power. As a result, in the case of medium and heavy elements beta particles are mainly scattered by the atomic nuclei, and in the case of hydrogen both by the electron shell and nuclei. The weaker the scattering the greater is the energy of beta particles.

The path length of beta particles in rocks containing a number of β emitters is characterized by the path length of beta particles having a maximum penetrating ability. The ionizing capacity of beta particles is less than that of alpha particles. In the air one beta particle causes 50 ion pairs to be formed per 0.01 m.

Gamma rays is ultra-short electromagnetic radiation with the wavelength from 0.1 nm and less. Gamma rays occur during spontaneous and induced nuclear transformations as well as in the deceleration and decomposition of charged particles and annihilation of pairs. Gamma rays observed during the decay of natural radioactive elements are characterized by:

(1) the wavelength λ (in m) from hundredth fractions of X** units (for hard gamma radiation) to X and more (for soft gamma radiation);

(2) the quantum mass

$$m = h\nu/c^2 = h/c\lambda = 2.21 \times 10^{-40}/\lambda$$

where h is the Planck constant equal to 6.62×10^{-34} J·s; c is the velocity of light in m/s; ν is the rate of emission of gamma quanta;

(3) the energy $E_\nu = h\nu$ varying from $(0.08\text{-}4.81) \times 10^{-13}$ J (0.05 to 3 MeV); also $E_\gamma = 1238/\lambda$ where λ is expressed in X units.

---

* Beta rays are also defined by symbols $_{-1}e^0$, $_{+1}e^0$, or $_{-1}\beta^0$ and $_{+1}\beta^0$.

** X unit (generally known as siegbahn or X-ray unit) equals $(1.00202 \pm 0.00003) \times 10^{-13}$.

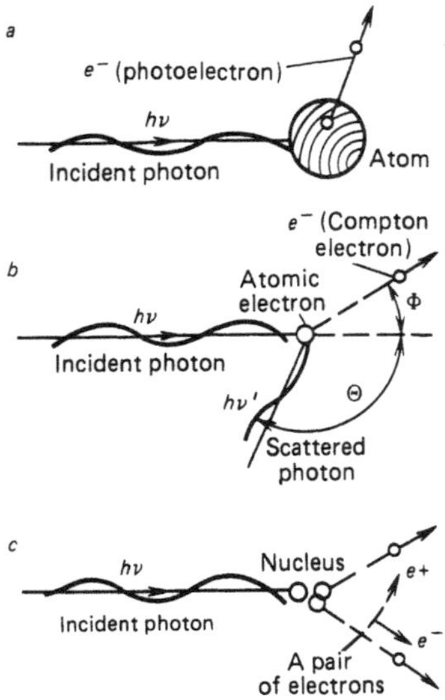

FIG. 136. Interaction of gamma-radiation with matter.

a—photoeffect; *b*—Compton effect; *c*—formation of electron-positron pairs

It should be pointed out that: (a) the maximum energy of gamma rays observed during the decay of natural β-emitters is about 4.17×10^{-13} J (2.6 MeV); during the transition of nuclei from one excited state to another the energy is within 0.80×10^{-13} J (0.5 MeV); (b) gamma quanta are mostly emitted by several groups with a definite energy of each quantum group.

The penetrating ability of gamma quanta is much more than that of beta particles since a gamma particle carries no charge (there is no interaction with the electric field of other charged particles) and its mass is less than that of alpha and beta particles.

The gamma radiation is attenuated in rocks due to: (1) internal conversion (for some radioactive elements); (2) photoeffect; (3) Compton (also known as Compton-Debye) effect; (4) pair production. Note that the attenuation of gamma radiation in the substance of a rock is approximately described (in a wide beam) by the relationship $I = I_0 e^{-\mu x}$, where I_0, I are the intensity of the original gamma radiation and of that after it has passed a thickness x of the rock layer; μ is a total attenuation coefficient consisting of gamma attenuation coefficients taking into account the photoeffect (τ), Compton effect (σ) and pair production ($\varkappa$): $\mu = \tau + \sigma + \varkappa$.

Given internal conversion characteristic of rocks containing heavy nuclei, gamma quanta are absorbed by the electron shell of the same atom with radiating electrons. Given a photoeffect (Fig. 136), gamma rays interact with the electron shell of another atom. The emerging photoelectron carries away a fraction of gamma radiation energy: $E_\beta = h\nu - E_0$, where E_0 is the electron bond energy in the atom. The process is probable if E_ν does not exceed 0.5 MeV.

The photoeffect also gives rise to X-ray radiation. The direction of the photoelectron scattering is governed by the gamma radiation energy. Hard gamma radiation* gives rise to photoelectrons whose orientation agrees with that of gamma radiation. Soft gamma radiation facilitates secondary beta radiation whose direction is mainly at right angles to that of the original rays.

The attenuation of gamma radiation intensity due to the photoeffect is described by a differential equation

$$-dI = \tau I dx$$

where dx is the thickness of a rock layer; τ is the total saturation coefficient due to the photoeffect.

The value of the coefficient is

$$\tau = a\left(\frac{1}{h\nu}\right)^{2.8} + Z^4 \frac{\delta_r N_A}{A}$$

where a is a constant; Z is the atomic number of the element; δ_r is the density of saturated rock; N_A is the Avogadro constant; A is a relative atomic mass.

Due to the proportionality of τ to Z to the fourth power and inverse power dependence on $h\nu$ the photoeffect is particularly pronounced in rocks containing heavy elements as well as upon rocks being irradiated by minor-energy photons.

Given the Compton effect, the gamma radiation interacts with the electron imparting to the latter a fraction of its energy to then propagate in the rock (see Fig. 13*b*) upon changing the original direction. This process in likely to occur for all gamma quantum energies and is principal for 0.2-3.0 MeV.

The coefficient of gamma radiation attenuation due to the Compton effect is

$$\sigma = \sigma_e \delta_r N_A Z/A$$

where σ_e is the attenuation coefficient calculated for one electron; $\delta_r N_A Z/A$ is the number of electrons in a unit volume of a rock; since Z/A is roughly constant for all elements, σ is proportional to the density of the substance.

The process of electron-positron pair production emerging from a photon in the field of atomic nuclei is most likely for rocks containing heavy atoms (e.g. Pb) irradiated by hard gamma rays having an energy not less than 1.02 MeV (see Fig. 136*c*). The gamma radiation attenuation due to it is determined by the coefficient $\varkappa$.

The Law of Radioactive Decay

Radioactive decay (alternatively known as decay, nuclear spontaneous reaction, radioactive disintegration, radioactive transformation, radioactivity) proceeds at the following rates established experimentally. During a time interval $T_{1/2}$ called a *half-life* (or half-value period, radioactive half-life) 50 per cent of radioactive atoms of a substance undergoes radioactive decay. This leads to a decrease of the intensity of radioactive radiation by 50%. Within a time interval $2T_{1/2}$ 50% of the atoms

---

* The gamma radiation energy of natural radioactive elements varies from 0.05 to 3 MeV; hard gamma radiation has a great, and soft radiation a small gamma radiation energy.

remaining intact by the time period $T_{1/2}$ decays, and the radiation intensity also diminishes twofold etc. The number of atoms dN decaying within a time interval $d\tau$ is proportional to $d\tau$ and the number of atoms, N, that remain intact by the given time moment τ. Therefore $dN = -\lambda N d\tau$ or $dN/N = -\lambda d\tau$. By integration we will find the law of decay:

$$N = N_0 e^{-\lambda\tau}$$

where N_0 is the number of atoms at the initial time moment $\tau = 0$.

The proportionality coefficient λ is called *a decay constant* (decay coefficient, disintegration constant, etc.):

$$\lambda = \frac{\ln(N_0/N)}{\tau} \tag{84}$$

Thus the decay constant is determined by the fraction of atoms that have decayed within a unit time.

By assuming in Eq. (84) $\tau = T_{1/2}$ and $N = N_0/2$, we have

$$T_{1/2} = \ln 2/\lambda = 0.693/\lambda$$

Radioactivity. Units of Its Measurement and Radioactive Element Concentrations

The decay of radionuclides* of a rock is characterized by its activity n. Activity is a quantity determining the number of decays in the radioactive component of a rock in a unit time. The unit of measuring activity is a bequerel (Bq = s^{-1} = decay/s) equal to the activity of a quantity of a radioactivity of a substance (in particular, radioactivity of a rock) having one spontaneous nuclear transition per second. 1 Bq = 0.27×10^{10} Ci (Ci is a nonsystem unit of radioactivity, defined as that quantity of any radioactive nuclide which has 3.7×10^{10} disintegrations per second as much as in 1 g of Ra), 1 Ci = 3.7×10^{10} Bq.

The radioactivity of rocks is also characterized by such quantities as *the specific mass activity*—Bq/kg, *the specific volumetric activity—Bq/m³* and *the surface activity*—Bq/m².

The gamma activity of rocks is still characterized by a nonsystem quantity—radium gamma equivalent with a unit of measurement expressed by a milligram-equivalent of Ra (mg-eqv Ra); 1 mg-eqv Ra is the activity of the substance, the gamma radiation of which has the same ionizing ability as has the radiation of 1 mg of ^{226}Ra (together with its decay products) after passing through a platinum filter 0.5 mm thick. Use is also widely made in geophysics of such quantities as: (1) the specific mass gamma activity having units of measurement mg-eqv Ra/g (or pkg-eqv Ra/kg = (kg-eqv Ra/kg) $\times 10^{-12}$—activity of 1 g of a rock in mg-eqv Ra (or 1 kg of a rock in kg-eqv Ra); (2) the specific volumetric activity having

* Nuclides are species of atoms differing in the composition of the nucleus, i.e. having different numbers of nucleons in the nucleus or involving different numbers of neutrons and protons.

units of measurement mg-eqv Ra/cm^3 (or nkg-eqv Ra/m^3 = (kg-eqv Ra/m^3) $\times 10^{-9}$.

When determining the radiation intensity of radioactive elements by pulsed methods (gamma spectrometric method, β-method, α-method) the intensity of gamma, beta and alpha radiation and Ra, Th and U content are measured in imp/s and %, and when determining the Ra, Th, U etc. content by using radiochemical, colorimetry, luminiscence, X-ray spectroscopy, isotope dilution, neutron activation and track method, in kg/kg (or g/g) and %.

The radioactivity of rocks containing definite radioactive elements (e.g. uranium enriched by radioactive minerals) is measured in % U or Th (for rocks enriched by thorium-bearing mineral) per kilogram of a rock. In this case the radioactivity of 1 kg of a rock is compared with that of 1 kg of the reference material containing a certain percentage of a radioactive element (e.g. 0.02% U) and is expressed in per cent U or other radioactive element per kilogram of the rock.

Sec. 61. **The Radioactivity of Elements and Minerals of the Lithosphere**

Most radioactive isotopes occurring in nature refer to U, Ac and Th families containing, respectively, 20, 15 and 13 genetically related radioactive and stable isotopes. The half-life of radioactive isotopes varies from millionth fractions of the second to many bullions of years. The longest lifetime is demonstrated: (a) in a uranium-radium series by UI (4.49×10^9 yrs), UII (2.522×10^5 yrs), I_0 (8×10^4 yrs), ^{226}Ra (1 590 yrs), RaD (22 yrs); (b) in the Ac series by AcU (7.13×10^8 yrs), Ra (3.43×10^4 yrs), Ac (21.7 yrs); (c) in the Th series by Th (1.39×10^{10} yrs), $MsTh_1$ (6.7 yrs), RaTh (1.9 yrs).

The transition from one to another radioactive, then a stable isotope occurs through alpha and beta decay. The U-Ra series contains twelve alpha, eleven beta and eleven gamma emitters. The Ac series contains ten alpha, seven beta and six gamma emitters. The Th series contains as few as eight alpha emitters, and six beta and gamma emitters. However, what is recorded is the alpha radiation of definite alpha emitters of the U-Ra family. The alpha radiation of the Ac family is much less intensive than that of the U-Ra family (less than 5% of the total alpha radiation of the U-Ra family members). This is due to the inappreciable actinouranium content in natural uranium. The U-Ra and Th family each contain four practically important beta emitters, the gamma radiation of the U-Ra family is mainly due to the presence of Ra(B + C). The gamma radiation of the Ac series is inappreciable compared with that of the U-Ra series. In the Th family principal gamma radioation is due to MsTh and Th (C′ + C″) with decay products.

In addition to radioactive elements of the U, Ac and Th families, radioactive properties are manifested by K, Ca, Ru, Zr, In, Sn, Tl, La, Nd, Sm, Lu, W, Re and Bi isotopes. These isotopes are long-lived, their half-life exceeds 10^9 yrs; they are genetically unrelated to other radioactive elements. They are characterized by beta decay or K capture and, sometimes, both processes excepting Sa, W and Bi that disintegrate forming alpha particles; the energy of the released particles varies over the range from (0.016-5.45) $\times 10^{-13}$ J (0.01 to 3 MeV): so does their relative abundance in nature vary appreciably, their content in a natural isotope mixture ranging from 0.0119 to 100%.

Radioactive isotopes represent a small fraction of the lithosphere. As roughly estimated by A. P. Vinogradov, the lithosphere (to a depth 16 km) contains 2.1×10^{-6} (1.4×10^{-7})* mass per cent of ^{235}U, 3×10^{-4} (1.98×10^{-5}) ^{238}U, 8×10^{-4} (7×10^{-5}) ^{232}Th, 10^{-10} ^{226}Ra, 10^{-10} ^{231}Pa, 6×10^{-14} ^{227}Ac, 3.1×10^{-4} (1.54×10^{-4}) ^{40}K, 6.4×10^{-3} ^{48}Ca, and 8.4×10^{-3} (1.9×10^{-3}) ^{87}Rb.

Optimal clarke values are currently taken to be 2.1×10^{-4}% for ^{238}U, 7.0×10^{-4}% for ^{232}Th, 1.8% for ^{40}K. These clarke values also known as crustal abundance values are not final since the radioactivity of the oceanic floor has been insufficiently studied.

Among the aforementioned isotopes the most abundant are Rb and Ca; the abundance of Th, K, U, Sn, Sa, Nd and Zr is less by an order of magnitude compared with that of Rb and Ca. The abundance of other long-lived radioactive isotopes is even less.

By taking into account the abundance of radioactive isotopes in the lithosphere and intensity of their decay it can be said that the greatest effect on the geological processes may be produced by U and Th with their disintegration products, ^{40}K and, in part, ^{87}Rb. Other radioactive elements, in view of their minor abundance and great half-life decay display an inappreciable total decay energy and cannot play an important role in the radioactivity of the lithosphere. Radioactive elements are nonuniformly distributed in individual spheres of the lithosphere and in rocks. The radioactivity of the spheres of the earth is less the nearer they are to the centre of the earth.

Radioactive elements occur in the lithosphere in the composition of minerals. More than 200 mineral species have been established to contain radioactive U, Th, Ra and K. A great number of minerals containing uranium are especially often occurred. U- and Th-bearing minerals are widely distributed but are very much scattered in the lithosphere and seldom occur as economic deposits.

The great number of minerals whose composition contains U and Th is accounted for by the ability of uranium to be found in the lithosphere in a tetra- and hexavalent form and easy isomorphic mixability** of tetravalent U, Th, rare earths, Zr and Ca at elevated temperatures. Uranium occurs in minerals as U^{4+} cations (in

* Parentheses contain the atomic percentage of radioactive isotopes.

** Under conditions of isomorphic replacement radioactive elements enter the crystal lattice of accessory minerals substituting ions close in radius without rupturing or altering the crystal lattice. For instance, the radioactivity of allanite (orthite) corresponding to the formula $(Ca, Ce)_4(Al, Mg, Fe)_6(OH)_3O_3 \times (Si_2O_7)_8$ is due to an isomorphic Th impurity. Monazite $(Ce, La)PO_4$ also contains up to 5-10% ThO_2 as an isomorphic impurity. This group includes xenotime YPO_4, often containing ThO_2 (up to 5%) and UO_2, sometimes ZrO_2. The same is also true of samarskite $(Y, Er)(Nb, Ta)_2O_6$ which contains from 4.02 to 16% U, pyrochlore $NaCaNb_2O_6$ containing from 0 to 11.4% U and 0 to 15.5% UO_2, zircon ($ZrSiO_4$), sphene (titanite), $CaTiO_2[SiO_4]$, apatite ($Ca_5[PO_4]_3(F, OH, Cl)$), baddeleyite (ZrO_2), nepheline $(Na, K)_2OAl_2O_32SiO_3$ etc.

a strongly reductional medium), U^{6+} cation (which is more stable in an acidic medium), UO_2^{2+} and $U_2O_7^{2-}$ anion. The high isomorphic mixability is due to the closeness of ionic radii of U^{4+}, Th^{4+}, Y^{3+}, Ca^{2+}, Pa^{5+}, TR (from La to Lu), Mn^{2+} The U^{4+} ionic radius is 0.105 nm. Uranium isomorphously replaces mainly zirconium (zircon, zitrolite) and yttrium (sphere, orthite), and thorium isomorphously replaces tetravalent thorium in ceric orthites and monazites. The number of thorium-bearing minerals is about twice less since it occurs in the lithosphere solely in a tetravalent form.

U and Th-bearing minerals fall into uranium mineral proper and thorium and uranium- and thorium-bearing ones. This division is made taking into account the content of and the fashion U and Th enter into minerals. The latter are classified as being uranium or thorium minerals if their composition constantly contain more than 30-45% U and Th.

U and Th occur in U- and Th-bearing minerals in relatively small and variable amounts. Here they are components of solid solutions, isomorphic impurities, microadmixtures of uranium and thorium minerals proper and crystals of rock-forming minerals and absorbed ones. U and Th minerals are lithophylic and oxyphylic.

Thorium proper and thorium-uranium minerals refer to oxide and silicate classes, sometimes the same minerals are concurrently U-bearing ones. For example, aldanite, uranothorianite and thorianite refer to anhydrous oxides; of these, uranothorianite represents also uranium minerals of this subgroup of the oxide class. Complex Ti, rare earth and Th oxides include a thorium mineral species smirnovite (thorutite) which, according to the U content, is classified as an U-bearing one. Thorite ThO_2SiO_2 or $ThSiO_4$, mackintoshite (Th, U)SiO_4H_2O, maitlandite (Pb, Ca)$_2Th_3U_4^{4+}Si_8O_{32}23H_2O$, uranothorite (Th, U)$SiO_4nH_2O$ etc. which can provide an example of thorium proper or uranium-thorium minerals of the silicate class also refer to uranium-bearing. The Th-bearing minerals are much more numerous. They are also, as a rule, concurrently uranium proper or uranium-bearing minerals.

U proper minerals which contain thorium are:

(1) simple tetra- or hexavalent uranium oxides represented, e.g. by uraninite, bröggerite, cleveite; (2) complex anhydrous U^{4+} oxides containing titanium and iron (U, Fe, Th titanates), e.g. lodochnikite, brannerite etc.

The specially investigated uranium minerals, which do not seem to contain Th, or if they do, then in minor amount, refer to : (1) simple oxides (U^{4+} + U^{6+})—pitchblende, uranium black; (2) hydrous uranium oxides U^{4+} (UO_2^{2+}) and uranates ($U_2O_7^{2-}$); (3) uranium sulphates (UO_2^{2+}); (4) uranium sulphate carbonates—(UO_2^{2+})—schroekingerite; (5) uranium carbonates (UO_2^{2+}); (6) complex carbonates (UO_2^{2+}); (7) arsenates (UO_2^{2+})—trögerite etc.; (8) uranium phosphates (U^{4+}); (9) uranium phosphates (UO_2^{2+})—autunite, torbernite etc.; (10) uranium vanadates (UO_2^{2+})—carnotite, tuyamunite etc.; (11) uranium molybdates (U^{4+} + U^{6+}); (12) uranium molybdates (UO_2^{2+}); (13) uranium silicates (UO_2^{2+})—uranophane and others.

Among minerals containing U^{4+} + U^{6+} and UO_2^{2+} whose composition does not appear to contain Th we can refer to: uranium (U^{4+} + U^{6+})—chalcosine, uranium (U^{4+} + U^{6+})—limonite, UO_2^{2+}—opal, UO_2^{2+}—calcite, UO_2^{2+}—allophane etc.

Th- and U-bearing minerals may include minerals of the following classes and groups: (1) simple oxides—baddeleyite; (2) complex titanium, rare earth, uranium U^{4+} and thorium oxides (rare earth and thorium titanates)—ufertite; (3) complex titanium, niobium, rare earth and uranium U^{4+} oxides (tantalo-niobate, niobo-tantalates, titano-tantalo-niobates)—fergusonite, euxenite, loparite, samarskite, pyrochlore, aeschynite, etc.; (4) zirconium silicates—zircon, nalgite, malacon, alvite etc.; (5) titanium silicates (titanosilicates)—rinkolite, lovchorrite, sphene; (6) rare earth silicates—orthite (allanite), thalleite; (7) silicates; (8) rare earth phosphates—xenotime and monazite; (9) phosphates—apatite. Uranium-bearing minerals also include uraniferous organic compounds carburan and thucholite. In addition to the aforementioned Th- and U-bearing minerals a number of radium-bearing minerals are known.

The U content in uranium minerals is characterized by the following data. The lower limit of U content in uranium minerals is 25-30% excepting some blacks poor in uranium. The highest U content (up to 70-77%) is shown by uranium oxides and hydroxides. Among oxygenous salts, a high U content (up to 60-70%) is manifested by simple salts with one uranyl (soddyite, rutherfordine, phosphuranylite, trögerite); the U content in double salts is generally as much as 40-60%.

According to the findings of R. V. Getseva and K. G. Savelyeva the Th content in thorium and in thorium and uranium minerals attains 45-90% and drops to 25-30% whereas the U concentration in uraniferous minerals does not exceed 22% and that of ThO_2 of thorium-bearing minerals is not more than 19%. It must be noted that the average U and Th contents in zircons are roughly equal. Sometimes Th is very much predominant in them, on rare occasions it represents the only radioactive element. The radioactivity of zircon crystals is not uniform even in one and the same rock. Thorium is predominant in ceriferous minerals (alanite, monazite), and uranium in yttrium-bearing rocks (xenotime).

The radioactive element (U and Th) concentration in individual rock-forming minerals varies from $n^* \times 10^{-7}$% (olivines, garnets) to $10n$% (monazites, uranium and thorium minerals).

Thorium and thorium-bearing minerals are primary minerals, whereas uranium and uraniferous minerals may be both primary and secondary ones. Primary materials originate in a magmatogenic, pegmatite, pneumatolytic and hydrothermal ways and secondary minerals through supergene processes. They form from primary minerals close to or at the surface of the lithosphere, in oxidation and cementation zones, with surface waters, oxygen, carbon dioxide and organic substances participating.

Primary thorium and uranium minerals include minerals of oxide and silicate (U^{4+} and $U^{4+} + U^{6+}$) class. Primary uraniferous and thorium-bearing minerals include minerals of classes (and groups) of complex oxides, rare earth silicates, rare earth phosphates.

Secondary uranium minerals are represented by hydroxides, uranates, phosphates, sulphate carbonates, carbonates, arsenates, phosphates, vanadates, molybdates and silicates of (UO_2^{2+}).

* Several first units.

TABLE 16. Classification of Hypogene and Supergene Minerals According to Their Radioactivity Degree

Mineral group	Minerals	Content, mass per cent		Th/U	Concentration clarke	
		U	Th		U	Th
Maximally radioactive (uranium and thorium minerals)	Hypogene: uraninite, pitchblende, thorianite etc.	56-85	20-40	<0.01 (uranium) 40-80 (thorium)	>10 000	>10 000
	Supergene: autunite, thorbernite, schrökengerite etc.	40-60	<0.01	<0.001	>10 000	-
Highly radioactive (rare accessories of felsic rocks)	Cyrtolite, zircon, thorium orthite	$(400\text{-}2\,000) \times 10^{-4}$	$(400\text{-}1\,000) \times 10^{-4}$	<1	100-1 000	10-100
	Orthite, monazite	$(600\text{-}2\,000) \times 10^{-4}$	$(10\,000\text{-}30\,000) \times 10^{-4}$	>10	100-1 000	100-1 000
Increasingly radioactive (most common accessories)	Sphene, apatite, magnetite	$(10\text{-}100) \times 10^{-4}$	$(30\text{-}200) \times 10^{-4}$	2-5	5-30	2-15
Normally radioactive (felsic rock minerals)	Biotite, hornblende	$(4\text{-}8) \times 10^{-4}$	$(8\text{-}18) \times 10^{-4}$	1.5-2.3	2-3	1-2
Weakly radioactive (main felsic rock minerals)	Quartz, potassium feldspars, acidic plagioclases	$(1\text{-}3) \times 10^{-4}$	$(2\text{-}8) \times 10^{-4}$	1.8-4.5	0.5-1	0.2-0.5
Low-radoactive (femic rock minerals)	Pyroxene, main plagioclase species	$(\leqslant 0.1\text{-}1.0) \times 10^{-4}$	$\leqslant(0.1\text{-}1.0) \times 10^{-4}$	2-5	<0.2	<0.2

Note. Concentration clarke is the ratio of an element content to its clarke (average abundance in the earth's crust).

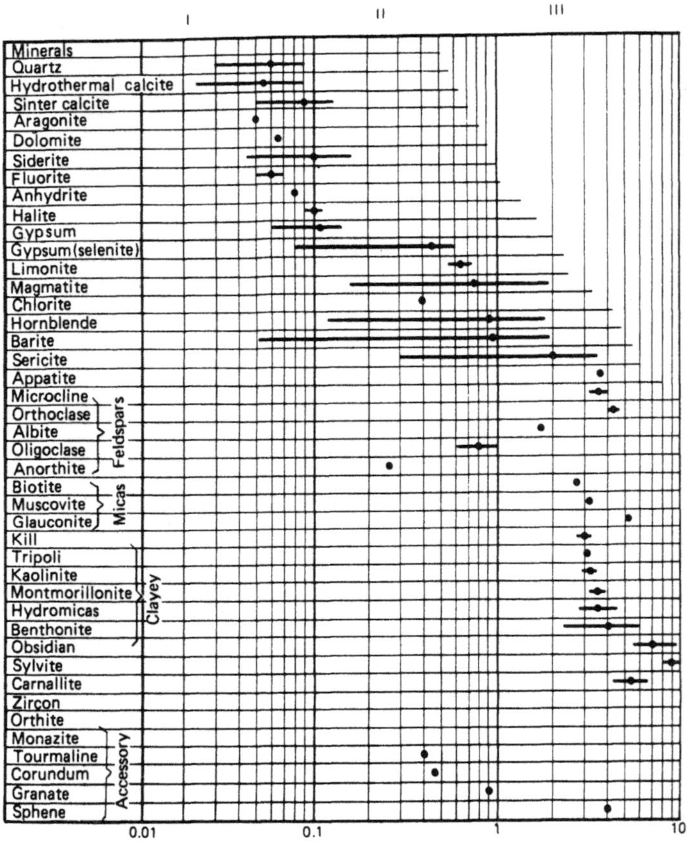

FIG. 137. Classification of minerals involved in the composition of sedimentary rocks (after V.V. Larionov, 1969).

Minerals having radioactivity: *I*—high; *II*—intermediate; *III*—higher; *IV*—anomalously high

Among uraniferous minerals, those whose composition contains (UO_2^{2+}) are primary ones as are minerals that have adsorbed (U^{4+} + U^{6+}) (chalcosine, limonite etc.). Radium-bearing minerals also refer to secondary supergene minerals.

The most important potassium-bearing minerals include sylvite KCl, carnallite $KClMgCl_26H_2O$ (14.1% K), potassium nitrate KNO_3 (38.5% K), feldspars, microcline ($KAlSiO_8$), orthoclase ($K_2OAl_2O_36SiO_2$), micas—biotite [$K_2OAl_2O_36(MgFe)O_2$] and muscovite ($K_2OAl_2O_36SiO_22H_2O$), nepheline $(NaK)_2OAl_2O_32SiO_3$, glauconite (K, Al, Fe hydrous silicate) etc.

Radioactive minerals most commonly contained in magmatic and sedimentary rocks divide, according to radiation intensity, into radioactive and moderately radioactive. Radioactive minerals include allanite, monazite, xenotime, thorite, pyrochlore, uraninite (minerals are listed in the order of decreasing their abundance).

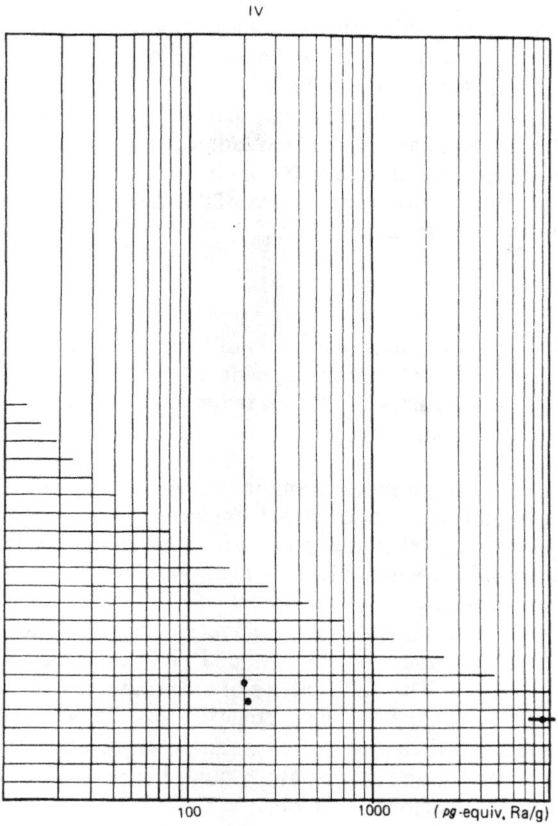

They occur more rarely than do such accessory minerals as zircon, sphene and apatite which are referred to moderately radioactive minerals.

The most abundant rock minerals fall into four groups according to their radioactivity. The first group comprises weakly radioactive principal rock-forming minerals: quartz, potassium feldspars, plagioclase, calcite, dolomite, anhydrite, rock salt, nepheline. The second group of minerals showing a normal or slightly increased radioactivity includes such rock-forming species as biotite, amphiboles, pyroxenes etc. The third group with increased radioactivity includes principal (commonly occurring) accessory and ore minerals: apatite, eudialyte, fluorite, ilmenite, magnetite and others. The fourth group having an increased radioactivity comprises less common accessory minerals: sphene, orthite, monazite, zircon, loparite etc. Table 16 shows U and Th contents in these minerals.

There is also a classification of sedimentary rock minerals according to their radioactivity values where they are likewise divided into four groups showing a low, intermediate, increased and high radioactivity. As believed by V. V. Larionov, the radioactivity of principal rock-forming minerals of the first group: quartz, calcite, dolomite, anhydrite, rock salt etc. (see Fig. 137) does not exceed 0.1 pkg-eqv Ra/kg.

The second group includes such accessory minerals as limonite, magnetite, tourmaline, corundum, garnet as well as such minerals as Na-K, feldspars (anorthite, aligoclase), hornblende, chlorite etc. Their radioactivity is within $0.1 < q_m < 1$ pkg-eqv Ra/kg. The third group includes clay minerals, micas, many feldspar minerals, potassium salts, sericite, apatite, obsidian and sphene, their radioactivity varies from 1 to 10 pkg-Ra/kg. The fourth group includes such accessory minerals as monazite, zircon, orthite having radioactivity $q_m > 10$ pkg-eqv Ra/kg. Their radioactivity exceeds that of group one minerals by a factor of 1 000.

Sec. 62. Magmatic Rocks

The division of radioactive minerals into groups and types of magmatic (plutonic) rocks and the radioactivity of uranium- and thorium-bearing minerals is governed by the conditions of rock formation and composition. The radioactivity of magmatic rocks is also conditioned by the latter factors.

Rocks of an ultrabasic and basic composition. Uranium and thorium rocks are, generally, absent from ultrabasic and basic rocks. What imparts radioactivity to these rocks are usually zircon, orthite (allanite), other accessory minerals and definite potassium-bearing minerals. Less commonly the bulk of uranium is concentrated in principal rock-forming minerals.

Uranium is present here in a dispersed state or as submicroscopic inclusions of uranium minerals, as to accessory minerals, the bulk of U and Th is isomorphously incorporated into their crystal lattice. The radioactivity of zircon and orthite in rocks under consideration is less than that of other magmatic rocks. Minerals of gabbroids, according to an increase in their radioactivity, can be arranged in the following series: pyroxene → hornblende → plagioclase → olivine → magnetite → biotite → titanite (sphene) → apatite → orthite → zircon.

Uranium- and thorium-bearing zircon and orthite are very much dispersed in basic and ultrabasic rocks. These rocks contain also minor amounts of radium present in talc, tremolite and serpentine.

Data on the average Ra, U, Th and K contents in ultrabasic and basic rock groups and types are presented in Table 17. The latter shows that ultrabasic and basic rocks contain microgram quantities of radioactive elements which fact is responsible for their very inappreciable radioactivity.

Intermediate rocks. Intermediate rocks may include: (1) uranium and thorium minerals—thorianite, thorite, uranothorite (in syenites and niobium-bearing syenites), brannerite, iriginite (in albitites); (2) of U- and Th-bearing minerals—delorenzite, pyrochlore, loparite, zircons, orthite, apatite (in syenites, nepheline syenites and occasionally in diorites). It must be noted that zircons of these rocks show much more radioactivity than those of basic rocks.

A definite fraction of radioactivity is also due to other rock-forming minerals. Nepheline-syenite minerals are arranged, for example, in this series in order of increasing radioactivity, and their U and Th content: nepheline → orthoclase → aegirine → arfvedsonite → biotite → ilmenite → eudialyte → apatite → zircon.

TABLE 17. Average Concentrations of Radioactive Elements (in %) in Magmatic Rocks

Rock group	Series	Rocks	K	U × 10^4	Th × 10^4	Th/U
Intrusive	Calc-alkali	Ultrabasic (dunite, pyroxenite etc.)	0.15	0.03	0.08	2.7
		Basic (gabbro, diabase etc.)	0.7	0.6	1.8	3.0
		Intermediate (diorite, quartz diorite)	1.8	1.8	6.0	3.3
		Acidic (granodiorite, plagiogranite, biotite granite, leucocrate alaskite granite)	2.3-4.0	2.1-7.0	8.3-40.0	5.6
	Alkaline	Melteigite, meimechite, ijolite	1.8	2.6	9.8	3.8
		Kimberlite (other than diamond-bearing)	—	2.6	8.8	3.4
		Kimberlite (diamond-bearing)	—	3.2	16.3	5.1
		Syenite	3.8	—	—	—
		Miaskite nepheline-syenite	4.5	4.1	7.2	1.8
		Agpaite nepheline-syenite	5.0	10.3	28.3	2.8
		Trachybasalt	2.6	2.4	8.0	3.3
		Trachyte, trachyliparite	4.8	3-8	30-50	4-5 to 10
Effusive		Basalt, diabase	1.0	0.7	2.3	3.2
		Andesite, andesite-porphyrite	1.7	1.2	4.0	3.3
		Dacite, dacite-porphyrite	2.3	2.5	10.0	4.0
		Liparite, quartz-porphyry	3.7	4.7	19.0	4.0

The concentration of uranium and uraniferous minerals in intermediate rocks is higher than in basic ones: the concentration of Ra, Th and K is also higher in intermediate rocks (see Table 17), which is responsible on the whole for the appreciable radioactivity of these rocks.

The most appreciable uranium amounts have been found for alkaline syenites and phonolites that form at the last magmatic differentiation stages. Generally, the U content of alkaline syenites is on the average several times greater than that of calc-alkaline granites.

Acidic-composition rocks. Uranium- and thorium minerals—uraninite (in soaa granites), brannerite, smirnovite, yttrocrasite, thorite and thorianite—are present in acidic magmatic rocks very rarely although in greater amounts compared with intermediate rocks. The accumulation of great amounts of uranium and thorium minerals in acidic rocks is accounted for by more considerable U and Th concentrations in the granitic magma.

Typical representatives of radioactive minerals of acidic magmatic rocks are considered to include zircons, sphene, orthite, monazite, xenotime and apatite; sometimes pyrochlore, ufertite and delorenzite are found in them. Moreover, zircons occur in granites (except leucocratic species) in greater amounts than in intermediate rocks and their radioactivity is higher.

An increase in the radioactivity of zircons and, it appears, of other accessory uranium- and thorium-bearing minerals and their greater concentration passing from ultrabasic to acidic rocks is due to the increased Th and U concentration in magmatic melts during their crystallization. The greatest content of uranium is shown by xenotime, zircon, monazite and apatite, typical granodiorites, calc-alkaline and soda granites; the U content in the remaining minerals of these rocks is inappreciable and varies from one rock to another.

Even though the radioactivity of these rocks is sometimes associated with their principal rock-forming minerals, the U and Th content as well as the radioactivity of granite minerals increases according to this series: quartz → orthoclase → plagioclase → hornblende → biotite → pyroxene → magnetite → fluorite → apatite → epidote → ilmenite → sphene → zircon → orthite → monazite.

The concentration of radioactive accessory minerals and, consequently, principal radioactive element (Ra, U, Th and K) is maximum in acidic magmatic rocks. This is responsible for the highest radioactivity of acidic magmatic rocks attaining the highest values in granites (see Table 17).

The average concentration of principal radioactive elements in granites of different regions varies comparatively little even though adjacent rock masses may sometimes manifest quite different radioactivity.

The U content in granites may be as great as $\geqslant(4\text{-}10) \times 10^{-6}$ kg/kg, the Th content $\geqslant(20\text{-}50) \times 10^{-6}$ kg/kg. An increased K content has been found in quartz porphyry, microcline granites (up to 6%) and leucitic nephrites (up to 7%). The U content in volcanic glasses varies over the range $(0.8\text{-}15) \times 10^{-6}$ kg/kg and is on the average 5.6×10^{-6} kg/kg. This value seems to represent the upper limit of the U content in acidic volcanic rocks.

Comparative radioactivity of magmatic rocks. The radioactivity of magmatic (plutonic) rocks is mainly connected with the presence of accessory uranium- and thorium-bearing minerals; it is partly due to uranium- and thorium-forming minerals and in a minor degree common rock-forming minerals.

The content of radioactive elements and radioactivity of intrusive rocks of a calc-alkali series and effusive rocks increases as we pass from ultrabasic to basic, intermediate and to acidic rocks (see Table 17, Fig. 138) in proportion to the increase in them of the silica and potassium content. In the case of separate rock masses of magmatic rock associations a direct relationship between the contents of U and Th, U and K, Th and K, U and SiO_2 shows itself more distinctly. K and SiO_2 as well as U, Th are concentrated in end products of magmatic differentiation which leads to the above correlations and increased gamma-activity with increasing the K and SiO_2 content in magmatic rocks. All magmatic rocks contain, on the average, microquantities of radioactive elements. An increased concentration (greater clarkes) is only common for acidic rocks (granites, liparites).

Particularly great is the difference in the radioactivity of granites having a high Ca content (plagiogranites) where U concentrations are $U \leqslant (1.5\text{-}3.1) \times 10^{-4}\%$ and Th concentrations are $Th \leqslant (5\text{-}15) \times 10^{-4}\%$ and a low Th content but a high K

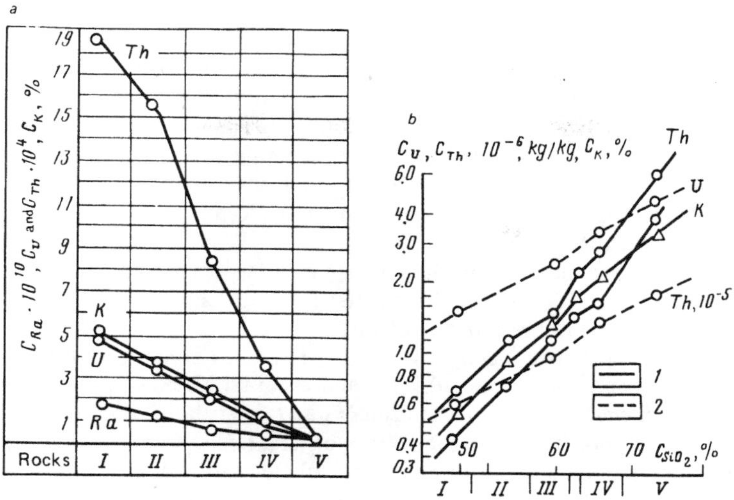

FIG. 138. Th, K, U and Ra contents in erupted rocks of the earth's crust (*a*) (after G.S. Smirnov, F.M. Kornilov and N.D. Tikhomirov) and accumulation of radioactive elements with increasing SiO_2 contents in calc-alkali volcanites of orogenic complexes (*b*).

a—rocks: *I*—alkaline; *II*—acidic; *III*—intermediate; *IV*—basic; *V*—ultrabasic; *b*: *1*—Kamchatka; *2*—Paleozoic folded regions (after A.A. Smyslov, Yu.M. Puzankov, V.A. Bobrov, A.D. Duchkov) *I*—basalts; *II*—andesito-basalts; *III*—andesites; *IV*—dacites; *V*—liparites

content. The latter species demonstrate typical U concentrations U $\geqslant$ (4-10) $\times$ 10^{-4}% and those of Th $\geqslant$ (20-50) $\times$ 10^{-4}%.

Magmatic intrusive and effusive rocks having an increased alkalinity show a higher radioactivity than do close to acidity rocks of the calc-alkaline series (see Table 17). This is characteristic of both ultrabasic alkaline rocks (meimechites, melteigites etc.) and acidic rocks (trachites, trachyliparites etc.).

Some areas of magmatic rocks have radioactive element content higher than their clarkes. These regions represent magmatic U and Th deposits which are not generally mined because of low contents of radioactive elements. Of these, the most interesting is presented by ones that are associated with magmatic formations of more recent origin.

A regular increase of the average radioactivity from rocks of ancient to more recent rocks seems to be observed for each petrographic series of magmatic rocks (see Fig. 140). The lack of a rigorous dependence is accounted for by the variability of accessory mineral contents.

Maximum radioactive element concentrations are confined to marginal portions of major intrusive bodies. Their distribution in porphyraceous dykes and small stocks seems to be uniform.

The radioactivity of effusive and endocontact rocks is greater than that of their intrusive analogs. The radioactivity of effusive rocks, q_r, is correlated with their crystallization water content w_{H_2O}.

Sec. 63. **Pegmatites**

Pegmatites contain more radioactive minerals than do magmatic rocks. What is more, the number of various uranium and thorium minerals increases appreciably. This is due to the fact that during the formation of pegmatites a basic magma contains greater amounts of radioactive elements accumulated as a result of previous magmatic differentiation stages than in a magma giving rise to acidic rocks. At the same time accessory minerals generally contain less U and Th than do identical minerals of granitoids.

A separation of uranium and thorium is observed in pegmatites. Uranium-forming volatile compounds goes to hydrothermal solutions, and most of thorium precipitates with a fraction enriched by silicon.

Of uranium and thorium minerals, unaltered pegmatites contain minerals of the anhydrous oxide group (except pitchblende and uranium blacks), uranium and thorium titanates, rare earth and thorium titanates and most thoro- and uranosilicates. Uranium- and thorium-bearing minerals of unaltered pegmatites include tantalo-niobates, niobo-tantalates, titano-tantalo-niobates, zirconium silicates, rare earth silicates, titanium silicates, rare earth silicates and such phosphates as apatite and loparite. Of organic compounds, thucholite and carburan occur here.

In addition, uranium and thorium silicates, uranates, simple carbonates, arsenates, phosphates, thoro- and uranosilicates occur in the oxidation zone of pegmatites. The most important minerals of the most common granite pegmatites are uraninite, thorite, samarskite, fergusonite, priorite, columbite, tantalite, pyrochlore, euxenite, zircon, monazite, xenotime, orthite etc.

Minerals that occur most generally in syenite and nepheline-syenite pegmatites are thorianite, uranothorianite, thorite, uranothorite, aeschynite, priorite, pyrochlore, samarskite, zircon, monazite, allanite, loparite. Uranium and thorium minerals and U- and Th-bearing minerals are found in pegmatites as phenocrysts that differ in size, as pockets, block masses and minor veins.

It has been established that even pegmatites may sometimes include large crystals of radioactive minerals the average radioactive element content of these rocks is low and irregular. At this, it presents quite a problem to estimate this content due to the heterogeneity and large grains of pegmatites.

Sec. 64. **Metamorphic Rocks**

Among metamorphic rocks, we refer to weakly radioactive (amphibolites, amphibolite shales, apodiabases, quartzites, marbles, calciphyres) and ones having a normal or slightly increased radioactivity (felsic gneisses, crystalline schists, porphyroids, metamorphosed sandstones) (Table 18). The different radioactivity of these varieties is accounted for by the difference in their original chemical composition. The rocks of the second group have an increased content of SiO_2, K_2O, CO_2 and H_2O. The radioactive element concentration of felsic gneisses and crystalline schists is also governed by the degree of their metamorphism—structure of mineral combinations. The less the average U and Th concentration in the rock shields and masses the higher is the degree of matamorphism of gneisses and crystalline schists (Fig. 139).

TABLE 18. Average Concentration of Radioactive Elements (in %) in Metamorphic and Ultrametamorphic Rocks

Facies	Rock	K	U × 10^4	Th × 10^4	Th/U
Granulite and eclogite	Eclogite	0.8	0.2	0.4	2.0
	Amphibolite	0.6	0.7	1.8	2.6
	Apodiabase porphyrite	0.8	0.9	2.7	3.0
	Magnesial-silicate schist	0.8	1.2	4.0	3.3
	Hypersthenic gneiss	1.6	0.6	2.1	3.5
	Plagioclase migmatite, charnockite	1.7	0.8	2.3	2.9
	Kordierite silimanite, gneiss and slate	2.6	1.3	4.2	3.2
Amphibolite and epidote-amphibolite	Amphibolite	0.8	1.0	4.0	4.0
	Marble, marmorized limestone	0.2	1.1	2.2	2.0
	Quartzite	0.6	0.8	3.1	3.9
	Metamorphosed sandstone	2.5	2.7	9.2	3.4
	Crystalline schist	3.1	2.6	10.0	3.8
	Feldspar gneiss	3.4	3.5	15.0	4.0
	Migmatite, granite gneiss	3.6	3.2	16.1	5.0

The effect of metamorphism on the radioactive element concentration in metamorphosed rock masses from an epidote-amphibolite to a granulite facies can be followed by referring to a decrease in U and Th concentration.

U and Th concentration of rocks, that formed as a result of metamorphism

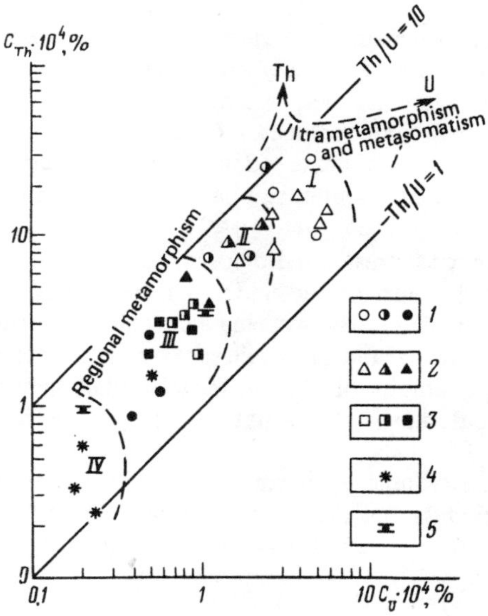

FIG. 139. U and Th contents in metamorphic rocks.

Metamorphism facies: *I*—greenstone-slate and epidote-amphibolite; *II*—amphibolite; *III*—granulite; *IV*—eclogite; rocks: *1*—gneisses; *2*—crystalline schists; *3*—amphibolites; *4*—eclogites; *5*—metamorphosed carbonates of volcanite of intermediate composition is invariably low irrespective of the metamorphism facies.

Thus, the U and Th concentration in rocks resulting from regional dynamothermal and contact metamorphism is nonuniform solely for varieties that formed under conditions of an amphibolite, epidote-amphibolite and greenstone slate facies. In the case of much more metamorphosed rocks (granulite facies) U and Th concentrations are almost levelled for all rock types. Ultrametamorphism and metasomatosis facilitate an increase in U and Th concentrations in their products, and the Th/U ratio of the latter may be both very small (<1) and very great (>10-20).

When considering the radioactivity of metamorphic varieties we must single out pneumatolytic and hydrothermal vein rocks. The bulk of uranium and a fraction of thorium migrate from the residual magma before pegmatite formation and are deposited in pneumatolytic and hydrothermal veins. Quite a number of types and varieties of uranium and thorium minerals as well as U- and Th-bearing minerals are confined to the latter. Of uranium minerals, pitchblende is most typical for rocks of low-temperature hydrothermal veins.

U- and Th-enriched quartz veins contain monazite or davidite (or pitchblende) and brannerite; in the latter case autunite and torbernite occur in the oxidation zone of veins.

In unaltered quartz and carbonate veins containing nickel and cobalt arsenides, native Ag and Bi—pentaelement formation, U accumulations occur as a rule as pitchblende. In oxidation zones of these formations uranates (curite), sulphates (zippeite and johannite), sulphate-carbonates (schrökengerite), carbonates (uranothalite), arsenates (trögerite, zeunerite and uranospinite), phosphates (torbernite, autunite and phosphuranilite) and silicates (uranophane, kasolite) as well as radioactive barite have been found. Thorite and thorogummite have been found in quartz-carbonate-thorite veins.

In unaltered chalcopyrite-pitchblende, polymetallic, gold ore and sulphide veins radioactive minerals are represented by nasturan (generally known as pitchblende) occurring here as collomorphic segregations and, less often, as separate veins. Minerals of oxidation zones of these veins are sulphates (zippeite), sulphate-carbonates (schrökengerite), silicates (uranophane, β-uranophane, cuprosklodowskite and kasolite). Uranium black and uranium sulphates are found in the cementation zone.

Nasturan has also been found to occur as streaks and dispersed impregnations in fluorite veins as well. A group of such veins formed under definite geological conditions having an increased U content is singled out as being an uranium fluorite formation. In fragmentation fissures of effusive and tufogenic rocks nasturan veins also occur, i.e. uranium formations. Uranium blacks, hydratized pitchblende, hydrosilicates have been found in hydroalumosilicate formations of hydrothermal origin.

U and Th concentrations in radioactive ores, in rocks of wallrock alteration zones and dispersion pattern rocks is much higher than in conventional rocks. The U content attains $4\,000 \times 10^{-4}\%$, and Th content $10\,000 \times 10^{-4}\%$, and Th/U ratio varies from 0.1 to 20 in ore bodies.

Sec. 65. **Sedimentary Rocks**

The radioactivity of sedimentary rocks is due to the presence in their composition of potassium, only uranium and thorium minerals, and only U- and Th-bearing minerals as well as adsorbed radioactive elements. It varies over a wide range.

Clastic Rocks

Placers. These have been found to contain uranium and thorium minerals, such as thorianite, brannerite, thorite, uranothorite etc., and such U- and Th-bearing minerals as euxenite, zircons, orthite, xenotime, monazite etc. These minerals are all primary and occur in placers as crystals and their fragments and as slightly rounded grains.

Due to the fact that during the disintegration of rocks enriched by radioactive minerals (for example, granites and their pegmatites) uranium is easily oxidized to form with the anions of acids soluble compounds and migrate in aqueous solutions, placers mainly contain thorium minerals. Owing to their high density, these latter are deposited and concentrated shortly after being ablated from the disintegration zone of the parental rocks forming together with other fragments economically important alluvial and sea-beach placers.

According to data available, the Th content in rocks of these deposits varies from 49×10^{-6} to $3\,500 \times 10^{-6}$ kg/kg, and U content from 12.7×10^{-6} to 119×10^{-6} kg/kg.

Clays and clay shales. Among sedimentary rocks (except potassium salts, some pyroschists and quartz conglomerates) clays demonstrate on the average the highest U, Th and K content and radioactivity (Table 19).

The relatively high radioactivity of clays and clay shales stems from the increased sorption of U, Ra, Th and K on clay particles, the presence of K, Th and hexavalent U minerals, formation under reductional conditions of tetravalent U minerals, origination of solid solutions (some radioactive minerals may combine with clay minerals to form solid solutions).

Appreciable adsorption of U ions is likely to occur from natural waters where U is present as: (1) readily soluble carbonate compounds , e.g. $Na_2UO_2(CO_3)_3$ dissociating into Na^+ ions and $[UO_2(CO_3)_3]^{2-}$; (2) urano-organic compounds; (3) $[UO_2(OH)_2]_n$ hydroxide colloidal solution or other colloids; (4) sulphate-uranyl UO_2SO_4—the least stable under natural conditions and others.

Sorbents fail to extract uranium from bottom water. Sorption starts at the stage of diagenesis when chemical reactions appreciably changing the composition and structure of the sediment proceed in silty waters rich in carbon dioxide. The saturation of silty waters by biogenic carbon dioxide causes uranyl and carbonate anions to be disintegrated. The disintegration of complexes resulted in an increased U cation concentration followed by their absorption by several sorbents concurrently. The sorbing minerals are clay minerals (montmorillonite, halloysite, kaolinite etc.), disperse hydromica and chlorite aggregates, altered biotites, sericite, glaukonite. The concentration of uniformly dispersed U in an alumosilicate material often may attain 2-3%.

TABLE 19. Concentration of U, Th and K (in %) in Sedimentary Deposits of the Continental Part of the Earth's Crust (Average for Continents)

Rock group	Rock	K	U × 10^4	Th × 10^4	Th/U
Terrigenic (sandy-argillaceous)	Conglomerate, gritstone	1.5	2.4	9.0	3.7
	Quartz conglomerate	1.2	6.3	31.0	5-10
	Sandstone, aleurolite	1.7	2.9	10.4	3.6
	Argillite, clay shale, clay	2.7	4.0	11.5	2.9
Siliceous	Flinty slate, quartzite	0.3	1.7	2.2	1.2
	Clay-siliceous shale	1.1	2.8	6.2	2.2
Carbonate	Limestone	0.3	1.6	1.8	1.1
	Marl	0.8	2.8	2.5	0.9
	Dolomite	0.4	3.7	2.8	0.8
	Bituminous limestone	0.3	7.8	11.9	1.5
Salt	Gypsum, anhydrite	0.02	1.0	1.0	1.0
	Rock salt	*n** -10*n*	0.9	1.0	1.1
Caustobiolith	Fossil coal	0.1	3.4	4.8	1.4
	Peat**	—	2.0-5.0	5.2	1.8
	Pyroschist	—	100-200	10-15	0.5

* *n*—several units.
** Concentration given as average for platform regions.

One of the causes of the accumulation of U in laminated alumosilicates and cryptograin pelites (having a cryptocrystalline structure) is the origination in them of a dispersed Ti hydroxide that behaves as it were an active sorbent extracting uranium from pore and sea waters.

Rocks containing uraniferous titanium hydroxide and laminated alumosilicates readily release U to the carbonate-bearing liquid phase. Uranium minerals may precipitate from U-bearing solutions that are thus formed. Cases may occur where, against the background of uniformly dispersed U concentration, the sparse grains of uranium minerals, explicitly secondary with respect to sorptional U accumulation, appear in clay cement minerals.

In this case adsorbing minerals are clay minerals as well as colloidal iron, aluminium, silicon, and magnesium oxides and organic substances. Adsorption occurs until definite, relatively small contents of radioactive elements in co-precipitating clay particles, colloidal oxides and organic materials.

The co-precipitation of uranium, uraniferous clay minerals and other compounds is particularly intensive if they occur in the immediate vicinity of decaying uranium deposits in oxidation and cementation zones of the primary uranium

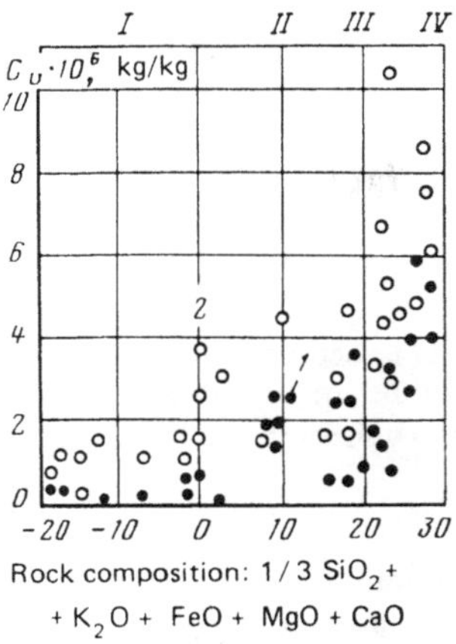

FIG. 140. U content and α-activity of various batholith rocks of South California (after E. Larsen and G. Fair).

1—U content; *2*—α-activity expressed in U equivalents; *I*—gabbro; *II*—quartz diorite; *III*—granodiorite; *IV*—granite

deposits. It is more common, however, that clay minerals and other compounds adsorb uranium released during the disintegration of various rocks.

Adsorption depends on the pH of natural waters: it is more appreciable, the less free oxygen and CO_3^{2-} ion the solution contains, the deeper and quieter the waters and the more dispersed the precipitating particles.

The precipitation of U compounds is most intensively in the reducing medium in the presence of organic substances (humic acids and carbonaceous residues) and sulphides. The organic matter adsorbs uranium from the environment sometimes reducing it to tetravalent compounds unsoluble in water.

The enhanced radioactivity of clays and clay rocks is also accounted for by a relatively high content in these rocks of K (up to 6.5%) which is found here not only in a mineral (microcline $KAlSi_3O_8$, orthoclase $KAlSi_3O_8$, anorthoclase (Na, K)$AlSi_3O_8$, nepheline (Na, K) × $AlSiO_4$, muscovite $KAl_2(AlSi_3O_{10}) \times (OH)_2$, phlogopite $KMg_3(Si_3AlO_{10}) \times (Fe, OH)_2$, biotite $K(Mg, Fe)_3 \times (Si_3AlO_{10})$ (OH, F_2), sericite (small flakes of muscovite), lepidolite $KLi_2Al(Si_4O_{10})(FeOH)_2$, illite $2K_2OMgO8Al_2O_312H_2O$, glaukonite $K_{<1}$ (Fe^{3+}, Fe^{2+}, Al, $Mg)_{2\text{-}3}Si_3$(Si, Al) × $O_{10}(OH)_2^n H_2O$ and others) but also in a sorbed form. Clays may contain also Ra-bearing minerals; the Ra content in these minerals is negligible (not more than 10^{-7} kg/kg). The content of radioactive elements in deep clay sediments may be as high as 50×10^{-12} g-eqv Ra/g and more (Fig. 140).

Continental and shoaly clay sediments are less radioactive (from 3×10^{-12} to 30×10^{-12} g-eqv Ra/g).

Sandstones and aleurolites. Sandstones may contain these primary minerals: pitchblende (streaks, lenses and bodies of other shapes), thorite, mackintoshite and other silicates. Uranium black, schrökengerite, sharpite, voglite and other carbonates, autunite and other phosphates, carnotite, tyuyamunite and other vanadates, soddyite and other silicates may sometimes be found in oxidation zones of sandstones and aleurolites. Secondary minerals appear as a result of pitchblende oxidation in rocks having a definite mineral complex.

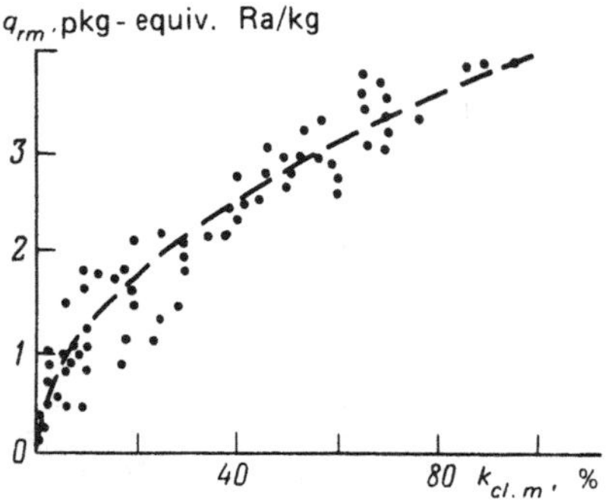

FIG. 141. The dependence of gamma-activity q_{rm} on the coefficient of mass clayiness $k_{cl.m}$ for sandy-clayey rocks

Pitchblende, uranium blacks, carnotite and tyuyamunite occur in considerable and sometimes commercial accumulations. Other minerals occur in negligible amounts and serve as pathfinder materials owing to the vivid coloration of secondary minerals.

Most commonly radioactive elements are contained in sandstones and aleurolites in the form of isomorphic impurity to heavy-fraction minerals. Radioactive elements are found in the clayey portions of these rocks and in an adsorbed form.

Sandstones and aleurolites contain on the average microquantities of U, Th and a little K (see Table 19). An increased radioactivity is displayed by their K-bearing polymictic varieties. A minimum content of radioactive elements is typically shown by well sorted pelagic, mainly quartz sandstones. An increased content of radioactive elements has been found for argillaceous sandstone varieties as well as varieties having organic impurities. The increase of the total (specific mass) radioactivity q_{rm} conditioned by the gamma radiation of all radioactive elements of the rock for sandy-argillaceous varieties with the growth of their clayiness $k_{cl.m}$ is shown in Fig. 141.

However, according to R. P. Gottich, for the Amu-Darya River basin intimate direct relationships are solely found between the content in sandy and argillaceous rocks of K and Th and their mass clayiness $k_{cl.m}$ whereas U concentration with increasing $k_{cl.m}$ remains practically unchanged.

The close intensive relationships have also been found by P. A. Kurochkin for terrigenous rocks between the total specific volumetric gamma activity q_rV and their volumetric exchange capacity Q, between q_rV and specific surface S_V, between q_{rV} and residual water saturation $k_{w.r}$ as well as between the specific volumetric thorium activity* $q_{rV\,Th}$ and the same quantities.

Often the clayiness also governs the porosity of sandstones, therefore inverse relationships are observed between their voids ratio k_r and total radioactivity q_{rV} (Fig. 142*b*). They have also been established between $q_{rV\,Th}$ and k_r (see Fig. 142*a*).

* Radioactivity conditioned by the gamma radiation of Th contained in a rock.

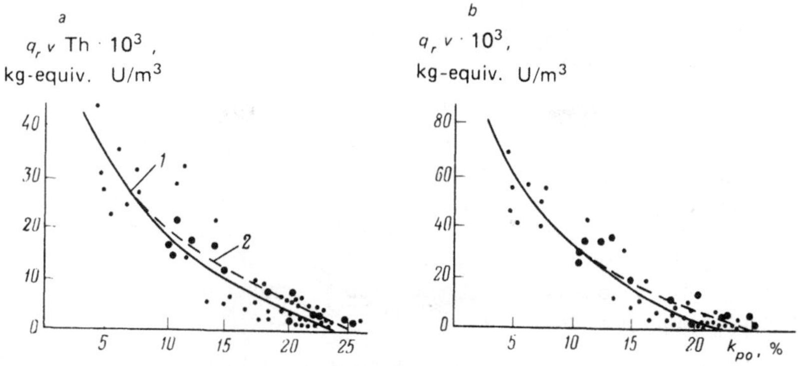

FIG. 142. Connections of Th concentration q_{rVTh} (*a*) and total concentration q_{rV} (*b*) on the open voids ratio k_{po} (after P.A. Kurochkin).

Deposits: *1*—Baklanovskoe; *2*—Savinsko-Moskudyinskoe

Inverse and sufficiently close relationships have been found between the total gamma activity q_rV and average radius of pore channels $r_{ch\ av}$ and permeability coefficient k_{perm} and not less close inverse connections: $q_{rV\ Th} = f(r_{ch\ av})$ and $q_{rV\ Th} = f(k_{perm})$.

Organogenic and Chemogenic Rocks

Carbonates. Pure sea limestones and dolomites are, as a rule, weakly radioactive. Their U content does not exceed 4×10^{-6} kg/kg. This radioactivity of organogenic limestones is governed by the oxidizing conditions of the medium where these sediments are formed (shells and other organisms inhibit shallow basins containing oxygen). The oxidizing environment does not facilitate uranium deposition.

The origination of chemogenic limestones and dolomites from sea and fresh waters is also possible under conditions that are not suitable for uranium adsorption and presipitation (weakly saturated by carbon dioxide, they contain carbonate ion-bearing turbulent rocks with a variable temperature). Uranium is not precipitated with magnesites, siderites, tuffs and other carbonates that originate in identical conditions. The clayiness enhances the radioactivity of carbonate rocks. So is the radioactivity of carbonate rocks related with their insoluble residue $C_{in.r}$ since the latter mainly contains a clay fraction.

Carbonate rocks contain microquantities of radioactive elements. The average radioactivity of limestones and dolomites as obtained by different workers (see Table 19) is very low and is less than that of clays.

In definite vertical sections of oil fields dolomitized limestones showing an increased radioactivity have been found. Appreciable and even economically important radioactivity is found in vanadium-rich sea limestones. Uranium is present here as uranium blacks and uranium-hearing organic substance. Carnotite, tyuyamunite occur in the oxidation zone.

Marls. The radioactivity of marls is on the average (see Table 19) higher than that of pure limestones and in some cases approaches the radioactivity of clays.

Potassium sediments. Potassium minerals—sylvine KCl, carnallite $KClMgCl_26H_2O$, kainite $KClMgSO_43H_2O$, polyhalite $K_2SO_4 \times MgSO_42CaSO_42H_2O$ and others where the K content varies from 12.9 (for polyhalite) to 52.5% (for sylvine) display a high radioactivity. This is what is responsible for a comparatively high radioactivity of potassium sediments. The U content in potassium salts is rather inappreciable (see what follows).

Halite, anhydrite, gypsum. Under conditions that are unfavourable for adsorption and precipitation these rocks much as carbonate sediments, form in an oxidizing environment. Uranium is concentrated here in residual brines. As to sediments formed from the latter, they are most commonly subject to weathering thus enriching the adjacent clastic rocks and are then leached from these. Gypsum, anhydrite, halite and other salts show an average U content 1×10^{-6} kg/kg U (see Table 19).

Phosphates. Phosphates are sometimes very much enriched by uranium. This is ascribed to their formation in the reducing medium, given appreciable amounts of organic material.

The U cations in carbonate fluopatite replace Ca ions by participating in the construction of the crystal lattice; alpha-emitters are uniformly distributed here. Ca phosphates extract uranium even from sea water, yet uranium minerals cannot originate since U and Ra are not practically extracted from natural phosphates by carbonate solutions. Ca phosphate, by strongly binding uranium and its decay products, limits the migration of radioactive elements. The U content in phosphates varies over the range $(10\text{-}1\,500) \times 10^{-6}$ kg/kg.

Fossil coals. The radioactivity of coals is, as a rule, inappreciable (see Table 19). However, highly radioactive varieties are encountered among rocks of this class. Uranium minerals of pitchblende and blacks and uranium combined with an organic substance convert to schrökengerite, carnotite, tyuyamunite and hydrous Zi and Ca carbonates in the oxidation zone. The most active U concentrator proved to be humic organic matter, and to a lesser degree sapropel material.

The radioactivity, q_{rm}, of coals and carboniferous rocks with a different degree of metamorphism grows with increasing the ash content. Natural bitumens concentrate negligible amounts of U.

Different radioactivity values are demonstrated by dry sedimentary rocks. A maximum average U and Th concentration and Th/U ratio has been found for clastic rocks, such as conglomerates, sandstones, clay shales (see Table 19). Smaller values of average concentration and Th/U ratio are attained by siliceous rocks (flinty slates, quartzites, clay-siliceous slates). Even smaller values of average concentrations and Th/U ratio are shown by such carbonate rocks as marls, limestones, bituminous limestones. The least U and Th concentrations and a relatively small Th/U ratio are characteristic of saliferous rocks. Dolomites and caustobioliths (coal, peat, pyroschists) are characterized by a comparatively high U content and a small Th content which is particularly true of the former.

A great amount of Th and U with Th is contained in rudaceous rocks (conglomerates, gritstones, sandstones) of eluvial, sea beach and other rock associations. By contrast, limestones, slates, sandstones that form in sea basins are mainly uraniferous and contain phosphate and organic matter.

The U and Th content, and, consequently, radioactivity of clastic rocks grows from conglomerates to sandstones, aleurolites and argillites.

Classification of Sedimentary Rocks According to Their Radioactivity

Rocks that have a low radioactivity include: well sorted and poorly cemented monomineral quartz sands, aleurolites, pure limestones, dolomites, rock salt, anhydrites, gypsums, most coals (humic coaly sediments) and oil-saturated rocks.

Rocks that have an increased radioactivity include: argillaceous varieties of sedimentary rocks, argillaceous sands, sandstones, aleurolites, some marl varieties, marlstones and dolomites and rocks with organic impurities.

Rocks that have a high radioactivity include: potassium salts, monazite and orthite sands, deep clays, globigerine muds and red clay.

Marine sediments have a higher radioactivity compared with lagoon and fluvial ones. A particularly high radioactivity has been established for bottom sediments of the Pacific and Atlantic coasts of the USA deposited in quiet waters. The Ra concentration in them is about 3 times that of metamorphic and sedimentary rocks of continents.

Sec. 66. The Liquid and Gaseous Phase

The liquid phase. U concentration in sea waters and atmospheric precipitation does nor exceed 10^{-6} kg/m^3 (g/l). The U content in the subsurface hydrosphere can be higher.

Maximum U concentrations (up to 2 kg/m^3) are found in acidic downpours from uranium ores enriched by sulphides. In semineutral waters from uranium ores U concentration does not exceed 10^{-2} kg/m^3 (g/l) and outside deposits and ore showings it is within 10^{-4} kg/m^3 (g/l) (generally not more than 10^{-5} kg/m^3). In the zone of intensive water exchange with the surface the U content in the solution is in direct proportion to its concentration in the solid portion of rocks and total mineralization of water.

The greatest effect of the formation of the chemical composition of subsurface waters and their uranium-bearing capacity is produced by the amount of atmospheric precipitation. Arid regions show the highest background U concentrations in ground water. As to barren rocks, weakly mineralized stratal waters are rich in U and poor in Ra. The U concentration in the solution dramatically drops to its values in salt water and brines and Ra concentration grows with attaining the total mineralization of stratal waters tenth fractions of kilogram-equivalents

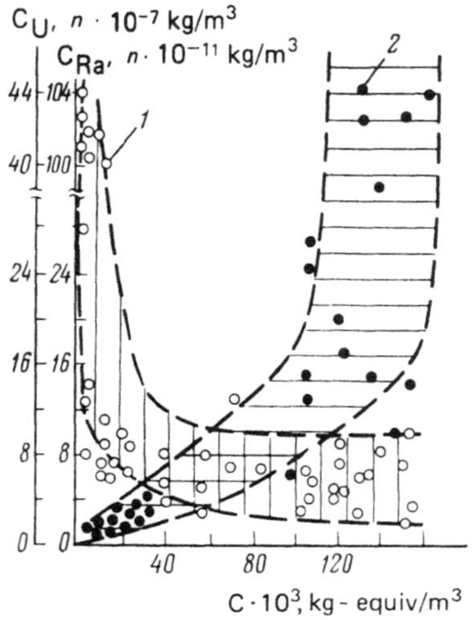

FIG. 143. Comparison of U and Ra content with mineralization C of subsurface waters of the Amu-Darya catchment area (after R.P. Gottich).

1—Ra; *2*—U

per 1 m³ (Fig. 143). On the whole water-soluble salts of ground water are rich in uranium compared with the mineral residue of stratal waters. Compared with an averaged U content in rocks the dry residue of groundwater is on the average richer in uranium, and poorer in stratal waters.

Close to the ground surface the U content in waters varies from 10^{-7} to 10^{-4} kg/m³ and more. The background U content (in barren rocks) attaining 10^{-5} kg/m³ has been found in waters to a depth of hundreds of metres. At depths measured in terms of kilometres the U content nowhere exceeds 10^{-6} kg/m³. With depth the maximum U content in waters tends to decrease (Fig. 144*b*). Internal waters generally contain no uranium altogether.

The removal of uranium from a solution in natural conditions may be due to: (1) the formation of hardly soluble compounds in the presence of V, P, As and Si, their anion forms, together with the cation uranyl constituent, give rise to these compounds; (2) sorption of the uranyl ion by Fe, Al, Ti hydroxides, organic substance of the coal and oil series, certain clay minerals and other natural sorbents.

High U concentrations are typically characteristic of waters with positive values of the Eh redox potential. The decrease of maximum U concentration with decreasing the Eh seems to be due to a decrease in the solution of its equilibrium concentration upon passing to conditions favourable for U to be found in a stable low-valency condition. On the whole oxygen waters show the increased U content, and hydrosulphuric waters, by contrast, are characterized by a very low U content. Uranium, however, may be present in waters without O_2 or H_2S either in amounts up to 10^{-4} kg/m³ or be absent altogether.

In any case, there seems to exist the substance capable of producing the required drop in Eh. The smaller the Eh values for U precipitation from stratal waters are

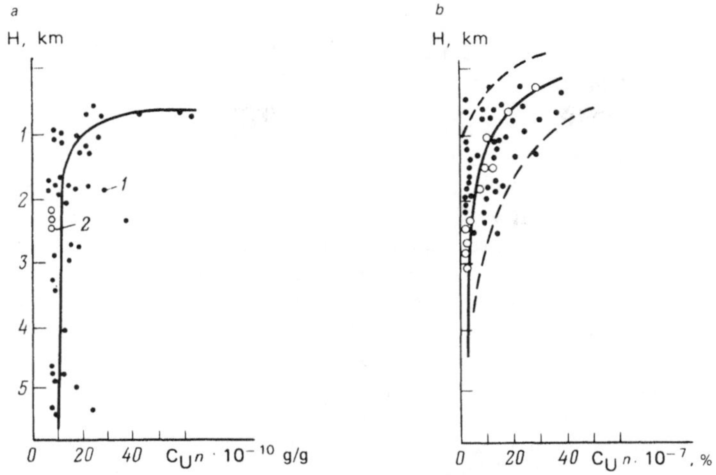

FIG. 144. Variation of U content in oils (*a*) and confined waters (*b*) with depth *H* (after R.P. Gottich).

Data: *1*—actual; *2*—averaged

needed the stronger the complex U ion forms are conditioned by the chemical composition of stratal waters and the less the U content of the waters.

The U content in numerous oil samples of various types does not exceed $n(10^{-7}\text{-}10^{-6})\%$. In a number of cases only certain heavy oils of a tarry and asphaltene composition have the increased quantities of uranium.

According to data obtained by measuring U concentration in liquid oil, its content is within $6 \times 10^{-9}\text{-}(0.9\text{-}1.2) \times 10^{-7}\%$. Note that U concentrators in oils are provided by asphaltenes and alcohol-benzene tars whose content increases with oil oxidation degree. The latter process occurs during seepage of waters containing oxidizing reagents, for example, surface water infiltration. That is why the U concentration in oil first decreased with depth to become then nearly constant. It has also been found that the U content in oil increases with its sulphur content, density and tarriness. A variation in U contents in oils with depth *H* in a vertical section is shown in Fig. 144*a*.

The gaseous phase. Radioactive elements come to the soil air as gaseous emanations of radon, thoron, actinon from rocks and, in part, natural waters. The radioactive element content in the soil air fluctuates both in space and in time. The radioactivity of the soil air containing radon is on the average 5.4×10^{-18} dec./(s·m^3).

Thus the contribution of the liquid and gaseous phase to the radioactivity of rocks is comparatively small.

Sec. 67. The Spatial Distribution and Migration of Uranium, Thorium and Potassium in the Earth Crust

It is necessary to study the spatial distribution and migration of uranium, thorium, potassium in the earth crust as well as the regional variation of total radioactivity within separate portions of it in order to solve such problems as geochemical and petrophysical zoning, identification of zones of distribution of radioactive ores, correlation of sedimentary rock masses, magmatic and metamorphic rock associations, determination of the contribution of radioactive heat to the total heat balance of the earth. By mapping and plotting vertical sections within separate regions of the USSR definite regularities have been established in regional distribution of uranium, thorium, potassium and total radioactivity of rocks. These are mainly continental areas or areas within the continent of Eurasia-Paleozoic folded systems.

The average content and distribution of radioactive minerals in rocks of South Kamchatka can be inferred from a schematic map showing average values and dispersions of distribution of the total radioactivity of rocks compiled by Yu. M. Puzankov, V. A. Bobrov and A. D. Duchkov. According to this map, the rocks having a relatively increased average radioactivity are found within the central part of the Sredinnyi massif where rocks are the most diversified. Maximum dispersion values refer to the southern and northern boundaries of the development region of deeply metamorphosed sediments of the Kolpakovskii complex. Minor values of average radioactivity and dispersion have been established for the remaining area of South Kamchatka excepting the regions of distribution of Neogene volcanites adjacent to the outcrops of metamorphosed rock associations of the foundation. Jointly with rocks of the Sredinnyi massif, volcanites form an extended band of increased radioactivity values coincident in general terms with the southern section of the Central Kamchatka geological and structural zone.

The distribution of K, U and Th in vertical sections and areas of sedimentary deposits of the oil and gas fields of the USSR have been studied in detail guided by F. A. Alekseev, R. P. Gottikh and others. These studies have been conducted due to a need to substantiate a method of gamma spectroscopy of bore holes for a paleographic reconstruction of the formation of paleobasins of sedimentation, correlation of complex cross sections of boreholes, the estimation of the clayiness of oil and gas traps.

Studies were made of regularities of the distribution of K, U, Th over the area of the Amu-Darya River drainage basin in sediments of the Carboniferous, Baltic sea basin in Ordovician and Silurian sediments, northwest fringe of the Caspian sea basin in sediments of the Vorobyev and Pasha age. Figure 145 presents a diagram of the distribution of K in the Neocomian (Barremian) over the area of the Amu Darya River basin. Let us consider general regularities in the distribution of K, U and Th within the limits of sediments of different age of the ancient-origin sedimentation basin.

Potassium. The distribution of K over the area of the sedimentation is conditioned by the number of ablation sources and composition of their rocks, specific

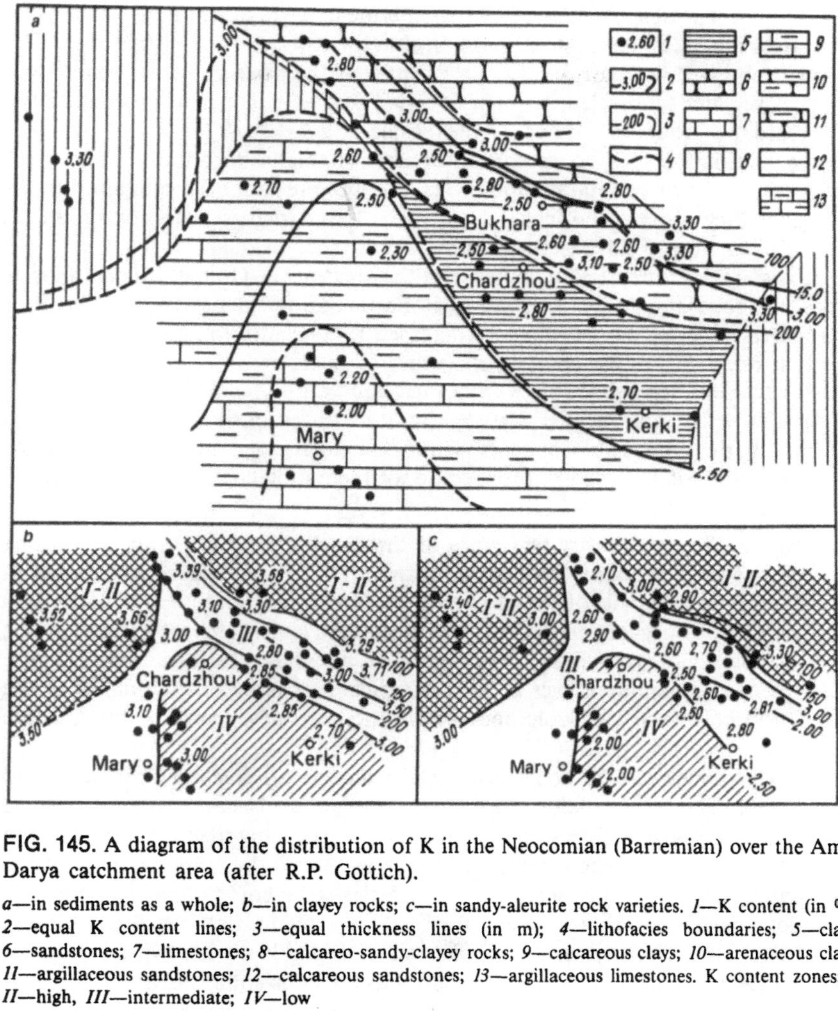

FIG. 145. A diagram of the distribution of K in the Neocomian (Barremian) over the Amu-Darya catchment area (after R.P. Gottich).

a—in sediments as a whole; *b*—in clayey rocks; *c*—in sandy-aleurite rock varieties. *1*—K content (in %); *2*—equal K content lines; *3*—equal thickness lines (in m); *4*—lithofacies boundaries; *5*—clays; *6*—sandstones; *7*—limestones; *8*—calcareo-sandy-clayey rocks; *9*—calcareous clays; *10*—arenaceous clays; *11*—argillaceous sandstones; *12*—calcareous sandstones; *13*—argillaceous limestones. K content zones; *I*, *II*—high, *III*—intermediate; *IV*—low

features of the distribution of clay minerals, governed by conditions of rock disintegration (climate, presence of vegetation, activity of tectonic movement) and hydrodynamic regime of the drainage basin. The distribution of K associated with a change of clay minerals is what governs the mineral composition of clays within the basin.

For Cretaceous sediments of the Amu-Darya River oil basin a drop in K concentration has been found, for example, from the marginal portions of the basin toward its centre (see Fig. 145) which is attributed to the presence of only one ablation source of detrital material and small degree of its disintegration. Several zones are identified showing different K contents: a littoral zone with an increased content, an abyssal zone with a low content and an intermediate zone separating them where K (feldspar and mica) concentration diminishes toward the centre of the basin. This zoning is observed both for the entire rock mass of the basin and for separate rock types (clayey and sandy-aleurite). The location and size of zones are conditioned by the sedimentation conditions of the basins.

The distribution of K concentration in clays of the basin can also attest to the distribution of their mineral composition. The more montmorillonite and the less hydromica they contain and, consequently, the lower is their K concentration the more distant are the clays from the ablation source.

Potassium is differently distributed over the areas of the Ordovician and Silurian deposits of the Baltic basin. This is due, in particular, to the presence of two ablation sources. Two regions showing an increased K concentration are observed here. The first one refers to the shallow, the other to relatively abyssal sediments, and two zones manifesting different K concentration can be identified in the area of the Silurian sediments. Peculiar features are to be traced in the K distribution over the area of the Paleozoic sediments within the northeast fringe of the Caspian depression which is governed by the specific sedimentation conditions and composition of clay minerals.

Uranium. We can identify two types of the distribution of U over the area of ancient-origin sedimentation basins. A maximum U concentration is observed in the first one, characteristic of basins showing a normal distribution and clarke content (0.3-0.5%) of organic matter for littoral shallow facies, and the minimum one for transition facies sediments. U concentration of abyssal facies increases again. This type of U distribution over an area is typical of Cretaceous sediments of the Amu-Darya basin and Paleozoic ones within the northwest fringe of the Caspian basin.

The second pattern of U concentration distribution observed in Ordovician and Silurian sediments of the Baltic basin shows the highest U concentration referred to comparatively abyssal facies that originate in the reducing medium and enriched by organic matter. The U concentration decreases toward the marginal parts of the basin and again increases in shallow littoral facies sediments failing to attain concentrations corresponding to abyssal facies. This U concentration distribution is typical both of sediments on the whole and of individual rock types.

Uranium is assumed to come to the basin as aqueous solutions. Its best adsorbents are organic matter, phosphorus, argillaceous minerals, Fe and Al hydroxides. A definite role in the distribution of uranium within the basin seems, however, to be played by the physicochemical conditions at different portions of the sea floor. The enrichment of sediments by U is associated with the sorption and formation of uranoorganic compounds and mineral formation. The course of all these processes is controlled by the physicochemical conditions in diagenesis, in particular, by the redox conditions. It must be noted that the highest U contents (up to $30 \times 10^{-4}\%$) are shown by sandy and aleurite rocks forming under the conditions of shallow littoral and shallow facies rather than for clays.

Given a normal organic matter content (0.3-0.5%), the U content increases in the series conglomerates → sandstones → aleurolites → argillites which is due to the increased concentration in them of adsorbing clay particles. Terrigenous sedimentary rocks of the oil and gas basins display an increased content of organic matter (up to 2.5-3.5%). With increasing the dispersity of abyssal sediments the content of organic matter, and, consequently, their uranium-bearing capacity increase.

In shallow facies, the concentration of organic matter is governed not only by the granulometric composition, but also by the accumulation of biomass in the sediments. The content of biomass in rocks of different granulometric composition varies appreciably and is governed by the rate of sedimentation, depth of the water reservoir, redox and other physicochemical conditions of sedimentation. Consequently, an important role in U distribution within sedimented deposits of ancient-origin basins is also played by the distribution of organic matter which gives an insight into the aforementioned regularities in the distribution of U and makes it possible to identify littoral, transition and relatively abyssal zones of sedimentary deposits of basins.

Thorium. Th-bearing minerals are accumulated in littoral zone rocks; the fraction that sorbs thorium is the clay fraction of rocks. If one ablation area is available, the enrichment by thorium of the emerging rock varieties occurs solely in the marginal portions of the basin. In this case the placers are likely to form.

As to rocks that are sufficiently remote from the ablation source, they show a fairly constant Th concentration. Given several distributive provinces that differ in the composition and transformation conditions of minerals, thorium accumulation proceeds nonuniformly. Data on the distribution of Th concentration over the areas of ancient-origin basins of sedimentation facilitate in locating the ablation sources of terrigenic material and shallow littoral zones.

Chapter XII

Neutron Activation

Neutrons are particles (their half-life is $T_{1/2} = 1.01 \times 10^3$ s) having a mass $1.0086654 \times 10^{-24}$ g decaying to form a proton, an electron and an antineutrino liberating energy (0.78 MeV). Since they have no charge and do not interact with the electrons of the shell and field of nuclei, neutrons easily penetrate the nuclei getting in touch with these latter. Neutrons fall into *fast neutrons* (having an energy $E = 2 \times 10^5$-2×10^7 eV), *intermediate* (0.5-2×10^5), *resonance* (100), *slow* (0.5), *thermal* (0.025) and *cold neutrons* (0.001). Neutrons having an energy from 0.3 to 0.5 to $n \times 10^2$ eV are often described as *suprathermal* ones. Energy groups of neutrons display a definite speed and other characteristics.

Neutron and gamma-neutron methods of borehole investigations use a source shifted along the borehole emitting a flux of multi- and/or single-energy neutrons through the water, often drilling column and cement to rocks. These sources may be provided, e.g. by polonium and beryl sources showing an energy spectrum from 0.1 to 13 MeV or neutron generators producing neutrons with an energy 14.1 MeV.

Sec. 68. The Interaction of Neutrons with Nuclei of Elements

Neutrons are scattered by and absorbed in a medium filling the bore hole, drilling column, cement and rocks.

Being scattered, neutrons lose energy, gradually passing to a thermal state having an energy of the order of 0.025 eV and speeds of propagation 2 200 m/s to be then captured by one of nuclei of chemical elements.

During scattering the trajectories of neutrons change as the latter collide with nuclei of elements of the medium and the kinetic energy of neutrons decreases. We distinguish elastic and inelastic scattering.

Elastic scattering. In this type of scattering the kinetic energy of the neutron-nucleus system remains constant both before and after the scattering act. However, if prior to scattering, in the laboratory frame of reference, the carrier of the kinetic energy of the system was provided by the neutron (the nucleus being considered stable), then following the scattering the kinetic energy is redistributed between the neutron and recoil nucleus according to their masses and angle of scattering.

The energy of the neutron, E_2, and its velocity, v_2, after its collision with a nucleus is found using these approximate equations:

$$E_2 = \frac{A_m^2 + 2A_m \cos\theta + 1}{(A_m + 1)^2} E_1 \tag{85}$$

$$v_2 = \sqrt{\frac{A_m^2 + 2A_m \cos\theta + 1}{(A_m + 1)^2}}\, v_1 \tag{86}$$

where E_1 and v_1 are, respectively, the energy and velocity of the neutron before the collision; A_m is the mass number of nuclei of the moderator on which neutron scattering occurs; θ is an angle between the original and subsequent trajectories of the neutron in the frame of reference of the centre of inertia.

The derivation of Eqs. (85) and (86) relies on the law of collision of elastic spheres.

By defining $(A_m - 1)/(A_m + 1) = \alpha$ we have

$$E_2/E_1 = [(1 + \alpha) + (1 - \alpha)\cos\theta]/2 \tag{87}$$

As can be seen from Eq. (87), the maximum ratio $E_2/E_1 = 1$ (minimum energy loss) is observed, given that $\theta = 0$ is the gliding collision when the energy of neutrons is equal before and after collision. The transfer of energy is maximally possible and the minimal value of E_2/E_1 takes place if $\theta = \pi$. In this case $E_2 = \alpha E_1$ and the maximum relative energy loss under conditions of single scattering is

$$\frac{E_1 - E_2}{E_1} = \frac{E_1 - \alpha E_1}{E_1} = 1 - \alpha$$

As a result, during neutron scattering on hydrogen for which $\alpha \approx 0$, the neutron can practically lose all its energy in a single collision act. The elastic scattering of fast neutrons is most likely to occur on element nuclei having a small relative atomic mass.

Inelastic scattering. Under conditions of inelastic scattering or a (n, n^*) reaction the nucleus that captured and then lost a neutron remains in an excited state. As it returns to the principal one, it emits a gamma quantum according to the schemes:

$$_{Z}X^{A_m} + {}_0n^1 \rightarrow {}_{Z}X^{A_m*} + {}_0\check{n}^1;$$

$$_{Z}X^{A_m*} \rightarrow {}_{Z}X^{A_m} + \gamma$$

where X^{A_m} and X^{A_m*} are, respectively, the nuclei of the original element X having the mass number A_m and atomic number Z and the nucleus of the same element in an excited state; $_0n^1$ and $_0\check{n}^1$ are, respectively, the neutrons absorbed and ejected by a nucleus with a lesser kinetic energy; γ is a gamma quantum.

The (n, n^*) reaction is most likely to occur in rocks containing heavy elements, given neutron energies from several keV to several MeV. Under conditions of inelastic scattering following several collisions, neutrons then scattered elastically.

As a result of an elastic and inelastic scattering the electrons lose energy and their velocity decreases.

Absorption. Absorption and sometimes reproduction of neutrons generally occurs during the (n, α), (n, p), (n, γ), $(n, 2n)$, (n, np) and other nuclear reactions.

At the first stage of nuclear reactions compound nuclei form composed by the original nucleus and a captured neutron. The kinetic energy contributed by the neutron is redistributed between the nucleons in these nuclei.

A compound nucleus lives during 10^{-12} to 10^{-14} s, until, as a result of a vast number of collisions between nucleons, energy sufficient to overcome nuclear forces does not concentrate at one or more of them following which the nucleus decays. The path of the decay is determined by the energy of the neutron and nucleus type—the neutron to proton ratio in the nucleus. The products of decay are one or more new elements or a nuclide of the element being irradiated, gamma quanta of a definite energy and one or more light particles (neutrons, protons, alpha particles).

The (n, α), (n, p) and other reactions that proceed accompanied by the ejection of two or more nucleons as well as fission reactions are most likely as rocks are being irradiated by high- and very high-energy neutrons (from 1 to 100 MeV and more).

The (n, α) and (n, p) reactions proceed according to the schemes:

$$_{Z}X^{A_m} + {}_0n^1 \rightarrow [_{Z}X^{A_m+1}]^* \rightarrow {}_{Z-2}X^{A_m-3} + {}_2He^4;$$

$$_{Z}X^{A_m} + {}_0n^1 \rightarrow [_{Z}X^{A_m+1}]^* \rightarrow {}_{Z-1}X^{A_m} + {}_1H^1$$

Examples of such reactions are:

$$_{14}Si^{28} + {}_0n^1 \rightarrow {}_{12}Al^{28} + {}_1H^1;$$

$$_{13}Al^{27} + {}_0n^1 \rightarrow {}_{11}Na^{24} + {}_2He^4;$$

$$_{17}Cl^{35} + {}_0n^1 \rightarrow {}_{15}P^{32} + {}_2He^4$$

The (n, p) and (n, α) reactions involving slow neutrons are observed solely for light elements (Li, B etc.):

$$_3Li^6 + {}_0n^1 \rightarrow {}_1H^{3**} + {}_2He^4;$$

$$_5B^{10} + {}_0n^1 \rightarrow {}_3Li^{7**} + {}_2He^4;$$

$$_{17}Cl^{35} + {}_0n^1 \rightarrow {}_{16}S^{35**} + {}_1H^1$$

The principal reaction to occur as a result of the interaction of rocks and slow neutrons is radiative capture in which the compound passes to the ground state emitting a gamma quantum:

$$_{Z}X^{A_m} + {}_0n^1 \rightarrow [_{Z}X^{A_m+1}]^* \rightarrow {}_{Z}X^{A_m+1} + \gamma$$

Here are some examples of such reactions in rocks:

$$_{11}Na^{23} + {}_0n^1 \rightarrow {}_{11}Na^{24**} + \gamma$$

$$_1H^1 + {}_0n^1 \rightarrow [_1H^2]^* \rightarrow {}_1H^2 + \gamma$$

$$_{12}Mg^{26} + {}_0n^1 \rightarrow {}_{12}Mg^{27} + \gamma$$

$$_{13}Al^{27} + {}_0n^1 \rightarrow {}_{13}Al^{28**} + \gamma$$

During a radiative capture the neutron to proton ratio increases often giving rise to radioactive* isotopes: $_{11}Na^{24}$, $_{12}Mg^{27}$, $_{13}Al^{28}$ etc.

---

* Radioactive isotopes are marked by two asterisks.

Sec. 69. The Ability of Rocks for Scattering and Absorbing Neutrons

The ability of rocks for scattering and absorbing (capturing) neutrons, i.e. their *scattering* and *absorbing activity*, are evaluated by the *macroscopic effective scattering cross sections* Σ_{scat} and *absorption (capture) cross sections* Σ_{capt} measured in m^{-1} and a number of other quantities. Neutron macroscopic effective cross sections are a measure of the probability of the outcome of a definite nuclear process under specified conditions. Macroscopic effective cross sections Σ_{scat} and Σ_{capt} are composed, respectively, of the *effective microscopic scattering cross section* σ_{scat} or *capture cross section* σ_{capt}, of all nuclei in 1 m^3 of rock.

When determining the effective microscopic scattering and absorption cross section let us imagine a layer of the element under consideration one atom thick (Fig. 146). Let us assume a unit area $S = 1\ m^2$ found at the above layer is transversed by a uniform neutron flux at right angles to its surface having a density 1 neutr./($m^2 \cdot s$) and that during 1 s C_{scat} of neutrons has been scattered or C_{capt} has been absorbed by nuclei at a unit area S. If N_S nuclei were at the unit area S, the number of C_{scat} or C_{capt} processes must be in direct proportion to the intensity of the flux I and the number of nuclei N_S. As to the proportionality coefficients, they will be provided by the microscopic scattering cross sections σ_{scat} during scattering and σ_{capt} during neutron capture. Note that σ_{scat} and σ_{capt} are governed by the nucleus type and energy of neutrons. Thus,

$$C_{scat} = \sigma_{scat} I N_S \quad \text{and} \quad C_{capt} = \sigma_{capt} I N_S$$

Consequently, σ_{scat} and σ_{capt} are the numbers of processes referred to one nucleus and one neutron. If C_{scat} and C_{capt} are the numbers of neutrons scattered on or captured by nuclei at $S = 1\ m^2$ within 1 s—neutr./($m^2 \cdot s$), I and N_S have, respectively, units neutr./($m^2 \cdot s$) and nucleus/m^2, the size of the effective cross sections will be provided by an area taken by one nucleus of the rock: m^2/nucleus. Since σ_{scat} and σ_{capt} values are in the range 10^{-26} to $10^{-30}\ m^2$ per nucleus, a unit equal to $10^{-28}\ m^2$/nucleus is adopted to measure them. This unit corresponds to 10^{-4} barns (1 barn = $10^{-24}\ cm^2$/nucleus).

Thus, for the calculation of the macroscopic effective cross sections of elements (native metals, carbon, sulphur etc.), it is obvious that these equations can be used:

$$\Sigma_{scat} = \sigma_{scat} N = \sigma_{scat} \frac{N_A \delta_s}{A}$$

$$\Sigma_{capt} = \sigma_{capt} N = \sigma_{capt} \frac{N_A \delta_s}{A}$$

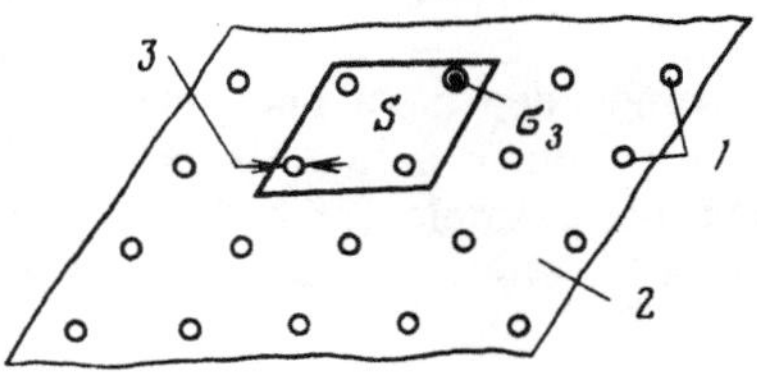

FIG. 146. Diagram illustrating the concept of a microscopic effective cross section.

1—atomic nuclei; *2*—plane over which atoms are distributed in a layer 1 atom thick; *3*—area of the cross section of nucleus where N is the number of nuclei in 1 m^3 of rock; A and δ_s are the relative atomic mass and density of the element, respectively; N_A is the Avogadro constant.

However, rocks generally consist of elements having different effective microscopic cross sections. If the chemical composition of a rock is presented in terms of individual elements, Σ_{scat} and Σ_{capt} are found using these equations:

$$\Sigma_{scat} = \frac{\delta_r N_A}{100} \sum_{i=1}^{i=n} \frac{m_{el.i}\sigma_{scat.i}}{A_i} \tag{88}$$

$$\Sigma_{capt} = \frac{\delta_r N_A}{100} \sum_{i=1}^{i=n} \frac{m_{el.i}\sigma_{capt.i}}{A_i} \tag{89}$$

where $\sigma_{scat.i}$, $\sigma_{capt.i}$, $m_{el.i}$ and A_i are microscopic effective scattering and capture cross sections, mass per cent content and relative atomic mass of the i-th unit element of rock; δ_r is the density of the rock.

If analysis of the solid portion of the rock is presented in terms of oxides of elements and analysis of the liquid portion in terms of elements, the calculation of its cross sections uses these equations:

$$\Sigma_{capt} = \frac{N_A}{100}\left[\delta_s(1 - k_p) \sum_{i=1}^{i=n} \frac{m_{m.i}\sigma_{capt.m.i}}{M_i} + k_p\delta_l \sum_{i=1}^{i=m} \frac{m_{el.i}\sigma_{capt.el.i}}{A_i}\right] \tag{90}$$

$$\Sigma_{scat} = \frac{N_A}{100}\left[\delta_s(1 - k_p) \sum_{i=1}^{i=n} \frac{m_{m.i}\sigma_{scat.m.i}}{M_i} + k_p\delta_l \sum_{i=1}^{i=m} \frac{m_{el.i}\sigma_{scat.el.i}}{A_i}\right] \tag{91}$$

where k_p is the voids ratio; δ_s and δ_l are densities of the solid and the liquid phase; $m_{m.i}$, $m_{el.i}$ are contents (in mass %) of the i-th molecule in the solid and the i-th element in the liquid phase of the rock; $\sigma_{scat.m.i.}$, $\sigma_{capt.el.i}$ are effective absorption cross sections of the i-th molecule of the solid and the i-th element of the liquid phase of the rock; $\sigma_{scat.m.i}$, $\sigma_{scat.el.i}$ are effective scattering cross sections of the i-th molecule of the solid and i-th element of the liquid phase of the rock; M_i is the molecular mass of the i-th molecule of the solid phase; A_i is the relative atomic mass of the i-th element of the liquid phase.

Thus, to calculate Σ_{capt} and Σ_{scat} we must know the chemical composition of the rocks, their density, given the natural moisture content or density of their solid and liquid portions, microscopic scattering and absorption cross-sections of molecules and elements that compose the rock.

The scattering and absorption cross sections strongly depend on neutron energy, so calculations of the nuclear scattering and absorption of a rock use averaged values of the effective cross sections of elements. For example, for hydrogen this averaging for fast neutrons employs special relationships to obtain $\sigma_{scat\ H} = 4.3 \times 10^{-28}\ m^2$.

The scattering cross sections of other elements much less depend on energy, that is why the averaged values are generally considered to be represented by cross sections for definite particular energy value, e.g. for the energy 1 MeV in calculating the cross sections of a rock for fast neutrons and 0.025 eV for thermal neutrons*. The averaged neutron effective capture cross sections for neutrons showing thermal and suprathermal energies can be determined from dependences of neutron capture cross sections on energy.

Sec. 70. Other Quantities Characterizing the Interaction of Rocks with Neutrons

Other elementary and complex variables determining the behaviour and intensity of the interaction between rocks and neutrons include the diffusion coefficient (diffusivity) D_0, slowing-down power, slowing-down coefficient, slowing-down length L_s, diffusion length L_d, slowing-down time τ_s or chronological age of neutrons and life time of thermal neutrons τ_d.

The slowing-down power, slowing-down coefficient, slowing-down length and time characterize more adequately than the energy loss in a single collision or macroscopic cross sections the ability of rocks for scattering and absorbing neutrons as these change their energy from high to thermal values. The slowing-down power, length, and time depend on the average energy loss during one collision and given a macroscopic scattering cross section. The slowing-down coefficient is also determined by the macroscopic capture cross section.

The diffusion length life time of thermal neutrons more accurately determine the intensity of their interaction with nuclei of rocks and are mainly governed by the macroscopic capture and scattering cross sections (diffusion length) or by the capture cross section alone (life time).

The diffusion coefficient of thermal neutrons. If $\vec{r}$ is the direction in a unit volume of a substance in which at a definite moment the density of n neutrons changes the most (their volume concentration), it seems to be true that their flux I over the interval dr across a surface dS normal to $\vec{r}$ is in direct proportion to grad n. Thence

$$IdS = -D\frac{dn}{dr}$$

where D is the diffusion coefficient; dn/dr is the value characterizing the maximum variation of neutron density in matter.

Suppose the distribution of neutrons in matter is independent of their energy and space orientation, then the neutron flux is a scalar quantity and the diffusion equation is true which for single energy neutrons appears as

$$\vec{I} = -D\,\overrightarrow{\text{grad}}\,n \qquad \text{(Fick's law)}$$

* Sometimes use is made of the scattering cross sections of thermal neutrons averaged over the Maxwell distribution.

where $\vec{I}$ is a vector quantity determining the number of neutrons crossing a unit area perpendicular to the direction of the neutron flux in a specified direction within a unit time; n is neutron density at a point of interest I.

Since the rate of diffusion of neutrons into matter characterizes their flux $\Phi = nv$. Fick's law can be presented as

$$\vec{I} = -D_0 \, \overrightarrow{\text{grad}} \, \Phi$$

where $D_0 = D/v$ is the diffusion coefficient for a neutron flux in linear units since the units of measuring I and Φ are the same (number of neutrons per 1 cm^2 per s); v is the constant velocity of neutrons.

The diffusion coefficient D_0 is inversely related to the macroscopic absorption and scattering cross sections. This quantity is also governed by the degree of anisotropy of scattering. Hence

$$D_0(T) = \frac{v(T)}{3[\Sigma_{capt}(T) + \Sigma_{tr}(T)]} \tag{92}$$

where $v(T)$, $\Sigma_{capt}(T)$ and $\Sigma_{tr}(T)$ are the speed of neutrons, macroscopic capture cross sections and transport scattering at the absolute temperature T.

For poorly absorbing substances

$$D_0(T) = \frac{v(T)}{3\Sigma_{scat}(T)[1 - (\cos\psi)_{av}]} = \frac{\lambda_{tr}(T)v(T)}{3}$$

where $\lambda_{tr}(T) = 1/\Sigma_{scat}(T)[1 - (\cos\psi)_{av}]$ is the average transport scattering length; $(\cos\psi)_{av}$ is the average cos scattering angle. Under condition of isotropic scattering $(\cos\psi)_{av} = 0$; for water and water bearing rocks $(\cos\psi)_{av} = 0.2$.

The calculation of D_0 (in cm^2/s or m^2/s) of poorly absorbing practically single-phase rocks generally employs an abridged equation

$$D_0 = v_{av}\left(3\sum_{i=1}^{i=n} n_i\sigma_{scat.i}\right)^{-1} \tag{93}$$

and a more rigorous equation

$$D_0 = \frac{v_{av}\lambda_{tr}(T)}{3} = \frac{1}{3} = \frac{1.128}{\Sigma_{tr}(T)} \tag{94}$$

where $v_{av} = 2.48 \times 10^5$ cm/s $= 2.48 \times 10^3$ m/s at $T = 293$ K; $\sigma_{scat.i}$ is the microscopic scattering cross section of thermal neutrons by the nucleus of the i-th element in cm^2 of m^2; $n = (6.023 \times 10^{23}/M)b\delta$ is the number of atoms of the i-th element in 1 cm^3 (m^3) of rock; b is the number of atoms of this element in one molecule; M, δ are the molecular mass (weight) and density of rock; $v(T) = 2.2 \times 10^5\sqrt{T/T_0}$ is the most probable velocity of thermal neutrons, cm/s or m/s, in rock having a temperature T, K; $T_0 = 293$ K is a temperature corresponding to 20 °C; the transport scattering length is $\lambda_{tr} = \lambda_{scat}/[1 - (\cos\psi)_{av}]$ where λ_{scat} is the average scattering length.

The slowing-down power and slowing-down coefficient. Separately taken macroscopic scattering cross sections and relative energy loss $(E_1 - E_2)/E_1$ for a single collision are insufficient for evaluating the capacity of rocks to slow down neutrons. The probability of neutrons being scattered on nuclei of elements can be great but in case little energy is lost in each collision, the rock will prove a poor slowing-down medium. Nor will the rock be a good moderator, given a high $(E_1 - E_2)/E_1$ but a small Σ_{scat}.

The slowing-down properties are more useful to evaluate in terms of a quantity known as the macroscopic slowing-down power of rocks equal to the product of the macroscopic scattering cross section Σ_{scat} with the average logarithmic energy loss ξ in a single collision.

$$\xi = (\ln E_1 - \ln E_2)_{av} = [\ln(E_1/E_2)]_{av} = 1 - \frac{(A_m - 1)^2}{2A_m} \ln \frac{A_m + 1}{A_m - 1}$$

where E_1 and E_2 are the neutron energy before and after a collision; A_m is the mass number of the element in question equal to the sum of the number of neutrons and protons in the nucleus. Given mass numbers $A_m > 10$, the following relationship is sufficiently accurate:

$$\xi = \frac{2}{A_m + 2/3}$$

which makes it possible even for $A_m = 2$ to determine ξ accurate to 3.3%.

For native elements

$$\Sigma_{scat}\xi = \frac{N_A \delta_s \sigma_{scat}}{A}\,\xi \simeq \frac{2N_A \delta_s \sigma_{scat}}{A(A + 2/3)}$$

and for A > 14 accurate to 5%

$$\Sigma_{scat}\xi \simeq \frac{2N_A \delta_s \sigma_{scat}}{A^2}$$

The slowing-down coefficient, i.e. the ratio of the slowing-down power $\Sigma_{scat}\xi$ to the average macroscopic absorption cross section $\Sigma_{capt.av}$ taking also into account the capacity of rock for absorbing fast neutrons more adequately characterize the slowing-down power of rocks.

The slowing-down, L_s, and **diffusion length,** L_d. Having in the general case undergone a sufficient number of collisions with nuclei of a rock, each of the neutrons is captured by them.

The trajectory of a separate neutron is made up of a number of segments representing the path of the neutron between two subsequent collisions with nuclei of the elements of the medium. These segments have different lengths and orientation due to which fact, before being captured, the neutron travels from its source along a straight line to a definite distance r^* whose approximate value is found from the equation

$$r^{*2} = r^2_{s.av} + r^2_{d.av} = 6(L_s^2 + L_d^2) = 6(\tau_s + L_d^2)$$

where $r^2_{s.av}$ is the rms distance along a straight line from the point of neutron origination to a point where they become thermal ones, m^2; $r^2_{d.av}$ is that from the last point to the absorption point, m^2; slowing-down and diffusion lengths, m, $L_s = r_{s.av}/\sqrt{6} = \sqrt{\tau_{th}}$ and $L_d = r_{d.av}/\sqrt{6}$; τ_{th} is the symbolical age of thermal neutrons, m^2.

Consequently, to determine slowing-down and diffusion lengths, we must know $r_{s.av}$ and $r_{d.av}$ or squares of these quantities.

Let us determine L_s. In so doing, let us calculate $r^2_{s.av}$ using this equation:

$$r^2_{s.av}(r_s, \tau) = \frac{\int_0^\infty r_s^2[4\pi r_s^2 q(r_s, \tau)]dr}{\int_0^\infty 4\pi r_s^2 q(r_s, \tau)dr}$$

$$= \frac{\int_0^\infty \text{quadratic distance from the source to a unit layer} \times \text{number of neutrons that attained thermal energy in this layer}}{\int_0^\infty \text{number of all neutrons that attained thermal energy (or corresponding lethargy or age), in the sphere of an infinite radius}}$$

$$= \frac{\text{sum of all quadratic slowing-down acts in a sphere of an infinite radius}}{\text{number of all neutrons that attained thermal energy (or corresponding lethargy or age) in the sphere of the same infinite radius}} \qquad (95)$$

where $4\pi r_s^2 dr$ is the volume dV of a unit spherical layer found at a distance r_s from a point source of neutrons where the slowing-down density is equal to q (r_s, τ).

The slowing-down density is the number of neutrons referred to a unit volume and time crossing during the slowing-down the specified value of lethargy or energy (in the particular case, thermal energy), consequently, having a definite age τ.

When solving Eq. (95) the main problem is to determine the explicit form function $q(r_s, \tau)$. In the case often occurring in nature where the hydrogen content is relatively small, the $q(r_s, \tau)$ function is determined in the age approximation on an assumption that neutrons are continuously losing energy as a result of a great number of collisions and the elastic scattering reaction cross section little depends on energy.

The basis for determining the q function is in this case provided by an equation for the rate of decreasing neutron density:

$$\frac{dn(r, t)}{\partial t} = -D_0 v \nabla^2 n(r, t) = -D_0 v \nabla^2 \left[\frac{\xi_v}{\lambda_{scat}} n(r, u)\right] \qquad (96)$$

where the diffusion coefficient for anisotropically scattering and poorly absorbing rock

$$D_0 = \frac{1}{3\Sigma_{scat}[1 - (\cos\psi)_{av}]} = \frac{\lambda_{scat}}{3[1 - (\cos\psi)_{av}]} = \frac{\lambda_{tr}}{3}$$

λ_{tr} is the transport free path; $(\cos \psi)_{av}$ is the average cos scattering angle in the laboratory frame of reference; v is the velocity of neutrons; the function determining neutron density,

$$n(r, t) = n(r, u)\frac{\partial u}{\lambda_{scat}} = \frac{\xi_v}{\lambda_{scat}} n(r, u) \tag{97}$$

since

$$du = d \ln \frac{E_0}{E} = \frac{\xi_v}{\lambda_{scat}} dt \tag{98}$$

Here $u = \ln(E_0/E)$ is lethargy or logarithmic decrement of energy; E, E_0 is the specified energy and energy neutron origination; $\lambda_{scat} = 1/\Sigma_{scat}$ is the average free path of the neutron relative to scattering—the average distance travelled by the neutron between two scattering acts.

By differentiating Eq. (97) we have

$$\frac{\partial n(r, t)}{\partial t} = \frac{\xi v}{\lambda_{scat}} \frac{\partial}{\partial u}\left[\frac{\xi v}{\lambda_{scat}} n(r, u)\right]$$

and, consequently,

$$D_0 v \nabla^2 \left[\frac{\xi v}{\lambda_{scat}} n(r, u)\right] = \frac{\xi v}{\lambda_{scat}} \frac{\partial}{\partial u}\left[\frac{\xi v}{\lambda_{scat}} n(r, u)\right]$$

However, since $vn(r, u) = \Phi(r, u)$, where $\Phi(r, u)$ is the neutron flux per unit lethargy interval, and $\lambda_{scat} = 1/\Sigma_{scat}$, then

$$D_0 \nabla^2 [\xi \Sigma_{scat} \Phi(r, u)] = \xi \Sigma_{scat} \frac{\partial}{\partial u} [\xi \Sigma_{scat} \Phi(r, u)],$$

taking into account that $\xi \Sigma_{scat} \Phi(r, u) = q$, we get

$$\nabla^2 q = \frac{\xi \Sigma_{scat}}{D_0} \frac{\partial q}{\partial u} \tag{99}$$

To solve Eq. (99), we introduce a new variable

$$\tau_0(u) = \int_0^u \frac{D_0}{\xi \Sigma_{scat}} du \tag{100}$$

known as the symbolic age or simply age, neutron age of Fermi age, following which we get a differential age equation

$$\nabla^2 q = \partial q / \partial \tau_0 \tag{101}$$

describing the continuous slowing-down of neutrons which in a nonabsorbing medium consisting of elements with sufficiently great mass numbers has this solution:

$$q = \frac{\exp(-r^2/4\tau_0)}{(4\pi\tau_0)^{3/2}} \tag{102}$$

By substituting Eq. (102) into Eq. (95) and integrating it we have

$$r_{s.av}^2(\tau) = \frac{\int_0^\infty \tau^4 \exp(-r^2/4\tau_0)dr}{\int_0^\infty r^2 \exp(-r^2/4\tau_0)dr} = 6\tau_0 \tag{103}$$

The age approximation yields correct values of $r_{s.av}^2(\tau)$, solely given small distances from the source, for $r \ll 6\tau_0/\lambda_{tr}$. Given great distances, the accuracy of determining $r_{s.av}^2$ decreases with decreasing the mass of nuclei of the medium, increasing the capture reaction cross section and given an appreciable dependence of cross sections on energy. It is recommended to calculate $r_{s.av}^2$ using the above method solely for rocks whose composition contains elements of a medium and great atomic mass—dense igneous rocks, dense sandstones, carbonates etc.

The age approximation is particularly inconvenient to apply under the aforementioned conditions to rocks with a high hydrogen content—loose water- and oil-saturated sands, sandstones, aleurolites, clays, limestones etc. as well as to dense rock species, e.g. gypsums. Neutrons lose energy in the aforementioned rocks not continuously, as a result of a great number of collisions but, rather, may totally lose this energy at a single, that is, the first collision. On the other hand, the reaction cross section of scattering on hydrogen is strongly dependent on energy.

D. A. Kozhevnikov, basing on the solution of a kinetic equation (under definite conditions), has derived an equation for this case:

$$\tau(u) = \frac{1}{3}\lambda^2(u)\left[\varepsilon_{01}(u) + \frac{u}{h_1[1-(\cos\psi)_{av}} + \zeta(u)\right] \tag{104}$$

Here $\lambda(u)$ is *the total free path** for elastic collisions with hydrogen nuclei of secondary (collided with protons) neutrons having lethargy u; $\varepsilon_{01}(u) = \nu_0^2(u)/h_{01}$, where $\nu_0(u) = \lambda_0/\lambda(u)$; λ_0 is the total free path of primary (not collided with protons) neutrons having zero lethargy; $h_{01} = \lambda_0/\lambda_{01}$ is the probability of virgin neutrons being scattered on hydrogen nuclei (λ_{01} is the mean free path with zero lethargy for elastic collisions with hydrogen); $u = \ln(E_0/E)$ is lethargy; $h_1(u) = \lambda(u)/\lambda_1(u)$ is the probability of scattering on hydrogen nuclei [$\lambda_1(u)$ is the free path of secondary (collided with protons) neutrons having lethargy u for scattering on hydrogen]; $(\cos\psi)_{av} = 2h_1/3$ is the average cos scattering angle of neutrons in the rock where only hydrogen scatters anisotropically

$$\zeta = \frac{2}{3[1-(\cos\psi)_{av}]} + \frac{1 - 16/9h_1 + 4/9h_1^2}{[1-(\cos\psi)_{av}]^2}$$

The age equation can be reduced to a form facilitating its interpretation in physical terms:

$$\tau = \frac{1}{3}\lambda_0\lambda_{01} + \tau_0 + \frac{1}{3}\lambda\lambda_1 + \frac{2}{9}\lambda_{tr}\left(\lambda_0 - \frac{2}{3}\lambda_{tr}\right) \tag{105}$$

* The total free path is $\lambda = 1/\Sigma = 1/\Sigma_{capt} + \Sigma_{scat} = \lambda_{scat}\lambda_{capt}/\lambda_{scat} + \lambda_{capt}$.

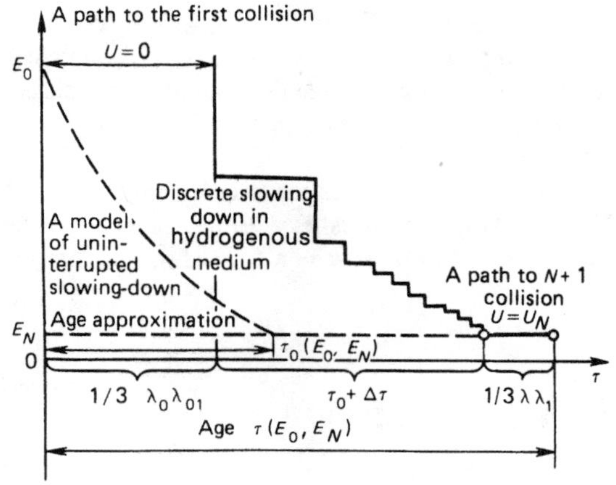

FIG. 147. Diagram illustrating the determination of the path length of neutrons

where

$$\lambda_{tr} = \frac{\lambda(u)}{1 - (\cos\psi)_{av}}$$

$$\tau_0 = \int_0^u \frac{\lambda^2(u)du}{3\xi[1 - (\cos\psi)_{av}]} = \int_0^u \frac{D_0}{\xi\Sigma_{scat}} du$$

As can be seen, the relationship permitting the calculation of neutron age in hydrogenous media includes the age equation (100) for rocks with heavy nuclei and, in addition, has extra terms. The first of these ($\lambda_0\lambda_{01}/3$) is a correction for the first path (Fig. 147), the third term ($\lambda\lambda_1/3$) is a correction for the last path, and terms $2\lambda_{tr}(\lambda_0 - 2\lambda_{tr}/3)/9$ allow for the discreteness of slowing-down in hydrogenous media.

The neutron age calculated from this equation is much greater than one found from the Fermi age equation [see Eq. (100) and Fig. 147].

The neutron age $L_s^2 = \theta$ with the initial energy E_0 to suprathermal energy E_{sth} is found from the relationship:

$$L_s^2 = \frac{1}{3\,[\sum_i n_i\sigma_{\Sigma_i}(E_0)]^2} + \int_{E_{sth}}^{E_0} \frac{dE'}{3\xi(E')\Sigma_\Sigma(E')\Sigma_{tr}(E')E'}$$

$$\approx \frac{1}{3[\sum_i n_i\sigma_{\Sigma_i}(E_0)]^2} + \frac{1}{3\xi\Sigma_\Sigma\Sigma_{tr}(E)} \ln\frac{E_0}{E_{sth}} \qquad (106)$$

Here $n_i = (6.023 \times 10^{23}/M)b\delta$ is the number of atoms of the i-th element in 1 m^3 (cm^3) of rock; b is the number of atoms of this element in one molecule of the mineral; δ is its density, kg/m^3 (g/m^3); $\sigma_{\Sigma_i}(E_0) = \sigma_{capt}(E_0) + \delta_{scat}(E_0)$ is the summed up microscopic cross section of the given element with neutrons of the initial energy (E_0), m^2 (cm^2); $\Sigma_\Sigma = \Sigma_{capt} + \Sigma_{scat}$; $\Sigma_{tr}(E) = 1/\Sigma_{scat}[1 - (\cos\psi)_{av}]$ is the transport cross section of neutron interaction with energy E with the nucleus of the same element, m^2 (cm^2); ξ is the average logarithmic decrement of increasing energy per one collision of a fast neutron with this nucleus; M is the molecular mass of the mineral, a dimensionless quantity.

The diffusion length L_d is bound as being a distance reckoned from the source at which the neutron flux* $\Phi = Qe^{-kr}/4\pi D_0 r$ diminishes e times. Here Q is the number of neutrons emitted by the source within 1 s; $k = \sqrt{\Sigma_{capt}/D_0}$; r is the distance from the source.

It has been shown in nuclear physics that

$$L_d^2 = \frac{1}{k^2} = \frac{D_0}{\Sigma_{capt}} = \frac{\lambda_{tr}\lambda_{capt}}{3} = \frac{1}{3[1 - (\cos\psi)_{av}]\Sigma_{scat}\Sigma_{capt}} \tag{107}$$

where λ_{capt} are mean free paths—the distance travelled by a slow neutron before being captured by matter; $\lambda_{capt} = 1/\Sigma_{capt}$.

Thence,

$$L_d = \sqrt{1/3\Sigma_{capt}\Sigma_{scat}[1 - (\cos\psi)_{av}]} \tag{108}$$

For poorly absorbing rocks containing atoms of great and medium mass numbers, under conditions of isotropic scattering $L_d = 1/\sqrt{3\Sigma_{capt}\Sigma_{scat}}$. Sometimes it is more reasonable, in place of the diffusion length, to use a quantity $r_{d.av}^2$ which is the rms displacement of a thermal neutron from a point where it became thermal before the point of capture.

Let us imagine a spherical layer r in radius and dr in thickness about a point source of thermal neutrons. The number of thermal neutrons captured in this layer within 1 s is equal to $4\pi r^2 dr\Phi\Sigma_{capt}$, where $4\pi r^2 dr$ is the volume of the layer.

Let us now determine the rms displacement $r_{d.av}^2$ from the source to the point of capture:

$$r_{d.av}^2 = \frac{\int_0^\infty r_d^2 \text{ (a number of neutrons captured at area } r \text{ in radius)}}{\text{(a number of neutrons emitted by the source)}}$$

$$= \frac{1}{Q}\int_0^\infty r_d^2 4\pi r^2 dr\Sigma_{capt}\Phi = \frac{1}{Q}\int_0^\infty \frac{r_d^2 4\pi r^2 \Sigma_{capt} Qe^{-\kappa r}}{4\pi r D_0}\, dr = 6L_d^2 \tag{109}$$

Equation (109) shows that the quadratic diffusion length L_d^2 (diffusion surface) equals 1/6 rms sum of elementary free paths of a thermal neutron from their source to the location of its absorption. The actual path of a neutron moving in a zigzag line is much greater.

---

* The relationship for Φ is derived following a solution of the diffusion equation $D\nabla^2\Phi - \Sigma_{capt}\Phi = 0$.

L_d can in the general case be found from the relationship

$$L_d(T_0) = \sqrt{D_0(T_0)\tau_d(T_0)} \tag{110}$$

where D_0 is the diffusion coefficient; τ_d is the average lifetime of thermal neutrons whose values are temperature-dependent.

The slowing-down time of neutrons. The slowing-down time or average chronological age of neutrons τ_s is the time taken by a neutron to slow down in a rock from energies of about 2-5 MeV to suprathermal energies (~1.46 eV).

Due to the fact that within a time interval $d\tau$ the neutron loses energy $-dE$, for the neutron to attain suprathermal energy time is needed

$$\tau_s = \int_0^{\tau} d\tau = \int_{E_{nth}}^{E_0} \frac{\lambda_{scat\Sigma}(E')}{\xi v_{nth}} \frac{dE}{E} = \frac{2}{\xi(E_{sth})} \frac{1}{\Sigma_{scat\Sigma}(E_{sth}) v_{sth}} \tag{111}$$

where τ_s is measured in s if $\lambda_{scat\Sigma}$ is taken in m, and E_0 and E_{sth} in J (or MeV).

Since the time needed for the first path is negligible, the latter equation is derived without taking it into account.

The average life time of the thermal neutron. The average life time of the thermal neutron or the average diffusion time τ_d is called the average time the neutron exists in the rock from the moment it attains thermal energy to its absorption by the nucleus.

The average life time τ_d in an infinite medium is equal numerically to the average distance λ_{capt} travelled by the neutron in a rock referred to its average velocity:

$$\tau_d = \frac{\lambda_{capt}}{v} = \frac{1}{\Sigma_{capt} v} = \frac{1}{N_A \sigma_{capt} v} = \left(v \sum_{i=1}^{i=m} n_i \sigma_{capt.i} \right)^{-1} \tag{112}$$

where v is the velocity of thermal neutrons, at 25 °C equal to 2.2×10^3 m/s; λ_{capt} is the mean free path of the neutron before it is absorbed.

Sec. 71. Minerals

The neutron activity of minerals and their ability of slowing down and absorbing the fast, then slow neutrons in the process of scattering is evaluated from their composition and such quantities as: the macroscopic capture cross section Σ_{capt}, scattering cross section Σ_{scat}, slowing-down length L_s and slowing-down time τ_d, the diffusion coefficient D_0, diffusion length L_d and diffusion time τ_d. The latter have been found using Eqs. (88), (89), (106), (111), (92) and (107), (112). This, however, been done not for all minerals.

The slowing-down length and time. The slowing-down length and time of neutrons in minerals to energy 1.46 eV has been shown by calculations to vary over a wide range [L_s = (5.7-106.5) $\times 10^{-2}$ m and τ_s = 0.6-297 μ/s. The principal neutron moderators are hydrogenous minerals. The hydrogen entering their composition has a great scattering cross section σ_{scat} (>20 b) increasing with diminishing neutron energies. Energy losses are also great as it collides with neutron nuclei.

Hence the greatest slowing-down power (minimum τ_s and L_s) of minerals having an appreciable content of crystallization and constitutional water. These include, for example, ascharite ($\tau_s = 0.9\ \mu/s$, $L_s = 8 \times 10^{-2}$ m), hydroboracite ($\tau_s = 0.6\ \mu/s$, $L_s = 6.5 \times 10^{-2}$ m), diaspore ($\tau_s = 1.9$, $L_s = 6 \times 10^{-2}$), hydrargillite ($\tau_s = 1.1$, $L_s = 5.7 \times 10^{-2}$), carnallite ($\tau_s = 1.4$, $L_s = 8.5 \times 10^{-2}$), gypsum ($\tau_s = 1.89$, $L_s = 7.7 \times 10^{-2}$), mirabilite ($\tau_s = 1.26$, $L_s = 6.7 \times 10^{-2}$), kaolinite ($\tau_s = 2.5$, $L_s = 7.9 \times 10^{-2}$), muscovite ($L_s = 12 \times 10^{-2}$), opal ($\tau_s = 7\ \mu/s$, $L_s = 15.6 \times 10^{-2}$ m) etc.

The poorest moderators (except native metals) whose τ_s and L_s have not been studied are sulphides; their L_s and τ_s show, respectively (32.3-85) $\times 10^{-2}$ m and 56.5-297 μ/s and have average values, respectively, 62.4×10^{-2} m and 140.8 μ/s. This is accounted for by a great relative atomic mass of metallic (Hg, Ni, Pb, Bi, Fe, Cu, Ag, Sb, Zn etc.) cations involved in their composition and their appreciable content in the mineral under which conditions the average logarithmical loss ξ during the collision of the neutron with a nucleus and the total energy loss during collisions of neutrons with metallic nuclei are small. Neutrons are relatively poorly slowed down by anhydrous chlorides, such as sylvite ($L_s = 106.8 \times 10^{-2}$ m, $\tau_s = 60\ \mu/s$), halite ($L_s = 47 \times 10^{-2}$ m, $\tau_s = 32.3\ \mu/s$), cerargyrite ($L_s = 53.5 \times 10^{-2}$ m, $\tau_s = 44.6\ \mu/s$). Minerals of other classes (including the most important rock-forming minerals of sedimentary rocks) show a medium capacity for slowing down neutrons, their L_s = (19-35) $\times 10^{-2}$ m, and τ_s = 21.5-47 μ/s.

The macroscopic capture cross section, lifetime, diffusion coefficient and diffusion length of thermal neutrons. The composition of minerals, energy of neutrons irradiating them and probability of a series of nuclear reactions with neutrons showing different energies are what ensures principally radiative capture. The probability of the reaction is particularly great for minerals containing Ti, Mn, Cl, V, Co, Nd, La, Ni, Cs, Sm, Sc and As and some other elements manifesting high effective microscopic capture cross sections.

The absorption of neutrons in lithium- and boron-bearing minerals seems to occur as a result of (n, α) and (n, γ) reaction. For nitrogen-, sulphur- and chlorine-bearing minerals not only (n, γ) but also (n, p) reactions are likely to occur. It is not uncommon that products of (n, γ) reactions include radioactive nuclei and hard gamma radiation having energies up to 9 MeV. Hard gamma radiation also occurs during the decay of radioactive isotopes. The neutron capture cross section Σ_{capt} of neutrons by minerals varies over a wide range: from 0.037 (graphite) to 2423 m^{-1} (boracite). The values of the diffusion time (lifetime) τ_d of thermal neutrons also show a wide range from 0.19 (boracite) to 12290 μ/s (graphite). So do the values of the diffusion coefficient D_0 and diffusion length L_d of neutrons in minerals vary appreciably: respectively from 0.045×10^5 (boracite) to $19.96 \times 10^5\ cm^2/s$ (sulphur) and from 0.05×10^{-2} (boracite) to 29×10^{-2} m (magnesite).

In terms of Σ_{capt}, L_d and τ_d, taking into account microscopic cross sections of mineral-forming elements minerals can be referred to six following groups:

Group I—very slightly active absorbers of thermal neutrons with $\Sigma_{capt} <$

$0.1\ m^{-1}$ and $\tau_d > 5\,000\ \mu/s$*—graphite, diamond and bismuth. These native elements do not practically absorb long-living thermal neutrons since predominant in their composition are elements having low ($n \times 0.001$ b for C) and decreased ($n \times 0.01$ b for Bi) effective microscopic cross sections. Here and elsewhere n is a numerical factor of several units.

Group II includes slightly active absorbers of thermal neutrons. This group comprises minerals showing Σ_{capt}, L_d and τ_d, respectively, in the range from 0.1 to $1\ m^{-1}$, $(12.5\text{-}30) \times 10^{-2}$ m, 500-5 000 μ/s. It is currently represented by 21 minerals (magnesite, periclase, spinel, quartz, albite, anorthite, nepheline, serpentine, montmorillonite, lead, anglesite etc.)

These minerals are compounds of elements (H, Ca, Na, Al, Zr, Si, S and others) having an intermediate ($n \times 0.1$ b), decreased ($n \times 0.01$ b for Mg), low ($n \times 0.001$ b for C) and very low ($n \times 0.0001$ b for O) microscopic capture cross section. Note that the composition of minerals ensures the aforementioned values of Σ_{capt}, L_d and τ_d.

Group III—averagely active absorbers of thermal neutrons: their Σ_{capt} over the range $1\text{-}10\ m^{-1}$. $L_d = (3.75\text{-}15) \times 10^{-2}$ m and $\tau_d = 50\text{-}500\ \mu/s$. It is represented by the greatest number of minerals (58). The composition of the minerals of this group rather often includes a comparatively appreciable amount of elements (Ba, K, Fe, Zn, Sr, Ti, La etc.) with increased absorption cross sections (σ_{capt}—n b); however, apart from these elements minerals contain elements having an intermediate (Cl, Sn, Zr, Si, Ca, Al, Na, H, P, S and others, $\sigma_{capt} = n \times 0.1$ b), lowered (Bi, Mg, N, Li, $\sigma_{capt} = n \times 0.01$ b), low (F, C, $\sigma_{capt} = n \times 0.001$ b) and very low (O, $\sigma_{capt} = n \times 0.0001$ b) capture cross sections. However, the composition and structure of minerals are what are responsible for their intermediate neutron absorption power. This group is represented, e.g. by galena, pyrrhotite, pyrite, corundum, anhydrite, gypsum, forsterite, fayalite, kaolinite and many other minerals.

Group IV—increasingly active absorbers of thermal neutrons. The group includes 42 minerals having Σ_{capt} varying from 10 to $100\ m^{-1}$, and L_d and τ_d, varying over the range, respectively, $(0.75\text{-}3.75) \times 10^{-2}$ m and 5-50 μ/s. These values of the quantities Σ_{capt} and τ_d for minerals are mainly due to the presence in their composition also of a definite combination of elements having very high ($\sigma_{capt} = n \times 100$ b for Pa and Er), high ($\sigma_{capt} = n \times 10$ b for Au, Ag, Co, Cl, Wi, Mn, Se), increased ($\sigma_{capt} = n \times$ barn for K, Pt, Cu, Li, Fe, Ti, Cr, Sb, Nb, As, Te), medium ($\sigma_{capt} = n \times 0.1$ b for P, S) and very low ($\sigma_{capt} = n \times 0.0001$ b for O) microscopic capture cross sections. This group includes, for example, native gold, silver, palladium, platinum, copper, iron, halite, sylvite, antimony, pyrolusite, brownite, manganite, hausmannite, cobaltite, bornite, chalcopyrite, covellite etc.

Group V are active absorbers of thermal neutrons. The group includes minerals having $\Sigma_{capt} = 100\text{-}1000\ m^{-1}$, $L_d = (0.075\text{-}0.75) \times 10^{-2}$ m and $\tau_d = 0.5\text{-}5\ \mu/s$. These minerals are only six in number: borax, tourmaline, cinnabar, argentite, cerargyrite and native tantalum. The composition of borax and tourmaline includes bo-

---

* L_d has not been practically investigated for this mineral group.

rax ($\sigma_{capt} = n \times 100$), and that of cinnabar contains mercury; the high activity of argentite is due to silver, that of cerargyrite is due to silver and chlorine, and, finally, native tantalum almost entirely consists of highly active tantalum.

Group VI includes extremely active neutron absorbers, minerals having $\Sigma_{capt} > 1000\,m^{-1}$, $L_d = (0.05\text{-}0.075) \times 10^{-2}\,m$ and $\tau_d < 0.5\,\mu/s$. The group is represented by boracite, ascharite, hydroboracite whose composition contains highly active boron. This group also includes highly active mercury.

The most important rock-forming minerals show an increased, medium and weak neutron activation power; their Σ_{capt}, L_d and τ_d have values, respectively, $0.1\text{-}100\,m^{-1}$, $(0.075\text{-}30) \times 10^{-2}\,m$ and $5\text{-}5000\,\mu/s$.

In the order of decreasing values of their average macroscopic cross sections mineral classes can be arranged in this series: borates ($\Sigma_{capt.av} = 1193\,m^{-1}$) → native metals (168) → chlorides (99) → sulphides (77) → oxides and salts (17.4) → phosphates (10.2) → native elements (nonmetals) (9.8) → nitrates (4.2) → carbonates (3.5) → sulphates (3.4) → silicates (1.92 without taking into account tourmaline, 13.8 taking it into account).

However, each mineral class is characterized by higher and lower $\Sigma_{capt.i}$ values compared with an average $\Sigma_{capt.av}$ value referred to individual mineral varieties enriched by strongly and weakly active elements. Among native metals, e.g. increased (compared with an average, $\Sigma_{capt.av}$ value) values of this quantity are shown by mercury and decreased ones by lead, bismuth, tin, zinc, antimony etc.

The greatest Σ_{capt} values are demonstrated, among chlorides, by cerargyrite whose composition includes chlorine and silver having a high neutron activation power. $\Sigma_{capt.i}$ values that are lower than the average one, $\Sigma_{capt.i}$, are shown by halite, sylvite and carnallite.

Among sulphides, higher than average $\Sigma_{capt.i}$ values are manifested by cinnabar and argentite, and lower macroscopic capture cross sections $\Sigma_{capt.i}$ by galena, bismuthine, sphalerite etc. Among oxides, increased $\Sigma_{capt.i}$ values are characteristic of minerals (pyrolusite, brownite) which have in their composition highly active Mg; and decreased values are shown by highly dense materials (massicot, zincite, cassiterite, periclase) that contain much Pb, Zn, Sn with a medium neutron absorbing power and Mg with a very low power.

As to carbonates, increased $\Sigma_{capt.i}$ values are shown by rhodochrosite containing much Mn, and decreased values by cerussite and magnesite that contain much lead and Mn. The value of $\Sigma_{capt.i}$ of such sulphate as kainite that has chlorine in its composition, and values lower than medium ones are demonstrated by such minerals as anglesite ($PbSO_4$) and kieserite ($MgSO_4H_2O$).

Such basic silicates as olivines and garnets show higher than average $\Sigma_{capt.av}$ values, zircons have lower values. Most laminated silicates have lower than average values of $\Sigma_{capt.i}$; high and increased $\Sigma_{capt.i}$ values are demonstrated by lepidolite, biotite, chloritoid and illite which seems to be due to K in their composition having an increased activity and, possibly, to boron. All reinforced silicates have lower than average values of $\Sigma_{capt.i}$ owing to their relatively small density and high content of very low oxygen activity. It is only potassium nepheline that has $\Sigma_{capt.i}$ which is higher than $\Sigma_{capt.av}$.

Sec. 72. The Solid Phase

The values of $\Sigma_{capt.s}$ and $\tau_{d.s}$ of the solid phase of rocks are governed by their mineral composition and structure. The power for absorbing neutrons of the solid phase $\Sigma_{capt.s}$ of, e.g. sandy and clay rocks that consists of quartz, feldspars, argillaceous minerals of ankerite, siderite and pyrite (according to findings obtained under the guidance of D. M. Srebrodol'skii) grows with increase of the content of argillaceous minerals, feldspars and calcite, since the neutron absorption power of these minerals is higher than that of quartz. Siderite and pyrite inclusions comparatively little change τ_{dsph} and $\Sigma_{capt.s}$ of sandstones since their content is small and constant even though τ_{dsph} is lower than that of the remaining components of this phase.

The effect of mass clayiness $k_{cl.m}$ on τ_{dsph} of sandy and clay rocks can be inferred from a comparison of these quantities (Fig. 148) and τ_{dsph} with the Al/Si ratio. The lesser the lifetime of thermal neutrons in the solid phase of sandy and clay and carbonate-argillaceous rocks and the more appreciable the neutron absorption activity of this component of the rock, the greater is $k_{cl.m}$ and Al/Si ratio determining the value of $k_{cl.m}$.

The presence of boron causes the lifetime of thermal neutrons in the solid phase of rocks to dramatically decrease.

Crystallization water produces the greatest effect on the values of the diffusion coefficient D_{0s} of thermal neutrons in the solid phase. It must be noted that with increasing its content, w_{cr}, the D_{0s} values decrease and, given $w_{cr} > 5\%$, become practically independent of other specific features of the mineral composition of the solid phase.

Sec.73. The Liquid and Gaseous Phase

Apart from water, the composition of the liquid phase of rocks includes, in definite combinations and concentrations, such salts as sodium and potassium chlorides, calcium and magnesium chlorides, carbonates, bicarbonates and sulphates of sodium and calcium etc. In case Σ_{capt} of sodium carbonate, sodium bicarbonate and

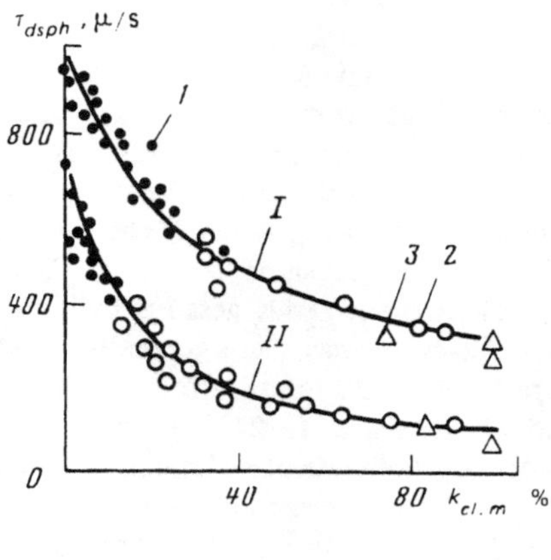

FIG. 148. Dependence of the lifetime of neutrons τ_{dsph} in the solid phase of rock on the coefficient of mass clayiness $k_{cl.m}$.

I—without taking into account the boron content; *II*—taking into account the boron content; *1*—sandstone; *2*—aleurolite; *3*—argillite.

TABLE 20. The Neutron Absorption Activity of Waters of Different Oil and Gas Deposits (after F.Ts. Kron and D.M. Srebrodol'skii)

Waters of sediments (name of oil or gas field)	Macroscopic capture cross section Σ_{capt}, m^{-1}	Lifetime of thermal neutrons in water τ_d, μ/s
Devonian (Tuimaza)	9.88	46
Triassic (Radchenkovskoye)	3.95	115
Devonian (Romashkinskoye)	12.98	35
Okobykaiskaya rock mass (Okhinskoye)	2.33	195
Jurassic (Iskine)	16.23	28

sodium sulphate are of the same order of magnitude as that of water, $\Sigma_{capt.w} = 2.22\ m^{-1}$, the macroscopic capture cross section of chlorides is much greater. For $MgCl_2$ Σ_{capt}, for example, equals $98.9\ m^{-1}$. Pore water concentration differs very much. The neutron absorption power of waters changes very appreciably. Calculations suggest that with decreasing NaCl concentration in water the neutron absorption power of the latter diminishes appreciably, the lifetime of neutrons in the solution increases, and the values of $\Sigma_{capt.w}$ and $\tau_{d.w}$ approach those for pure water.

Waters of different composition and mineralization of oil and gas deposits, as follows from what has been said above, also differ in terms of values of $\Sigma_{capt.w}$ and $\tau_{d.w}$ (Table 20).

The neutron absorption power of liquid hydrocarbons that compose oil is less than that of confined waters. It grows from pentane having $\Sigma_{capt.o} = 2.002\ m^{-1}$ to *n*-undecane having $\Sigma_{capt.o} = 2.271\ m^{-1}$ due to the packing of the constituents of this series whereas the lifetime of neutrons $\tau_{capt.o}$ in oil hydrocarbons, as their density increases, drops from 227 μ/s (pentane) to 200.15 μ/s (*n*-undecane). The neutron absorption power of oil under atmospheric conditions also diminishes with increasing the density from $2.802\ m^{-1}$ for $\delta_{oil} = 10^3\ kg/m^3$ to $2.223\ m^{-1}$ for $\delta_{oil} = 0.78\ kg/m^3$; accordingly its $\tau_{capt.o}$ increases from 162 to 204.5 μ/s which is close to the values of $\Sigma_{capt.w}$ and $\tau_{d.w}$ of pure water having $\Sigma_{capt.w} = 2.208\ m^{-1}$ and $\tau_{d.w} = 206\ \mu/s$. Confined oil is generally heated, compressed and gas-saturated. Its $\Sigma_{capt.o}$ and $\tau_{capt.o}$ values differ from ones under atmospheric conditions.

The microscopic absorption cross section of dry air at the atmospheric pressure p and room temperature t is very small (about $0.0003\ m^{-1}$), whereas the lifetime of thermal neutrons in it is great (~ 1.5 s). Given the same p and t, $\Sigma_{capt.g}$ (~ 0.0028-$0.0067\ m^{-1}$) is negligible and $\tau_{d.g}$ of methane-series geseous hydrocarbons is very great (162 000-68 000 μ/s). These gases, however, are generally found in confined conditions, e.g. under a pressure 50 MPa and temperature 100 °C. At this, $\Sigma_{capt.g}$ increases to attain $1.6\ m^{-1}$ and $\tau_{d.g}$ decreases to 278 μ/s their values approaching those of Σ_{capt} and τ_d of rock-forming minerals of sedimentary rocks.

However, macroscopic capture cross sections of compressed and heated natural gases are less than Σ_{capt} of confined waters and oils whereas $\tau_{d.g}$ values are greater.

This is particularly true of strongly mineralized chlorine-bearing waters. The $\tau_{d.g}$ values of gases are calculated from the equation:

$$\tau_{d.g} = \frac{3.67 \times 10^{-3} \beta_g T}{v_{av} A p \sum_{k=1}^{k=n} q_k \sum_{i=1}^{i=m} \sigma_{capt.i} n_i} \tag{113}$$

where p and T are the pressure and absolute temperature in the stratum; β_g are the compressibility factor of gas for the given stratal pressure; $v_{av} = 2.48 \times 10^3$ m/s is the average velocity of thermal neutrons; q_k is the relative volume content of the k-th gas; A is the relative atomic mass of gas; $\sigma_{capt.i}$ is the microscopic capture cross section; $n_i = 6.023 \times 10^{23}\ b_i \delta_{g.i}/M_i$ is the number of atoms of the i-th gas in 1 cm^3 of their mixture; b_i is the number of atoms of this gas in one molecule; $\delta_{g.i}$ is the density of the i-th gas; M_i is the relative molecular mass of the gas.

Sec. 74. **Rocks**

The macroscopic capture cross section $\Sigma_{capt.r}$ and values of $1/\tau_{d.r}$ and $1/D_{0r}$ that are reciprocals of the lifetime $\tau_{d.r}$ and diffusion coefficient D_{0r} of neutrons in a clayey oil and water- or gas- and water-saturated rock can be found from Eq. (42)

$$a_r = (1 - k_p - k_{clv}) a_s^* + k_{clv} a_{cl} + k_p k_w a_w + k_p (1 - k_w) a_o (\text{or } a_g) \tag{114}$$

where $a_r = \Sigma_{capt.r} = 1/\tau_{d.r} = 1/D_{0r}$ is the macroscopic capture cross section in an oil and water- or gas- and water-saturated clayey rock; $a_s^* = \Sigma_{capt.s}^* = 1/\tau_{d.s}^* = 1/D_{0s}^*$ is the same but in a solid non-clayey rock component; $a_{cl} = \Sigma_{capt.cl} = 1/\tau_{d.cl} = 1/D_{0cl}$ is the same but in a clayey component of the solid phase; $a_w = \Sigma_{capt.w} = 1/\tau_{d.w} = 1/D_{0w}$ is the same but in water; $a_o = \Sigma_{capt.o} = 1/\tau_{d.o} = 1/D_{0o}$ is the same, in oil; $a_g = \Sigma_{capt.g} = 1/\tau_{d.g} = 1/D_{0g}$ is the same in gas; k_p, k_{clv}, k_w are the voids ratio, clayiness and water saturation coefficients, respectively.

For clean rocks practically containing no clay, Eq. (114) takes on a simplified form. To calculate $\tau_{d.r}$ it has, e.g. this form:

$$\frac{1}{\tau_{d.r}} = (1 - k_p) \frac{1}{\tau_{d.s}^*} + \frac{k_p k_w}{\tau_{d.w}} + k_p (1 - k_w) \frac{1}{\tau_{d.o} (\text{or } \tau_{d.g})} \tag{115}$$

where $\tau_{d.s}^*$ is the lifetime of neutrons in the solid phase having no clayey component.

The composition of the non-clayey solid phase much affects the values of $\tau_{d.r}$, $\Sigma_{capt.r}$ and D_{0r}. A comparison of the lifetimes of thermal neutrons in sandstones, dolomites and limestones having identical k_p and pore water mineralization calculated by D.M. Srebrodol'skii, F.Ts. Denisik, A.A. Polyachenko suggests that their greatest values are shown by sandstones, lower ones by dolomites and still lower ones by limestones. This is good in agreement with the lifetime of thermal neutrons in quartz, dolomite and calcite.

According to the calculations of the above workers, with increasing the clayiness of quartz sandstone, the lifetime of thermal neutrons in their water-, gas- and gas-saturated varieties diminishes. Compared with quartz sandstone, the lifetime of

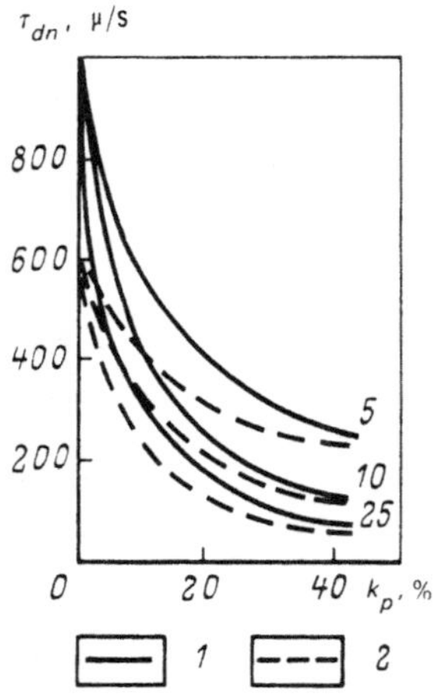

FIG. 149. Connection between the lifetime τ_{dr} of thermal neutrons in water-saturated rocks and their voids ratio k_p (after F.Ts. Denisik and D.M. Srebrodol'skii).

1—quartz sandstone; *2*—limestone

these neutrons is also less in calcareous water-, oil- and gas-saturated quartz sandstone. This regularity is valid for NaCl solution concentration in sandstone 50, 100 and 300 g/l.

Due to the fact that the mean lifetime of thermal neturons in the solid phase is generally higher than in water or oil, the growth of the voids ratio of water- and oil-saturated quartz sandstone, dolomite and sandstone lead to a decrease of $\tau_{d.r}$ (Fig. 149) and increase of $\Sigma_{capt.r}$ of these rocks. The higher the pore water mineralization (Fig. 149) and the higher NaCl concentration, also values $\tau_{d.r}$ are lower, and higher the value Σ_{capt}. However, the lifetime $\tau_{d.o}$ of thermal neutrons in oil is greater than that in mineralized water since the composition of oil practically has no NaCl ions, that is why with increasing the oil saturation of sandstones and limestones as well as other rocks of a specified porosity the $\tau_{d.r}$ values decrease. A.V. Avdeeva has calculated a relationship $\tau_{d\,m.r}/\tau_{d\,o.r} = f(k_o)$ the derivation of which has taken into account the presence in the voids space of an oil-saturated rock of residual water using the dependence $k_{w.r} = f(k_p)$. According to this dependence with

TABLE 21. The Effect of Elements Intensively Absorbing Thermal Neutrons on Their Lifetime τ in Sedimentary Rocks

Element	Element content, g/t			τ*, μs	Element	Element content, g/t			τ*, μs
	Sandstones	Clays	Carbonates			Sandstones	Clays	Carbonates	
B	4.0	150	12	300	Eu	0.7	1	-	427
Cd	0	0.37	0.03	429	Gd	0.7	5	-	390
Sn	3.7	5	1.4	400	Dy	2.6	4	0.8	427

* τ—taking into account the average element content in sedimentary rocks.

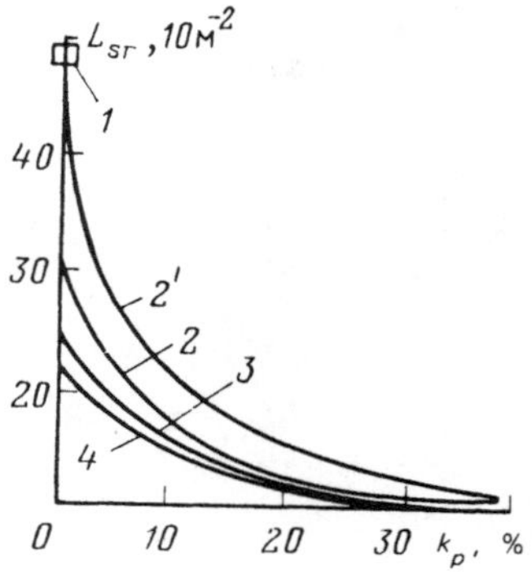

FIG. 150. Dependence of the path of slowing-down of neutrons L_{sr} on the voids ratio k_p of rocks (design) (after A.I. Kedrov).

1—rock salt; *2* and *2'*—sandstones; *3*—dolomite; *4*—limestone. For curve *2'* F_0 = 14 MeV, for others 5 MeV

increasing the oil saturation of rocks the $\tau_{dm.r}/\tau_{do.r}$ ratio decreases since upon substituting mineralized confined water by oil having a lesser neutron activation power $\tau_{d.ro}$ increases.

Increased gas saturation results in greater $\tau_{d.r}$ values, since the lifetime of neutrons even in strongly compressed and heated gases is longer than in oil and water.

The diffusion coefficient is also inversely related to the voids ratio k_p. Its values are comparatively little dependent on the mineralization of confined waters but are sufficiently strongly dependent on the composition of the solid phase. The greatest D_{0r} values, given definite k_p, are shown by sandstones in whose composition quartz is predominant, lesser D_{0r} values are demonstrated by limestone and still lesser ones by dolomite.

Note that the $\tau_{d.r} = f(k_p)$ and $D_{0r} = f(k_p)$ relations established from chemical analyses of the solid phase of a number of sedimentary rock samples of a definite origin, age and conditions of occurrence of strata having the identical composition of pore waters not only differ from ones obtained earlier for rocks of a homogeneous composition of the solid phase (the solid phase represented either by quartz, or calcite or dolomite only) but also fail to correspond to each other. Each of the dependences calculated from the results of chemical analyses of the solid phase of samples taken from a particular stratum displays its own slope rate of the $\tau_{d.r} = f(k_p)$ straight line.

This is accounted for by the heterogeneity of the composition of the solid phase of individual rock varieties in the strata. For example, sandstones may prove to include not only clay minerals but calcite and even minor amounts of elements having very high capture cross sections (boron, lithium, cadmium, samarium, europium, gadolinium and dysprosium). Limestones and dolomites, in addition to calcite and dolomite, also prove to contain argillaceous minerals, pyrite, siderite and other elements intensively absorbing neutrons. The composition and areal distribution of impurities differ for strata referred to different ages. They are also differently distributed in cross sections. Particularly many impurities are manifested, for example, by low-porosity clayey and calcareous sandstones that occur in oil and gas traps together with well-sorted weakly clayey and calcareous high-porosity sandstones.

TABLE 22. The Limiting Values of Quantities Characterizing the Capacity of Rocks for Activation by Neutrons

Rock	Macroscopic cross section, m^{-1}		Length, m		Mean path of neutron, before absorption $\lambda \cdot 10^2$	Mean lifetime of thermal neutron τ_{dr}, μs	Diffusion coefficient $D_{or} \cdot 10^{-5}$, cm^2/s
	of capture $\Sigma_{capt.r}$	of scattering $\Sigma_{scat.r}$	slowing-down, $L_{sr} \cdot 10^2$	diffusion, $L_{d.r}$			
Sedimentary and metamorphic rocks derived from them							
Conglomerate	0.7-3.2	25.6	27	17	-	1100	-
Sand:							
dry	0.9	81	-	6.8	-	506	-
moist	0.4-5.0*	-	-	-	-	-	-
Sandstone:							
gas-saturated	-	-	-	-	-	760-320	-
oil-saturated**	-	-	-	-	-	968-220	-
moist	-	-	-	-	-	1050-84	0.7-2.52
Sandstone***	0.866	-	-	-	-	-	-
Clay:							
dry	-	-	-	14.8	197	890	-
moist	-	-	-	4	75	340	-
Clay	0.5-1.2	19-60	28	4-18.7	75-197	910-340	1.007-1.9
Shale	1.0-1.5	-	-	-	-	-	-
Gneiss***	0.7-1.1	-	-	-	-	-	-
Limestone	0.7-3.6*	-	-	-	-	615-83	0.65-2.3
Dolomite	0.45-3.5*	44	-	12.8	217.4	988-924	-
Marl	0.8-3.0	-	-	-	-	600-530	1.46
Anhydrite	1.3	27-34.8	26-27****	8.97-9.8	84	350-380	-

Gypsum	1.7-1.8	151	11	3.6-3.7	55-56	268-250	-
Rock salt	50-75	31-44	-	1.0-1.2	1.41	6.4-6.07	-
Coal:							
brown	1.9	233.5	-	2.7	-	240	-
hard	0.154-1.6	211	-	3.1	-	284	-
Anthracite	0.108-0.9	137	-	4.9-5.2	114	520-501	-
			Magmatic rocks				
Dunite (4 samples)	1.703	-	-	-	-	-	-
Gabbro (27 samples)	2.147						
Norite (11 samples)	1.28						
Diorite (13 samples)	1.43						
Syenite (24 samples)	1.64						
Granodiorite							
(11 samples)	1.133						
Granite							
(average of 155 samples)	1.162						
Granite	0.7-1.1						
Basalt	2.10-2.22						
Diabase							
(average of 6 samples)	1.722						

* Maximum values are extremely rare.

**k_o = 0.9 at water mineralization 50 g/l; for other mineralizations k_o = 0.8.

*** No detailed characteristic of the rock is available.

**** L_{sr} is calculated without allowing for the first path.

Data illustrating the contents of elements intensively absorbing thermal neutrons and their effect on the lifetime of the latter in sedimentary rocks are listed in Table 21. The lifetime of thermal neutrons in the solid phase of sedimentary rocks (sandstones, limestones and dolomites) in the calculations were taken to be equal on the average to 430 μs if the rocks did not contain elements having high absorption cross sections. The lifetime decreases if a sedimentary rock contains an element intensively absorbing thermal neutrons whose content is an average one compared with that in sandstones, clay and carbonate rocks (see Table 21).

The slowing-down length $L_{s.r}$ (Fig. 150) and diffusion length $L_{d.r}$ are also inversely related to the voids ratio k_p. For example, it can be seen from Fig. 150 that the slowing-down length $L_{s.r}$ is much more dependent on the voids ratio (i.e. on the volumetric content of the liquid phase) than on the mineral composition of the solid phase of rocks.

A distinct differentiation of rocks in terms of their $\Sigma_{capt.r}$, $L_{s.r}$, $\tau_{s.r}$, $L_{d.r}$ and $\tau_{d.r}$ has not as yet been established (Table 22). What has been best studied is $\tau_{d.r}$ of sedimentary rocks. The highest $\tau_{d.r}$ values (615-1100 μs) refer to dry high-porosity or moist low-porosity sedimentary rock varieties, and the least ones (84-87 μs) refer to the same maximum-moisture rocks, particularly, under conditions of their saturation by strongly mineralized water. Rock salt displays a very low (6.4-6.7 μs) $\tau_{d.r}$ value. The range of variation of $\tau_{d.r}$ of igneous rocks is not reported in literature. As to the average $\Sigma_{capt.r}$ values, they increase as we pass from their acidic ($\Sigma_{capt.r} = 1.16\ m^{-1}$) to basic ($\Sigma_{capt.r} = 2.15\ m^{-1}$) to ultrabasic ($\Sigma_{capt.r} = 1.7\ m^{-1}$) varieties. Consequently, $\tau_{d.r}$ values decrease in the same direction.

Chapter XIII

Elasticity

Elasticity is the property of rocks to change in shape and size and frequently in volume due to the external action (stress, volumetric force, heating etc.) to then recover completely or partially the original configuration after the forces have been removed and their action has not exceeded the elastic limit.

A change in size and shape of a rock (deformation) results from the mutual displacement of its constituent elements: atoms, molecules or ions of crystal lattice nodes, solid-phase mineral grains or molecules or ions of the liquid and gaseous phase. Rock deformation is counteracted by internal forces, of interaction between their different particles.

Owing to this fact a rock under load is in a state of stress. If a force, varying in magnitude and direction, is acting in its minor region, then in its immediate environment particle displacements varying in value and direction take place as do states of stress of the rock in hand. Thus rock particles of the region under consideration begin to vibrate. These variations in the state of stress and particle vibrations expand to the adjacent and subsequent unit volumes, i.e. a longitudinal and transverse elastic waves pass across the rock. In so doing, due to the absorption and scattering of elastic energy by the rock the vibrations propagating in the rock decrease in amplitude, the vibration process being attenuated. The capacity of rocks for deformation and transfer of elastic vibrations at a definite rate and to a definite distance as well as for absorption and scattering of elastic energy is characterized by a number of petrophysical quantities connected with their other characteristics and temperature and pressure.

Sec. 75. **Minerals**

Elasticity modulus and coefficient. The phase and material homogeneity of individual minerals sometimes permit us to consider their polycrystalline varieties as being quasi-homogeneous and isotropic continuous media, and their single crystals as being anisotropic bodies. The quasi-homogeneity of polycrystalline minerals shows itself in the identity of elastic properties of their individual volumes each of which is much greater than that of their average-size crystal. As to their isotropism, it is revealed in the identity of properties under study in any orientation of individual volumes: anisotropic single crystals of minerals show elastic properties that differ in different directions from the point specified in the crystal.

The elasticity of homogeneous, isotropic and anisotropic continuous media has been sufficiently studied, the result being summed up essentially as follows. If a

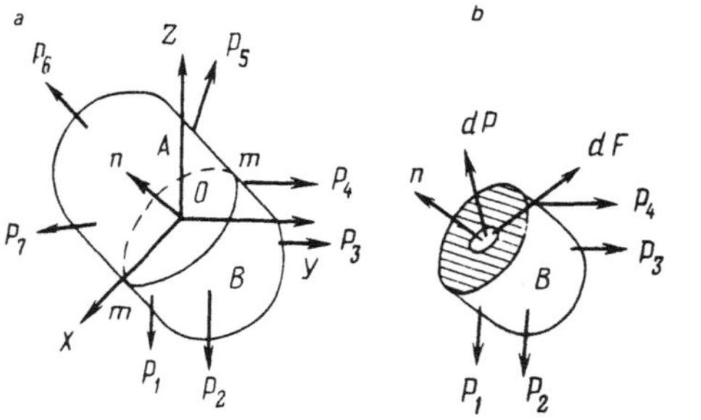

FIG. 151. Diagram illustrating the state of stress of a mineral sample being acted on by the forces (*a*) and the concept of strain (*b*)

sample of a medium, e.g. a mineral, is acted on by balanced forces $\vec{P}_1$-$\vec{P}_7$ (Fig. 151), it deforms and passes to a state of stress. Let us divide a mineral sample into portions *A* and *B* by a plane *mm* passing through point *O*. Portion *B* can be considered to be in equilibrium due to the action of forces $\vec{P}_1$-$\vec{P}_4$ and internal forces continuously distributed over the surface *mm*. The latter forces are characterized by strain, a force acting on a unit area of the surface at which they are operating.

In the general case (see Fig. 151*b*) the stress is nonuniformly distributed over the cross section *mm* in which case its magnitude in any point is determined by the limiting ratio $\Delta p/\Delta F = p_n^*$.

The stress vector $\vec{p}_n$ acting at the particular point on any sloping elementary area in the mineral is completely defined by six independent components σ_x, σ_y, σ_z, τ_{xy}, τ_{yz} and τ_{zx} of stress acting at three unit areas perpendicular to the coordinate axes and forming with a sloping area a tetrahedron (Fig. 152). To get $\vec{p}_n$ we write the conditions of the equilibrium state of the tetrahedron taking into account a theorem on the area of projection (sloping area of tetrahedron). Then the components of vector $\vec{p}_n$ prove equal to

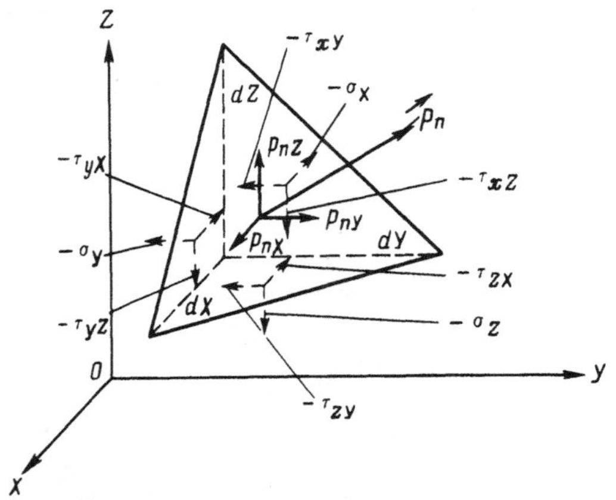

FIG. 152. Components of the strain vector $\vec{p}_n$

* The subscript *n* of the stress vector acting on a unit area *F* implies that the outer normal to *F* has an orientation *n*.

$$p_{nx} = \sigma_x \cos(n, x) + \tau_{yx} \cos(n, y) + \tau_{zx} \cos(n, z)$$
$$p_{ny} = \tau_{xy} \cos(n, x) + \sigma_y \cos(n, y) + \tau_{zy} \cos(n, z)$$
$$p_{nz} = \tau_{xz} \cos(n, x) + \tau_{yz} \cos(n, y) + \sigma_z \cos(n, z)$$

where (n, x), (n, y) and (n, z) are angles made by a normal to the sloping unit area in the mineral with the coordinate axes.

Since $\tau_{yx} = \tau_{xy}$, $\tau_{zy} = \tau_{yz}$ and $\tau_{zx} = \tau_{xz}$, p_n is known provided that the above components have been determined. The components σ_x, σ_y and σ_z are normal, and τ_{xy}, τ_{yz} and τ_{zx} are tangential to unit areas that are perpendicular to the coordinate axes. The stress components are connected with one another and volumetric forces through equilibrium equation. In order to get these equations, we mentally cut out of the mineral a unit parallelepiped. It is in equilibrium provided that

$$\frac{\partial \sigma_x}{\partial x} + \frac{\partial \tau_{yx}}{\partial y} + \frac{\partial \tau_{zx}}{\partial z} + \delta_m g_x = 0$$

$$\frac{\partial \tau_{xy}}{\partial x} + \frac{\partial \sigma_y}{\partial y} + \frac{\partial \tau_{zy}}{\partial z} + \delta_m g_y = 0$$

$$\frac{\partial \tau_{xz}}{\partial x} + \frac{\partial \tau_{yz}}{\partial y} + \frac{\partial \sigma_z}{\partial z} + \delta_m g_z = 0$$

where g_x, g_y and g_z are projections of mass force g onto the coordinate axes; δ_m is the density of the material of the parallelepiped (mineral).

The relative displacement of adjacent particles of the mineral, P and Q, (its deformation) is generally characterized by a relative displacement e_{PQ} of a section connecting the aforementioned particles:

$$e_{PQ} = \frac{L_1 - L}{L} = l^2 \frac{\partial u}{\partial x} + m^2 \frac{\partial v}{\partial y} + n^2 \frac{\partial w}{\partial z} + mn\left(\frac{\partial v}{\partial z} + \frac{\partial w}{\partial y}\right) + nl\left(\frac{\partial w}{\partial x} + \frac{\partial u}{\partial z}\right) + lm\left(\frac{\partial u}{\partial y} + \frac{\partial v}{\partial x}\right)$$

where L and L_1 are the lengths of line sections connecting P and Q in a non-loaded and a loaded state: l, m, n are direction cosines of section PQ in a non-deformed state; u, v and w are particle displacements in directions x, y and z (projections of its displacement).

If segment PQ prior to deformation was in parallel to the OX axis, then $l = 1$, $m = n = 0$, $e_{PQ} = \partial u/\partial x = e_x$ and provides a relative extension of the line section along the X axis.

In like manner we get the relationships $e_y = \partial v/\partial y$ and $e_z = \partial w/\partial z$ which represent relative extensions of line segments before deformation parallel to the Y and Z axes.

It can also be proved that the values (displacements)

$$\frac{\partial u}{\partial y} + \frac{\partial v}{\partial x} = e_{xy}; \quad \frac{\partial v}{\partial z} + \frac{\partial w}{\partial y} = e_{yz} \text{ and } \frac{\partial w}{\partial x} + \frac{\partial u}{\partial z} = e_{zx}$$

express (in radians) the deformation (decrease) of the angle between mutually perpendicular straight lines passing through the point of interest, O.

Thus, a parallelepiped having sides before deformation parallel, respectively, to the X, Y and Z axis, and equal to Δx, Δy, Δz and a volume $\Delta V = \Delta x \Delta y \Delta z$ as a result of overall extension will have edges of a length

$$\Delta x\left(1 + \frac{\partial u}{\partial x}\right), \quad \Delta y\left(1 + \frac{\partial v}{\partial y}\right) \quad \text{and} \quad \Delta z\left(1 + \frac{\partial w}{\partial z}\right)$$

and volume

$$V + \Delta V = \Delta x \Delta y \Delta z\left(1 + \frac{\partial u}{\partial x}\right)\left(1 + \frac{\partial v}{\partial y}\right)\left(1 + \frac{\partial w}{\partial z}\right)$$

where ΔV is the increment of the volume, V, of the parallelepiped due to its extension.

Hence, by neglecting squares and products of derivatives of displacements, we get

$$\frac{\Delta V}{V} = \theta = \frac{\partial u}{\partial x} + \frac{\partial v}{\partial y} + \frac{\partial w}{\partial z} = e_x + e_y + e_z = \operatorname{div} \vec{U}$$

where $\vec{U}$ is the displacement vector. Note that $\theta = \Delta V/V$ is said to be a volumetric expansion in the vicinity of a given point.

Under conditions of the most general state of stress where not only normal but also tangential stresses act at the faces of a unit parallelepiped near the given point, the edges of the parallelepiped upon passage from a non-stressed to a state of stress extend by values e_x, e_y and e_z, and the tangential stresses (strains) change the right angles made by the adjacent faces e_{xy}, e_{yz}, e_{zx}.

In the general case rock deformation at a given point is quite definite if we know the three values of extension, e_x, e_y and e_z and three displacements, e_{xy}, e_{yz}, e_{zx} said to be deformation components.

For anisotropic media, given stresses less than the limit of elasticity, according to the generalized Hooke's law, the components of stress tensor and deformation are connected through the following relationships:

$$\left.\begin{aligned}
\sigma_x &= C_{11}e_x + C_{12}e_y + C_{13}e_z + C_{14}e_{yz} + C_{15}e_{zx} + C_{16}e_{xy}\\
\sigma_y &= C_{21}e_x + C_{22}e_y + C_{23}e_z + C_{24}e_{yz} + C_{25}e_{zx} + C_{26}e_{xy}\\
\sigma_z &= C_{31}e_x + C_{32}e_y + C_{33}e_z + C_{34}e_{yz} + C_{35}e_{zx} + C_{36}e_{xy}\\
\tau_{yz} &= C_{41}e_x + C_{42}e_y + C_{43}e_z + C_{44}e_{yz} + C_{45}e_{zx} + C_{46}e_{xy}\\
\tau_{zx} &= C_{51}e_x + C_{52}e_y + C_{53}e_z + C_{54}e_{yz} + C_{55}e_{zx} + C_{56}e_{xy}\\
\tau_{xy} &= C_{61}e_x + C_{62}e_y + C_{63}e_z + C_{64}e_{yz} + C_{65}e_{zx} + C_{66}e_{xy}
\end{aligned}\right\} \tag{116}$$

and, vice versa, six deformation components are linear functions of six independent stresses, and proportionality coefficients in this case are quantities called elasticity coefficients as distinct from proportionality coefficients in Eqs. (116). Of 36 moduli of elasticity C_{11}, C_{12}, ..., C_{66} in Eq. (116) only 21 moduli are independent. This number of moduli characterizes the elasticity of triclinic syngony single crystals. The elasticity of single crystals of other syngonies is determined by a lesser number of moduli of elasticity due to their higher symmetry: 13 for monoclinic syngony;

9 for orthorhombic; 7 for trigonal, 6 or 7 for tetragonal, 5 for hexagonal and 3 elasticity moduli for cubic syngony.

Independent moduli of elasticity in cubical crystals are considered to be provided, for example, by C_{11}, C_{12} and C_{44} and are determined as a result of measurements: (a) of the relative decrease of the length of the crystal upon its compression by a known stress: (b) of the relative decrease of the transverse dimension of the crystal, given the same state of stress; (c) of the displacement due to the spalling stress normal to the cubical axis.

Moduli and coefficients of elasticity of the most important rock-forming minerals (quartz, calcite, biotite, olivine, magnetite and many more) have been studied to establish interconnections between elasticity characteristics of single crystals of different structure, estimates of moduli of elasticity of the solid phase from moduli of elasticity and concentration of its component minerals.

Moduli C_{ij} of single crystals differ very much in value and sometimes in sign. They also differ in sign, value and number for minerals of different structure. The same is true also of elasticity coefficients. This mirrors the dependence of characteristics of elasticity of single crystals on their characteristics of elasticity of single crystals on their actual crystal structure, very much complex processes of various microscopic interactions in the crystal lattice.

According to K.S. Aleksandrov and T.V. Ryzhova, the values of elasticity coefficients and moduli and therefore of forces of bonds in the crystal structure are governed by ionic radii, valency and coordination numbers* of crystal cations. They increase with increasing the valency and coordination number and decreasing the ionic radius of the cations. This rule makes it possible to account for the change in the elasticity of single crystals of isomorphic series of minerals where this variation is connected with the substitution of cations. For example, the decrease in the elasticity of garnets upon substitution of Mg^{2+} (ionic radius $r = 0.074$ nm) by Fe^{2+} ($r = 0.08$ nm) or growth of moduli and coefficients of elasticity in the plagioclase series with increasing the anorthite component content attributed to the substitution of the Na^{+} ions by the Ca^{2+} ions. If in place of Na and Ca, as is the case in plagioclases, K characteristic of structures of soda potash feldspars appears in the crystal lattice of a mineral, then the values of elasticity characteristics decrease compared with the respective characteristics of plagioclases. Some more examples illustrating this rule can be provided.

If several cations are found in the crystal lattice of a mineral, the above rule will enable us to determine which of them are responsible for the strength of bonds in different directions.

A total study of the dependence of values of C_{ij} on pressure and temperature has been conducted solely for quartz in the pressure intervals from atmospheric to 210 MPa and temperature range from 25 to −195.8 °C. C_{ij} values of quartz increase with decreasing the temperature, in this case the greater the pressure, the higher the rate of increase.

* The coordination number is the number of the nearest neighbours of a definite atom (ion) in a crystal.

Any two simple moduli of elasticity will suffice to describe the elasticity of homogeneous and isotropic polycrystal minerals. In this case the equations of the generalized Hooke's law reduce to this form:

$$\left.\begin{aligned}
\sigma_x &= \frac{\nu E}{(1+\nu)(1-2\nu)}\theta + \frac{E}{(1+\nu)} e_x = \lambda\theta + 2\mu e_x \\
\sigma_y &= \frac{\nu E}{(1+\nu)(1-2\nu)}\theta + \frac{E}{(1+\nu)} e_y = \lambda\theta + 2\mu e_y \\
\sigma_z &= \frac{\nu E}{(1+\nu)(1-2\nu)}\theta + \frac{E}{(1+\nu)} e_z = \lambda\theta + 2\mu e_z \\
\tau_{yz} &= \frac{E}{2(1+\nu)} e_{yz} = \mu e_{yz} \\
\tau_{zx} &= \frac{E}{2(1+\nu)} e_{zx} = \mu e_{zx} \\
\tau_{xy} &= \frac{E}{2(1+\nu)} e_{xy} = \mu e_{xy}
\end{aligned}\right\} \qquad (117)$$

since in Eq. (116)

$$C_{11} = C_{22} = C_{33} = \lambda + 2\mu$$

$$C_{44} = C_{55} = C_{66} = \mu$$

$$C_{12} = C_{21} = C_{13} = C_{31} = C_{32} = \lambda$$

the remaining C_{ij} ($i = 1, 2, ..., 6$ and $j = 1, 2, ..., 6$) are equal to zero. In Eqs. (117) the following symbols are presented: $E = C_{11}$ is *the modulus of linear expansion* (*Young's modulus*); ν *is the ratio of the transverse contracting strain or Poisson ratio*; λ and μ *are Lamé's coefficients* which are thus related with E and ν:

$$\lambda = \frac{\nu}{(1+\nu)(1-2\nu)} E \quad \text{and} \quad \mu = \frac{1}{2}\frac{E}{1+\nu}$$

Young's modulus provides a ratio of an acting longitudinal-isolated tension stress σ_x (or σ_y and σ_z) applied to a material to the resulting strain parallel to the tension e_x (or e_y and e_z). The elongation strain appears, for example, if a pair of straining forces P_1 (or P_2 or P_3) acting parallel to the X (Y or Z) axis (Fig. 153) is applied to the centres of parallel faces of a mineral (rock) bar.

In each of these cases a cube isolated in a non-loaded sample with edges $a_1 = a_2 = a_3$ (see Fig. 153*a*) is deformed in the state of stress such that its edges parallel to the force applied get relative increments e_x or e_y and e_z proportional to the appearing stress σ_x (σ_y or σ_z) and the length of edges perpendicular to them decreases. Relative decrements of these edges of the cube are equal and proportional to the longitudinal deformation e_x (or e_y or e_z).

If the lengths of the sides of the cube deformed by a pair of strains P_1 are equal to a_1', a_2', a_3', and by the other two pairs of strains P_2 and P_3 equal to a_1'', a_2'', a_3'' and a_1''', a_2''', a_3''', then in conformity with what has been said above, we can write the following:

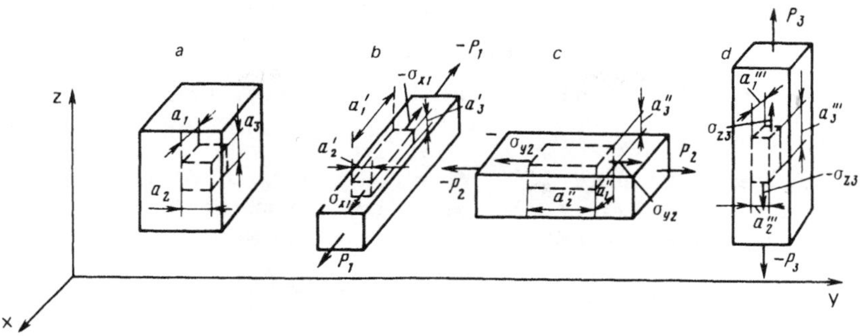

FIG. 153. Diagram illustrating a simple expansion of a bar of a mineral.
a—bar of mineral before application of expanding effort; *b-d*—the same bar, but the pairs of forces parallel to *x*, *y* and *z* axes act on the parallel faces, respectively, perpendicular to *x*, *y* and *z* axes

$$e_{x1} = \frac{a_1' - a_1}{a_1} = \frac{1}{E}\sigma_{x1}$$

$$e_{y1} = e_{z1} = \frac{a_2' - a_2}{a_3} = \frac{a_3' - a_3}{a_3} = -\nu e_{x1} = -\frac{\nu}{E}\sigma_{x1}$$

$$e_{y2} = \frac{a_2'' - a_2}{a_2} = \frac{1}{E}\sigma_{y2}$$

$$e_{x2} = e_{z2} = \frac{a_1'' - a_1}{a_1} = \frac{a_3'' - a_3}{a_3} = -\nu e_{y2} = -\frac{\nu}{E}\sigma_{y2}$$

$$e_{z3} = \frac{a_3''' - a_3}{a_3} = \frac{1}{E}\sigma_{z3}$$

$$e_{x3} = e_{y3} = \frac{a_1''' - a_1}{a_1} = \frac{a_2''' - a_2}{a_2} = -\nu e_{z3} = -\frac{\nu}{E}\sigma_{z3}$$

where e_{x1}, e_{x2} and e_{x3} are deformations of edges of the cube parallel to the X axis upon sequential application of strains σ_{x1}, σ_{y2} and σ_{z3}; e_{y1}, e_{y2} and e_{y3} and e_{z1}, e_{z2} and e_{z3} are deformations of the cube's edges parallel to the Y and Z axes, respectively.

If strains σ_{x1}, σ_{y2} and σ_{z3} are applied to the cube (see Fig. 153*a*) simultaneously, deformations of its edges oriented similarly relative to the coordinate axes are summed up, and deformations e_{x123}, e_{y123}, e_{z123} of the cube's edges are found from relationships:

$$e_{x123} = e_{x1} + e_{x2} + e_{x3} = \frac{1}{E}\sigma_{x1} - \frac{\nu}{E}\sigma_{y2} - \frac{\nu}{E}\sigma_{z3} = \frac{1}{E}[\sigma_{x1} - \nu(\sigma_{y2} - \sigma_{z3})]$$

$$e_{y123} = e_{y1} + e_{y2} + e_{y3} = \frac{1}{E}[\sigma_{y2} - \nu(\sigma_{x1} + \sigma_{z3})]$$

$$e_{z123} = e_{z1} + e_{z2} + e_{z3} = \frac{1}{E}[\sigma_{z3} - \nu(\sigma_{x1} + \sigma_{y2})]$$

By adding together the equations obtained we get deformation

$$\Delta = e_{x123} + e_{y123} + e_{z123} = [(1 - 2\nu)(\sigma_{x1} + \sigma_{y2} + \sigma_{z3})]/E$$

For $P_1 = P_2 = P_3 = P$ and, consequently, for $\sigma_{x1} = \sigma_{y2} = \sigma_{z3}$

$$\Delta = 3(1 - 2\nu)\sigma/E$$

As has been shown above, $\Delta = \Delta V/V = e_x + e_y + e_z$, i.e. provides a relative increment of the volume of the cube upon its overall elongation by equal strains normal to the opposite faces of the cube. Thus we can write

$$\frac{\Delta V}{V} = \frac{3(1 - 2\nu)}{E}\sigma = \beta\sigma = \frac{1}{K}\sigma,$$

where $K = E/3(1 - 2\nu)$ is the *modulus of overall expansion* (or compression); $\beta = 1/K$ is the *coefficient of compressibility.*

The modulus K defines the proportionality of the relative volumetric expansion $\Delta V/V = e_x + e_y + e_z$ or compression of the applied strain σ_x, σ_y, σ_z, provided they are equal and in the absence of other components of strain. This corresponds to the overall hydrostatic pressure.

$$\sigma_x = \sigma_y = \sigma_z = \sigma = -K(\Delta V/V) = -K(e_x + e_y + e_z)$$

The theory of elasticity considers another simple quantity, $G = \mu = E/2(1 + \nu)$. *The shear modulus* G or Lamé's coefficient μ establishes a ratio of the tangential stress acting individually to the corresponding deformation. E.g., between τ_{yz} and e_{yz}, at this, $G = C_{44}$: $\tau_{yz} = \mu e_{yz} = Ge_{yz} = C_{44}e_{yz}$. The moduli E, G and K have a stress difference, their SI units of measurement are N/m^2 or Pa, ν is a dimensionless quantity. The quantity $\beta = 1/K = 3(1 - 2\nu)/E$ is said to be the coefficient of overall compression or compressibility. Its units of measurement are reciprocals of units of measurement of K.

The quantities E, ν, $G(\mu)$, K are interconnected through definite relationships (Table 20). This makes it possible, given two simple characteristics, to find the remaining ones.

The values of E, G, K and coefficients ν and β of polycrystalline minerals can be established by averaging moduli and coefficients of elasticity of their single crystals and from data of their measurements in laboratory.

The values of E, G, K and ν have been obtained for the most important rock-forming minerals of sedimentary, igneous and metamorphic rocks as well as for ore minerals and pure metals. They vary over a fairly large range: $E = (0.1\text{-}10) \times 10^{11}$ Pa, $G = (0.1\text{-}4) \times 10^{11}$ Pa, $K \approx (0.25\text{-}2.5) \times 10^{11}$ Pa (few measurements of this quantity have been made). It must be noted that design and experimental values of K are in fairly good agreement.

In terms of E, G and K, minerals can be referred to groups having high, increased, average and low values of the quantities under consideration. The limits of variation of E, G and K are listed in Table 23.

Group I includes such a native metal as iron, nonmetals such as diamond, sulphides such as pyrite and marcasite, AX_2 type oxides (rutile, cassiterite etc.), AX type (periclase), A_2X_3 type (corundum, hematite), uraniferous and thorium-bearing

TABLE 23. The Limits of Variation of Elasticity Moduli (E, G and K) of Group of Minerals Separated by Their Values

Group	Characteristic of group of minerals according to their moduli of elasticity	E	G	K
			$\cdot 10^{-11}$, Pa	
I	High	2->4	0.6->0.8	1.0->1.25
II	Higher	1.2-2.0	0.4-0.6	0.75-1.0
III	Medium	0.5-1.0	0.2-0.4	0.5-0.75
IV	Low	<0.5	<0.2	<0.5

oxides (uraninite and thorianite), compound oxides except some chromite varieties (spinel, magnetite, ilmenite etc.), most nesosilicates (forsterite, grossular, garnet, zircon etc.), rare varieties of alkali silicates (beryll, tourmaline).

Group II includes such native metals as platinum, palladium, tantalum, and arsenides as cobaltine and smaltite, and also AX type oxides as zincite, circonium-bearing oxides as baddeleyite, phospates as apatite, most inosilicates such as pyroxenes (diopside, augite, diallage etc.), layer silicates as serpentine.

Group III is represented by metals—silver and antimony, nonmetals—graphite and selenium, many sulphides and sulphsalts (galenite, sphalerite, pyrrhotite, chalcopyrite etc.), anhydrous sulphates (anhydrite, barite etc.), anhydrous carbonates except magnesite (calcite, siderite, dolomite etc.), ribbon silicates (hornblende, tremolite etc.), layer (sheet) silicates of mica and brittle mica groups [muscovite, phlogopite (also known as bronze or brown mica), biotite, chlorites], most tectosilicates: quartz group minerals (quartz and chalcedony), feldspars (albite-anorthite), soda potash feldspars (orthoclase, microcline), feldspathoids (nepheline, leucite, analcime), nitrates-sodium saltpeter.

Finally, Group IV comprises metals—gold, lead, bismuth and tantalum, some sulphides—arsenopyrite, cinnabar, argentite, chalcosine, bismuthine, known A_2X oxides—ice, hydrous sulphates (gypsum, epsomite), carbonates—magnesite, layer silicates—talc and chrysocolla, chlorites—halite, sylvite, cerargyrite.

This regularity in the variation of values of the moduli of elasticity of main rock-forming minerals of sedimentary, igneous, metamorphic rocks is accounted for by N.B. Dortman and other workers. The weighted mean relative atomic weight of main rock-forming minerals (olivines, pyroxenes, feldspars, quartz etc.) relatively little changes. They are characterized by a covalent interatomic bonding, regular variation of the packing density of atoms in a mineral governed by atomic radii and structure and responsible for the elasticity of these compounds. The packing density of atoms in silicates, for example, and stronger covalent bonding between principal structural units (silicon hydrogen or aluminium silicon hydrogen tetrahedra) increase in the series tectosilicates—inosilicates—nesosilicates which leads to increased values of moduli of elasticity of minerals passing from the first to the last term of the series. A still greater packing density of atoms, strength of covalent bonding of structural units (aluminium oxygen tetrahedra) are shown by such oxides as corundum and spinel also having a nesosilicate structure. Due to this their elasticity moduli have greater magnitudes compared with nesosilicates. The highest elasticity is demonstrated by diamond owing to the perfect and strong tetrahedra of carbon. Since the density of rock-forming minerals is conditioned by the same features as moduli of elasticity, there is direct correlation between these quantities.

The change in values of moduli of elasticity of sulphides, iron oxides, other ore minerals and native metals is mainly associated with a change of their weighted mean relative atomic mass. These minerals are characterized by a covalent-metallic or metallic type of atomic bonding, minor dependence of the structure and properties of the atomic radii and packing density of these properties in crystals and a much more intimate connection between the structure and properties and their weighted mean relative atomic mass. The lesser the moduli of elasticity, the greater is the atomic mass and, consequently, the density of ore minerals.

Owing to this, values of moduli of elasticity of this mineral group decrease with increasing their density, but they increase with growth of their hardness. However, the values of moduli of elasticity of these minerals are greater for a specified hardness compared with ones for principal petrogenic minerals.

The values of the coefficient ν vary over a lesser range from 0.08 (for quartz) to 0.34-0.35 (for zircon, gypsum, sphalerite). Most generally it varies from 0.24 to 0.26.

Moduli E, G and K and coefficients ν, β of minerals are governed by temperature and pressure. The variation of the compressibility of principal rock-forming minerals due to pressure and temperature has been studied. It has been shown to increase with pressure and decreases with temperature.

The speed of propagation of elastic waves. Let us isolate in a sample of quasi-homogeneous and isotropic mineral near the point M whose coordinates are x, y, z, a unit parallelepiped with sides dx, dy and dz, volume $dV = dxdydz$ and mass $\delta_m dV$ where δ_m is the mineral's density.

Since the product of the mass of a unit parallelepiped with acceleration along one of the axes (e.g. $\partial^2 u/\partial t^2$ along the X axis) equals a sum of projections along any of the axes of all forces acting on the parallelepiped (e.g. the sum $(\partial\sigma_x/\partial x + \partial\tau_{xy}/\partial y + \partial\tau_{zx}/\partial z)dV$) on the X axis, then for a quasi-homogeneous and isotropic mineral the equation of movement of point M in strains will take on this form

$$\left.\begin{aligned} \frac{\partial\sigma_x}{\partial x} + \frac{\partial\tau_{xy}}{\partial y} + \frac{\partial\tau_{zx}}{\partial z} &= \delta_m \frac{\partial^2 u}{\partial t^2} \\ \frac{\partial\tau_{xy}}{\partial x} + \frac{\partial\sigma_y}{\partial y} + \frac{\partial\tau_{yz}}{\partial z} &= \delta_m \frac{\partial^2 v}{\partial t^2} \\ \frac{\partial\tau_{zx}}{\partial x} + \frac{\partial\tau_{yz}}{\partial y} + \frac{\partial\sigma_z}{\partial z} &= \delta_m \frac{\partial^2 w}{\partial t^2} \end{aligned}\right\} \qquad (118)$$

or a vector form

$$\operatorname{div} \vec{P} = \delta_m \frac{\partial^2 \vec{U}}{\partial t^2}$$

where u, v and w are projections of the displacement of point M on X, Y and Z axes; $\vec{P}$ is the stress vector; $\vec{U}$ is the displacement vector.

Equations (118) can be transformed into equations of movement of point M in displacements when taking into account regularities mirrored in Eq. (117) concerning strains and deformations:

$$\left.\begin{aligned} \delta_m \frac{\partial^2 u}{\partial t^2} &= (\lambda + \mu) \frac{\partial \theta}{\partial x} + \mu \nabla^2 u \\ \delta_m \frac{\partial^2 v}{\partial t^2} &= (\lambda + \mu) \frac{\partial \theta}{\partial y} + \mu \nabla^2 v \\ \delta_m \frac{\partial^2 w}{\partial t^2} &= (\lambda + \mu) \frac{\partial \theta}{\partial z} + \mu \nabla^2 w \end{aligned}\right\} \qquad (119)$$

where dilatation, increase in volume near point M is

$$\theta = \operatorname{div} \vec{U} = \frac{\partial u}{\partial x} + \frac{\partial v}{\partial y} + \frac{\partial w}{\partial z} = \frac{\Delta V}{V}$$

$$\nabla^2 = \frac{\partial^2}{\partial x^2} + \frac{\partial^2}{\partial y^2} + \frac{\partial^2}{\partial z^2}$$

Equations (119) reduce to two relationships*:

$$\frac{\partial^2 \theta}{\partial t^2} = \frac{\lambda + \mu}{\delta_m} \nabla^2 \theta \qquad (120)$$

$$\mu \nabla^2 \omega = \delta_m \frac{\partial^2 \omega}{\partial t^2} \qquad (121)$$

A substitution of $\omega = \operatorname{rot} \vec{U}$ yields

$$\frac{\partial^2 \operatorname{rot} \vec{U}}{\partial t^2} = \frac{\mu}{\delta_m} \nabla^2 \operatorname{rot} \vec{U}$$

Equation (120) describes the law of propagation of longitudinal compression waves. The variable θ represents dilatation near the specified point.

---

* To get Eq. (120), we must differentiate each of Eqs. (119), respectively, with respect to x, y, z, add together the equations obtained and represent the results in a vector equation form. Equation (121) is derived as follows. Differentiate the first of Eqs. (119) with respect to y, the second with respect to x, subtract the second from the first derivative. This yields

$$\delta_r \frac{\partial^2}{\partial t^2}\left(\frac{\partial v}{\partial x} - \frac{\partial u}{\partial y}\right) = \mu \nabla^2 \left(\frac{\partial v}{\partial x} - \frac{\partial u}{\partial y}\right)$$

In like manner we derive two other similar equations integrating the three equations obtained to form one vector equation.

The integral of Eq. (120) is a function of x, y, z, t. At t = const this function determines the pattern of distribution of θ in the volume of the mineral under study. Assuming x, y, z to be constant, we can determine the pattern of variation of mineral unit volume with time.

Whereas Eq. (120) corresponds to compression and dilatation, Eq. (121) describes torsion and shear. By integrating Eqs. (120) and (121), we can show that the $(1/r)F_1(r - v_P t)$ and $(1/r)F_2(r - v_S t)$ form functions satisfy differential equations (120) and (121) if

$$v_P = \sqrt{(\lambda + 2\mu)/\delta_m} \quad \text{and} \quad v_S = \sqrt{\mu/\delta_m}$$

where r is the distance from point M to the origin of coordinates.

As also follows from the form of these functions, v_P and v_S are, respectively, speeds of propagation of elastic waves—compression and dilatation and torsion or shear waves.

Thus longitudinal and transverse waves propagate in quasi-homogeneous and isotropic minerals at speeds:

$$v_P = \sqrt{\frac{E(1-\nu)}{\delta_m(1+\nu)(1-2\nu)}} = \sqrt{\frac{\lambda + 2\mu}{\delta_m}} = \sqrt{\frac{K + 4/3G}{\delta_m}} \tag{122}$$

$$v_S = \sqrt{\frac{E}{\delta_m 2(1+\nu)}} = \sqrt{\frac{G}{\delta_m}} = \sqrt{\frac{\mu}{\delta_m}} \tag{123}$$

Since mineral elasticity characteristics are governed by temperature and pressure, so do speeds of wave propagation in minerals depend on the same factors.

Speeds of propagation of longitudinal elastic waves in minerals vary from 2 (gold) to 18 (diamond) km/s, and of transverse waves from 1.2 to over 7.1 km/s.

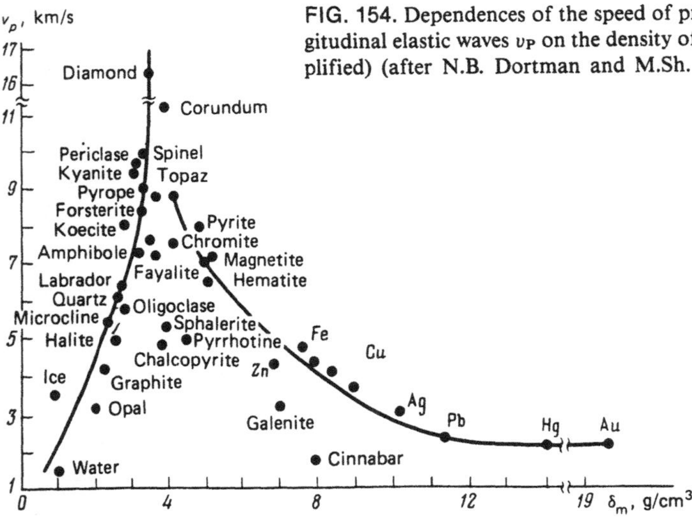

FIG. 154. Dependences of the speed of propagation of longitudinal elastic waves v_P on the density of minerals δ_m (simplified) (after N.B. Dortman and M.Sh. Magid)

This has been found from calculating v_P and v_S using Eqs. (122) and (123), values moduli E (or K), G, coefficient ν and mineral densities δ_m and has been validated by experimental findings. In terms of variations of v_P and v_S values minerals fall into principal rock-forming minerals (Type I), ore and native metals (Type II).

The speed of propagation of elastic waves and density of Type I minerals much as their moduli E and G increase with increasing atomic packing density of their crystals, that is why a straightforward correlation is observed between v_P and δ_m (see Fig. 154). An inverse relationship has been bound between v_P and δ_m (see Fig. 154), for Type II minerals. This is accounted for by decreased values of modulus of elasticity E and increased mineral density with increasing the far from being uniform relative atomic and weighted mean atomic mass of their constituent element(s); the packing density of atoms in crystals of these minerals has a relatively minor effect.

Sec. 76. **The Solid Phase**

Moduli E_s, G_s, K_s, coefficient ν_s and speeds of propagation of elastic waves, v_{P_s} and v_{S_s}, in an isotropic solid component of polymineral rocks are evaluated from their mineral composition, values of moduli of elasticity G_{m_i}, K_{m_i}, densities of constituent minerals, δ_{m_i}. Fundamentals of the theory and procedures of such an estimate (calculation) that reduced to obtaining average values of moduli of elasticity G_s, K_s and δ_s and calculation of other moduli, elasticity coefficients, speeds of propagation of elastic waves using the appropriate equations for a homogeneous and isotropic medium are presented in Belikov's work (see Bibliography).

If G_{m_i} and K_{m_i} are considered in the first approximation as being independent, the averaging can be carried out, in particular, using these relationships:

$$\ln G_s = \sum_i V_{m_i} \ln G_{m_i}$$

$$\ln K_s = \sum_i V_{m_i} \ln K_{m_i}$$

where V_i is the specific volume of the i-th mineral of the solid component of rocks with G_{m_i} (or K_{m_i}) determining its contribution to ln G_s (or K_s) of the mixture. At this, summation is carried out relative to the number of minerals of the solid component of the given rock.

When analysing the design values of speeds of propagation of waves and moduli of elasticity of the solid component of magmatic rocks it will be pointed out that their maximum values are demonstrated by ultrabasites ($v_{P_{s.des}} = 7.5\text{-}8.5$ km/s, $v_{S_{s.des}} \approx 4\text{-}5$ km/s, $E_{s.des} \approx 1.2\text{-}2$ MPa, $G_{s.des} \approx 0.5\text{-}0.8$ MPa, $K_{s.des} \approx 1.0\text{-}1.3$ MPa) whereas their lowest values are shown by granitoids. These latter have $v_{P_{s.des}} \approx 5\text{-}6.2$ km/s, $v_{S_{s.des}} \approx 3.5\text{-}3.6$ km/s, $E_{s.des} \approx 0.8\text{-}0.9$ MPa, $G_{s.des} = 0.32\text{-}0.37$ MPa, $K_{s.des} = 0.475\text{-}0.58$ MPa. The values of elasticity characteristics of basic and intermediate rocks are completely consistent with specific features of the mineral composition of the solid component of this rock series and are found between those for ultrabasites and granitoids.

Predominant among ultrabasites are minerals with relatively greater moduli of elasticity: olivine, diopside, augite and more whereas principal rock-forming minerals of granitoids are quartz, feldspars having comparatively low values of moduli of elasticity. The composition of the solid phase of other varieties of the rock groups under study includes minerals having both increased and low E, G, K values. The elasticity of the solid component of sedimentary rocks has been less studied using this method. According to the values of elasticity characteristics of calcite and quartz the elasticity of the solid phase of limestones somewhat exceeds that of granitoids, the elasticity of sandstones approaching them.

Elasticity characteristics of the solid component of the investigated metamorphic rocks are fully determined by their mineral composition and vary over about the same range as those of magmatic rocks. Very high values of elasticity characteristics are demonstrated by eclogites, in the composition of which garnet is predominant having high values of E, G and K.

Minor values of elasticity characteristics have been found in quartzites practically entirely composed of quartz. Design values of elasticity characteristics are in good agreement with experimental values of these quantities obtained for low-porosity air-dried rocks showing the same mineral composition and for rocks composing a definite type of sedimentary deposits. In the latter case $v_{P_{s.ex}}$ is taken to be represented by the value $v_P = 1/\tau$, given $k_p = 0$, found from experimental correlation relationships of the $\tau = a + bk_p$ type where τ is the time interval; a and b are constants typical of a definite sediment type.

Sec. 77. **The Liquid and Gaseous Phase**

It is solely waves associated with a variation in the volume of the liquid or gaseous component of a rock that propagate in the liquid or gaseous phase, viz. longitudinal compression or dilatation. We will present a wave equation illustrating variations, $\Delta\varphi$, of the potential, φ, of an acoustic wave in time and space satisfactorily describing the propagation of elastic waves in the above rock components:

$$\frac{d^2\varphi}{dt^2} - v_P^2 \Delta\varphi = 0 \quad \text{or} \quad \Delta\varphi = \frac{1}{v_P^2}\frac{\nabla^2\varphi}{dt^2} \tag{124}$$

where $\Delta\varphi = v_{vb}$ are vibrational speeds of propagation of particles in the media in question much less than the speed of sound; $v_P = \sqrt{(\partial p/\partial\delta)_s}$ is the speed of propagation of elastic waves in the liquid and gaseous components; p and δ are pressure and density of the same media; s is a symbol of entropy (due to the adiabaticity of the process it is constant).

Equation (124) has been derived as a result of a joint solution under definite conditions of lianized motion equations

$$\frac{\partial v_{vb}}{\delta t} + \frac{\nabla p'}{\delta} = 0$$

continuity equations

$$\frac{\partial\delta'}{\delta t} + \delta_0 \operatorname{div} v_{vb} = 0$$

and static state equations

$$p' = \left(\frac{\partial p_0}{\partial \delta_0}\right)_s \delta'$$

where p', δ' are the pressure in the liquid or gas and their density in the acoustic wave; p_0 and δ_0 are the same but in a static state.

Equation (124) applies to media that relatively little change the energy of an elastic wave. It is satisfied by solutions describing two waves propagating along the X axis in opposite directions.

In an ideal gas and liquid where a sound is not attenuated the speed of propagation of elastic waves is

$$v_P = \sqrt{\nu p_{0av}/\delta} = \sqrt{1/\delta\beta_a}$$

where ν is Poisson's ratio, δ is the density of the medium, β_a is adiabatic compressibility, p_{0av} is average pressure in a static state.

Speeds of propagation of longitudinal elastic waves have also been studied experimentally on samples of distilled water, NaCl solutions, confined waters, oils, kerosene, air, methane, isooctane and other materials under general laboratory and

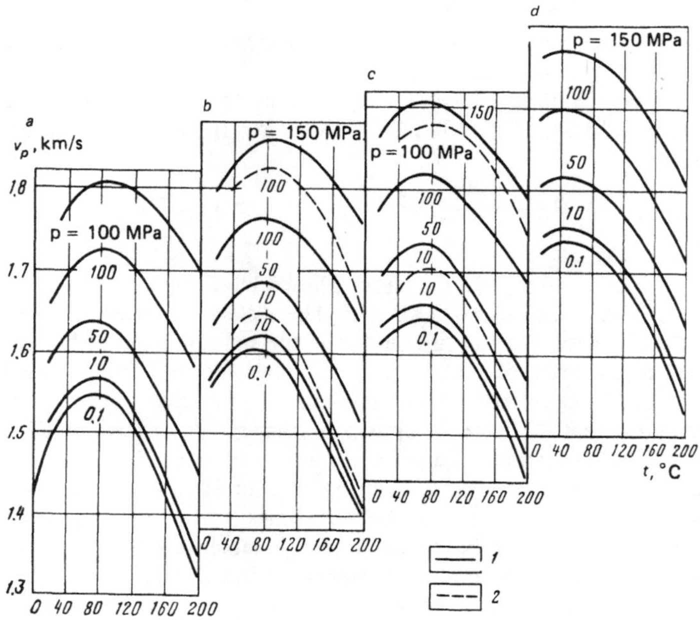

FIG. 155. Dependences of the speed of propagation of longitudinal ultrasonic waves v_P in water on temperature t, pressure p and mineralization C [after N.G. Kulikova et al. (*1*) and V.P. Lukov (*2*)]

a—for distilled water; for NaCl solution concentration C (in g/l) and confined water of mineralization (in g/l): *b*—58.5 and 60; *c*—117 and 120; *d*—234 other, occurring in the Earth's bowels, thermobaric conditions (temperature interval 0-200 °C and pressure interval 0.1-150 MPa) (Fig. 155).

At a temperature 20 °C and pressure 0.1 MPa the speed v_P in water (1500 m/s) exceeds that in oil (1300 m/s) and both much exceed v_P in air (330 m/s) or methane (~490 m/s).

Values of v_P in water rapidly increase with growth of its mineralization and pressure (see Fig. 155*b*). With increasing water temperature, they gradually attain peak magnitudes, then drop. With increasing water mineralization, maximum v_P values are attained at ever lower temperatures (see Fig. 155*b*, *d*). For other liquids (kerosene, oil) v_P continuously decreases with increasing the temperature.

The speed of sound in salt water, in subsurface layers, can be determined at different temperatures, e.g. by using the relationship

$$v_{P_w} = 1445.5 + 3.92t - 0.024t^2$$

with v_{P_w} increasing approximately by 18 m/s per km of depth.

Another equation determining v_{P_w} of water for mineralization C (in kg/m^3), pressure p (in MPa) and temperature t (in °C):

$$v_{P_w} = \frac{(1 + 2.2 \times 10^{-3}t - 1.65 \times 10^{-5}t^2)(1 + 5.5 \times 10^{-4}C)}{710(1 - 1.2 \times 10^{-3}p)} \tag{125}$$

All this, taken together, points to a substantial effect on v_{P_w} of not only the composition but also thermobaric conditions.

The volumetric modulus of elasticity of sea pore water also changes with the temperature, pressure and salinity of water. There exists a definite relationship between k_w and v_{P_w}.

Sec. 78. **Rocks**

In the general case rocks are aggregates whose phases have distinctly different elasticity varying from one rock variety to another. The skeletal particles of a rock also show a specific elasticity. They occupy in space the same volume as has the whole medium. As to the mass of the skeleton, it equals the mass of the solid component of a rock. The skeleton is mainly composed by relatively large grains of the solid component that differ in composition, shape, interconnection. The same is also true of carbonate, quartz, swelled clayey and other cement types weakly connected with grains. Its intergrain cement-bearing communicating voids contain a liquid, gaseous or both components of rocks. Sometimes the liquid and gaseous component is practically immovable in the voids being unable to migrate through them under the effect of minor and variable increments of strains that occur during the passage of elastic waves through a rock.

Nor does the liquid component filling the voids space come to the voids space of such a rock. Minor and time-variable variations of strains in rocks induced by acoustic waves have not irreversible deformations. The deformations are proportional to stages, and the losses of energy and the changes observed in the rocks, having low porosity, are closely approaching those of ideally elastic waves. Voids of such rocks are considered closed. Examples of rocks having closed voids can be provided by low-porosity unaltered varieties of sedimentary, magmatic and metamorphic rocks occurring below the upper fissured zone; as to clays and rocks of nonexploited oil fields under conditions of a closed elastic regime represent high-porosity varieties of these media. According to F. Gassman such rocks will be called differentially elastic with a perfect bond between components.

However, there exist rocks containing such amounts of the liquid and gaseous phase as to appreciably affect their elasticity, having less strong bonds between the phases; they divide into rocks having a relatively small and practically nonexisting interphase connection. The former (gravel, quicksands, worked oil- and water- and gas-saturated sands, sandstones, aleurolites, loose limestones and dolomites) will be termed differentially elastic rocks with an imperfect connection between the phases; their liquid and solid components shift with respect to one another, these displacements being interconnected.

In the practically total absence of connection between the components filling the voids space the material filling this latter readily circulates in the voids space, and minor increments of strains of the filling component of the voids space. Variations of strains act only on the solid material and skeletal particles of the rock.

The connection between the phases is affected by the viscosity of the liquid and diameter of channels. The simple structure of the voids space, given other favourable factors, provides a minor degree of interphase connection. Until strains at which the configuration of the voids space begins to change and the movement of gases and liquids and gases appears, the elastic properties of differentially elastic rocks do not qualitatively differ from similar properties of perfectly elastic media.

Sec. 79. Differentially Elastic Homogeneous, Isotropic and Anisotropic Rocks with a Perfect Interphase Connection

Model Theories

Moduli and coefficients of elasticity. According to F. Gassman, the components of differentially elastic rocks (solid phase, gas or liquid) behave as if they were perfectly elastic, isotropic and homogeneous rocks to which all the above laws of the theory of elasticity in a differential form can be applied. If the skeleton of a rock, given definite shape, pore size and measuring device, can be considered as macroscopically isotropic and homogeneous, it can also be considered, within certain limits, to be perfectly elastic.

The elastic properties of such a rock, given a perfect connection between its phases and laboratory temperatures and pressures, are characterized, as is the case of perfectly elastic media, by two moduli (coefficients) of elasticity. As a characteristic of elasticity let us select moduli of overall compression K_{s+fill} and shear G. Clearly, moduli of shear of the skeleton of a two- and single-phase rock are equal; and modulus of overall compression K_{s+fill} for a two-phase rock (solid phase + filling component) having a perfect connection between the components is governed not only by moduli of overall compression of phases and skeleton of the rock and its voids ratio and differs from K_{s+sk}.

When determining the dependence of K_{s+fill} on the aforementioned characteristics of elasticity and voids ratio, F. Gassman proceeded from the following relationships:

(1) for a two-phase rock as a whole

$$\frac{\Delta V}{V} = -\frac{1}{K_{s+fill}}\Delta\sigma = -\frac{1}{K_{s+fill}}(\Delta\sigma_{fill} + \Delta\sigma_{sk}) \tag{126}$$

(2) for the material filling the voids space

$$\frac{\Delta V_{fill}}{V_{fill}} = -\frac{\Delta\sigma_{fill}}{K_{fill}} \tag{127}$$

(3) for the skeletal particles of the rock

$$\frac{\Delta V}{V} = \frac{\Delta V_{sk}}{V_{sk}} = -\left(\frac{\Delta\sigma_{fill}}{K_s} + \frac{\Delta\sigma_{sk}}{K_{sk}}\right) \tag{128}$$

(4) for the solid phase of the rock

$$\frac{\Delta V_s}{V_s} = -\left(\frac{\Delta\sigma_{fill}}{K_s} + \frac{\Delta\sigma_{sk}}{K_s(1-k_p)}\right) \tag{129}$$

In these equations V_{fill} is the volume of the voids filling component in the volume of the rock $V = V_{sk}$ whose size is sufficiently great with respect to the cross section of the voids; ΔV_{fill}, ΔV_s and $\Delta V = \Delta V_{sk}$ are the change in the volume of the filling component, solid phase and skeleton of the rock resulting from the action on the rock of a variation of the hydrostatic strain $\Delta\sigma_{fill}$ and on its skeleton of the variation of the hydrostatic strain $\Delta\sigma_{sk}$; K_{fill}, K_s and K_{sk} are, respectively, moduli of overall compression of the voids space of the rock, its solid phase and skeleton.

Since $V_s = (1-k_p)V$, Eq. (129) takes on this form:

$$\frac{\Delta V_s}{V} = -\left[\frac{(1-k_p)\Delta\sigma_{fill}}{K_s} + \frac{\Delta\sigma_{sk}}{K_s}\right] \tag{130}$$

where $\Delta V_s/V$ is the relative variation in the volume of the solid phase caused by strains σ_{sk} and σ_{fill}.

Equations (126)-(129) allowing for relationships $V_{fill} = k_p V$ and $\Delta V = \Delta V_s + \Delta V_{fill}$ the last of which expresses a perfect connection between the phases of the rock make it possible to determine the modulus of overall compression of a differentially elastic two-phase rock having a perfect interphase connection:

$$K_{s+fill} = K_s\frac{K_{sk}+Q}{K_s+Q}$$

where

$$Q = \frac{K_{fill}(K_s - K_{sk})}{k_p(K_s - K_{fill})} \tag{131}$$

Analysis of Eq. (131) shows that

(1) if K_{fill} is so small that it can be neglected, then $K_{s+fill} = K_{sk}$, i.e. the elasticity of a rock is governed by the elasticity of its skeletal particles (as is the case of

porous and permeable gas-saturated rocks in the open air in the practical absence of a connection between its phases);

(2) given $k_p = 0$, $K_{s+fill} = K_s$, i.e. as the modulus of overall compression the modulus of overall compression of the solid material of the rock can be adopted; the case of low-porosity and permeable homogeneous and isotropic rocks, e.g. magmatic, below the upper fissured zone is close to this case;

(3) if the compressibility of the pore filling component approaches that of a solid material, $K_{s+fill} = K_s = K_{fill}$ (the case of two-component rocks one of solid materials of which can be described as being the voids space filler);

(4) if the compressibility of a solid material compared with that of the pore filling component can be neglected ($K_s = \infty$), then $K_{s+fill} = K_{sk} + K_{fill}/K_r$.

Since differentially elastic macroscopically homogeneous and isotropic rocks having a perfect connection between the components behave very much like ideally elastic rocks, the relationships between strains σ_i and deformation e_i for them appear as follows:

$$\Delta\sigma_i = -\sum_{j=1}^{6} C_{ij}e_j \begin{Bmatrix} i = 1, 2, ..., 6; \\ j = 1, 2, ..., 6 \end{Bmatrix} \tag{132}$$

where

$$C_{ij} = C_{ji}; \quad C_{11} = C_{22} = C_{33} = \lambda_{s+fill} + 2\mu;$$

$$C_{44} = C_{55} = C_{66} = \mu; \quad C_{12} = C_{21} = C_{31} = C_{23} = C_{32} = \lambda_{s+fill}$$

where λ_{s+fill} is found from the relationship between K_{s+fill} and μ.

The elastic properties of differentially elastic homogeneous anisotropic rocks can be described, e.g. according to F. Gassman, using the following relationships whose derivation relies on the theory of elasticity:

$$C_{ij} = C_{skij} + \frac{\alpha^*}{D^*} b_i b_j \tag{133}$$

where

$$b_i = \epsilon_i \frac{1}{3K_s}(C_{ski_1} + C_{ski_2} + C_{ski_3})$$

$$\epsilon_1 = \epsilon_2 = \epsilon_3 = 1; \quad \epsilon_4 = \epsilon_5 = \epsilon_6 = 0$$

$$\alpha^* = \frac{1}{k_p[(1/K_{fill}) - (1/K_s)]}$$

$$D^* = 1 + (\alpha^*/3K_s)(b_1 + b_2 + b_3)$$

C_{ij} and C_{skij} in Eqs. (132) and (133) are characteristics of elasticity of a two-phase anisotropic rock, respectively, in six different orientations and skeletal particles oriented in the same directions; ϵ are auxiliary quantities.

The speed of propagation of elastic vibrations. The propagation of longitudinal and transverse elastic waves in differentially elastic homogeneous and isotropic two-phase rocks seems to satisfy the following equations:

$$(K_{s+fill} + 4/3G)\nabla^2\theta = [(1 - k_p)\delta_s + k_p\delta_{fill}]\frac{\partial^2\theta}{\partial t^2} = (\delta_{sk} + \delta_{fill}k_p)\frac{\partial^2\theta}{\partial t^2} \quad (134)$$

$$G\nabla^2\omega = [(1 - k_p)\delta_s + k_p\delta_{fill}]\frac{\partial^2\omega}{\partial t^2} = (\delta_{sk} + \delta_{fill}k_p)\frac{\partial^2\omega}{\partial t^2} \quad (135)$$

where θ and ω are, respectively, the divergence and curl (rotation) of the displacement vector $\vec{U}$; δ_{sk}, δ_s and δ_{fill} are densities of the skeleton, solid and liquid phase; $G = G_{sk}$ is the modulus of shear of the skeleton.

The speeds of propagation of longitudinal and transverse waves in differentially elastic homogeneous and isotropic rocks, respectively, will be these:

$$v_{P_{s+fill}} = \sqrt{\frac{K_{s+fill} - 4/3G}{(1 - k_p)\delta_s + k_p\delta_{fill}}} = \sqrt{\frac{K_{s+fill} - 4/3G}{\delta_{sk} + \delta_{fill}k_p}} \quad (136)$$

$$v_{S_{s+fill}} = \sqrt{\frac{G}{(1 - k_p)\delta_s + k_p\delta_{fill}}} = \sqrt{\frac{G}{\delta_{sk} + \delta_{fill}k_p}} \quad (137)$$

For $G = G_{sk} = 0$ (noncemented rocks)

$$v_{P_{s+fill}} = \sqrt{\frac{K_{s+fill}}{(1 - k_p)\delta_s + k_p\delta_{fill}}} \quad (138)$$

where K_{s+fill} is found from Wood's equation

$$\frac{1}{K_{s+fill}} = \frac{k_p}{K_{fill}} + \frac{1 - k_p}{K_s} \quad (139)$$

According to M.R. Willie, A.R. Gregory and L. Gardner, if different-grained adjacent mineral fragments are submerged in a liquid, and the volume of this medium somewhat exceeds that of the solid phase and its pore filling component, minor pressure will cause a simple compression of this medium, since the pressure exerted on the skeleton of a model is also equal to zero.

Moduli of overall compression of the components of a model are

$$K_i = \delta_i v_i^2/(1 + q_i)$$

where δ_i and v_i are, respectively, the density and speed of propagation of elastic waves in the particular component of the medium;

$$q_i = \frac{4G_i}{3K_i} = 2\left(\frac{1 - 2\nu_i}{1 + \nu_i}\right)$$

A substitution of values of q_i of individual phases and the whole medium into Eq. (139) yields

$$\frac{1 + q_{s+l}}{v_{s+l}^2} = \left[\frac{(1 - k_p)(1 + q_s)}{\delta_s v_s^2} + \frac{k_p}{\delta_l v_l^2}\right][k_p\delta_l + (1 - k_p)\delta_s] \quad (140)$$

For liquids whose $q_s = 0$ and $q_{s+l} = 0$, Eq. (140) appears as Eq. (139).

The best approximation to experimental data for water-saturated sandstones is achieved for $q_{s+l} = 0.6$. For weakly cemented rocks the values of speeds are ap-

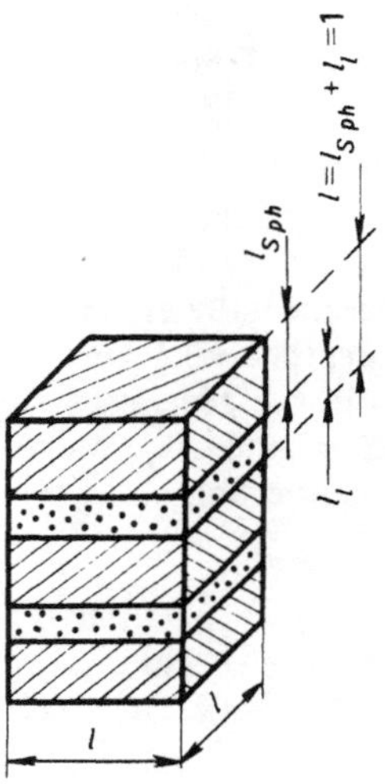

FIG. 156. Model of a bedded medium

proaching ones obtained by using Wood's equation. Equation (140) is also in agreement with equations proposed by Gassman, White and Sengbush for saturated rocks when there is no pressure on the skeletal particles.

The aforementioned researchers believe that maximally moist rock acted on by a pressure with uniformly distributed voids can be modelled by alternating phases (Fig. 156), the first of which is the solid phase l_s in height, its length and width is l, followed by the next, liquid, phase which has a height l_l and similar length and width.

Given time intervals τ_s and τ_l of passage of a longitudinal elastic wave through the solid and liquid phase, respectively, an elastic pulse will pass through a unit structure volume of the medium under study $l_{s+l} = l_s + l_l = 1$ in length and l^2 in cross section within a time interval $\tau_{s+l} = \tau_l + \tau_s$. Since time intervals τ_{s+l}, τ_s and τ_l are in direct proportion to the lengths l_{s+l}, l_s and l_l travelled by an elastic wave through the solid, liquid phase and their population and are in inverse proportion to the speeds of propagation $v_{P_{s+l}}$, v_{P_s}, v_{P_l} of the wave in the same unit volume of a model, one can write

$$\frac{l_{s+l}}{v_{P_{s+l}}} = \frac{l_s}{v_{P_s}} + \frac{l_l}{v_{P_l}},$$

taking into account that

$$k_p = \frac{l_l l^2}{l^2(l_s + l_l)} = l_l; \quad l_s = 1 - l_l = 1 - k_p,$$

one gets

$$\frac{1}{v_{P_{s+l}}} = \frac{1 - k_p}{v_{P_s}} + \frac{k_p}{v_{P_l}} \tag{141}$$

or

$$\tau_{s+l} = (1 - k_p)\tau_s + k_p\tau_l \tag{142}$$

Equations (141), (142) find an extensive use in acoustic soundings of boreholes even they are based on a model which is far from representing actual rocks failing to take into account the elasticity of the skeletal particles and effect of temperature and pressure. Equation (142) agrees with the experimentally established one solely for dense water-saturated rocks occurring at a depth 2.5-3.5 km.

Since experimental $\tau_{s+l} = f(k_p)$ relationships in the general case depart from ones calculated from the average time equation $\tau_{s+l} = A + Bk_p$, Wachholz has proposed an exponential equation to describe this connection

$$\tau_{s+l} = A + Bk_p + Ck_p^2 + Dk_p^3 \tag{143}$$

For $k_p = 0$ its coefficient $A = \tau_s$; $B = \partial\tau_{s+l}/\partial k_p$. Given $k_p = 1$, $A + B + C + D = \tau_l$; $B + 2C + 3D = \partial\tau_{s+l}/\partial k_p$. These relationships determine coefficients C and D. A substitution of the found values of A, B, C, D into Eq. (143) yields

$$\tau_{s+l} = \tau_s + (\tau_l - \tau_s)(3k_p^2 - 2k_p^3) + (k_p - 2k_p^2 + 3k_p^3)\left.\frac{\partial\tau_{s+l}}{\partial k_p}\right|_{k_p=0} + (k_p^3 - k_p^2)\left.\frac{\partial\tau_{s+l}}{\partial k_p}\right|_{k_p=1} \tag{144}$$

Partial derivatives from τ_{s+l} with respect to k_p, given $k_p = 1$ and 0, are not known in Eq. (144). Wachholz has determined these derivatives. By substituting the partial derivatives in Eq. (144) by relationships one gets

$$\frac{\tau_{s+l}}{\tau_s} = 1 - \frac{k_p}{2} - 2k_p^2 + \frac{3}{2}k_p^3 + \left(\frac{k_p}{2} - k_p^2 + \frac{k_p^3}{2}\right) \times \left\{\frac{\delta_l}{\delta_s} + \frac{a(1 - 2\nu_s)}{1 - \nu_s^2} + \frac{a(1 + \nu_s)}{2(1 - \nu_s)(1 - 2\nu_s)\left[1 + \dfrac{3a}{2(1 - 2\nu_s)\left(\dfrac{\delta_s(1 + \nu_s)\tau_l^2}{3\delta_l(1 - \nu_s)\tau_s^2} - 1\right)}\right]}\right\} + (k_p^2 - k_p^3)\left(\frac{\tau_l}{\tau_s}\frac{\delta_s}{2\delta_l} + \frac{3\delta_l(1 - \nu_s)\tau_s}{2\delta_s(1 + \nu_s)\tau_l}\right) + \frac{\tau_l}{\tau_s}(2k_p^2 - k_p^3) \tag{145}$$

Equation (145) is the most valid for nonclayey water-saturated sands and sandstones. The quantity a is a function of the shape of voids, material of skeletal particles and the state of stress of the latter. The form of the function is established experimentally. To do this, we get relationships between a and pressure Δp on samples of definite water-saturated media or sediments. For a rock having a structure approximated by a packing of spheres of equal diameter

$$a = 1.4\left(\frac{1}{3} + \frac{2}{3}\sqrt[3]{\frac{E_s(1 - \nu_s^2)^2}{3\Delta p}}\right) \tag{146}$$

When determining the speed of propagation of compression waves in a liquid-saturated medium including spheres of identical diameter with a cubic packing, J.E. White and R.L. Sengbush calculated a modulus of elasticity for the unit element of such a packing (cube involving portions composing one sphere). Since a displacement of the face of a unit element causes a deformation of hard grains, liquid and skeleton of a rock, the modulus of elasticity K_{s+l+sk} must be composed of two moduli, K_{s+l} and K_{sk}. According to Eq. (139),

$$K_{s+l} = \frac{1}{[(1 - k_p)/K_s + k_p/k_l]}$$

and, according to Hertz's law, the modulus of elasticity of the skeleton is

$$K_{sk} = [\pi(\delta_s - \delta_l)gE_s^2z/16(1 - \nu_s^2)^2]^{1/3} \tag{147}$$

where z is the depth of occurrence of a sphere in a row of superimposed spheres having a cubic packing; g is gravitational acceleration: and, since $v_{P_{s+l}}^2 \approx K_{s+l+sk}/\delta_{s+l}$, then

$$v_{P_{s+l}} = \left\{ \frac{\dfrac{1}{(1 - k_p)/K_s + k_p/K_l} + \left[\dfrac{\pi(\delta_s - \delta_l)gE_s^2z}{16(1 - \nu_s^2)^2}\right]^{1/3}}{(1 - k_p)\delta_s + k_p\delta_l} \right\}^{1/2} \tag{148}$$

In this case use has been made of a model of a rock approaching well sorted and weakly cemented clastic rocks, the elasticity of the skeletal particles and depth of occurrence of rocks, z, (or pressure acting on them) have been taken into account.

For a rock containing a solid, liquid and gaseous phase and found at the Earth's surface (such are, for example, denudation sands) the relationships to determine the speed have been proposed by I.S. Berzon et al. These authors have accepted an average modulus of overall compression

$$K_{s+l+g} \approx v_{PO_{s+l+g}}^2 \delta_{r(s+l+g)} \tag{149}$$

where $v_{PO_{s+l+g}}$ is the speed of propagation of elastic waves in a three-component medium; $\delta_{r(s+l+g)} = \delta_s f_s + \delta_l f_l + \delta_g f_g$, where δ_s, δ_l, δ_g and f_s, f_l and f_g are densities and volumetric contents of the solid, liquid and gaseous phase in a unit volume of a rock. Since

$$\frac{1}{K_{s+l+g}} = \frac{f_s}{K_s} + \frac{f_l}{K_l} + \frac{f_g}{K_g} \tag{150}$$

$$K_s = v_s^2\delta_g, \quad K_l = v_l^2\delta_l \quad \text{and} \quad K_g = v_g^2\delta_g \tag{151}$$

the equation for the speed of propagation of elastic transverse waves in a three-component medium appears as

$$v_{PO_{s+l+g}} = v_s = \frac{1 + [(f_s + f_l)/f_s]}{\sqrt{\left(1 + \dfrac{f_l\delta_l}{f_s\delta_s}\dfrac{f_g\delta_g}{f_s\delta_s}\right)\left[1 + \dfrac{f_l\delta_s}{f_s\delta_l}\left(\dfrac{v_s}{v_l}\right)^2 + \dfrac{f_g\delta_g}{f_s\delta_s}\left(\dfrac{v_s}{v_g}\right)^2\right]}}$$

where $f_g + f_l = k_p$.

The calculation of the speed of propagation of elastic waves in a three-component rock at a depth z used Eq. (148) reduced to a form:

$$v_{Pz_{s+l}} = \sqrt{(v_{P0_{s+l}})^2 + \frac{5.78 \sqrt[3]{z(\delta_s - \delta_{fill}) \dfrac{E^2}{(1-\nu^2)^2}}}{\delta_s f_s + \delta_{fill} f_{fill}}} \tag{152}$$

where $v_{Pz_{s+l}}$ and $v_{P0_{s+l}}$ are speeds of propagation of elastic waves, respectively, in a two-phase rock at a depth z and at the surface of the earth; δ_{fill}, f_{fill} are the density and volumetric content of the filling component.

The passage from Eq. (152) for a two-component rock to an equation for a three-component rock is given in Kobranova's book (see Bibliography).

$$v_{Pz_{s+l+g}} = \sqrt{(v_{P0_{s+l+g}})^2 + \frac{5.78 \sqrt[3]{z\left[\delta_s - \delta_g - \dfrac{(\delta_l - \delta_g)f_l}{k_p}\right] \dfrac{E^2}{(1-\nu^2)^2}}}{\delta_s f_s + \delta_l f_l + \delta_g f_g}}$$

F. Gassman has provided a solution of the problem for spheres of identical diameter having a hexagonal packing; such a medium is anisotropic, and its elasticity is determined by five moduli of elasticity. These latter can be calculated by using Eq. (133), having previously determined moduli of elasticity of the skeletal particles of the rock. Use is made for this of Eq. (132) showing a dependence between strains and deformations, having preliminarily determined from Hertz's law the deformation of coming in contact with elastic spheres at likely strains of the skeleton and taken into account the transverse isotropy of these rocks.

Knowing moduli of elasticity we set up motion equations in displacements from which to derive this relationship to find the speed of propagation of elastic waves:

$$v_{s+fill} = \frac{1}{\delta_r}\left(\frac{\alpha^*}{D^*} b_{fill}^2 + 16C_{sk}\right)^{1/2} \tag{153}$$

where

$$\delta_r = (1 - k_p)\delta_s + k_p\delta_l$$

$$\frac{\alpha^*}{D^*} = \frac{1}{\dfrac{k_p}{K_l}\,\dfrac{3(1-\nu_s)}{E_s}\left[(1-k_p) - \dfrac{12(1-2\nu_s)C_{sk}}{E_s}\right]}$$

$$b_{fill} = 1 - \frac{24(1-2\nu_s)C_{sk}}{E_s}$$

$$C_{sk} = \frac{1}{24\sqrt{2}}\,\frac{2\pi E_s(\delta_s - \delta_l)gz}{(1-\nu_s^2)^2}$$

Here for a hexagonal packing $k_p = 0.26$.

C. Brandt has analytically determined $v_{P_{s+l}}$ in randomly packed spherical particles of four different sizes. He obtained the relationship between the volume of the aggregate, V, and pressure p following Hertz's law, yet gaps between spheres contained a fluid. The $V = f(p)$ relationship enabled the modulus of overall compression, K_{s+l}, to be established. This modulus and average density of the aggregate, δ_{s+l}, permitted the calculation of the speed

$$v_{P_{s+l}} = \left[\frac{3g}{\delta_{s+l}}\left(-V\frac{dp}{dV}\right)\left(\frac{1-\nu_s}{1+\nu_s}\right)\right]^{1/2} \tag{154}$$

where

$$-V\frac{dp}{dV} = K_{s+l} = \frac{E_s}{3(1-2\nu_s)}$$

C. Brandt further extended his theory approximately to cemented granular media and three-phase aggregates to obtain complex equations for calculating $v_{P_{s+l}}$ and $v_{P_{s+l+g}}$ allowing for the pressure.

Thus v_P is also affected by the degree of gas saturation, particularly where rocks are being acted on by pressure. The modulus of elasticity of gas is determined not only by pressure but also by a compression technique.

If the appreciable fraction of voids is filled by argillaceous cement, clay swelling must be taken into account.

E.A. Kozlov's theory proposed for a water-saturation aggregate of perfectly elastic partially cemented fragments of a random shape and size is based on the same equation (148) as is the theory of White and Sengbush. However, both the rock's model and path of determination of K_{s+l} and K_{sk} are different. According to E.A. Kozlov,

$$K_{s+l} = \frac{1}{\beta_s(1-k_p) + \beta_l k_p} \tag{155}$$

where

$$\beta_s = 1/v^2_{P_s}\delta_s; \quad \beta_l = 1/v^2_{P_l}\delta_l$$

and

$$K_{sk} = \frac{dp}{dV_{sk(p)}/V_{Psk(p)}} \tag{156}$$

is determined without allowing for the displacement of particles or deformation induced by squeezing out of free and bound water. In addition, E.A. Kozlov believes that the own weight of the rock mass causes only a volume of the skeletal particles, $V_{sk(p)}$ $(p = 0)$ isolated out of the mass to vary.

We will omit the original, multi-stage transformations for getting a relationship to determine K_{sk}, and will present it jointly with the final equation for $v_{P_{s+l}}$.

According to E.A. Kozlov,

$$K_{sk} = -V_{sk(p)}\frac{dp}{dV_{sk(p)}} = \frac{(1-k_p)mE_s - 0.5[(1-C)/\beta_l k_p]N^2}{L(M-N)C\delta_l/\delta_s} \tag{157}$$

where m is the ratio of the areas of horizontal projections of grains contacts to the area of the horizontal projection of their total surface in a unit volume of a model [for clays $m = 1.0\text{-}1.43k_p$ and for sands $m = 7.34\,(k_{p(p=0)} - k_p)^{2.17}$; k_p,

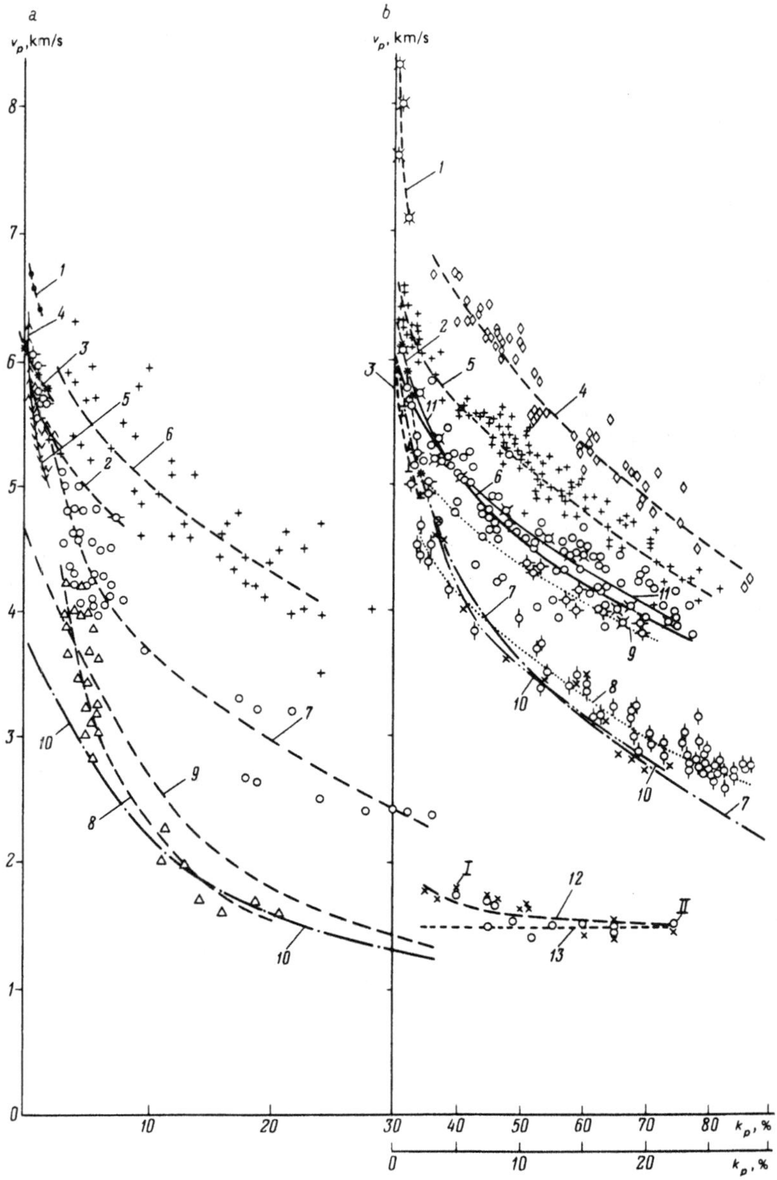
a
v_p,km/s
b
v_p, km/s
0
1
2
3
4
5
6
7
8
10
20
30
40
50
60
70
80
k_p, %
0
10
20
k_p, %
I
II
11
12
13

$k_{p(p=0)}$ are voids ratios at a depth z and at a pressure $p \to 0)$]; C is a coefficient characterizing the degree of grains cementation (at $C = 0$ cement fills the entire volume of the voids—dense sedimentary rocks; given $C \to 1$ and $m \to 0$, rocks are weakly cemented); $L = 1 - k_{p(p=0)} - 2\nu_s m[1 - k_{p(p=0)}]$; $M = 1 - k_p - 2\nu_s m(1 - 2k_p)$; $N = (1 - m)(1 - k_p) + 2\nu_s m k_p$.

A substitution of the value of K_{sk} from Eq. (157) into Eq. (148) in place of its value obtained according to Hertz [see Eq. (147)] yields Kozlov's equation

$$v_{P_{s+l}} = \sqrt{\frac{1}{\delta_r}\left[\frac{1}{k_p\beta_l + (1 - k_p)\beta_s} + \frac{(1 - k_p)mE_s - 0.5[(1 - C)/\beta_m k_p]N^2}{L(M - N)C\delta_l/\delta_s}\right]} \tag{158}$$

where $\delta_r = k_p\delta_l + (1 - k_p)\delta_s$.

This equation can also be used in the case of a heterogeneous filling component in which case β_l must be substituted by a compressibility coefficient of the filling component $\beta_{fill} = \Sigma\beta_{fill.i}f_i$, where f_i are volume fractions of the constituents of the filling component, and $\beta_{fill.i}$ are their compressibilities. Accordingly, equations determining densities δ_r, δ_l and compressibility β_l will change.

Sand grains are approximated by spheres, in this case $C \approx 1$, $m \approx 0$, and v_P can be found using this equation:

$$v_{P_{s+l}} = \sqrt{\frac{1}{\delta_r}\left\{\frac{1}{k_p\beta_l + (1 - k_p)\beta_s} + 7.4\,\frac{1 - k_p}{k_p(1 - \delta_l/\delta_s)}\sqrt[3]{\frac{E_s^2(\delta_s - \delta_l)z}{(1 - \nu_s^2)^2}}\right\}} \tag{159}$$

For sand having spherical grains and voids completely filled by clay material,

$$v_{P_{s+l}} = \sqrt{\frac{1}{\delta_r}\left\{\frac{1}{(k_{pPs}/K_{s+lP1}) + (1 - k_{pPs})\beta_s} + 7.73\,\frac{1 - k_{pPs}}{k_{pPs}}\sqrt[3]{\frac{E_s^2\delta_r z}{(1 - \nu_s^2)^2}}\right\}} \tag{160}$$

where k_{pPs} is the voids ratio of the sandy fraction of the rock; K_{s+lP1} is the volume

---

FIG. 157. Dependences of the speed of propagation of longitudinal elastic waves v_P on the total voids ratio k_p for air-dry (*a*) and moist (*b*) rocks.

a: *1*—gabbros; *2*—granitoids (after B.I. Belikov, K.S. Aleksandrov, T.V. Ryzhova); *3-5*—small-, medium- and large-grained granites, respectively (after T.S. Lebedev, P.A. Burtnoi, V.A. Korchin); *6*—pure limestones (Borovskaya area, borehole 14) (after B.E. Feldman, V.P. Tyurin); *7, 8*—sandstones and argillites (Central Donets Basin) (after A.A. Golubev, V.N. Oparin, G.Ya. Rabinovich); *9, 10*—sandstones and clay rocks (Ob River Region) (after N.A. Tuezova, L.M. Doroginitskaya, R.G. Demina, N.I. Bryuzgina).

b: *1-3*—respectively, pyroxenites, syenites, granites (after M.Sh. Magid); *4-6*—dolomites, limestones, sandstones and aleurolites, given confined water mineralization 250 g/l, and in confined conditions (Perm Kama River Region) (after V.P. Potapova, I.S. Zhukova, B.K. Fedoseev, V.G. Chumakov); *7*—polymictic sandstones in general laboratory conditions of the Ob River Region (after N.A. Tuezova, L.M. Doriginitskaya, R.G. Demina, N.I. Bryuzgina); *8, 9*—polymictic sandstones of West Siberia, at an effective pressure equal to zero and 40-45 MPa (after G.P. Stavkin); *10, 11*—sandstones at an effective pressure equal to zero and 80 MPa (after G.I. Petkevich and T.V. Verbitskii); *12*—littoral sand sediments under *in situ* conditions (pressure 1.37 MPa) (points *I*) and under laboratory conditions (points *II*); *13*—bottom water (pressure 1.37 MPa) (after Brandt) modulus of elasticity of clay; $k_{pPs}k_{pPl}$ is the overall porosity of the clayey rock; its density is $\delta_r = k_{pPs}\delta_{Pl} + (1 - k_{pPs})\delta_s$.

To get probable types of the $v_{P(p)} = f(z)$ relationship use also was made of experimental compression curves $k_{p(p)} = f(p)$ for clean clays and clean sand, where p is the pressure of the overburden.

There exist other theories of elasticity of differentially elastic rocks having a strong interphase connection. They characterize the elasticity of these media depending on the specific content and elasticity of individual rock phases and pressure.

Experimental Investigations

Experimental studies also suggest that the speed of propagation of elastic vibrations and other characteristics of elasticity in the sediments and rocks under investigation are largely governed by temperature. Laboratory investigations have shown that the speeds of propagation of longitudinal elastic vibrations, Young's modulus and other characteristics are the most affected by the porosity of rocks. v_P drops and τ grows with increasing k_p values. Given a specified k_p, the speed of propagation of elastic waves increases in this sequence: air-saturated clayey rocks-sandstones-granitoids-limestones-gabbros (Fig. 157*a*). This corresponds to an increase in the density and, consequently, elasticity of their solid component. The same is true of a comparison of $v_P = f(k_p)$ relationships of water-saturated rocks (see Fig. 157*b*). Here for specified k_p values, v_P values increase from seabed sand deposits toward sandstones, dolomites and pyroxenites (see Fig. 157, curves *12, 7, 5, 4* and *1*). But v_P values are much higher than those of similar-type air-saturated rocks, particularly, when v_P determinations have been carried out for p_{eff} or p (see Fig. 157, curves *5, 9, 11*).

The speed of propagation of longitudinal waves, v_P, in sandstones having different cement types, argillites (Fig. 158), granites, diorites and other rocks until k_w values 30-40% slowly increases or even somewhat decreases, to rapidly increase at $k_w > 30$-40% approaching the limiting values. Note that $v_P = f(k_w)$ and

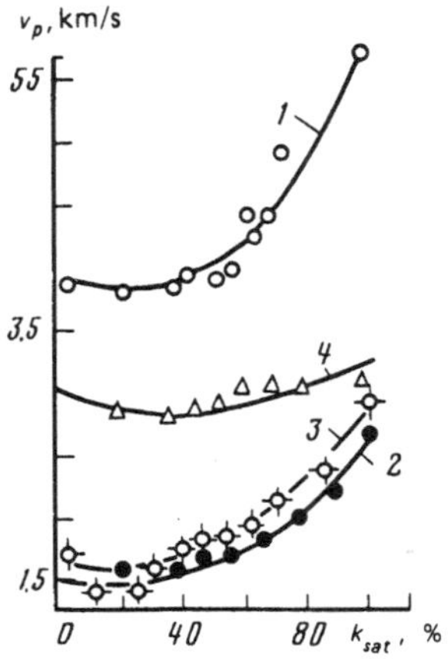

FIG. 158. Dependences of the speed of propagation of longitudinal elasticity waves v_P on the coefficient of water saturation of rocks $k_{w.s}$ (after N.A. Tuezova, L.M. Doroginitskaya, R.G. Demina, N.I. Bryuzgina).

Sandstone containing (in %) calcite cement: *1*—44; *2*—12; *3*—10; *4*—argillite

$v_P = f(k_{oil})$ (k_{oil} is coefficient of oil saturation of rock) are the same whether rocks are saturated by water or oil. The saturation mostly affects v_P of rocks acted on by pressure. This is accounted for by the increased density and elasticity of their filling component (water and air) with growth of k_w and by the fact that with increasing the pressure intergrain contacts become more compact.

According to experiments

$$v_P = (v_{Ps} - v_{Pmin})e^{-Bk_p} + v_{Pmin} \quad (161)$$

where v_{Ps} is the speed of propagation in the solid phase of the given sedimentary rock type; v_{Pmin} is the minimum speed of propagation for the same rock—that for k_{pmax} in nonloaded rocks; B is the coefficient determining the dependence of v_P on the sizes and shapes of voids. By using this equation $v_P = f(k_p)$ dependence graphs have been plotted (Fig. 159).

The variation of v_P values induced by pressure is governed not only by the composition of principal rock-forming minerals of a rock but also by the composition of their cement (Fig. 160) owing to their various compressibility β_r. Given k_p = const, the highest relative variations of the values of the speed of propagation of longitudinal waves in sandstones and aleurolites containing clayey and micaceous cement, and the lowest ones have opal and siliceous cement.

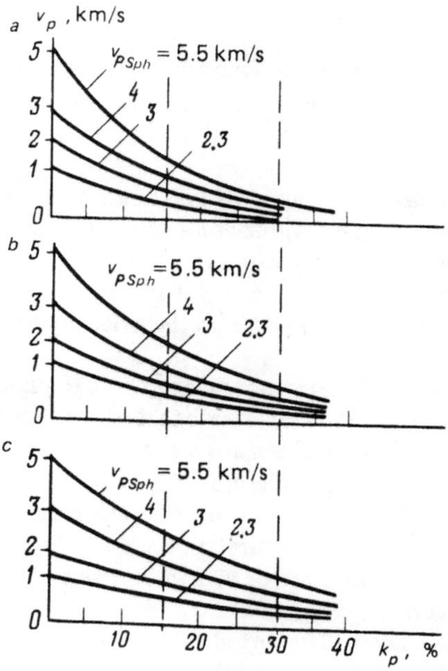

FIG. 159. Dependence of v_P in sedimentary rocks on k_p (after M.L. Ozerskaya).

a—0.08; *b*—0.06; *c*—0.04

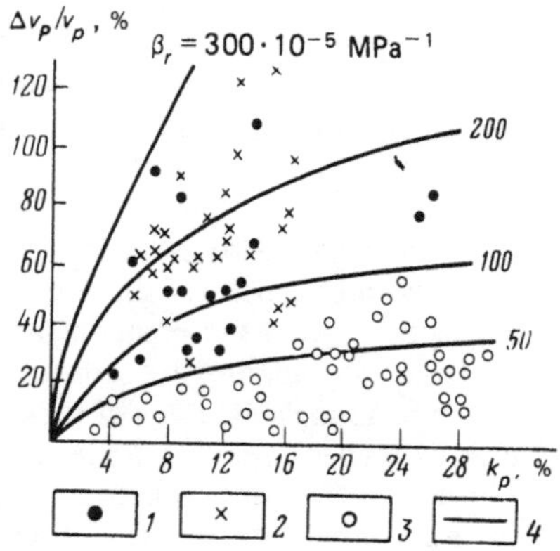

FIG. 160. Relative variation $\Delta v_P/v_P$ of the speed of propagation of longitudinal elastic waves in sandstones and aleurolites with increasing the overall pressure from 0.1 to 150 MPa and depending on the initial voids ratio k_p and cement composition (after G.M. Avchyan).

Sandstones saturated by binormal NaCl solution. Cement: *1*—clayey-micaceous; *2*—clayey carbonate; *3*—opal and siliceous; *4*—calculated graphs

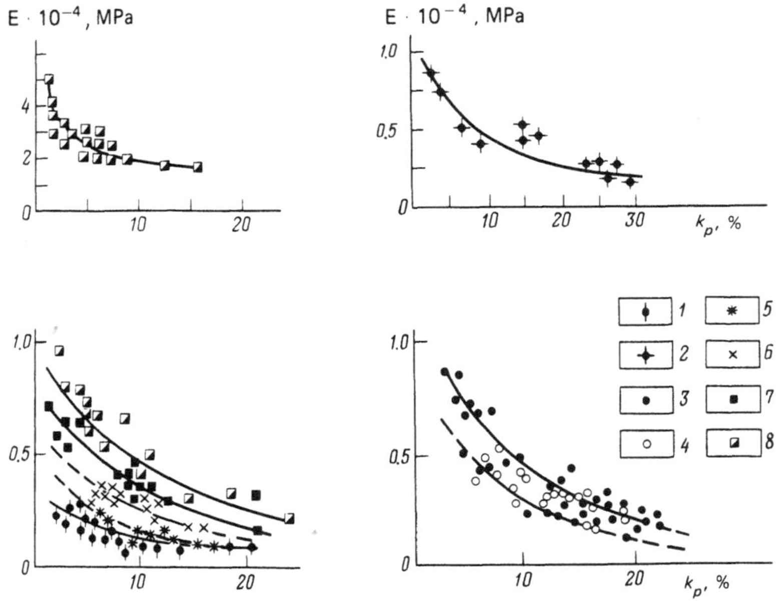

FIG. 161. Dependence of Young's modulus E on k_p (after I.S. Finogenov).
1—aleuritic clay; *2*—clay rocks; *3*—sandstones; *4*—aleurolites; *5*—clayey marls; *6*—marls; *7*—marlstones; *8*—limestones

The modulus E also increases with decreasing k_p (Fig. 161) and is governed by the density and elasticity of their solid phase. Given a definite voids ratio, the higher the density of the solid component of the rock, the greater is the modulus.

This is also validated by a direct correlation between E and hardness P_{sh} for sandstones. Young's modulus of rocks is the higher, the greater is their density that is closely connected with hardness.

Since the values of elasticity characteristics and many more petrophysical quantities taken separately are determined by the voids ratio, one can observe, even though not very close, correlations (Fig. 162). They are closer for a specified density of the solid phase and quantities characterizing the structure of rocks (S_V, Md etc.), since in this case determined are the mineral composition and structure of the series of samples of identical-type rocks for which the correlation is to be found.

Special studies have established the pattern of variation of v_P and other elasticity characteristics with growth of overall pressure, p, temperature, t, and, to a lesser degree, with simultaneously increasing these latter. The speed of propagation of longitudinal elastic waves in samples of magmatic, metamorphic and sedimentary rocks increases with p. Note that at first at pressures from an atmospheric one (0.1 MPa) to 50-100 MPa v_P varies dramatically, then, over the entire interval to 250-500 MPa, slowly, often tending to the limit which is often called a final one. This v_P value is used in average time equations (141) and (142).

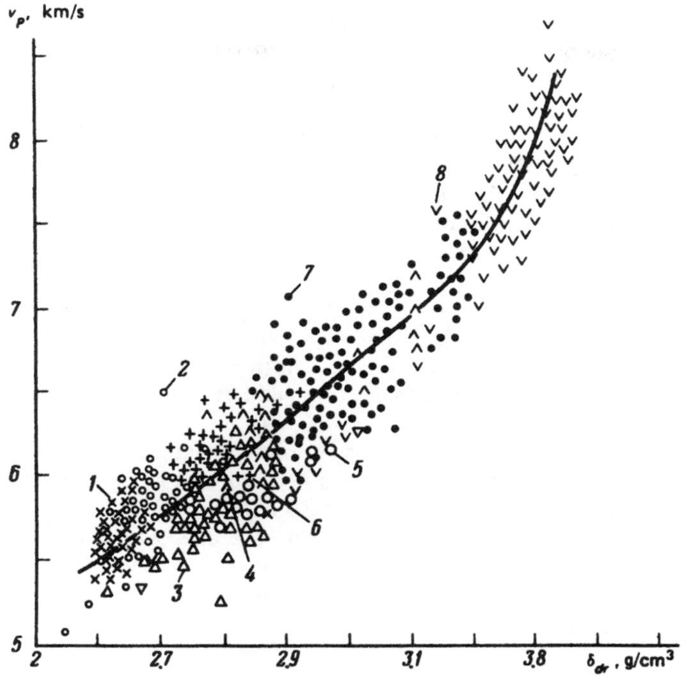

FIG. 162. Connections between v_P and δ_{dr} for intrusive and metamorphic rocks (after N.B. Dortman and M.Sh. Magid).

1—granites; *2*—biotite and biotite-amphibolite gneisses; *3*—kyanite-granite-biotite gneisses; *4*—biotite-amphibole gneisses and amphibolites; *5*—plagioclase granulites; *6*—diorites and gabbro diorites; *7*—gabbro-norites; *8*—hyperbasites

This pattern of variation of v_P depending on p is accounted for by the closing of voids mainly at pressures 50-100 MPa and by appreciable growth of elasticity forces at contacts between other crystals owing to the increase of the number and area of contacts upon compression. However, the pattern of increasing v_P with pressure p is determined for magmatic and metamorphic rocks by the composition of their rock-forming minerals, and for sedimentary rocks, additionally, by the composition of their cement. The v_P values increase with pressure in the series acidic-intermediate-basic-ultrabasic-alkaline rocks.

Under conditions of a concurrent action of increasing temperature t and pressures p on different types of magmatic and metamorphic rocks complex $v_P = f(p, t)$ and $v_S = f(p, t)$ relationships have been found. The anisotropy coefficient, particularly appreciable for definite types of metamorphic rocks little changes with temperature at a specified pressure, however, it does appreciably decreases with increasing pressure.

The volumetric deformation of sedimentary, relatively high-porosity water-saturated rocks, according to V.M. Dobrynin, is generally determined by effective pressure p-p_p (p, p_p are the overall and pore pressure) and temperature t. It must be noted that the pore pressure of the liquid opposes the overall pressure acting on the solid phase of the rocks.

The speed of propagation of longitudinal waves in sedimentary rocks grows with p_{eff}, however, it does so the less, the lesser is their initial elasticity (voids ratio).

The speed of propagation of elastic waves, $v_{P(p-p_p)}$, in a reservoir bed at a depth h can be evaluated using the relationship

$$v_{P(p-p_p)} = v_P \left[\frac{(p - p_p)_H}{(p - p_p)_{min}} \right]^{1/n}$$

where $(p - p_p)_{min}$ and $(p - p_p)_H$ are effective pressures at the minimum depth of occurrence of the bed and one equal to H, and v_P is the speed at a minimum pressure; n grows with increasing cementation of sandstones and decreases with growth of the depth of their layer.

With increasing the temperature a decrease of v_P in an oil-saturated reservoir bed is more appreciable than in a water- and gas-saturated bed. The greatest growth of v_P (up to 60-80%) with increasing pressure to 150 MPa is referred to high-porosity sandy clayey rocks, and in the presence of siliceous and opal cement v_P values change with pressure less (by 25-40%). The v_P values of carbonate rocks change still less (by 15-25%). The less the difference induced by the type of the filling component, the higher is the pressure.

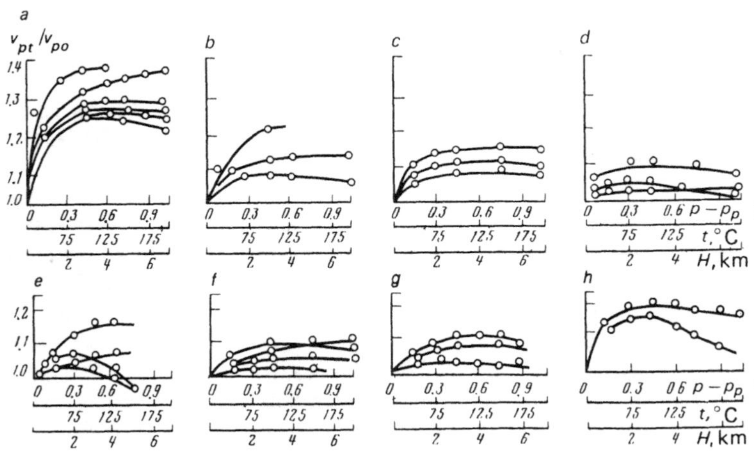

FIG. 163. Dependences of the relative speed of longitudinal waves v_{Ppt}/v_{P_0} in water-saturated sedimentary rocks (different samples) for a concurrent effect of overall p, stratal p_{st} pressure and temperature t (after G.M. Avchyan, A.A. Matveenko, Z.B. Stefankevich).

a—highly porous clay sandstones and aleurolites; *b*—highly porous nonclayey quartz sandstone; *c*—sandstone and aleurolite with argillaceous and carbonate cement; *d*—low-porosity sandstone and aleurolite with high content of clayey and carbonate material; *e*—dolomite; *f*—limestone; *g*—marl; *h*—clay

The v_P values increase with concurrently increasing p, p_p and t to values corresponding to a depth 3 km (Fig. 163), i.e. the speed of propagation of elastic waves as determined in laboratory conditions is underestimated and must be corrected. Given greater values of p, p_p and t, corresponding to depths 3-5 km, v_P values are markedly affected by temperature which fact leads to a drop in v_P which is particularly true of dolomite, marl, argillaceous sandstones, clays (see Fig. 163).

The value of v_{P_H} at a depth of bed occurrence H can be found using the equation

$$v_{P_H} = v_{P0} K_{p-p_p} K_p K_t$$

where v_{P0} is the speed of propagation of elastic waves determined in laboratory at atmospheric pressure and room temperature; K_{p-p_p} is a correction coefficient representing a partial variation of the speed v_P following a variation of $p_{eff} = p - p_p$ from 0 to $p - p_p$; K_p is the same but at changing of stratum pressure p_p from 0 to p_p is the same and following temperature variation from 0 to t.

Methods of determining K_{p-p_p}, K_p and K_t are presented in works by G.M. Avchyan, V.M. Dobrynin, B.N. Kulikov, A.A. Matveenko and Z.B. Stefankevich.

According to M.L. Ozerskaya, the speed of propagation of longitudinal elastic waves increases with depth:

$$v_P = v_{Pmax} - (v_{Pmax} - v_{Pmin})e^{-0.45H}$$

where v_{Pmax} is the v_P value at a depth H; v_{Pmin} is that at $H = 0$. The pattern of variation of v_P with depth is illustrated in Fig. 164.

The v_S values also increase with increasing pressure and decrease with increasing temperature. The v_S values for magmatic rocks increase passing from their least dense acidic varieties to alkaline ones demonstrating the maximum density because of a minimum silica content.

Young's modulus, moduli of shear, overall compression and Poisson's ratio increase with increasing pressure p. This is due to an increased area of grains contacts as a result of the pressure-induced closing of intergranular and intercrystalline voids.

For granites, gabbros, diorite, labradorite, basalt, marble and low-porosity limestone Young's modulus values continuously decrease with temperature. For dense sandstones and quartzite until 500 °C they decrease gradually and then dramatically, having reached a minimum at 575 °C, increase again. This latter fact is accounted for by quartz alteration.

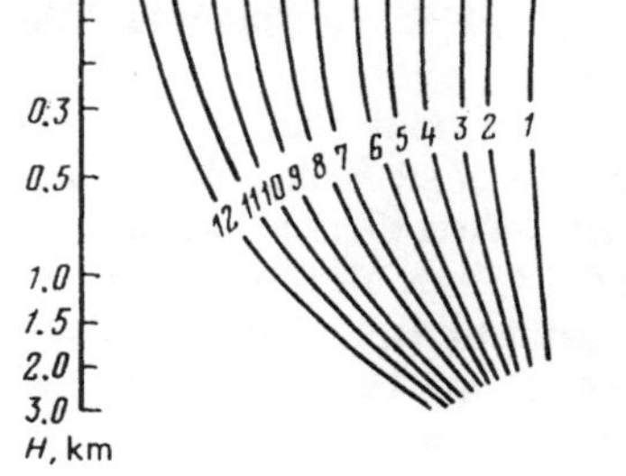

FIG. 164. Dependence of v_P velocity of propagation of longitudinal elasticity waves on depth H of occurrence of sedimentary rocks (according to M.L. Ozerskaya).

Ticks at the abscissa axis correspond to the origin of coordinates for each curve. Curves code: v_{Pmin} (in km/s): *1*—5.4; *2*—5.0; *3*—4.6; *4*—4.2; *5*—3.8; *6*—3.4; *7*—3.0; *8*—2.6; *9*—2.2; *10*—1.8; *11*—1.4; *12*—1

Sec. 80. Elasticity Characteristics of Rock Groups and Types

The values of the best studied elasticity characteristics of rocks vary respectively within these limits: v_p = 0.2-8.8 km/s, E = (0.01-2.35) × 10^{11} Pa and ν = 0.04-0.6 (Table 24).

In the order of decreasing limiting values of the speeds of propagation of elastic waves and Young's modulus, rocks groups can be arranged in this series: magmatic (v_p = 2.52-8.77 km/s, E = (0.1-2.3) × 10^{11} Pa)—metamorphic (v_p = 1.45-7.6 km/s, E = (0.05-1.91) × 10^{11} Pa)—sedimentary rocks (v_p = 0.3-7.6 km/s, E = (0.01-1.75) × 10^{11} Pa). The least elastic are sedimentary rocks. Particularly small v_p or E values are demonstrated by high-porosity, minor-density varieties, and high values by low-porosity relatively dense samples of sandstones, limestones, dolomites and other sedimentary rocks. High v_p values have also been found for rock salt, and v_p and E values for some anhydrite varieties.

Magmatic rock varieties are the most elastic including ones referred to a family of ultrabasic intrusive rocks having a high content of mafic minerals (olivine, pyroxenes, hornblende) that demonstrate low porosity, high density and v_p and E values. The v_p and E values of dunite and peridotite are, e.g. found in the range 7.4-8.8 km/s and (0.18-2.35) × 10^{11} Pa, respectively. The limiting high values of these quantities are less for basic and still less for intermediate and acidic rocks. The lowest v_p and E values are demonstrated by granite, granodiorite, quartz diorite owing to a relatively high quartz content and increased porosity. The elasticity of effusive rocks is less than that of intrusive rocks. For basalts, e.g. the speed of propagation of elastic waves can drop to 2.5 km/s, since they sometimes show a rather high porosity.

As to metamorphic rocks, the least limiting values of v_p and E are manifested by acidic-composition ortho- and paraschists [v_p = 1.5-6.6 km/s, E = (0.05-1.09) × 10^{11} Pa], phyllites [E = (0.09-0.7) × 10^{11} Pa], and the greatest values by amphibolites [v_p = 6.5-7.2 km/s, E = (0.12-1.35) × 10^{11} Pa], hornfelses (v_p = 5.39-6.21 km/s, E = 0.6 × 10^{11} Pa), eclogites [v_p = 6.6-8.0 km/s, E = (1.23-1.91) × 10^{11} Pa]. Intermediate v_p and E values are shown by such metamorphic rocks as serpentinites, marbles, quartzites (gneisses). Comparatively few values of the shear modulus G have been obtained, they are sometimes twice as small compared with the respective values of Young's modulus.

The $v_{p\parallel}$ and $E_\parallel$ values of layered rocks are greater than $v_{p\perp}$ and $E_\perp$, measured, respectively, parallel to the layers or at right angles to these latter. The coefficient of anisotropy K_{v_p} of longitudinal waves attains 1.2-1.3 and generally exceeds K_{v_s} of transverse waves, and the coefficient of anisotropy of Young's modulus, K_E, varies from 1.1 to 2.0 for most rocks.

The values of Poisson's ratio vary from 0.04 to 0.6, generally being in the range 0.18-0.35. Essentially quartz rocks, such as some varieties of aleurolites, sandstones, quartzoids, granites, quartz-porphyry quartz phyllites, quartz diorites have demonstrated the values of ν = 0.05-0.15. Minor ν values (0.04) have also been found for porous basalts. Sedimentary rock (clays, aleurolites, sandstones, rock salt, coal) and metamorphic rock (clay shales, amphibolites) varieties have been discovered that demonstrate ν = 0.4-0.6. These high ν values seem to refer to very low-porosity samples containing many mineral species having high ν values.

TABLE 24. Limiting Values of Quantities Characterizing Elasticity of Primary Sediments and Rocks (after various authors)

Primary sediment or rock	Speed of propagation of elastic waves, km/s		Young's modulus $E \times 10^{-11}$ Pa	Shear modulus**, $G \times 10^{-11}$ Pa	Poisson's ratio ν
	longitudinal* v_P	transverse v_S			
Primary sediments, derived sedimentary, metamorphic rocks and ores					
Breccia	1.45-5.6	—	—	—	—
Conglomerate	1.45-5.6	—	—	—	—
Gravellite	1.7-5.4	—	—	—	—
Sand	0.2-2.0	0.4-0.5	—	—	—
Sandstone	0.8-4.5	—	<0.3->1.0	—	0.06-0.6
Sandy shale	1.45-5.18	—	0.05-0.69	—	0.12-0.21
Quartzitic sandstone	—	—	0.4-1.0	0.26-0.47	0.07-0.3
Quartzite	5.3-6.35	—	—	—	—
Fine aleuritic mud	1.46-1.68	$(0.6)_{av}$	—	—	—
Aleurite clayey mud	1.50-1.54	—	—	—	—
Fine aleuritic calcareous foraminiferous mud	1.42-1.55	—	—	—	—
Loess	0.3-0.6	—	0.3-0.7	—	—
Aleurolite	0.8-4.0	—	0.01-0.69	—	0.05-0.6
Mud:					
clayey	1.49-1.51	—	—	—	—
clayey lime	1.44-1.50	—	—	—	—
Moist clay	1.1-2.5	—	—	—	—
Clay	0.3-3.0	$(0.59)_{av}$	$(0.3)_{av}$	—	0.25-0.45
Argillite	0.9-4.8	—	0.02-0.4	—	0.08-0.34
Shale					
clay shale	1.78-4.74	—	0.05-0.59	—	0.1-0.43
roofing slate	—	—	0.43-1.09	—	0.09
slate	2.3-6.65	$(2.86)_{av}$	0.24-0.72	—	0.17
Quartz, sericite, graphite phyllite	—	—	0.09-0.7	—	0.06-0.28
Para- and orthogneiss	3.5-7.5	3.43-3.61	0.5-0.58	0.24	0.31-0.32
Chalk	1.7-4.2	up to >3.0	—	—	—
Organogenic limestone	3.0-3.3	—	0.13-0.75	—	0.11-0.2
Limestone	1.0-7.0	2.75-3.59	0.13-1.75	0.12-0.71	0.18-0.31
Marble	3.75-6.94	2.02-3.86	0.23-0.98	0.28-0.36	0.22-0.4
Dolomite	0.9-7.6	—	0.15->1.5	0.1-0.4	0.17-0.38
Marbled dolomite	—	—	0.71-0.87	0.34	0.21-0.28
Dolomitic marble	—	—	0.85	—	—
Marl	1.3-4.5	—	0.1-0.135	—	0.11-0.23
Anhydrite	1.5-6.0	—	0.6-1.0	(0.28)	0.27-0.3
Gypsum	1.5-4.6	—	0.12-0.47	—	0.27-0.3

Primary sediment or rock	Speed of propagation of elastic waves, km/s		Young's modulus $E \times 10^{-11}$ Pa	Shear modulus**, $G \times 10^{-11}$ Pa	Poisson's ratio ν
	longitudinal* v_P	transverse v_S			
Rock salt	3.5-5.5	—	0.14-0.32	(0.15)	0.11-0.44
Carnallite	4.4-6.5	—	—	—	—
Sylvinite	4.4-6.5	—	0.09-0.135	—	0.26-0.31
Ores:					
ferruginous quartzite (magnetite)	5.96/5.81	3.36-3.48	0.41-1.19	—	0.18-0.34
ferruginous quartzite (hematite)	5.53/	2.86-3.32	1.03	—	0.33
hematite	7.04/	3.12-3.42	1.42-2.0	0.27-0.87	—
martite	—	—	0.05-1.25	—	0.1-0.35
Coal:					
brown	2.6-2.7	—	1.8	—	0.31
hard	1.7-2.6	—	0.05-0.1	—	0.43
Anthracite	2.5-3.5	—	—	—	0.14-0.16
Magmatic and derived metamorphic rocks and ores					
Dunite	6.0-8.77	3.68-4.83	0.89-1.95	0.48-0.74	(0.27-0.40)
Peridotite	7.8-8.7	—	0.18-2.35	—	0.24-0.31
Pyroxenite	7.8-8.7	—	0.3-4.55	—	0.25-0.37
Serpentinite	3.7-6.84	2.71-3.83	0.34-0.74	—	0.27-0.35
Gabbro	5.8-7.4	3.42-3.84	0.34-1.22	0.03-0.4	0.16-0.3
Anorthosite	5.7-7.21	3.56-3.81	0.825	(0.328)	0.26
Labradorite	6.17-7.05	—	0.65-1.75	—	0.32-0.7
Diorite	5.78-6.13	3.06-3.55	0.34-0.87	0.28-0.37	0.10-0.33
Quartz-diorite	5.13-6.7	3.4-3.8	0.34-0.75	0.21-0.366	0.16-0.23
Syenite	5.8-6.45	3.0-5.4	0.72-1.05	—	0.21-0.23
Granite	3.7-6.51	2.7-3.8	0.2-0.85	0.16-0.28	0.08-0.34
Nepheline-syenite	5.5-6.45	2.6-3.45	0.28-0.86	(0.251)	0.26
Ijolites:					
leucocratic	5.8-6.45	2.7-3.05	—	—	—
mesocratic	6.15-6.80	—	—	—	—
Ijolite-urtite	5.8-6.9	—	—	—	—
Basalt	2.52-6.4	2.72-3.20	0.1-1.2	0.27-0.34	0.04-0.36
Andesite	5.19; 5.61	2.73	0.2-0.82	—	0.16-0.3
Diabase	6.14-6.94	3.49-3.85	0.21-1.29	—	0.2-0.3
Porphyrite	4.32-5.31	—	0.89	—	0.23
Quartz-porphyry	—	—	0.15-0.83	—	0.08-0.26

Primary sediment or rock	Speed of propagation of elastic waves, km/s		Young's modulus $E \times 10^{-11}$ Pa	Shear modulus**, $G \times 10^{-11}$ Pa	Poisson's ratio ν
	longitudinal* v_P	transverse v_S			
Liparite	5.0	3.0	—	—	—
Pegmatite	—	—	0.12-0.9	—	0.16; 0.23
Tuff	1.89-2.38	—	0.014-0.87	0.017-0.034	0.08-0.11
Ash tuff	2.16; 3.35	0.83	0.25	—	0.24
Chlorite slate, quartz-chlorite schist, quartz-talc schist	4.8-6.97	—	0.09-0.45	—	0.18-0.4
Amphibolite	6.5-7.35	3.9-4.3	0.12-1.35	0.47	0.16-0.46
Hornfels	5.99-6.21	—	0.6-1.25	—	0.1-0.27
Eclogite	5.2-8.0	3.7-4.61	1.23-1.91	0.49-0.63	0.21-0.28
Gneiss:					
amphibole, biotite, garnet-biotite, biotite-amphibole, garnet-biotite-amphibole, muscovite-biotite	5.2-6.35	—	0.085-0.84	0.24	0.18-0.32
diorite	—	—	0.68-0.7	—	0.08-0.11
syenite	—	—	0.55	0.24	—
pyroxene	—	—	0.25-0.7	—	0.23-0.33
Ores:					
magnetite	3.4-6.05	—	0.68-0.7	—	0.27
pyrrhotite	—	—	0.45	—	0.31
sulphide	—	—	0.34-1.07	—	0.34
zinc (marmatite)	5.48	2.72-2.98	—	—	—
banded lead-zinc	5.74/5.34	2.37-2.58	—	—	—
lead (bleischweif)	3.47/3.22	1.65-1.98	—	—	—
lead	3.8-4.61	—	—	—	—
cinnabar	2.2/2.3	1.25-1.33	—	—	—
molybdenum (molybdenite)	/3.34	1.52-2.27	—	—	—
antimonite	/2.74	1; 1.86-5	—	—	—
fahlore (boulangerite)	/2.30	1.32-1.46	—	—	—
apatite	—	—	0.43-1.47	—	0.24-0.31
apatite-nepheline	—	—	0.21-1.11	—	0.1-0.39
scheelite	—	—	0.59	—	0.19-0.25

* The numerator contains v_P measured parallel to lamination, and the denominator those perpendicular to it.

** Parenthesized are calculated values.

Sec. 81. Absorption of Elastic Waves

The energy of elastic waves is wasted in a rock irreversibly transforming to thermal energy and scattering in various directions due to heterogeneities of rocks (voids, impregnations, fissures etc.). Conversion of elastic energy to thermal energy is accounted for by the friction of vibrating particles. As a result of absorption and scattering of elastic energy the waves decrease in amplitude and the vibration process decays.

The amplitude of elastic waves varies with a distance x from a point source of vibrations in conformity with the law

$$A = A_0 \left(\frac{x}{x_0}\right)^{-n} e^{-\alpha(x - x_0)}$$

Here A_0 is the amplitude at a distance x_0 from the centre of the source of vibrations located on a line connecting its centre and point x where the amplitude A is determined; n is the (amplitude) *index of divergence of the wave from the source* governed by the type of wave, structure and degree of material homogeneity of rock; α is the (amplitude) *absorption coefficient.*

According to Yu.V. Riznichenko, n and α values have a straightforward physical meaning corresponding to their names during the propagation of a plane wave:

(a) in a homogeneous isotropic, perfectly elastic medium $A = A_0$ where $A_0 = \text{const}$, $n = 0$, $\alpha = 0$ (there is no divergence or absorption of wave);

(b) in a homogeneous isotropic absorbing medium $A = A_0 e^{-\alpha x}$, where α is a coefficient characterizing the absorbing properties of the medium, $x_0 = 0$ (the origin of reckoning the distance coincides with the centre of the source), $n = 0$ (there is no divergence). The experimentally established decrease of the amplitude dA/A at a minor line segment dx is in direct proportion to its length:

$$dA/A = -\alpha dx$$

Both scattering and absorption of elastic waves due to their imperfect elasticity and heterogeneity are encountered in actual rocks. The different capacity of rocks for absorbing the elastic energy of longitudinal, transverse and surface waves is characterized, respectively, by absorption coefficients α_P, α_S and α_R. In practically single-phase monocrystals and in a homogeneous liquid phase of rocks (e.g. natural water)

$$\alpha_P = \frac{2}{3} \frac{\omega^2 \eta}{v^3{}_P \delta_r}$$

where η is a viscosity coefficient, Pa·s; $\omega = 2\pi f$ is the frequency; δ_r is the density of the medium; v_P is the speed of propagation of longitudinal elastic waves.

Coefficients of absorption α_P and α_S in actual rocks are found by using the relationships:

$$\alpha_P = \frac{\omega^2}{2\delta_r v_P^3} \left[\left(\frac{4}{3}\eta_1 + \eta_2\right) + \frac{\lambda T \beta^2 \delta_r^2 v_P^4}{c_{pV}^2} \left(1 - \frac{4}{3}\frac{v_S^2}{v_P^2}\right)\right],$$

$$\alpha_S = \frac{\eta_1 \omega^2}{2\delta_r v_S^3}$$

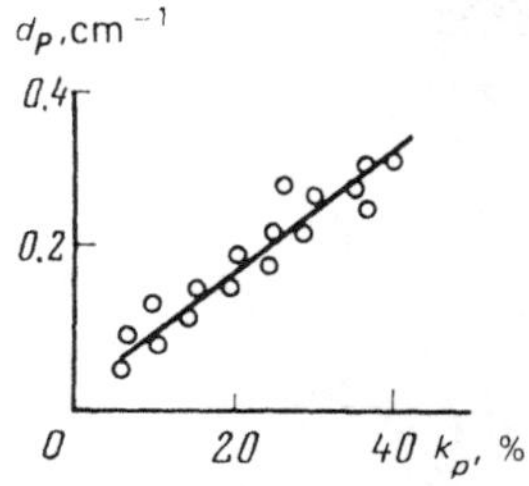

FIG. 165. Dependence of the coefficient of absorption of the longitudinal wave α_P on k_p for carbonate rocks (after I.P. Dzeban)

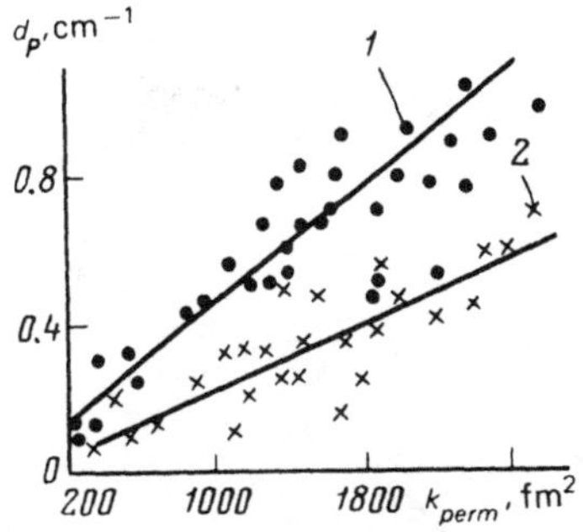

FIG. 166. Connection of the coefficient of absorption of ultrasound waves α_P with the coefficient of permeability k_{perm} (after A.A. Bolotov, B.A. Belinskii and L.L. Sinii).

1—sandstone; *2*—aleurolite

where $\eta_1$and η_2 are volumetric viscosities of the phases; λ is the coefficient of thermal conductivity of the rock; c_{pV} is its specific isobaric thermal capacity for a constant volume; β is the compressibility of the rock.

Thus, α_P and α_S values are determined by quantities characterizing the elasticity of rocks (v_P, v_S, β), their texture (density δ_r), thermal properties (λ, c_{pV}), as well as temperature T, frequency and viscosities η_1 and η_2.

As far as multi-phase permeable rocks are concerned, the absorption of elastic energy is accounted for, in addition, by viscous interphase friction (friction between the liquid and solid components of rocks) during their relative motion induced by the action of an acoustic field followed by interphase heat exchange.

Experiments substantiate the increase of α_P values with decreasing δ_r since a direct correlation has been found between α_P and the voids ratio k_p (Fig. 165), permeability coefficient k_{perm} (Fig. 166) which are inversely related with density. Growth of α_S and α_P values has been found with increasing the clayiness of rocks (Fig. 167) and frequency of sound (Fig. 168). Given a specified frequency, the highest α_P values have been found for gas-saturated sands. They are lower for oil-saturated sands and still lower for their water-saturated varieties. As to α_S values, given identical conditions, they are arranged in a reverse order (see Fig. 168). The design values show a sufficiently good fit to the experimental ones and are consistent with concepts of a viscous inertia and thermal mechanism of absorption of elastic waves.

The pattern of the dependence of the absorption coefficient on frequency is believed to be affected by the cementation of rocks. Over a frequency range 30-110 kHz in water-saturated sand the α_P values increase in direct proportion to $\sqrt{f}$. By contrast, in cemented rocks α_P is in proportion to frequency.

For $f = 1$ MHz the coefficient α_P varies from 0.8 to 5 m^{-1} for fresh magmatic rocks. For weathered granites and granosyenites it is greater, from 11 to 14 m^{-1}.

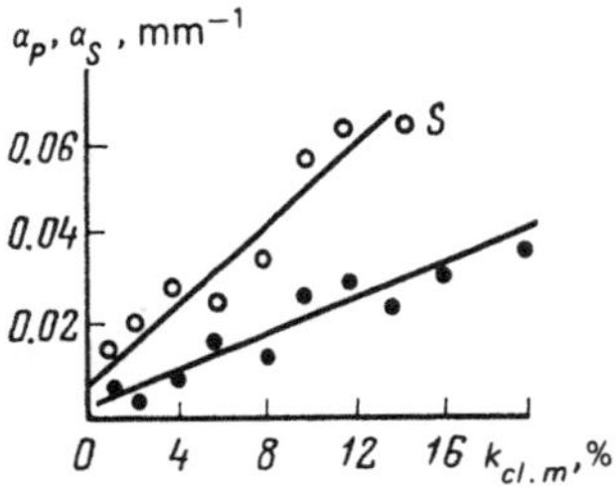

FIG. 167. Dependences of the coefficients of absorption of longitudinal and transverse waves α_P and α_S on the coefficient of mass clayiness $k_{cl.m}$ (after I.P. Dzeban)

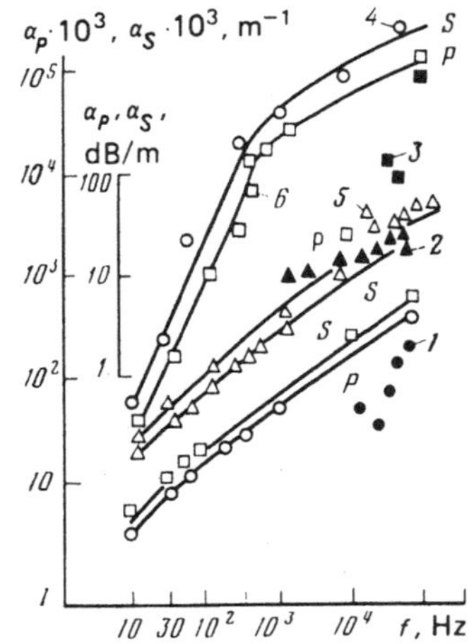

FIG. 168. Dependences of α_P and α_S in sands on the sound frequency f (after B.N. Ivakin, E.V. Karus, O.L. Kuznetsov)

Experimental (*1-3*) and design (*4-6*) values for sands upon their saturation by: *1*, *4*—water; *2*, *5*—oil; *3*, *6*—gas

Still greater values for the same rocks are shown by coefficients α_S = 18-27 m^{-1}. In water-saturated sands, given a frequency 80-90 kHz, it attains 1.5 m^{-1}; in oil-saturated sand, at the same frequencies, it equals 4.3 and 5.6 m^{-1}.

Coefficients α_P, α_S and α_R have been insufficiently studied, therefore the above data incompletely characterize likely limits of their variations for rocks.

Sec. 82. Differentially Elastic Rocks Showing an Imperfect Connection Between Phases

Relationships between strains and deformations for a unit volume of a macroscopically homogeneous and isotropic rock having voids sizes much less than the size of the unit volume and filled by a viscous material displacing relative to the skeletal particles of the rock appear as follows:

$$\left.\begin{aligned}
\sigma_x &= \lambda_{s+fill}\theta + 2\mu_{s+fill}e_x + Q\theta_{fill}\\
\sigma_y &= \lambda_{s+fill}\theta + 2\mu_{s+fill}e_y + Q\theta_{fill}\\
\sigma_z &= \lambda_{s+fill}\theta + 2\mu_{s+fill}e_z + Q\theta_{fill}\\
\tau_{yz} &= \mu_{s+fill}e_{yz}\\
\tau_{zx} &= \mu_{s+fill}e_{zx}\\
\tau_{xy} &= \mu_{s+fill}e_{xy}\\
s &= Q\theta + R\theta_{fill}\\
\theta &= e_x + e_y + e_z
\end{aligned}\right\} \qquad (162)$$

where Q is a proportionality coefficient in the relationship between expansions θ_{fill} and θ, respectively, of the voids filling component and of the solid aggregate of particles (if the pressure of the filling component is zero, $\theta_{fill} = -Q\theta/R$); R is the pressure acting on the filling component at which a known amount of it is impressed into a unit volume of the rock without being changed; s is a scalar quantity determining the state of stress of the filling component.

Having determined from Lagrange's equations the forces acting on the solid and liquid component in a unit volume of the rock having an imperfect connection between its components, we can, representing these forces in terms of strain gradients and substituting into the equations obtained the strain values from Eq. (162), find motion equations for the particular two-component medium. After their transformation taking into account that the pore fluid is viscous we get:

$$\nabla^2(M_{s+fill}\theta + Q\theta_{fill}) = \frac{\partial^2}{\partial t^2}(\delta_{11}\theta + \delta_{12}\theta_{fill}) + \frac{\mu^* k_{p.o}^2}{k_{perm}} \frac{\partial}{\partial t}(\theta - \theta_{fill}) \qquad (163)$$

$$\nabla^2(Q\theta + R\theta_{fill}) = \frac{\partial^2}{\partial t^2}(\delta_{12}\theta + \delta_{22}\theta_{fill}) - \frac{\mu^* k_{p.o}^2}{k_{perm}}(\theta - \theta_{fill})$$

where $M_{s+fill} = \lambda_{s+fill} + 2\mu_{s+fill}$; λ_{s+fill} and μ_{s+fill} are Lamé's coefficients of a two-component medium; μ^* is a dynamic shear viscosity; $\delta_{11} = \delta_{dr} - \delta_{12} = (1 - k_{p.o})\delta_s - \delta_{12}$, $\delta_{22} = \delta_{fill} - \delta_{12} = k_{p.o}\delta_l - \delta_{12}$ is the specific masses, respectively, of the solid component not participating in the motion of the fluid and fluid not participating in the motion of the solid component, called dynamic density coefficients; $\delta_{12} = -\delta_d$ is a coefficient of mass connection between the fluid and solid component—skeleton of the rock (defined as an "apparent mass", this quantity, probably, due to the difference of forces of inertia of the skeleton and fluid, is called an attached mass).

Having definitely transformed Eqs. (163), writing solutions of the equations obtained for plane harmonic waves as $e = C_1 \exp[j(\omega t - kx)]$ and $\theta_{fill} = C_2 \exp[(j(\omega t - kx)]$, ($C_1$, C_2 and k are constants) and substituted into the transformed equation (163), we get two equations characterizing the speeds of propagation of longitudinal waves in two-phase rocks having an imperfect connection between the components. One of them is meant for a wave propagating with a great speed, v_{PI}, called a type one wave, the other for a wave propagating with a lesser speed v_{PII}—a type two wave. This theory makes it also possible to obtain for the particular medium absorption coefficients of type one and two.

In the general form these equations are all complicated and cumbersome. In the limiting cases they become much more simplified. Suppose that the Q-factor of the medium is $q > 5$ ($q = \pi/\vartheta$; ϑ being an absorption decrement) in so doing,

$$v_{PI} = \sqrt{\frac{H(\nu_c + \sigma_L - 2\nu_l\delta_K)^2 + (\omega_c/\omega)^2}{\delta_{s+l}(\nu_c + \nu_l^2)(\nu_c + \sigma_L - 2\nu_l\sigma_K) + (\omega_c/\omega)^2}}, \qquad (164)$$

where

$$\nu_l = \frac{\delta_l}{\delta_{s+l}}; \quad \delta_l = \frac{\delta_{12} - \delta_{22}}{k_{p.o}}; \quad \delta_{s+l} = \delta_{11} + 2\delta_{12} + \delta_{22};$$

$$\nu_c = \frac{\delta_c}{\delta_{s+l}}; \quad \delta_c = \frac{\delta_{22}}{k_{p.o}^2} = \frac{k_{p.o}\delta_l - \delta_{12}}{k_p^2}; \quad \sigma_L = \frac{L}{H};$$

$$\sigma_K = \frac{K}{H}; \quad L = R/k_p^2; \quad H = \lambda + 2\mu + R + 2Q;$$

$$K = (Q + R)/k_p; \quad \omega_c = \frac{\mu^*}{k_{perm}\delta_2} = \frac{\mu^*}{k_{perm}(\delta_{12} + \delta_{22})};$$

μ^* is dynamic shear viscosity.

As can be seen from Eq. (164), given $\omega \to 0$.

$$v_{\mathrm{PI}_0} = \sqrt{H/\delta_{s+l}}$$

and given $\omega \to \infty$,

$$v_{\mathrm{PI}\infty} = \sqrt{\frac{H}{\delta_{s+l}} \frac{\nu_c + \sigma_L - 2\nu_l\sigma_K}{\nu_c + \nu_l^2}} = \sqrt{\frac{L\delta_{s+l} + H\delta_c - 2\delta_l K}{\delta_{s+l}\delta_c - \delta_l^2}}$$

The equation for v_{PII}, referred to low frequencies for $\omega/\omega_{av} \ll 1$ has this form

$$v_{\mathrm{PII}} = \sqrt{\frac{2(LH - K^2)}{\delta_{s+l}H} \frac{\omega}{\omega_c}} \tag{165}$$

A consideration of Eqs. (164) and (165) shows that the speeds of propagation of longitudinal elastic waves, v_{PI} and v_{PII} in two-phase rocks saturated by viscous liquid with an imperfect connection between the components are governed not only by the ratio of the components and their elasticity characteristics but also by the quantities (μ^*, k_{perm}) of the hydrodynamic interaction between the components of the media and elasticity vibration frequency ω.

To simplify matters, N.V. Nikolaevskii divided two- or three-phase rocks, according to the values of the quantities $\varepsilon = \beta_s K_{sk}(1 - k_p)$ (β_s being the compressibility of the solid component, K_{sk}—modulus of the overall compression of the skeletal particles), into little compacted (soft) for which $\varepsilon \ll 1$ and dense (cemented) rocks having $\varepsilon \to 1$. Both for soft and dense rocks, following from equations approaching in form to Eq. (164), he has derived relationships to determine speeds v_{PI} and v_{PII} as well as absorption coefficients α_{PI} and α_{PII} depending on elasticity characteristics and hydrodynamic connection of the components and frequency of propagating elastic vibrations. For soft rocks, type one longitudinal waves and given very low vibration frequencies ω or very high resistance of the medium [$\zeta = \delta_r\omega a_0/\mu^* \to 0$ (a_0 is the initial coefficient of the hydraulic resistance of rock)]

$$v_{\mathrm{PI}} = \frac{1}{\sqrt{\beta_{p.o}\delta_{p.o}}}\left[1 - \frac{1}{8}\zeta^2\left(\frac{R_1}{R_2} + 3\right)\left(\frac{R_1}{R_2} - 1\right)\right] \tag{166}$$

$$\alpha_{\mathrm{PI}} = \omega\sqrt{\frac{\beta_r\delta_{p.o}}{2}}\sqrt{\left(\frac{1 + R_1^2\zeta^2/R_2^2}{1 + \zeta^2}\right)^{1/2} - \frac{1 + R_1\zeta^2/R_2}{1 + \zeta^2}} \tag{167}$$

where $\beta_r = (1 - k_{p.o})\beta_s + k_{p.o}\beta_l$; $\delta_{p.o} = (1 - k_{p.o})\delta_s + k_{p.o}\delta_l$; $k_{p.o}$ are the average, initial stationary compressibility, density and voids ratio: β_s, β_l, δ_s, δ_l are, respectively, the compressibility and density of the solid and liquid component of the rock:

$$R_1 = \frac{k_{p.o}\delta_r^{2\infty}}{\delta_s\delta_l} A_{fill}; \quad R_2 = \frac{k_{p.o}\delta_{r\infty}\delta_{p.o}}{\delta_s\delta_l} A_{fill}$$

$$A_{fill} = 1 - \beta_s K_{sk} + \frac{\beta_l k_{p.o}}{\beta_s(1 - k_{p.o})}$$

Here K_{sk} is the modulus of overall compression of the entire aggregate of solid particles (skeletal particles) of rock; $\delta_{r\infty}$ is rock density, given its negligible filtration resistance or very high vibration frequency. For a very low ω or very high resistance of a porous medium and $\zeta = \omega\delta_r\alpha_0/\mu^* \rightarrow 0$ in the limit

$$v_{PI} \rightarrow v_{P0} \rightarrow 1/\sqrt{\beta_r\delta_{r0}}$$

where $\delta_{r0} = (1 - k_{r0})\delta_s + k_{r0}\delta_l$ is rock density, given elastic vibration frequency $\omega \rightarrow 0$. If ω values are very great, and rock filtration resistance values $\zeta \rightarrow \infty$, the limiting values of

$$v_{PI} \rightarrow v_{PI\infty} = 1/\sqrt{\beta_r\delta_{r\infty}},$$

where

$$\frac{1}{\delta_{r\infty}} = \frac{1 - k_{p.o}}{\delta_s} + \frac{k_{p.o}}{\delta_l}$$

Given a fixed frequency ω and negligible ($\zeta \rightarrow \infty$) or very large ($\zeta \rightarrow 0$) filtration resistance values, $\alpha_{PI} \rightarrow 0$, and for intermediate values α_{PI} attains peak values.

For the same soft rocks and type two longitudinal waves

$$v_{PII} = \sqrt{\frac{2}{b\delta}} \left(\sqrt{1 + \frac{1}{\zeta^2}} + 1\right)^{-1/2} \tag{168}$$

$$\alpha_{PII} = \omega \sqrt{\frac{b\delta}{2}} \left(\sqrt{1 + \frac{1}{\zeta^2}} - 1\right)^{-1/2} \tag{169}$$

where $b = (\lambda_1 + 2\lambda_2)^{-1}$; λ_1 and λ_2 are elasticity moduli of solid particles repacking; $\delta = \delta_{11} + \delta_{22} + 2\delta_{12}$; $\zeta = f_s/f_0$ is a dimensionless frequency; f_s, f_0 are the current and initial frequency, respectively.

Equation (163) has also been solved with respect to the speed of propagation of a plane monochromatic transverse wave in soft waves. The speed of a transverse wave propagating in soft rocks is also determined by the relative frequency and density, hydrodynamic characteristics of the medium under consideration and its structure.

The limiting values of the speed of propagation of a type two longitudinal wave for $\zeta \rightarrow \infty$ and $\zeta \rightarrow 0$ are, respectively, $v_{PII\infty} = (b/\delta)^{-1/2}$; $v_{PII0} \rightarrow 0$, and the limiting values of $\alpha_{PII\infty}$ and α_{PII0} (absorption coefficients) of the same waves, respectively, vanish and tend to infinity.

Thus, as far as soft rocks are concerned, longitudinal compression waves and particles repackings and a transverse wave propagate. The speed of propagation of a type one longitudinal wave v_{PI} is mainly governed by the compressibilities,

β_s and β_l, of the solid and liquid components and values of $k_{p.o}$ of the medium, whereas type two longitudinal waves propagating in soft rocks are mainly characterized by quantities λ_1 and λ_2. These waves characterize solid particles repacking (they can be observed in dry sand at atmospheric pressure); they demonstrate more pronounced attenuation compared with a type one wave.

Due to this fact, solely type one waves practically propagate in slightly cemented water-saturated rocks, and type two waves only propagate in dry slightly cemented rocks. For these rocks $\mu^* \to 0$, $\delta_{fill} \to \delta_{air} \to 0$, i.e. $\zeta \to \infty$, $\alpha_{PII} \to 0$, $\upsilon_{PII} \to C = 1/\sqrt{b\delta_s}$. Here $\alpha_{PII} = \sqrt{\omega\mu^* b/2a_0}$, i.e above the water table $\upsilon_{PII} = 1/\sqrt{b\delta_s}$, and below it the speed of propagation is found form Eq. (168). As can be seen from Eqs. (164) and (168), the speed of type one waves depends on the vibration frequency.

Analysis of the elasticity of hard (cemented) porous rocks without and with taking into account unbalanced heat exchange also suggests that two longitudinal and one transverse wave propagate in these rocks. However, in cemented water-saturated rocks the particles repacking longitudinal wave is rapidly, at short distances, attenuated and thus needn't be considered. A slower speed of the type one (pressure) wave propagating in cemented rocks is conditioned by the compressibility of the solid and liquid component and voids ratio, k_p, of rocks. It is also associated with characteristics of hydro- and thermal interaction between such quantities as translucence, permeability coefficient, viscosity coefficient, coefficients of thermal conductivity and thermal capacity. However, at low frequencies, type one waves move in phase in the liquid and solid component whereas at high frequencies the motion of the fluid begins to lag in phase, and, as a result, υ_{PI} becomes a complex quantity, and dynamic quantities, density (less than static one) and viscosity, appear.

A second longitudinal wave propagates in gas-saturated cemented rocks without attenuation, at a speed depending solely on Lamé's coefficients of the skeletal particles and density of the solid component. Its values approaches to that in a continuous solid medium. The first (slower in strongly cemented rocks) wave (sometimes known as an air or a gas wave) is, in fact, a type one wave, and its small speed of propagation is accounted for by the great compressibility of gas. The skeletal particles, as it propagates, practically remain immovable, being perfectly hard.

Given a drop-like fluid in a rock, the variation of the pore pressure mainly generates a faster wave, that is why slower waves originate in gas-saturated cemented rocks.

The limiting values of speeds in a fluid-saturated rock for $\omega \to 0$ and for $\omega \to \infty$ are found from the relationships:

$$\upsilon_{PI0} = \sqrt{\frac{\left[(n + k_p)\frac{(1 - k_{p0})\alpha_s\alpha^* T_0}{\beta_s C}\beta_s K\right]^2 + \frac{(1 - k_{p0})}{b}\theta_0\left[1 + \frac{(1 + k_{p.o})}{c}\alpha_s^2 T_0 b K^2\right]x}{\delta_0\theta}} \tag{170}$$

where $\theta_0 = \beta_r[1 - T_0(\alpha^2/c\beta_r)]$; $x = [1 - (\alpha_r^2\beta_r/b)]$; $b = (\lambda_1 + 2\lambda_2)$; $\alpha^* = n\alpha_s + k_p\alpha_l$; $c = nc_s + k_pc_l$; T_0 is the initial stationary absolute temperature; n is translucence of the rock; α_s, α_l and c_s, c_l are coefficients of thermal conductivity and thermal capacity of the solid and liquid phase of the rocks; K is the modulus of overall compression;

$$v_{PI\infty} = \frac{1}{\sqrt{\delta_l b}}\sqrt{\frac{(r + \varepsilon^* f) + \sqrt{(r + \varepsilon^* f)^2 - 4\varepsilon^* r[(1 - k_p)/n](\delta_s/\delta_l)}}{2\varepsilon^*[(1 - k_p)/n](\delta_s/\delta_l)}} \tag{171}$$

where

$$\delta^* = \delta_s + \frac{n}{k_p}\delta_l; \quad r = \frac{k_p(1 - k_p)\delta^*}{\delta_l n}$$

$$\varepsilon^* = \frac{\theta^*}{b}; \quad \theta^* = \beta_r\left[1 - \frac{n^2\alpha_s^2 T_0}{(1 - k_p)v_{Ps}\beta_r} - \frac{k_p\alpha_l^2 T_0}{v_{Pl}\beta_r}\right]$$

$$f = \frac{1 - k_p}{n}\left(1 - \frac{\beta_s}{\theta^*} nk_w\right)$$

(without taking into account heat exchange).

Calculations using Eqs. (170) and (171) can show that the limiting speeds are much greater, given that rocks are water- and not oil-saturated; v_P of elastic waves in oil-bearing rocks grow with frequency, whereas in water-saturated rocks they do not.

Sec. 83. Differentially Elastic Isotropic and Anisotropic Rocks Lacking Interphase Connections

In such rocks (highly porous rocks having communicating voids and good permeability air-saturated in the open air) two elastic waves, through the skeletal particles and the voids filling component, propagate independently. Due to this mechanism, the motion of phases in such a rock is described by two equations, for the skeletal particles and for the voids filling component. Stresses act on each phase separately, their motions being characterized by these equations:

$$\frac{\partial^2\theta}{\partial t^2} = \frac{\lambda_{sk} + 2\mu_{sk}}{\delta_{sk}}\nabla^2\theta = \frac{\lambda_{sk} + 2\mu_{sk}}{\delta_s(1 - k_p)}\nabla^2\theta$$

$$\frac{\partial^2\theta_{air}}{\partial t^2} = \frac{K_{air}}{\delta_{air}k_p}\nabla^2\theta_{air}$$

where the expansion (dilatation) is

$$\theta_{air} = \frac{\partial u_{air}}{\partial x} + \frac{\partial v_{air}}{\partial y} + \frac{\partial w_{air}}{\partial z}$$

the Laplacian operator for the Cartesian x, y, z coordinate system is

$$\nabla^2\theta_{air} = \frac{\partial^2\theta_{air}}{\partial x^2} + \frac{\partial^2\theta_{air}}{\partial y^2} + \frac{\partial^2\theta_{air}}{\partial z^2};$$

$\delta_{sk} = \delta_s(1 - k_p)$, δ_s, δ_{air} are the densities of the skeleton, solid phase, air, respectively.

A solution of these equations yields

$$v_{Pair} = \sqrt{\frac{K_{air}}{k_p\delta_{air}}} \quad \text{and} \quad v_{Psk} = \sqrt{\frac{\lambda_{sk} + 2\mu_{sk}}{\delta_{sk}}} = \sqrt{\frac{\lambda_{sk} + 2\mu_{sk}}{\delta_s(1 - k_p)}}$$

For anisotropic rocks these equations are sometimes written as

$$v_{Pair} = \sqrt{K_{air}/R\delta_{air}}$$

where K_{air} and δ_{air} are the modulus of overall compression and density of the voids filling component, air; R is the structure coefficient of rock;

$$v_{Pski} = \sqrt{C_{ski}/\delta_{sk}} \tag{172}$$

where C_{ski}, δ_{sk} are the elasticity coefficient, and the density of the skeleton [C_{ski} can be found using Eq. (172), given v_{Psk} and δ_{sk})].

In theory, the speed of propagation of waves in the skeletal particles has been calculated for a hexagonal packing—equal diameter spheres. In this case the speed of propagation of vertical compression waves is

$$v_{sk} = 860\left[\frac{2\pi E_s^2 g}{(1 - \nu_s^2)^2 k_p^3 \delta_s}\right]^{1/6} z^{1/6} \tag{173}$$

where z are the coordinates of the centres of spheres (or depth of occurrence of the bed).

The presence of two waves in differentially elastic rocks lacking or having an imperfect interphase connection as well as the general theory of the dependence of the speed of propagation of these waves on pressure have been first validated by works of N.R. Peterson on glass balls, quartz sands and Pleistocene coastal sands.

Early attacks of waves, via the water (or air) and skeletal particles, the early arrivals via water (or air) anticipating those via the skeletal particles. With increasing the load the speeds of waves in the air and skeletal particles increase, the speed growing more rapidly in the skeleton than in the air. Theoretical speeds in these media are lower compared with the experimental ones, owing to a different effect of the structure than it is assumed in theory.

Bibliography

Main Literature

Belikov, B.P., Aleksandrov, K.S., Ryzhova, T.V. Elastic Properties of Rock-forming Minerals and Rocks. Moscow, Nauka, 1970.
Kobranova, V.N. Physical Properties of Rocks (Petrophysics). Moscow, Gostoptekhizdat, 1962.
Kozhevnikov, D.A. Neutron Characteristics of Rocks and Their Applications in Oil and Gas Prospecting Geology. 2nd ed., Moscow, Nedra, 1978.
Leontyev, E.I. Modelling in Petrophysics. Moscow, Nedra, 1978.
Petrophysical Quantity Determinations from Rock Samples. Ed. by V.N. Kobranova, B.I. Izvekov, S.L. Patsevich, M.D. Shvartsman. Moscow, Nedra, 1977.
Physical Properties of Minerals and Rocks, Given High Thermodynamic Parameters. A Handbook. Moscow, Nedra, 1978.
Physical Properties of Rocks and Mineral Deposits (Petrophysics). A Geophysicist's Handbook. Moscow, Nedra, 1984.

Additional Literature

Adsorption and Porosity. Moscow, Nauka, 1976.
Analysis of Oil and Gas Reservoir Beds of Oil and Gas Fields of West Siberia Using Geophysical Methods. Ed. by E.I. Leontyev, L.M. Doroginitskaya, G.S. Kuznetsov, A.Ya. Malykhin, Moscow, Nedra, 1974.
Bolotov, A.A., Belinskii, B.A., Sinii, L.L. The Effect of the Structure of Rocks on Their Acoustic Properties. In: *Prospecting Geophysics*, Vol. 45, Moscow, Nedra, 1971.
Borehole Nuclear Geophysics. A Geophysicist's Handbook. Moscow, Nedra, 1978.
Bulatova, Zh.M., Volkova, E.A., Dubrov, E.F. Acoustical Logging. Leningrad, Nedra, 1970.
Dakhnov, V.N. Geophysical Methods of Determining Reserved Properties and Oil and Gas Saturation of Rocks. 2nd ed., Moscow, Nedra, 1985.
Dubinin, M.N. Adsorption and Porosity. Moscow, 1972.
Friedrichsberg, D.A. A Course of Colloidal Chemistry. Leningrad, Khimiya, 1974.
Gorbatov, G.P., Shibanov, A.S. Modern Methods of Mineralogical Analysis. In: *Electrical Characteristics of Minerals, Their Measurements and Applications in Mineralogical Investigations.* Moscow, Nedra, 1969.
Gottikh, R.P. Radioactive Elements in Oil and Gas Geology. Moscow, Nedra, 1980.
Grigorov, O.N. Electrokinetic Phenomena (A Course of Lectures). Leningrad, LGU, 1973.

Ivakin, B.N., Karus, E.V., Kuznetsov, O.L. Acoustical Method of Investigating Boreholes. Moscow, Nedra, 1978.

Kerkis, E.E. Methods for Studying Filtrational Properties of Rocks. Leningrad, Nedra, 1975.

Komarov, V.A. Electrical Logging Using a Method of Induced Polarization. Leningrad, Nedra, 1980.

Kormil'tsev, V.V. Transition Processes During Induced Polarization (Theory and Applications in Geophysics). Moscow, Nauka, 1980.

Kozlov, E.A. Concerning Speeds of Propagation of Longitudinal Waves in Terrigenic Clastic Deposits.—*Izv. AN SSSR*, Ser. geophys., No. 8, 1962.

Lisitsin, A.K. Uraniferous Properties of Subsurface Waters. Novosibirsk, Nauka, 1975.

Magnetism and Formation Conditions of Erupted Rocks. Ed. by D.M. Petcherskii, V.N. Bagin, S.Yu. Brodskaya, Z.V. Sharonova, Moscow, Nauka, 1975.

Nikolaevskii, V.N., Basniev, K.S., Gorbunov, A.T. Mechanics of Saturated Porous Media. Moscow, Nedra, 1970.

Ozerskaya, M.L., Semenova, S.G. Recommended Procedures for Using a Method of Graph Analysis of Parametric Cross Sections of Physical Properties of the Regolith and Their Applications in Solving Geological and Physical Problems. Moscow, VNIIGeofizika, 1976.

Parkhomenko, E.I. Electrical Properties of Rocks. Moscow, Nauka, 1965.

Petkevich, G.I., Verbitskii, T.Z. Acoustical Studies of Rocks in Oil Boreholes. Kiev, Naukova dumka, 1970.

Physical Properties of Rocks of the West Siberian Oil and Gas Province. Ed. by N.A. Tuezova, L.M. Doroginitskaya, R.G. Demina, N.I. Bryuzgina. Moscow, Nedra, 1975.

Pinneker, E.V. Problems of Regional Hydrogeology. Moscow, Nauka, 1977.

Problems of Homo- and Heterogeneity of Minerals. Moscow, Nauka, 1973.

Rats, M.V. Structure Models in Engineering Geology. Moscow, Nedra, 1973.

Rzhevskii, V.V., Novik, G.Ya. Fundamentals of Rock Physics. Moscow, Nedra, 1978.

Shapiro, D.A. Physico-chemical Phenomena in Rocks and Their Applications in Oil Prospecting Geophysics. Moscow, Nedra, 1977.

Vendel'shtein, B.Yu. Analysis of Cross Sections of Oil and Gas Drill Holes Using a Method of Self-potentials. Moscow, Nedra, 1966.

Vonsovskii, S.V. Magnetism of Microparticles. Moscow, Nauka, 1973.

Subject Index

To the Reader

Mir Publishers would be grateful for your comments on the content, translation and design of this book. We would also be pleased to receive any other suggestion you may wish to make.

Our address is:
Mir Publishers
2 Pervy Rizhsky Pereulok
I-110, GSP, Moscow, 129820
USSR

SPRINGER NATURE

GPSR Compliance

The European Union's (EU) General Product Safety Regulation (GPSR) is a set of rules that requires consumer products to be safe and our obligations to ensure this.

If you have any concerns about our products, you can contact us on ProductSafety@springernature.com

In case Publisher is established outside the EU, the EU authorized representative is:

Springer Nature Customer Service Center GmbH
Europaplatz 3
69115 Heidelberg, Germany

Kolibri 360 GmbH
Ferdinand-Jühlke-Straße 7
99095 Erfurt, Deutschland
produktsicherheit@kolibri360.de